ENERGY IN ARCHITECTURE
The European Passive Solar Handbook

EDITED BY

JOHN R. GOULDING   J. OWEN LEWIS   THEO C. STEEMERS

D1341390

Commission of the European Communities

This publication has been prepared in the Third Solar R+D Programme of the Commission of the European Communities Directorate-General XII for Science, Research and Development, within the SOLINFO Action coordinated by the Energy Research Group, University College Dublin.

Publication arrangements have been made under the VALUE Programme (specific programme for the Dissemination and Utilization of Community research results) within the Research Dissemination: Energy Efficient Building project of the Commission of the European Communities Directorate-General XIII for Telecommunications, Information Industries and Innovation

Editorial Advisory Group:
Patrick Achard, Valbonne
Alex Lohr, Köln
Albert Mitja i Sarvise, Barcelona
Martin de Wit, Eindhoven

Design by W.H. Hastings, Dublin

Graphical illustration and pagemaking by
W.H. Hastings, John Kelly, Ciaran O'Brien and Pierre Jolivet, Dublin

The European Passive Solar Handbook (Preliminary edition) 1986 of which this publication is a revised edition was edited by Patrick Achard & Renaud Giquel

Additional assistance by:
Judith Stammers, London
Professor Dermot O'Connell, Loughlin Kealy, Mary Rigby, Kay Dunican and Shane O'Toole, University College Dublin

The Editors would like to thank all those throughout Europe who have put so much time and effort into the preparation oft his book

The support of the VALUE Network is particularly appreciated:
Professor Andre De Herde, Louvain-la-Neuve.
Alex Lohr, Köln.
Dr. Maria del Rosario Heras, Madrid.
Alexandros N Tombazis, Athens.
Dott. Ing. Paolo Oliaro, Torino.
Joep Habets, Amsterdam.
Professor Eduardo Maldonado, Porto.
Dr. N V Baker, Cambridge.
T Vest Hansen and Ivor Moltke, Tåstrup.
Jerome Adnot and Eric Durand, Paris.

ENERGY IN ARCHITECTURE
The European Passive Solar Handbook

Produced & coordinated by
The Energy Research Group,
School of Architecture,
University College Dublin.
Richview, Clonskeagh,
IRL-Dublin 14.

Publication No. EUR 13446 of the Commission of the European Communities,
Scientific and Technical Communication Unit, Directorate-General
Telecommunication, Information and Innovation, Luxembourg.

Reprinted 1993.

Typeset by
the Energy Research Group, School of Architecture, University College Dublin.

Printed and bound in Great Britain by
Dotesios Limited, Trowbridge, Wiltshire.

Published by
B.T. Batsford Limited, 4 Fitzhardinge Street, London W1H 0AH

for the
COMMISSION OF THE EUROPEAN COMMUNITIES
Directorate General XII for Science,
Research and Development
within the SOLINFO Action of the
Solar Energy Applications to Buildings Programme
managed by Theo C. Steemers at the Commission
and
Directorate General XIII for Telecommunications,
Information Technology and Innovation
within the Research Dissemination: Energy Efficient Building project of the
VALUE programme (specific programme for the Dissemination and Utilisation
of Community research results).

A catalogue record for this book is available from the British Library.

ISBN 0 7134 69188

The Editors wish to particularly thank the following authors for their contributions:

Chapter 1 'Strategy' - by Alex Lohr, Büro für Energiegerechtes Bauen, Köln.

Chapter 2 'Climate and Design' - by Professor J.K. Page, Initiative Director, Cambridge Interdisciplinary Environmental Centre.

Chapter 3 'Passive Solar Urban Design' - by Prof. A. Dupagne and L. Mattelig - LEMA - University of Liège.

Chapter 4 'Thermal Comfort' - based on the Preliminary Edition of The European Passive Solar Handbook 1986 with revised sections 4.1-4.13 by Prof. Ole Fanger and Henrik N. Knudsen, Laboratory of Heating and Air Conditioning, Technical University of Denmark, Lyngby, and section 4.14 by Dr. Ian D. Griffiths, Energy and Indoor Climate Research Group, Department of Psychology, University of Surrey, Guildford.

Chapter 5 'Passive Solar Heating' - based on the Preliminary Edition of The European Passive Solar Handbook 1986 with additional material from Dr Martin De Wit, University of Eindhoven and Alex Lohr, Büro für Energiegerechtes Bauen, Köln.

Chapter 6 'Passive Cooling' - based on material from the 'Horizontal Study on Passive Cooling' 1990, prepared within the CEC DG XII Building 2000 Project by M. Antinucci, B. Fleury, J. Lopez d'Asiain, E. Maldonado, M. Santamouris, A. Tombazis, S. Yannas, and material commissioned from A. Mitja, J. J. Escobar, J. Esteve, C. Torra, Departament d'Industria i Energia, Generalitat de Catalunya, and J. A. Cusidó, J. Jorge, J. Puigdomènech, Escola Tècnica Superior d'Arquitectura des Vallés, Universitat Politècnica de Catalunya.

Chapter 7 'Daylighting' - by Dr Marc Fontoynont, Ecole Nationale des Travaux Publics de l'Etat, Vaulx en Velin, Lyon.

Chapter 8 'Control Systems' - by Dr A.H.C. van Paassen, Coordinator of CEC Concerted Action - 'Control of Passive Buildings', Technical University of Delft, Department of Mechanical Engineering, Laboratory of Refrigeration and Indoor Climate Technology.

Chapter 9 'Atrium Design' - based on contributions by Dr. John Littler, Polytechnic of Central London, Dr. Nick Baker, University of Cambridge/The Martin Centre of Architectural and Urban Studies, Professor Gabi Willbold-Lohr, University of Aachen and Alex Lohr, Büro für Energiegerechtes Bauen, Köln.

Chapter 10 'Design Guidelines' - By Alex Lohr, Büro für Energiegerechtes Bauen, Köln.

Appendices 1 to 8 - by Professor J.K. Page, Initiative Director, Cambridge Interdisciplinary Environmental Centre.

The LT Method and the Daylight Factor Meter were prepared by Dr N.V. Baker, University of Cambridge/The Martin Centre of Architectural and Urban Studies.

Method 5000 was prepared by Michel Raoust, DIALOGIC S.A., Paris with additional support from Agence Francaise pour le Maitrise de l'Energie (AFME).

The work of the European Passive Solar Working Groups (coordinated by Ralph Lebens and David Clarke, London), which prepared the first draft of the preliminary Editions, and the Revisers and Editors of that edition, is also gratefully acknowledged.

ENERGY IN ARCHITECTURE
The European Passive Solar Handbook

Contents

APPENDICES

PREFACE

This handbook represents an important strand in the Commission of the European Community's programme to promote the design and construction of energy-efficient, passive solar buildings in Europe.

It has evolved from an earlier 'Preliminary' edition printed in a limited run in 1986 which has provided a significant amount of material for this edition. However, the enlargement of the Community with the accession of Greece, Portugal and Spain together with progress in research has prompted a revised edition of the handbook with a new balance of emphasis on building design for northern and southern European climates combined with extensive rewriting and increased coverage.

In addition to new sections on climate, passive solar urban design, natural cooling, control systems and atrium design, certain other topics including daylighting and thermal comfort are covered in greater depth.

The format of this revised edition has been redesigned, separating illustrated introductory texts from design information and providing two distinct but logically consistent volumes, 'Energy Conscious Design - A primer for European architects' and 'Energy in Architecture - The European Passive Solar Handbook'. This approach has allowed the content of the handbook to be tailored more appropriately in its complexity and presentation to the different needs of building designers who wish to inform themselves about passive solar design and those engaged in the task of designing or constructing passive solar building. It is anticipated that both volumes will be translated into other EC languages.

Energy in Architecture - The European Passive Solar Handbook provides a significant body of information for European building designers, drawing on recent European R+D and the knowledge of European experts. Revised and specially-commissioned chapters contribute to a balanced treatment of passive solar design issues for application throughout the enlarged Community.

For those less familiar with the subject, the companion volume 'Energy Conscious Design - A Primer for Architects' is strongly recommended.

Dr. W. PALZ
Division Head "Renewable Energies"
Commission of the European Communities

INTRODUCTION

To make a building is to create a system linked to its surrounding environment and subject to a range of interactions affected by seasonal and daily changes in climate and by the requirements of occupants varying in time and in space.

Some twentieth century buildings seek to deny these inevitable interactions and subdue them with expensive heating, cooling and lighting equipment. A more climate-sensitive approach is proposed here which recognises and responds to seasonal and daily changes in the environment for the well-being and comfort of the occupants. The relationship between people, their living place and the environment is re-examined and resolved in an architecture which dynamic interaction.

In recent years these issues have most frequently been addressed in the design process after the building form has been fixed, and we have become used to thinking of heating, cooling and lighting devices as add-on equipment to be sized and placed in more or less completed buildings. While this may be a pragmatic or convenient approach, it diminishes the opportunity to design, at a more holistic level, buildings which can respond to the environment by virtue of their form and the intelligent use of materials with minimal reliance on machinery. Rediscovery of this design skill adds a dimension to the design process which offers sound parameters as generators of architectural form.

To achieve this calls for a knowledge of climate and an awareness of the available technologies which can be employed in building, combined with an understanding of what constitutes human comfort and discomfort and how these conditions can be affected by changes in climate. These issues are relevant to all buildings and locations whether the predominant need is for heating, cooling or daylighting. Initial design decisions will focus on the location of the building, its basic form, the arrangement of the spaces, the type of construction and the quality of the environment to be provided, resulting in an architectural response of high quality which is in harmony with its environment.

In most situations it is necessary to provide some additional heating or cooling at certain times. Similarly, daylighting cannot meet all lighting requirements and therefore these auxiliary inputs and their control must be addressed once the contributions by natural means and the patterns of use are known.

The design and construction of a building which takes optimal advantage of its environment need not impose any significant extra cost, and compared to more highly-serviced buildings it may be significantly cheaper to operate.

There are two major strategies, depending on the regional climate and the predominant need for heating or cooling:

- in cold weather - maximize 'free' heat gains, provide good heat distribution and suitable storage within the building and reduce heat losses while allowing for sufficient ventilation;
- in warm weather - minimize heat gains, avoid overheating and optimize cool air ventilation and other forms of natural cooling.

To the above must be added a daylighting strategy. The availability of daylight is influenced by latitude and climate. The use of natural light to replace electrical light is particularly important in large buildings with a low surface-to-volume ratio, where unwanted internal heat gains caused by artificial lighting can be considerable and often may require the use of mechanical air-conditioning.

The design of an isolated building and its immediate environment, where it is unaffected by neighbouring buildings, is one matter. However, more often, one must consider the negative and positive aspects of building in urban locations. Clusters of buildings create their own microclimates through shading, shelter, wind deflection, and the emission of heat. Depending on the climate, some of these aspects may be turned to advantage while others may have to be minimized. The interactions between buildings are by their nature very complex and while some design tools and guidelines exist to help the designer to understand the phenomena involved and predict how the building will perform, the development of more elaborate, often computer-based tools continues.

In order to provide conditions of thermal comfort a knowledge of the range of people's comfort tolerances is needed. Thermal comfort is affected by such factors as temperature, humidity, airflow (draughts), the level of physical activity, the amount of clothing being worn and even the weight of the individual concerned. Comfort is to some extent subjective and consequently a capacity for the individual to have some control over his or her environment is desirable. The section on Comfort characterises the range of comfort conditions while the section on Behaviour gives pointers to the way people respond in buildings and how occupant behaviour may affect the building's performance.

This book attempts to demonstrate the benefits of an approach to the design of buildings and their immediate surroundings which takes advantage of natural phenomena instead of fighting the influences of nature with expensive and often environmentally-destructive heating, cooling or lighting equipment and the energy they consume. The overall goals to which the book is directed are improved thermal and visual comfort in more environmentally-benign buildings, and the synthesis of these objectives in good architectural design. What is proposed is a fundamentally more thorough approach to building design which, while adopting additional performance parameters, offers the possibility of exciting new architectural design opportunities.

Sᴛʀᴀᴛᴇɢɪᴇs

Contents

1.1

Iɴᴛʀᴏᴅᴜᴄᴛɪᴏɴ

It is becoming clear that buildings contribute significantly to the serious environmental problems of the planet.

The close connection between a building's energy use and environmental damage arises because we still look to technical solutions to meet our energy-related needs. In so doing, we rely on mechanical systems to solve climate-related heating, cooling and lighting problems induced by inadequate building design. It is time to think again about our approach.

By changing our design approach, a building can be created which not only meets the occupants' needs for thermal and visual comfort but also requires less energy to run and consequently has a reduced impact on the environment. It is to be expected that clients will increasingly seek out competent, environmentally-sensitive architects who are able to integrate environmentally-responsive concepts into the final design. This chapter describes briefly the strategies which can be used by the architect to develop such a building.

1.2 INTEGRATION OF ENVIRONMENTALLY-RESPONSIVE CONCEPTS IN THE DESIGN PROCESS

As matters stand at present, the building design process reflects too closely the fragmentation of the design professions, with broad questions of building function and appearance seen as the province of the architect, and questions of appropriate interior environment seen as the province of the services engineers. In this way of working, issues of the comfort of building users have been addressed by the later addition of mechanical systems.

Heating performance, for instance, has too often been defined in a narrow way by considering what goes on inside the building's skin. In an environmentally-responsive approach, the skin itself is the subject of design: a building envelope is created which modifies the exterior microclimate as well as the environmental conditions inside the building. Heating performance, therefore, becomes another part of the responsibility of the architect.

Thermal performance, however, is of course only one of the aspects of performance which has to be considered. Spatial, acoustic and lighting performance all have to be taken into account. These will have different priorities according to building type. The architect has to choose how to use environmentally-sensitive concepts and integrate them with other, equally-important design goals

1.3 PERFORMANCE OBJECTIVES

In designing a building, it is of fundamental importance for the designer to have a clear idea of objectives. The goal of meeting the user's need for a comfortable environment involves achieving and maintaining acceptable levels of interior comfort, day and night, all year round. Consequently, he or she should have a good working knowledge of the occupants' requirements and of the basic architectural concepts which may be used to meet these requirements in an environmentally-responsive way.

Firstly, the designer has to establish a list of performance goals. This will include a definition of the thermal conditions acceptable for various occupants' activities at different times of day during each season of the year. The factors to be considered will include indoor air temperature, temperature variations, air movement, humidity, air changes, surface temperatures and lighting levels.

At a later stage, the designer should identify the key energy-using processes which can help achieve the performance goals. He or she needs to consider, for instance, the heating, cooling, ventilation, lighting, humidification or dehumidification, cooking, washing and bathing requirements.

Realistic energy performance targets can be developed which permit substantial cuts in fossil fuel consumption and allow for use of local renewable sources where appropriate. These performance targets may have to be reconsidered and refined throughout the design process. They are guiding forces for competent design. For example, if they are kept in mind, then overheating and cooling problems may be solved by introducing innovative shading devices rather than energy-consuming cooling equipment. Increasingly such targets will be influenced by external recommendations or even mandatory requirements.

1.4 THE PROCESS OF INTEGRATION

Architects are accustomed to establishing goals or parameters for their designs. They use these goals to measure the success of various solutions or compromises during the design process. The relative importance of individual goals has to be determined by the architect for each building.

The process by which different goals are taken into account during design can be loosely illustrated by the diagram in Figure 1.1. Here, the design process is depicted as a spiral, and the design parameters as radial beams. The end of each radial beam marks the achievement of the particular goal.

As the architect proceeds with the design work, he or she moves from one design parameter to another trying to integrate the requirements related to each parameter in a creative way. As the spiral unfolds, the amount of detailed knowledge which is needed to move along the path increases. The designer has to acquire this information, obtaining specialist external advice where this is necessary. The higher the design goals or the more complex the building, the sooner this specialized information has to be integrated.

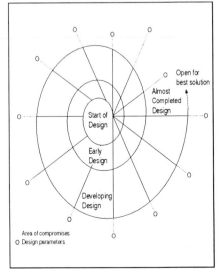

Figure 1.1. Environmentally-sensitive design process.

THE VARIOUS DESIGN PHASES

The best opportunities for improving a building's energy performance occur early in the design process when basic decisions are made concerning the site, orientation, configuration and passive solar strategies. If the designer does not realize the potential for energy saving during the initial phase the opportunity will be lost to make significant savings by relatively simple adjustments to the design. Increasingly sophisticated or costly efforts are needed to save energy at successive stages.

The Early Phases of Environmentally-Sensitive Design

By using environmentally-sensitive design strategies, the potential of the building envelope to control the heat and light entering the building can be realized. The natural energy flows brought about by the sun, wind and temperature differences can be organized so that heating is provided in winter, cooling in summer and lighting all year round. As a general rule, designs which put natural solutions before mechanical/artificial measures are most likely to create least environmental damage. Therefore, the possibility of using solar heating should be considered before introducing fossil fuel-based systems. The natural lighting should be designed before the electric lighting systems. Cooling should be by natural means alone; use of mechanical cooling systems should be avoided as far as possible.

As indicated earlier, the decisions which architects make during their earliest design considerations have a major impact upon the final energy consumption of the building. At the early stages, the following are important:

Securing Solar Access

Where architects are asked to advise their clients on the selection of a good site, they should be conscious of the benefits of solar access.

In northern Europe, ideally the siting should be such that the sun's rays fall on the south facade of the building during the heating season. Future building developments and growing trees on and around the site must be taken into account.

Conceptual Layout and Thermal Zoning of the Building

Unheated spaces such as storage or service areas and garages and other rooms requiring lower temperatures such as bedrooms, staircases, entrances, draught lobbies and corridors where possible should be placed on the north side of the building and most of the rooms requiring heating can then be positioned on the south side, only a few being located to the east and west.

Creating a Compact Building Form

Loss of heat through the building envelope can be reduced by creating a compact building form. The smaller the area of outside wall per heated volume, the less energy will be required to operate the building. In this respect, terraced housing or apartment buildings are energy savers compared to single-family detached dwellings. In addition, in a low-volume building less energy is required for heating the fresh air supply, while still allowing sufficient fresh air to avoid discomfort, or health problems.

Integrating Performance Requirements into the Design of Facades

Designing the envelope of an environmentally-sensitive building is a demanding task for the architect. It is the area where the inter-relationship between the given external conditions and the required internal environment is determined. Consequently, each facade must be designed according to different outside conditions in such a way that conditions inside the building

Figure 1.2 (A to H)
Performance of a solar facade in winter and summer
(Architects: Gabi and Alex Lohr, Cologne. 1987)

Figure 1.2 A

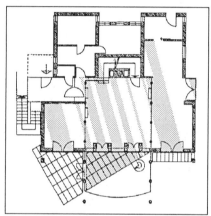

Figure 1.2 B – 21 January 10.°° hrs

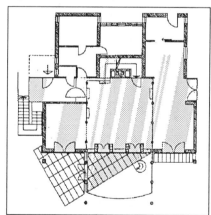

Figure 1.2 C – 21January 12.°° hrs

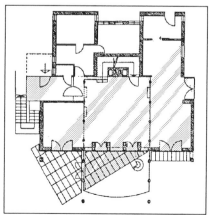

Figure 1.2 D – 21 January 14.°° hrs

Figure 1.2 E

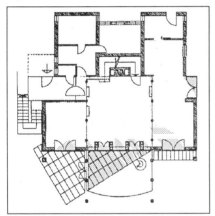

Figure 1.2 F – 21 June 10.°° hrs

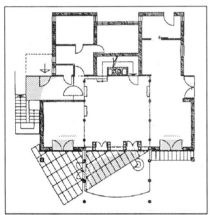

Figure 1.2 G – 21 June 12.°° hrs

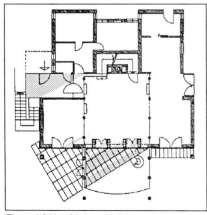

Figure 1.2 H – 21 June 14.°° hrs

remain within the occupants' comfort range. For example, if one may simplify for the sake of clarity, in an environmentally-sensitive building the total area of glazing may be similar to that in conventional buildings. One might expect however, that distribution among the various facades might be different. In the northern part of Europe, for instance, most of the glazed area would be placed on the south facade to improve solar gains; with fewer, smaller windows located on the other sides. Use of glazing materials with good insulating properties can prevent down-draughts associated with the flow of cold air inside the glazed south facade. Fixed overhangs can often provide enough shade on sunny summer days to make movable shading devices unnecessary. An example of the performance of a south facade in winter and summer is illustrated in Figure 1.2 (A to H).

Daylighting

Besides the south, west, north and east faces, another facade has to be taken into account - the roof. Roof openings are of particular interest where an even level of daylighting is required. Unfortunately, their potential in this respect is often ignored. In designing roof openings, however, it must be remembered that good architectural solutions are needed to avoid overheating.

Any building which is used mainly during the day provides an opportunity to design facades and roof in a way that both improves daylighting and saves energy. This is because in such buildings (schools and offices are examples), sunlight is available at times when it is needed most.

Architects have become accustomed to looking to other building professionals for help in evaluating the performance of a (conventional) design. With environmentally-sensitive buildings, decisions about the building shape, orientation and the responses of facades to different climatic conditions require the use of new skills. The acquisitionof these skills will give the architect greater control of the design and performance of the final building. To help with the assessment, various "evaluation tools" or "analysis tools" are becoming available. They are no substitute for good working knowledge and experience but are a means of quantifying the effect of energy-related design decisions.

Today, most analysis tools focus on one particular design parameter only, such as daylighting for instance. To obtain a clear picture of the impact on other parameters such as overheating, the designer has to use a variety of analytical tools. The dependencies and interrelationships of all design parameters may be obscured within this procedure. It may, for instance, be difficult to see whether a particular design modification which leads to better daylighting also creates an unacceptable amount of overheating. In the near future, however, new analysis tools will be available which permit evaluation of more than one design parameter at a time.

It takes time and experience to develop a basic understanding of the extent to which decisions in different areas affect the performance of the final design.

Environmentally-sensitive design involves both analysis and intuition in an iterative process which is intertwined with the creation of form and space. In today's world, a major challenge for architects is to link building form with environment-related performance goals.

CHAPTER TWO

BASIC CLIMATIC DATA

2.1

INTRODUCTION

This Chapter considers how to obtain and use the quantitative climatic data essential for the rational design of passive solar buildings. Design requires a seasonal analysis of the relationship between the patterns of natural energy supply, and the patterns of energy demand arising from the impacts of climate. The demand for both heating energy or for natural cooling is strongly influenced by the outdoor air temperatures around the building. Designers have first to consider the general climate of the region in which they are to build, and then to assess its modification, both by terrain and by urban impacts. They finally need to consider the detailed modification of climate within the site. These modifications may be very considerable, for example the impacts of overshadowing, or the redirection of the wind by built form.

The supply of renewable energy of climatic origin includes short wave radiation from the sun, long wave thermal radiation from the atmosphere, and wind energy. The most important energy source in the context of this book is the short wave radiation from the sun. Interseasonal analysis is important. Summer overheating is a common

problem in many areas of the European Community. Solar radiation in the overheated season exerts an adverse effect, impacting on cooling energy demands: therefore the positive radiative cooling potential of the atmosphere is also important. The downward flux of long wave radiation from the atmosphere is less than the outward loss of long wave radiation back to the atmosphere, so there is a net energy loss from buildings due to this cause. This loss is greatest under cloudless conditions. This exchange creates a potential night-time cooling resource, important in many parts of the European Community region.

The wind is a potential energy supply source of considerable European significance, though this aspect is not discussed in this text [1]. Wind, however, is also of considerable importance as a means of keeping buildings acceptably cool through natural ventilation in hot summer weather. Wind direction exerts a significant impact on the outdoor site microclimate, and hence influences building energy demands, for both cooling in summer and for heating in winter.

Contents

BASIC CLIMATIC DATA REQUIREMENTS

The key climatic requirements are:-

1. Solar radiation data, including knowledge of the solar geometry,
2. Wind speed and wind direction data,
3. Air temperature data,
4. Long wave radiation.

The generalized macroclimatic data obtainable from various sources have to be modified as discussed later to allow for the detailed effects of terrain and site layout

Data for heating season design assessment

The emphasis on the energy supply side in heating in this Chapter is on solar radiation data. A rather broader climatic perspective has to be taken to estimate heating demands. The influence of outdoor temperature on the monthly heating energy demand is usually assessed through the use of the monthly degree day concept. The sources of European degree day data are discussed in this Chapter. Wind exerts a significant effect on winter space heating energy demands, influencing both heat losses at the external surfaces of buildings, especially windows, and also the rates of internal ventilation. The losses at the outside surfaces of buildings are by long wave radiation

exchanges and by convection. The convection losses depend on the temperature difference between the air and the external surface and on the wind speed, which determines the surface conductance for forced convection. If the building is very well insulated, then, in the absence of energy recovery from the outgoing heated air, the ventilation losses in heated buildings become the key path of climatic heat loss.

Data for hot weather design assessment

Designers must also address the issues of hot weather building design [2]. There has been a tendency to overlook this aspect of design performance analysis in the development of European solar buildings. A solar building that performs well in winter, has also to provide an acceptable summer living environment. Under overheated conditions, the impacts of solar radiation create a demand for cooling, which, by appropriate design, may be offset by enhanced natural ventilation, and also more effective use of long wave radiative losses. Important regulation of solar overheating may be established through layout and landscape, as well as through detailed building design. Climatic data for assessing summer design performance therefore are needed in addition to data for assessing winter design performance

CLIMATIC DATA REQUIREMENTS FOR DESIGN ASSESSMENT

Selecting the design method

The first climatic decision a designer has to make is to select the design methods to be adopted for quantitative design. This decision defines the essential climatic data inputs required to support the selected method. Design assessment techniques of different complexity are used. This influences the level of detail needed in the supply of climatic data. Additionally, at early stages of design, quantitative data may be usefully applied in a qualitative manner, for example to decide on an orientation, or to structure a layout before detailed analysis is carried out. As climate involves directional concepts, it is useful in layout studies to construct directional design aids, compiled from the various data sources that are described later in this book.

The formal quantitative assessment of the performance of passive solar buildings may be carried out at two levels:-

1   Detailed simulation,
2   Simplified correlation models.

Data for thermal simulation studies

Simulation techniques are especially useful at the research level. This detailed approach requires the availability of hourly values of the relevant climatic observations for all the heat transfer variables over a significant period of time, say a year. The key variables are global and diffuse horizontal radiation, screen dry and wet bulb temperatures, 10m high wind speeds and directions, cloud cover and sunshine duration. The humidity is normally a derived variable, obtained from the observed dry and wet bulb temperatures. Long wave radiation data is not normally available, so it has to be derived from either the dew point temperature and the cloud cover, or from the air temperature and the cloud cover [3] [4]. Accurate validated techniques are also needed for the estimation of hourly slope radiation from the horizontal radiation data on the data tapes. The isotropic approximation for estimating sky diffuse irradiation is not satisfactory.

The concept of test reference years, TRYs, has been widely used in many European countries for simulation studies. A test reference year is a set of selected hourly climatic data used for system simulation to compare the dynamic responses of different building energy systems. In the past, the basis for the selection of the test reference years has varied from European country to country. One important climatic achievement of the CEC Solar Energy R&D programme has been the production of test reference years, TRYs, for 29 locations in seven Member States of the Community on a common basis [5]. A further development has been to produce short test reference years [6] [7]. The use of full test reference years consumes quite a lot of computing time. So there are advantages in preparing, in a statistically validated way, more compact dynamic weather descriptions to speed up such simulation assessments. Such compacted test reference years are known as short reference years. SRYs are, like Test Reference Years, used as input data for computer simulation of heating and cooling loads in buildings, solar energy systems, energy consumption and indoor climate. Offering a substantial reduction in computer time, SRYs give nearly the same results as TRYs. They are, however, less suited for the determination of extreme temperatures, or simulation of solar systems with long storage times. It is assumed in this chapter that, in general, most designers will not be using simulation methods for design assessment, so the details of the TRYs are not further discussed. However, full details of the CEC studies on Test Reference Years may be obtained from H. Lund [6].

Climatic design data for simplified design methods

Simulation is complex and difficult to use for practical design problems and is inappropriate in the initial phases, so a number of simplified design methods for passive solar buildings based on correlation studies have been developed. In a correlation technique, one uses the results of detailed simulation to develop simplified relationships for design purposes which take account of building performance, especially

the impacts of the dynamic effects of energy storage in the fabric. As Balcomb [8] has indicated, in general one attempts to relate performance assessment to environmental inputs, often described using dimensionless quantities as inputs. Success is more likely, if the chosen correlating parameters preserve some essence of the overall physics governing the energy balances. Widely used correlation techniques include the Solar Load Ratio technique, where the ratio of the monthly solar gain to the monthly heating load is used as the basis of the correlation [9]. Method 5000 [10] provides another example for passive solar building design which is described in Appendix 13.2 of this book. The impact of thermal capacity is introduced in a dimensionless way in the Portuguese FIT method [11]. The f-chart method represents another correlation method, but, in this case, for active systems [12].

The issue in each case is what input climatic data is essential to the selected design method? The supply side of the analysis usually needs values of the monthly mean daily solar radiation transmitted through single and double glazing systems. Obstructions have to be considered. On the demand side of the analysis, the climatic impact of screen air temperature in simplified methods is normally evaluated through the use of monthly heating degree days, or rather less satisfactorily through the use of monthly mean temperatures.

The wind influence is only implicit and not explicit in the above assessment techniques. It is implicit in the choice of the design ventilation rate, and also in the choice of U-valves, through the choice of the external surface conductances. However, for well insulated parts of the structure the influence of wind speed on energy demand is relatively small. The impact on window heat losses is more significant, especially for single glazing. The wind driven ventilation losses are crucial, and, in general, there is inadequate knowledge of the actual ventilation rates achieved in solar buildings on sites of different relative exposure in areas of different basic windiness.

EUROPEAN COMMUNITY CLIMATIC DATA SOURCES

The Commission of the European Communities has initiated and funded a number of special studies on European climate data, including Atlases, intended to assist with renewable energy design and systematic energy conservation design. The majority of these studies are now complete. Their systematic use ensures that the basic quantitative data needed for passive solar building design climatic decision-making are available on a Community-wide basis. This section and the following one concentrate on reviewing these macroclimatic data sources. Additional data is usually available in publications covering more limited areas [13], for example those published by the Member States' meteorological services or energy departments of governments [14]. Therefore designers should additionally explore the availability of relevant national data from national sources, which are not systematically considered in this Handbook. For example, a bibliography covering sources of European wind data may be found in the European Wind Energy Atlas [1].

2.5

THE CEC EUROPEAN SOLAR RADIATION ATLASES

Radiation Data Sources

There are two CEC European solar radiation Atlases, one for horizontal surfaces [15], the other for inclined surfaces [16], together with supporting books [17] [18]. There is also the EUFRAT climatic data handbook [19] [20] which contains statistical data on radiation as well as temperature .

European Solar Radiation Atlas, Volume 1, Horizontal Surfaces

European countries both within the Community and outside the Community were involved in providing data for production of the Atlases. A common time base, 1966-1975, was used throughout. The horizontal Atlas also provides data for Cyprus, Malta, Syria, Lebanon, Jordan, Israel and Turkey . Additionally, maps are provided for Saudi Arabia.

Observed daily global radiation data were processed systematically to provide the monthly data in the horizontal surface Atlas. Tables were prepared, on a month by month basis, of global mean daily irradiation, G; monthly mean maximum daily global irradiation, GMAX; monthly mean minimum daily radiation, GMIN; extraterrestrial daily radiation averaged over the month, G_O; monthly mean daily sunshine, S; monthly mean astronomical daylength, S_O. Two dimensionless ratios were also included, the ratio of the mean monthly global radiation to the extraterrestrial irradiation, G/G_O, often referred to as the KT value and the relative duration of bright sunshine, S/S_O. Table 2.1 provides an example of the published Table for Hamburg, and for Langeoog.

The number of stations observing global solar radiation in Europe is limited. There are far more stations with bright sunshine observations. In order to get better

geographic coverage of radiation data, a systematic method for using the observed sunshine data to estimate daily horizontal irradiation was used. To broaden the data in the Atlas, a detailed study was made of the monthly relationships between daily values of G/G_O and daily values of S/S_O using observed daily data, entered in the standard Angstrom linear regression formula:-

$$G/G_O = a + b\, S/S_O$$

where a and b are monthly regression coefficients for a specific site.

This process provided [for each site where observed data were available monthly] values of the regression coefficients a and b. These values of a and b are published in both Atlases. The next stage in the mapping process was to use these derived values of a and b in conjunction with the observed sunshine data for a large number of

additional sites to estimate, from monthly mean daily sunshine data, values of the monthly mean daily global irradiation, again on a month by month basis. Tables for three hundred and forty sites were produced. The GMAX values for sites without observations were estimated by setting $S/S_O = 1$. This produces estimates of maximum radiation for these sites, which are usually somewhat higher than those found at nearby observing sites. The mean values however agree closely. These Tables were then used, as the basis of the production of monthly horizontal surface mean daily radiation maps for Europe. Figure 2.1 shows the horizontal radiation map for December. The mean daily horizontal radiation in December in Southern Spain is 2.4 kWh/d compared with 0.2 kWh/d in Northern Scotland.

A *Site with radiation observations*

GERMANY, FEDERAL REPUBLIC of								STATION		HAMBURG		
Latitude 53 38' N				Longitude 10 0' E					Altitude 14 M			
GLOBAL RADIATION G in WH M-2 (WRR)							SUNSHINE DURATION S in 1/10H					
10 years means (1966 - 1975) of monthly means of daily sums												

	JAN	FEB	MAR	APR	MAY	JUN	JUL	AUG	SEP	OCT	NOV	DEC	ANN
G	521	1232	2231	3553	4688	5437	4820	4340	2786	1489	671	401	2680
GMAX	1260	2645	4435	6231	7783	8130	7499	6606	4863	3043	1596	959	4597
GMIN	114	238	426	707	1220	1479	1783	1358	845	284	109	80	724
GO	1982	3512	5866	8479	10558	11513	11018	9249	6800	4290	2387	1569	6450
G/GO	.026	0.32	0.38	0.42	0.44	0.47	0.44	0.47	0.41	0.35	0.28	0.26	0.38
S	13	22	38	54	68	82	71	73	48	30	16	12	44
SO	79	96	117	138	157	167	162	146	125	104	84	73	121
S/SO	0.16	0.23	0.32	0.39	0.43	0.49	0.44	0.50	0.38	0.29	0.19	0.16	0.33

B *Site where radiation is estimated using Angstrom's equation*

GERMANY, FEDERAL REPUBLIC of								STATION		LANGEOOG		
Latitude 53 45' N				Longitude 7 29' E					Altitude 5M			
GLOBAL RADIATION G in WH M-2 (WRR)							SUNSHINE DURATION S in1/10H					
10 years means (1966 - 1975) of monthly means of daily sums												

	JAN	FEB	MAR	APR	MAY	JUN	JUL	AUG	SEP	OCT	NOV	DEC	ANN
G	542	1265	2501	4103	5056	5822	5287	4631	2980	1608	705	435	2919
GMAX	1668	2759	4854	7027	8547	9324	8922	7114	5291	3289	1847	1179	5164
GMIN	373	768	1345	2032	2427	2763	2533	2217	1560	940	450	295	1479
GO	1963	3492	5848	8467	10532	11511	11015	9239	6784	4271	2368	1551	6436
G/GO	0.28	0.36	0.43	0.48	0.48	0.51	0.48	0.50	0.44	0.38	0.30	0.28	0.41
S	10	24	38	57	67	78	70	72	48	30	15	12	44
SO	79	96	117	138	157	167	163	146	125	104	84	73	121
S/SO	0.13	0.25	0.33	0.41	0.43	0.47	0.43	0.49	0.38	0.28	0.18	0.16	0.33

Table 2.1. Tables of solar radiation data for Hamburg and Langeoog from the CEC European Solar Radiation Atlas, Volume 1, Horizontal Surfaces. Refer to text for the meaning of symbol.

The Atlas also contains diagrams showing the annual mean course of global radiation for 124 stations, for which observed data then existed, with the standard deviation of the individual daily values (extracted from the 10 year monthly means) plotted.

Finally frequency distributions of horizontal surface global radiation for 49 stations were compiled. These give the mean monthly cumulative frequency distribution of daily irradiation above given thresholds, and also mean monthly maximum number of consecutive days with daily irradiation above given thresholds.

The horizontal surface Atlas contains no information about horizontal diffuse radiation.

CEC EUROPEAN SOLAR RADIATION ATLAS VOLUME II, INCLINED SURFACES

A substantial research effort went into the development of the CEC European Solar Radiation Atlas, Vol 2, Inclined Surfaces. All data provided are estimates. There are few centres where measured inclined surface radiation data are available. Obviously, if such measurements are made, the results have to be limited to surfaces of specific tilt and orientation. Solar designers however require data for any slope and orientation they may wish to specify. Slope values cannot be estimated unless the beam and diffuse components can be separated. As part of the task of producing the Atlas, the diffuse horizontal radiation had to be estimated for each site.

The modelling method evolved had to accept the limitations of the input data base. This input data base was the data in the CEC European Solar Radiation Atlas, Vol I. The input data used to develop by numerical modelling the inclined surface Atlas were:-

1. Monthly mean daily global radiation on a horizontal surface, G.
2. Monthly mean values of the daily duration of bright sunshine, S.
3. Monthly mean values of the monthly maximum values of daily global radiation on the horizontal surface, GMAX.
4. Monthly mean maximum duration of bright sunshine, SMAX.
5. Values of the monthly Angstrom regression co-efficients, a and b, taken out at the daily levels.

In the CEC Inclined Surface Radiation Atlas, two slope tables were developed for each site with observed horizontal data, one for cloudless days, one for monthly means. Such observed monthly mean daily and clear day global radiation data of appropriate quality were available for 102 sites in Europe together with sunshine data.

Table 2.2 is a sample table from the Atlas. The top part of the table provides estimates of monthly mean daily slope global G_m and diffuse irradiation D_m for a range of slopes and orientations. The bottom part of the table provides estimated clear day data of daily slope global, G_c and diffuse irradiation, D_c for days of high irradiation for the same slopes and orientations. Values for intermediate slopes and orientations can be estimated from the tables by interpolation. Attention is drawn to the fact that different computation dates are used for monthly mean values in the top half of each table and the clear day values in the bottom half of each table. The clear day computation dates were selected to allow for the typical date of occurrence of the maximum horizontal surface radiation values within any particular month.

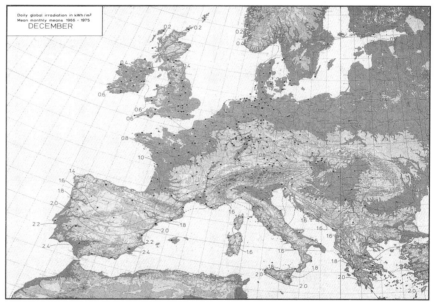

Figure 2.1. Horizontal surface daily mean irradiation map for Europe for December. Period 1966-75. Source CEC European Solar Radiation Atlas, Vol. 1, Horizontal Surfaces.

Surface	Monthly D/G %												
	J	F	M	A	M	J	J	A	S	O	N	D	Year
10° South	59	49	51	52	52	54	56	54	52	52	56	60	53
Latitude South	35	32	39	47	51	55	56	51	43	37	34	35	46
90° South	27	28	38	51	62	68	67	57	43	32	27	26	48
90° East/West	62	52	54	56	56	59	60	58	55	56	61	63	57
90° North	100	100	100	96	85	83	86	94	100	100	100	100	92
Ireland, Valentia 51°56'N - relatively cloudy climate													
Surface	Monthly D/G %												
	J	F	M	A	M	J	J	A	S	O	N	D	Year
10° South	41	40	41	37	40	34	29	32	35	33	42	36	36
Latitude South	30	32	36	46	42	37	32	33	32	28	31	25	34
90° South	27	31	38	47	61	63	56	48	38	29	28	22	41
90° East/West	50	50	50	49	53	48	46	47	47	47	51	48	49
90° North	100	100	100	95	88	77	84	92	100	100	100	100	91
France, Carpentras 44°5'N - relatively sunny climate													

Table 2.3. Estimated ratio of monthly mean daily diffuse to monthly mean daily global irradiation on various unobstructed surfaces, expressed as a percentage.

9

BUNDESREPUBLIK DEUTSCHLAND   **HAMBURG**

LATITUDE : 53° 38'N   Longitude : 10° 0'E   Altitude : 14m

Estimated monthly means of daily global radiation G_m and diffuse radiation D_m on inclined planes (1966 - 75)

Units : $kWh.m^{-2}$

		JAN	FEB	MAR	APR	MAY	JUN	JUL	AUG	SEP	OCT	NOV	DEC	MEAN
10° South	G_m	0.62	1.32	2.48	3.79	4.86	5.57	4.96	4.60	3.05	1.71	0.81	0.50	2.86
	D_m	0.40	0.76	1.37	2.07	2.64	2.87	2.88	2.42	1.66	0.94	0.48	0.31	1.57
30° South	G_m	0.79	1.60	2.81	4.00	4.88	5.49	4.93	4.80	3.37	2.04	1.02	0.67	3.04
	D_m	0.40	0.76	1.37	2.06	2.60	2.82	2.83	2.43	1.67	0.94	0.47	0.31	1.56
Latitude South (53°38')	G_m	0.89	1.75	2.88	3.81	4.41	4.84	4.40	4.51	3.36	2.19	1.15	0.79	2.92
	D_m	0.37	0.70	1.28	1.90	2.38	2.58	2.57	2.27	1.56	0.88	0.43	0.29	1.44
60° South	G_m	0.90	1.75	2.83	3.68	4.20	4.58	4.17	4.34	3.29	2.18	1.16	0.80	2.83
	D_m	0.36	0.68	1.24	1.84	2.29	2.48	2.47	2.20	1.51	0.85	0.42	0.28	1.39
Vertical South	G_m	0.83	1.56	2.33	2.72	2.87	3.01	2.81	3.12	2.60	1.90	1.08	0.77	2.13
	D_m	0.29	0.54	1.00	1.46	1.80	1.97	1.91	1.77	1.22	0.69	0.33	0.22	1.10
Vertical SE/SW	G_m	0.67	1.28	2.05	2.65	3.00	3.25	2.94	3.09	2.36	1.59	0.86	0.60	2.03
	D_m	0.28	0.53	0.98	1.44	1.79	1.96	1.90	1.73	1.19	0.68	0.33	0.21	1.09
Vertical E/W	G_m	0.37	0.80	1.50	2.26	2.83	3.21	2.82	2.69	1.82	1.04	0.48	0.31	1.68
	D_m	0.26	0.50	0.91	1.37	1.75	1.94	1.86	1.64	1.10	0.62	0.31	0.20	1.04
Vertical NE/NW	G_m	0.24	0.48	0.97	1.62	2.23	2.60	2.27	1.94	1.20	0.62	0.29	0.18	1.22
	D_m	0.23	0.45	0.83	1.27	1.66	1.85	1.76	1.49	1.00	0.57	0.28	0.18	0.97
Vertical North	G_m	0.22	0.43	0.79	1.25	1.77	2.09	1.85	1.47	0.96	0.54	0.27	0.17	0.99
	D_m	0.22	0.43	0.79	1.22	1.60	1.80	1.69	1.41	0.96	0.54	0.27	0.17	0.93

INPUT DATA

	JAN	FEB	MAR	APR	MAY	JUN	JUL	AUG	SEP	OCT	NOV	DEC	MEAN
Daily sunshine hours	1.3	2.2	3.8	5.4	6.8	8.2	7.1	7.3	4.8	3.0	1.6	1.2	
Angstrom a + b	0.71	0.77	0.77	0.77	0.77	0.74	0.72	0.72	0.74	0.75	0.76	0.72	0.74
Normalisation factor	1.06	1.02	1.00	1.01	1.02	1.03	1.02	1.05	1.02	0.98	0.99	1.01	

Estimated clear day means of daily global radiation G_c and diffuse radiation D_c on inclined planes (1966 - 75)

Units : $kWh.m^{-2}$

		JAN	FEB	MAR	APR	MAY	JUN	JUL	AUG	SEP	OCT	NOV	DEC	MEAN
Date in month		29	26	29	28	29	21	4	4	4	4	4	4	
10° South	G_c	1.77	3.43	5.16	6.76	8.09	8.39	7.77	7.05	5.50	3.76	2.26	1.53	5.13
	D_c	0.65	0.82	1.04	1.47	1.64	1.88	1.89	1.98	1.37	0.93	0.56	0.44	1.23
30° South	G_c	2.64	4.70	6.20	7.31	8.21	8.40	7.81	7.43	6.35	4.89	3.40	2.53	5.83
	D_c	0.80	0.98	1.20	1.61	1.75	1.98	1.98	2.11	1.54	1.10	0.70	0.56	1.36
Latitude South (53°38')	G_c	3.31	5.56	6.59	7.07	7.42	7.47	6.99	7.01	6.53	5.56	4.26	3.36	5.93
	D_c	0.90	1.10	1.32	1.72	1.84	2.03	2.02	2.16	1.65	1.22	0.80	0.65	1.45
60° South	G_c	3.41	5.65	6.53	6.83	7.03	7.05	8.61	6.73	6.41	5.60	4.39	3.49	5.81
	D_c	0.91	1.11	1.34	1.73	1.86	2.03	2.02	2.15	1.66	1.24	0.82	0.66	1.46
Vertical South	G_c	3.35	5.22	5.33	4.88	4.53	4.43	4.21	4.66	5.01	4.97	4.29	3.59	4.53
	D_c	0.87	1.10	1.35	1.70	1.85	1.97	1.91	1.98	1.60	1.21	0.81	0.65	1.42
Vertical SE/SW	G_c	2.52	4.13	4.76	5.00	5.12	5.06	4.75	4.86	4.69	4.14	3.28	2.64	4.25
	D_c	0.76	1.01	1.30	1.69	1.88	2.02	1.96	2.00	1.56	1.15	0.73	0.56	1.39
Vertical E/W	G_c	1.11	2.25	3.35	4.23	5.02	5.06	4.68	4.30	3.53	2.56	1.59	1.04	3.23
	D_c	0.54	0.82	1.16	1.59	1.87	2.02	1.95	1.92	1.41	0.96	0.57	0.40	1.27
Vertical NE/NW	G_c	0.44	0.86	1.63	2.59	3.54	3.64	3.35	2.81	1.91	1.11	0.55	0.33	1.90
	D_c	0.42	0.67	0.98	1.39	1.70	1.84	1.76	1.66	1.17	0.77	0.46	0.32	1.10
Vertical North	G_c	0.40	0.63	0.90	1.45	2.25	2.30	2.20	1.74	1.09	0.70	0.43	0.31	1.21
	D_c	0.40	0.63	0.90	1.26	1.57	1.68	1.61	1.47	1.05	0.70	0.43	0.31	1.00

INPUT DATA

	JAN	FEB	MAR	APR	MAY	JUN	JUL	AUG	SEP	OCT	NOV	DEC	MEAN
Daily sunshine hours	6.4	8.6	10.5	12.5	14.4	15.5	14.7	13.4	11.4	8.7	7.1	6.0	
Angstrom a + b	0.71	0.77	0.77	0.77	0.77	0.74	0.72	0.72	0.74	0.75	0.76	0.72	0.74
Normalisation factor	1.00	1.01	0.98	1.00	1.02	1.01	0.99	1.03	0.99	1.00	0.95	1.01	

Table 2.2. Representative table from the CEC European Atlas, Volume II, Inclined Surfaces, showing tabular data available for 102 European sites. The Atlas also contains mapped values.

LATITUDE : 48º 46'N   Longitude : 2º 1'E   Altitude : 168m

Estimated monthly mean daily transmitted solar radiation fluxes through vertical clear single and double glazing for unobstructed sites (1966 - 75) (Glazing algorithms from method 5000) Units : kWh.m^{-2}

Surface Azimuth	Glazing type	JAN	FEB	MAR	APR	MAY	JUN	JUL	AUG	SEP	OCT	NOV	DEC	MEAN
South	Single	1.02	1.68	2.03	2.15	2.00	2.04	2.17	2.24	2.38	2.04	1.29	0.85	1.82
	Double	0.85	1.39	1.64	1.70	1.55	1.57	1.68	1.75	1.91	1.68	1.08	0.72	1.46
SSE/SSW	Single	0.96	1.58	1.98	2.19	2.08	2.16	2.28	2.30	2.36	1.93	1.21	0.80	1.82
	Double	0.79	1.30	1.61	1.75	1.63	1.67	1.78	1.82	1.91	1.59	1.00	0.66	1.46
SE/SW	Single	0.78	1.36	1.84	2.21	2.23	2.39	2.50	2.36	2.26	1.70	1.01	0.65	1.78
	Double	0.64	1.12	1.50	1.79	1.78	1.90	1.99	1.90	1.84	1.40	0.83	0.54	1.44
ESE/WSW	Single	0.59	1.10	1.63	2.13	2.28	2.54	2.61	2.32	2.05	1.42	0.79	0.49	1.66
	Double	0.48	0.90	1.33	1.73	1.84	2.05	2.11	1.89	1.67	1.15	0.64	0.39	1.35
East/West	Single	0.42	0.84	1.37	1.94	2.21	2.53	2.56	2.15	1.75	1.11	0.58	0.34	1.48
	Double	0.33	0.68	1.11	1.57	1.79	2.06	2.08	1.75	1.42	0.89	0.46	0.27	1.20
ENE/WNW	Single	0.30	0.62	1.10	1.67	2.02	2.36	2.35	1.88	1.42	0.83	0.41	0.25	1.27
	Double	0.24	0.49	0.88	1.34	1.63	1.91	1.90	1.51	1.14	0.65	0.33	0.19	1.02
NE/NE	Single	0.26	0.49	0.87	1.38	1.76	2.08	2.03	1.56	1.11	0.63	0.33	0.21	1.06
	Double	0.20	0.38	0.68	1.10	1.41	1.67	1.63	1.24	0.87	0.49	0.26	0.17	0.84
NNE/NNW	Single	0.25	0.45	0.74	1.15	1.50	1.77	1.71	1.29	0.90	0.56	0.31	0.21	0.90
	Double	0.20	0.35	0.58	0.90	1.18	1.40	1.35	1.02	0.71	0.45	0.24	0.16	0.71
North	Single	0.25	0.44	0.72	1.05	1.38	1.61	1.55	1.17	0.86	0.55	0.30	0.20	0.84
	Double	0.20	0.35	0.57	0.83	1.08	1.26	1.21	0.92	0.68	0.44	0.24	0.16	0.66

Estimated clear day means of transmitted solar radiation fluxes through vertical clear single and double glazing for unobstructed sites (1966 - 75) (Glazing algorithms from Method 5000) Units : kWh.m^{-2}

Surface Azimuth	Glazing type	JAN	FEB	MAR	APR	MAY	JUN	JUL	AUG	SEP	OCT	NOV	DEC	MEAN
South	Single	4.11	4.50	4.11	3.43	2.77	2.67	2.76	3.10	3.79	4.13	3.97	4.06	3.61
	Double	3.46	3.73	3.31	2.68	2.30	2.01	2.08	2.40	3.02	3.39	3.33	3.44	2.91
SSE/SSW	Single	3.79	4.19	4.09	3.63	3.05	2.95	3.04	3.30	3.84	3.94	3.67	3.73	3.60
	Double	3.18	3.48	3.33	2.87	2.35	2.26	2.33	2.58	3.09	3.25	3.07	3.14	2.91
SE/SW	Single	3.05	3.55	3.89	3.88	3.60	3.54	3.65	3.59	3.78	3.50	3.00	2.88	3.49
	Double	2.54	2.94	3.20	3.14	2.87	2.81	2.90	2.89	3.09	2.90	2.49	2.40	2.85
ESE/WSW	Single	2.21	2.77	3.45	3.84	3.92	3.93	4.03	3.62	3.48	2.89	2.21	1.93	3.19
	Double	1.82	2.28	2.85	3.16	3.20	3.20	3.29	2.96	2.87	2.38	1.82	1.58	2.62
East/West	Single	1.40	1.96	2.81	3.49	3.85	3.92	4.00	3.35	2.95	2.18	1.45	1.10	2.71
	Double	1.13	1.59	2.30	2.87	3.16	3.22	3.29	2.74	2.42	1.78	1.17	0.87	2.21
ENE/WNW	Single	0.78	1.23	2.06	2.86	3.41	2.52	3.59	2.81	2.27	1.47	0.84	0.53	2.12
	Double	0.60	0.97	1.66	2.34	2.79	2.89	2.94	2.29	1.84	1.18	0.66	0.40	1.72
NE/NE	Single	0.45	0.74	1.34	2.11	2.71	2.85	2.88	2.13	1.57	0.92	0.50	0.32	1.55
	Double	0.35	0.58	1.06	1.69	2.19	2.30	2.34	1.71	1.25	0.72	0.39	0.25	1.24
NNE/NNW	Single	0.39	0.59	0.87	1.42	1.95	2.08	2.09	1.50	1.06	0.65	0.44	0.31	1.11
	Double	0.31	0.47	0.68	1.12	1.54	1.65	1.66	1.18	0.83	0.51	0.35	0.25	0.88
North	Single	0.38	0.59	0.77	1.10	1.57	1.71	1.70	1.19	0.88	0.62	0.43	0.31	0.94
	Double	0.31	0.47	0.61	0.86	1.22	1.32	1.32	0.93	0.70	0.49	0.34	0.25	0.73

Table 2.4. Representative table from CEC European Solar Radiation Atlas, Vol. II, Inclined Surfaces giving daily irradiation incident on and directly transmitted through single and double clear glazing on unobstructed sites (1966 – 75)

MONTHLY MEAN VALUES

J	F	M	A	M	J	J	A	S	O	N	D

Incident radiation kWh.m^{-2} per day

1.23	2.05	2.57	2.82	2.69	2.74	2.95	2.97	3.06	2.53	1.57	1.03

Transmitted single glazing kWh.m^{-2} per day

1.02	1.68	2.03	2.15	2.00	2.04	2.17	2.24	2.38	2.04	1.29	0.85

Transmittance single glazing

0.83	0.82	0.79	0.76	0.74	0.74	0.74	0.75	0.78	0.81	0.82	0.83

Transmitted double glazing kWh.m^{-2} per day

0.85	1.39	1.64	1.70	1.55	1.57	1.68	1.75	1.91	1.68	1.08	0.72

Transmittance double glazing

0.69	0.68	0.64	0.60	0.58	0.57	0.57	0.59	0.62	0.66	0.69	0.70

CLEAR DAY VALUES

J	F	M	A	M	J	J	A	S	O	N	D

Incident radiation kWh.m^{-2} per day

4.94	5.50	5.29	4.64	3.92	3.79	3.92	4.25	4.93	5.15	4.79	4.82

Transmitted single glazing kWh.m^{-2} per day

4.11	4.50	4.11	3.43	2.77	2.67	2.76	3.10	3.79	4.13	3.97	4.06

Transmittance single glazing

0.83	0.82	0.78	0.74	0.71	0.70	0.70	0.73	0.77	0.80	0.83	0.84

Transmitted double glazing kWh.m^{-2} per day

3.46	3.73	3.31	2.68	2.10	2.01	2.08	2.40	3.02	3.29	3.33	3.44

Transmittance double glazing

0.70	0.68	0.63	0.58	0.54	0.53	0.53	0.56	0.61	0.64	0.70	0.71

Table 2.5. The computation of estimated daily transmittance values for single and double, south-facing vertical glazing for Paris/Trappes, derived from the data in the CEC Radiation Atlas.

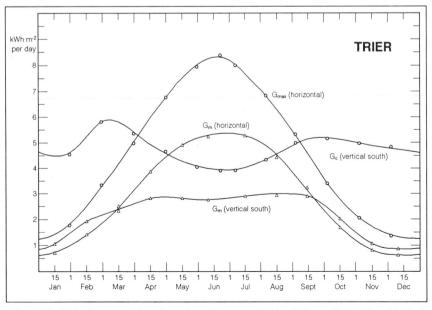

Figure 2.2. Plotted values of clear day and monthly mean daily irradiation on horizontal and vertical south surfaces at Trier. The horizontal values are observed. The vertical values are predicted. Source CEC European Solar Radiation Atlas, Part II, Inclined Surfaces.

The corresponding daily beam irradiation on any slope is found by subtracting the diffuse irradiation from the global irradiation.

At the base of each part of the tables, key input data is given. The horizontal global radiation values are not given, because they are the values provided in the CEC European Solar Atlas, Vol 1, Horizontal Surfaces. The normalization factor is the arithmetic adjustment that was applied to the modelled values in that specific month to match the daily sum of the predicted hourly horizontal global irradiation values to the actual observed horizontal daily irradiation values.

The estimates of the daily slope diffuse radiation include a ground reflected component based on a ground albedo of 0.2, added to the sky component. The Atlas explains how to adjust the data for other values of the ground reflectance.

The tables then were used in conjunction with the horizontal surface radiation maps to construct monthly mean slope irradiation maps. These plot isolines of daily slope irradiation. Unobstructed sites are assumed, and the detailed effects of obstruction by terrain are not considered.

Designers can achieve a better idea of the annual pattern of variation at any particular site by plotting out the tabulated results graphically. Figure 2.2 shows such a diagram for Trier.

For certain inclined surfaces, in certain months, the monthly variation of the slope irradiation may proceed in the opposite direction to the variation in the monthly mean horizontal irradiation. For example, at Trier, as Figure 2.2 shows, the estimated monthly irradiation values on a vertical south slope on clear days (days of low cloudiness) reach a maximum value at the beginning of March, then fall to reach a mid summer minimum around 21 June. They then start to rise again to reach a secondary maximum at the beginning of October before falling to a mid winter minimum. The risk of solar overheating of heavily glazed south facing structures at this latitude is thus greatest in September/early October, when high irradiation may coincide with reasonably high outdoor temperatures. The reasons for these seasonal patterns of irradiation are discussed later.

It is also instructive for design understanding to calculate the ratio of diffuse to global radiation for different slopes for a station close to the site under consideration. Table 2.3 provides such data for two European sites, one a relatively cloudy maritime site, the other a relatively sunny inland site.

The following salient features should be noted:-

1. The annual proportion of diffuse radiation is considerable and is greater in the cloudier climate on all surfaces. It is therefore advantageous to try to harness diffuse radiation as well as beam radiation.

2. The annual proportion of diffuse radiation is least on surfaces with a tilt equal to latitude. This tilt optimizes the annual input of direct beam energy.

3. The lowest proportion of diffuse radiation on vertical south surfaces is found in the middle of winter.

4. The proportion of diffuse radiation on vertical south surfaces is high in midsummer. This is partly the consequence of relatively high mid day solar altitude, when the sun's rays strike the facade very obliquely reducing the beam radiation. More importantly, in this season the sun swings over from the north to reach its lowest profile angle on the south facade at solar noon. The value of direct beam summer shading by overhangs, as opposed to general sky and ground reflected energy shading by diffusing or reflecting devices, fitted over the plane of the window, like traditional venetian blinds, is thus limited.

The Atlas also contains monthly mean daily values and maximum clear day values of the energy transmitted per unit area through vertical single and double clear glazing for the same 102 sites. Table 2.4 presents a typical table.

Transmittance is the ratio of the transmitted energy to the incident energy. The daily mean transmittances of single and double glazing depends on the month and the orientation. As an example, Table 2.5, using the data from Table 2.4, sets out the calculation of the daily transmittance values for south facing vertical glazing for a site located at Paris/Trappes. The fall off in transmittance with the high summer sun and its increase with the low winter sun is evident.

It is immediately seen that the glazing radiative transmission losses, even with totally clean windows are quite considerable, and vary with season.

The European Inclined Surface Solar Radiation Atlas thus provides an important tool for preliminary analysis of the passive solar building design. It can help the designer decide on choice of facade slope and orientation. However it does not provide a direct tool for considering detailed site layout with its consequent solar obstruction. The geometric and microclimatic energy analysis of solar irradiation impacts, are discussed later. Designers however should start by preparing a clear macroclimatic statement concerning the un-obstructed radiation on their site, concentrating particularly on the facade orientations of greatest significance.

2.7
PROJECT EUFRAT

The EUFRAT project [19] [20] is another important climatic data project sponsored by the CEC as part of its Solar Energy R & D programme. While this project has specifically aimed to produce additional climatic data needed for the design of active solar systems, much of the data is still very useful for passive solar design purposes. The publication contains systematically computed data on the frequency distribution of hourly solar radiation on different slopes and on temperature degree days. The data are mounted on a Personal Computer Data Base on diskettes readable on a personal computer (IBM or Compatible). A list of the European sites included in the data

base is given in Appendix 7. In addition to systematic tables for the statistical study of hourly solar radiation, the EUFRAT publication also contains plotted Cumulative Frequency Curves (CFC) of the hourly means of irradiance incident on inclined slopes for a range of 30 European sites, derived from long term hourly data files.

2.8
WINTER DESIGN TEMPERATURES

Peak heating demands, which influence the size of heating plant required, are determined using the outdoor design temperature. Normally, account is taken of the thermal inertia of buildings, and so extreme hourly minimum values of temperature are not used to determine design temperatures. Averaging over one or two days may be used. Different methods for deciding on the design temperatures are used in different European countries [21] [22]. In Germany, for example, the two day mean of the lowest air temperature occurring 10 times in twenty years is used in DIN 4701 for calculating heating needs of buildings. Table 2.6 provides some representative values to provide a European-wide view. In general, winter design temperatures rise as one moves west towards the Atlantic. Temperatures at low levels are higher south of the Alps. Altitude is always important.

Accumulated temperature differences (Heating Degree Days)

Clearly degree day data is important in assessing the impact of variations of monthly outdoor air temperatures on the demand side of active and passive system design analysis. Sometimes, less satisfactorily, monthly mean or winter mean temperatures are used instead in energy estimates. A climatic atlas of Europe giving temperature and precipitation data was published in 1970 [23].

Until recently, rather little European coordination had been achieved on degree day data, in contrast with solar radiation and wind energy data. Different methods

Country	Place	Winter design temperature, deg C.
Belgium	Brussels	-7
Denmark	Copenhagen	-7
France	Brest	-2
	Paris	-4
	Lyon	-10
	Strasbourg	-15
	Marseille	-2
Germany	Hamburg	-12
	Trier	-10
	Berlin	-14
	Munich	-16
Italy	Rome	-1
Ireland	Dublin	-3
	Shannon	-2
Netherlands	Amsterdam	-7
Portugal	Lisbon	4
Sweden	Stockholm	-13
UK high thermal inertia buildings:		
	Plymouth	0
	Belfast	-2
	Glasgow	-2
	Cardiff	-2
	Birmingham	-3
	Edinburgh	-4
	London	-2
DIN 4071 values		

Table 2.6; Approximate winter design temperatures (Sweden): Source mainly CIBSE Guide, Vol A, Design Data, [22]

for assessing degree days have previously been used in different EC Member States. The number of degree days at any site has to be extracted using an appropriate base temperature. In the past, there have been national variations in the selection of fixed base reference temperatures. It is now agreed that fixed base temperatures provide an unsatisfactory approach. The correct base temperature to be used in any design situation depends on the interaction of three factors, the comfort thermostat setting required, the internal casual heat gains, and the thermal loss characteristics of the structure. The balance temperature at which the monthly casual gains just balance the monthly thermal losses with the internal temperature at the thermostat setting, provides guidance on the correct choice of base temperature. This new approach avoids the limitations of the older single base temperature approach. The improved analytical design methods now being used, require the number of degree days in each month to be taken out to different base temperatures according to the balance of internal heat gains and building thermal losses. This data is now referred to as the accumulated temperature

difference, ISO [24]. It is essential therefore to provide such data for a range of base reference temperatures. For example, a recent U.K. design handbook [25] presented data for the U.K. degree days at 2 degree centigrade intervals of base temperature between 0 deg. C. and 22 deg. C. This data was extracted from the original temperature data in the U.K. Meteorological Office's data bank. Very often, however, statistical methods are used to estimate degree days, based on the use of the standard deviation of the monthly mean temperature about its long term monthly mean in different months.

In the ISO method [26], the balance temperature, t_b is calculated from the formula:-

$$t_b = t_{th} - h_m (Q_i + Q_s)/Q_e \text{ where:}$$

t_{th} is the average thermostat set point.

h_m is the monthly mean utilization factor of the internal gains.

Q_i is the mean monthly incidental heat gain per day from people, appliances, water heating & lighting.

Q_s is the mean monthly incidental heat gain per day from solar energy.

Q_e is the building heat loss per unit temperature difference per day (Fabric plus infiltration losses).

As the thermal insulation of structures is improved, the balance temperature becomes lower. The ISO Standard provides guidance on the estimation of the utilization factor for the combination of internal gains and solar gains. However, in many correlation models for passive solar design, the solar radiation gains are input separately, for example as the solar load ratio in the SLR method. Here the specific aim is to estimate the fractional solar contribution. A simpler form of the balance temperature formula is then used to establish the base temperature.

$$t_b = t_{th} - Q_i /Q_e$$

The symbols have the same meaning as above. It is essential in establishing the base temperature to use the same form of the equation as that actually used to develop the correlation method adopted.

EUFRAT data on temperature and degree days.

The EUFRAT publication [20] and data disc contain temperature and degree day data and provide a truly comparable European basis for the first time. A validated statistical method, described in the publication, has been used to produce the data.

EUFRAT degree day data has been prepared for over 50 locations in Europe. The tables provided give monthly mean screen temperatures, Ta, together with the associated standard deviations of the monthly means about their long term monthly mean, S. These monthly means and standard deviations provide the inputs for the subsequent degree day calculations. The degree day data has been computed to base temperatures of 20,18,16, and 14 deg. C. Table 2.7 provides an example of the data for 5 locations. Note the site at Santis in Switzerland at 2469m is very high, and heating is needed throughout the year.

The impact of site micro-climate and changes in site elevation on heating degree days needs more investigation. In the U.K. the mapped annual values are increased by 2 degree days C for each metre of altitude increase [22]. Hitchin [27] confirmed Chandler's estimate that the annual number of degree days in central London was about 10% less than at London, Heathrow. Urban heat island effects clearly exert an influence, but it is difficult at present to make accurate estimates of the impact of urban effects on degree days.

EUFRAT computer software.

The following pieces of software have been produced, including programs on diskettes and the corresponding user documentation:-

- EUFRAT 2.0 for hourly radiation file processing (FORTRAN; micro-computer + main-frame) (Author LNETI, Portugal)
- CFC and utilizability data base + user's routines: data from EUFRAT first phase (IBM-PC compatible + user's reading routines: BASIC or PASCAL) (Author EUFRAT Program participants + ETSII Seville).

The estimation of diurnal swings of outdoor temperature.

Knowledge of the typical swings of outdoor ambient temperature is sometimes required, for instance to assess the summer performance of buildings, or to check that the thermal capacity of a passive solar heated building is adequate to deal with the expected swings of indoor temperature on sunny days in winter [28]. For example, the admittance method for assessing indoor temperatures in unheated naturally ventilated buildings requires knowledge of both the mean radiative inputs and peak radiation inputs on the sunny days likely to cause overheating, as well as the mean daily sunny day outdoor design temperature and its diurnal swing about the mean [29].

It is fairly easy to get statistical summaries of outdoor dry bulb temperatures in the form given in Table 2.8 . It is simple to obtain the monthly mean daily swing of outdoor temperature by subtracting the mean monthly daily minimum temperatures from the mean monthly daily maximum temperature. However the range of temperature on hot sunny days cannot be

NETHERLANDS DeBILT
LAT.:52.06N LONG.:05.11E ALT.: 0m PERIOD 19310 - 60

MONTH	JAN	FEB	MAR	APR	MAY	JUN	JUL	AUG	SEP	OCT	NOV	DEC	YEAR
Ta	1.7	2.0	5.0	8.5	12.4	15.5	17.0	16.8	14.3	10.0	5.9	3.0	9.3
S	2.2	3.6	2.4	1.7	2.0	1.9	1.9	1.8	2.3	2.2	2.2	3.5	
DD(20)	567	504	465	345	235	137	98	103	173	310	423	527	3887
DD(18)	505	448	403	285	173	76	38	42	111	248	363	465	3157
DD(16)	443	392	341	224	111	20	2	2	52	186	302	403	2478
DD(14)	381	336	278	164	49	0	0	0	4	124	243	341	1920

BELGIUM Uccle
LAT.:50.48N LONG.:04.21E ALT.: 104m PERIOD 19310 - 60

MONTH	JAN	FEB	MAR	APR	MAY	JUN	JUL	AUG	SEP	OCT	NOV	DEC	YEAR
Ta	2.2	2.6	6.0	9.2	13.6	16.0	17.5	17.3	14.7	10.3	6.2	3.3	9.9
S	3.0	3.9	2.9	2.0	2.0	2.0	2.0	2.0	2.3	2.3	2.0	3.5	
DD(20)	551	487	434	324	199	123	86	91	161	300	414	517	3687
DD(18)	489	431	372	264	136	62	28	32	100	238	354	455	2961
DD(16)	427	375	310	204	74	11	1	1	41	176	294	393	2307
DD(14)	365	319	247	144	15	0	0	0	1	114	234	331	1770

LUXEMBURG Luxemburg
LAT.:49.37N LONG.:06.03E ALT.: 330m PERIOD 19310 - 60

MONTH	JAN	FEB	MAR	APR	MAY	JUN	JUL	AUG	SEP	OCT	NOV	DEC	YEAR
Ta	0.3	1.0	4.9	8.5	12.8	15.7	16.7	17.4	13.8	9.0	4.6	1.3	8.8
S	3.0	3.9	2.9	2.0	2.0	2.0	2.0	2.0	2.3	2.3	2.0	3.5	
DD(20)	610	532	468	345	223	131	107	89	187	341	462	579	4074
DD(18)	548	476	406	285	161	71	46	30	126	279	402	517	3347
DD(16)	486	420	344	225	99	16	4	1	66	217	341	455	2674
DD(14)	424	364	282	165	37	0	0	0	11	155	282	393	2113

SWITZERLAND Zurich
LAT.:47.23N LONG.:08.34E ALT.: 569m PERIOD 19310 - 60

MONTH	JAN	FEB	MAR	APR	MAY	JUN	JUL	AUG	SEP	OCT	NOV	DEC	YEAR
Ta	-1.1	0.3	4.5	8.6	12.7	15.9	17.6	17.0	14.0	8.6	3.7	0.1	8.5
S	3.0	3.9	2.9	2.0	2.0	2.0	2.0	2.0	2.3	2.3	2.0	3.5	
DD(20)	654	551	480	342	226	126	84	99	181	353	489	616	4201
DD(18)	592	495	418	282	164	65	26	39	120	291	429	554	3475
DD(16)	530	439	356	222	102	12	0	2	60	229	368	492	2812
DD(14)	468	383	294	161	40	0	0	0	7	167	308	430	2258

SWITZERLAND Santis
LAT.:47.15N LONG.:09.21E ALT.: 2496m PERIOD 19310 - 60

MONTH	JAN	FEB	MAR	APR	MAY	JUN	JUL	AUG	SEP	OCT	NOV	DEC	YEAR
Ta	-9.0	-9.0	-6.6	-4.1	0.4	3.6	5.6	5.5	3.5	-0.6	-4.5	-7.6	-1.9
S	3.0	4.6	3.4	2.9	2.1	2.1	2.1	2.0	2.3	2.7	2.5	3.4	
DD(20)	899	812	824	723	607	492	446	449	495	638	735	855	7975
DD(18)	836	756	762	663	545	432	384	387	435	576	674	793	7243
DD(16)	775	700	700	602	483	372	322	325	374	514	615	731	6513
DD(14)	713	644	638	542	421	311	260	263	314	452	555	669	5782

Table 2.7.: EUFRAT monthly mean temperature data, standard deviations of monthly means about the long term mean and derived degree day data for 5 sites to base temperatures of 20,18,16 and 14 deg.C. Source EUFRAT climatic data handbook.

	JAN	FEB	MAR	APR	MAY	JUN	JUL	AUG	SEP	OCT	NOV	DEC	YEAR
PLYMOUTH (Mount Batten): 1941-70													
Average Daily													
Max	8.0	7.9	9.9	12.3	14.9	17.6	19.0	19.0	17.5	14.8	11.1	9.2	13.4
Min	3.8	3.2	4.3	6.1	8.2	11.1	12.7	12.7	11.6	9.4	6.4	4.9	7.9
Mean	5.9	5.5	7.1	9.2	11.5	14.3	15.9	15.9	14.5	12.1	8.7	7.1	10.7
Average Monthly													
Max	11.9	11.9	14.5	17.5	21.0	23.7	24.7	23.6	21.8	19.0	14.9	12.9	26.0
Min	-3.6	-3.0	-1.6	0.7	2.8	6.3	8.0	7.8	5.5	2.2	-0.5	-2.1	-5.1
Absolute													
Max	13.9	15.0	19.4	22.2	26.1	27.8	28.9	30.6	27.2	23.3	16.7	14.4	30.6
Min	-8.9	-8.3	-7.0	-1.7	-0.6	-1.7	6.1	3.9	2.8	-1.7	-3.9	-5.7	-8.9
LONDON (Kew): 1941-70													
Average Daily													
Max	6.1	6.8	9.8	13.3	16.8	30.2	21.6	21.0	18.5	14.7	9.8	7.2	13.8
Min	2.3	2.3	3.4	5.7	8.4	11.5	13.4	13.1	11.4	8.5	5.3	3.4	7.4
Mean	4.2	4.5	6.6	9.5	12.6	15.9	17.5	17.1	14.9	11.6	7.5	5.3	10.6
Average monthly													
Max	11.8	12.2	16.5	20.1	24.3	26.4	28.4	26.9	23.7	19.9	14.9	12.5	29.7
Min	-4.5	-3.4	-2.2	-0.6	3.0	7.0	9.3	8.5	5.7	1.4	-1.6	-3.5	-5.8
Absolute													
Max	14.4	16.1	22.0	25.6	30.0	32.8	33.9	31.1	30.0	25.6	17.4	15.0	33.9
Min	-9.7	-9.4	-5.9	-1.7	-1.2	3.5	6.5	6.2	2.1	-2.8	-5.0	-7.2	-9.7

Table 2.8: Typical meteorological summary of observed long term dry bulb screen temperatures. Source Page and Lebens [25].

obtained by subtracting the average monthly minimum temperature from the average monthly maximum temp-erature, as the two data sets are derived from different days in the daily record, the maxima being associated with hot days, and the minima being associated with cold days.

Table 2.9 compares diurnal temperature swing data for clear and overcast days at Plymouth, a maritime site, and London Kew, an urban site towards the edge of London, with the corresponding observed monthly mean values, using data from reference 25. It will be noted that the monthly mean outdoor temperature swing on clear days is about twice the average temperature swing. Further, the amplitude of the swing varies with time of year. This is to be expected, because it is the diurnal cycle of

solar radiation that drives the diurnal temperature cycle.

On overcast days the swing is very small, especially at the maritime site. It will also be noted that clear days in winter are relatively cold days. Such specially abstracted data was available for five U.K. sites. Figure 2.4 plots some closely allied data for London, Kew. It brings out some of key factors concerning the temperature regimes in the regions of Northern Europe strongly influenced by proximity to the westerly oceans and seas.

The monthly mean maximum temperatures provide some guidance on hot weather maximum design temperatures. The problem of characterizing swings of temperature on such days on a European wide basis remains, in the absence of systematic published data. In order to generalize the clear day temperature swing data for hot weather design purposes, regressions were sought, using the U.K. data, between the monthly observed clear day swings and the monthly values of GMAX given in the European Solar Radiation Atlas, Vol 1, Horizontal Surfaces. Very good correlations were found between clear day temperature swing and the European Atlas values of GMAX. The results are given in Table 2.10.

The regression curves found are plotted out in Figure 2.4. The lowest summer swings were found at Plymouth, a maritime site where land and sea breeze effects are likely to moderate swings. Kew and Manchester Airport have similar patterns with a greater swing. The swings are greatest at Glasgow and Aberdeen. Glasgow Airport temp-eratures are probably influenced by night time air drainage from the foothills of nearby mountains, which will tend to increase the diurnal range. There are also foothills to the Highlands close to Aberdeen Dyce Airport, again probably leading to night air drainage, and consequently increased diurnal swings. Night air drainage seems to increase the swing by 2 to 3 degrees. Land and sea breezes appear to reduce the swing by about 3 degrees.

It is not known to what extent these relationships would hold for other parts of Europe. However examination of hot weather temperature data given in various

	JAN	FEB	MAR	APR	MAY	JUN	JUL	AUG	SEP	OCT	NOV	DEC
PLYMOUTH (Mount Batten): 1941-70												
Clear skies Max	2.3	7.0	11.9	15.6	17.6	20.2	21.1	22.7	20.8	17.3	9.9	4.8
Clear skies Min	-1.9	0.5	3.1	7.7	8.3	12.6	12.3	14.0	11.7	8.3	5.1	0.6
Difference	4.2	6.5	8.8	7.9	9.3	7.6	8.8	8.7	9.1	9.0	4.8	4.2
Monthly Mean Max	7.4	7.6	9.0	11.0	13.5	16.4	17.9	18.1	16.7	14.0	10.3	8.3
Monthly Mean Min	5.5	5.1	5.6	6.8	9.0	11.8	13.4	13.6	12.3	10.7	7.9	6.4
Difference	1.9	2.5	3.4	4.2	4.5	4.6	4.5	4.5	4.4	3.3	2.4	1.9
Overcast skies Max	7.3	6.2	8.6	9.6	11.2	13.8	14.8	16.3	15.1	12.8	11.3	9.6
Overcast skies Min	6.2	5.2	7.2	8.3	10.2	12.3	13.6	15.2	14.1	11.9	18.2	7.8
Difference	1.1	1.0	1.4	1.3	1.0	1.5	1.2	1.1	1.0	0.9	1.1	1.8
LONDON AREA OUTER REGION												
London Clear skies Max	3.2	5.9	11.2	16.1	20.6	24.8	26.8	25.1	21.2	15.4	8.1	3.4
Kew Clear skies Min	-1.8	-0.8	1.0	4.7	8.6	12.5	13.9	12.3	10.4	6.6	2.5	-0.1
Difference	5.0	6.7	10.2	11.4	12.0	12.3	12.9	12.8	10.8	8.8	5.6	3.5
Monthly Mean Max	6.1	6.9	9.1	11.9	15.7	19.2	20.7	20.2	17.9	14.4	9.4	6.8
Monthly Mean Min	3.8	3.7	4.3	6.1	8.8	11.7	13.7	13.4	11.6	9.6	6.3	4.7
Difference	2.3	3.2	4.8	5.8	6.9	7.5	7.8	6.8	6.3	4.8	3.1	2.1
Overcast skies Max	5.8	5.9	6.7	8.8	11.9	14.4	16.6	16.3	15.4	13.1	9.3	6.5
Overcast skies Min	4.1	3.9	4.8	6.4	9.4	12.4	14.2	14.4	13.0	10.7	7.0	4.4
Difference	1.7	2.0	1.9	2.4	2.5	2.0	2.4	1.9	2.4	2.4	2.3	2.1

Table 2.9.: Monthly values of daily maximum and minimum temperatures for clear days, average days and overcast days for two U.K. sites. Source of data: Page and Lebens [25].

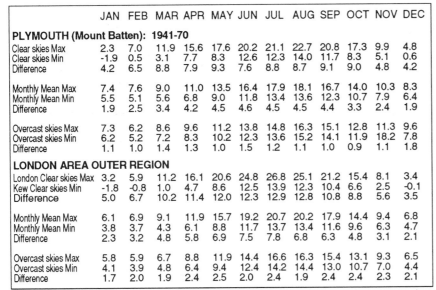

Figure 2.3. Diurnal variations in shade air temperature at Kew for three classes of daily global irradiation: smooth curves-days of high radiation, dotted curve-average days and curve with circles-days of low irradiation, for the months of March, June, September and December. Source "Solar Energy, a UK Assessment", UK ISES [41].

SITE	a_0	a_1	a_2	Correlation coefficient
Plymouth	0.67	2.78×10^{-3}	-2.28×10^{-7}	0.934
London Kew	0.53	2.85×10^{-3}	-1.68×10^{-7}	0.990
Manchester	0.45	3.46×10^{-3}	-2.56×10^{-7}	0.990
Glasgow	1.51	4.53×10^{-3}	-3.68×10^{-7}	0.969
Aberdeen	2.09	4.05×10^{-3}	-3.51×10^{-7}	0.937

Table 2.10.: Regression equations linking the monthly diurnal swing of temperature on clear days to the values of GMAX, the monthly mean maximum daily global radiation on horizontal planes given in the CEC European Solar Radiation Atlas, Vol. 1, Horizontal Surfaces [15].

papers to a workshop on passive cooling [2], reveals observed outdoor diurnal temperature ranges on hot days as follows:-

1.   31 - 13 = 18 deg. C reported for Senlis, France

2.   36 - 24 = 12 deg. C in July at Giarra, a coastal site in Sicily, Italy.

3.   36 - 21.5 = 14.5 deg. C in August at Seville, Spain, a site with probably some air drainage from nearby hills.

4.   36 - 24 = 12 deg. C in July in Rome.

These figures are therefore broadly in line with the values given by Figure 2.4.

Estimation of hourly temperatures
In clear weather, the lowest temperatures occur just before dawn. Screen temperatures reach a peak about two hours after solar noon in winter and about three hours after solar noon in summer, i.e. around 1600 hours, when adjusted for European summer time. Statistically similar patterns are found for monthly mean and overcast day mean hourly values of temperature. Given knowledge of the maximum and minimum temperature, hourly temperature variations can be simulated by using two separate sinusoidal interpolations, one covering the period between the minimum and maximum, and the other the period between the maximum and the minimum. Figure 2.5 illustrates the principle. The detailed methodology is given in the CIBSE Guide together with supporting tables for easy calculation [22].

2.9
HUMIDITY DATA

Humidity is crucial for comfort in hot weather, therefore the availability of humidity data is important for bioclimatological design analysis in summer. There is currently no collated data on humidity for the CEC region. Data however may be found in individual national summaries. The best parameter with which to work is the vapour pressure. The mean values of the vapour pressure are relatively stable across the course of the day. The relative humidity in contrast varies strongly across the course of the day, as it is so dependent on ambient temperature. The highest values of the relative humidity tend to occur around dawn. The lowest values tend to occur in the middle of the

afternoon, when the air temperatures are highest. The midday drop in the mean relative humidity tends to be considerable in summer, and very much less in winter, when the diurnal swings are smaller. Figure 2.6 shows some representative data for the U.K. The relative stability of the vapour pressure data stands in marked contrast to that of the relative humidity data.

Summer dry and wet bulb design temperatures are tabulated in many engineering guides for example the CIBSE Guide [22]. These figures can be used to derive the vapour pressure for design hot days. If the hourly course of the daily temperature is estimated, it then becomes possible to estimate the associated variations in relative humidity, hour by hour, using standard methods.

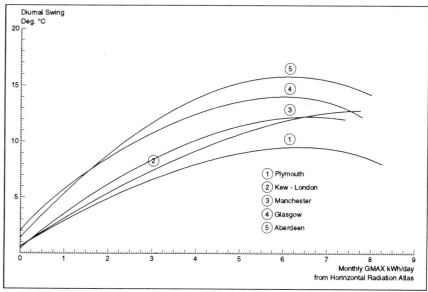

Figure 2.4. The relationship between the observed clear day diurnal swing of temperature, and the CEC European Solar Radiation, Atlas, Vol 1, Horizontal Surface values of GMAX for five U.K. sites.

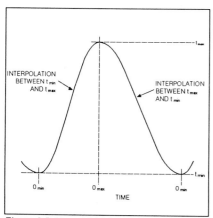

Figure 2.5 Sinusoidal interpolations of outdoor temperature between minimum temperature and maximum temperature and between maximum temperature and minimum temperature: Source CIBSE Guide [22].

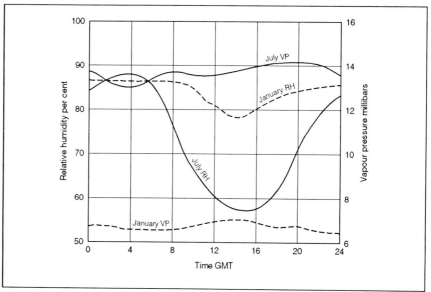

Figure 2.6. The mean diurnal variation of relative humidity and vapour pressure at London, Kew in January and July. Averages for 30 year period, 1886-1915 Source Lacy [30].

2.10

LONG WAVE THERMAL RADIATION DATA

There are relatively few measurements available of long wave radiation, in spite of its thermal importance, (especially for natural cooling). However the heat exchange by long wave thermal radiation can be indirectly assessed from the screen air temperature and the cloud cover. For inclined surfaces, the thermal energy exchange with the ground has also to be taken into account. One needs to be able to make some estimates of ground temperature conditions as part of the assessment process. Appendix 8 sets out in detail a methodology for the estimation of the long wave energy exchanges on unobstructed sites, giving particular attention to cloudless condition, when the cooling effects are greatest. Knowledge is lacking on long wave exchanges on obstructed sites. For example, heated buildings opposite another building will modify the long wave radiative exchanges, but practicable methods for assessing such impacts have yet to be developed. The basic assessment problem bears some relationship to the daylighting assessment problem, because both problems rely on knowledge of sky radiance distributions. Better characterization of the long wave radiance of the sky is needed, before practical design assessment procedures can be developed. Another gap in current knowledge is lack of representative ground surface temperatures to adopt for design purposes under different weather conditions. Appendix 8 provides further details.

2.11

WIND AND TEMPERATURE

Because the surface layers always exert such a strong influence on wind flows and directions, all wind measurements tend to be microclimatologically influenced. Detailed wind flows are therefore best considered in the context of microclimate. The assessment of the site wind climate using the CEC European Wind Atlas is discussed in Section 2.14 onwards. Wind however interacts with other climatic design variables.

In mid winter, easterly weather in Northern Europe tends to bring cold dry air, and westerly weather tends to bring warmer damp air from the Atlantic and is often associated with considerable cloudiness. In summer the temperature and wind direction issues are reversed and easterly winds blowing from the centre of the continent tend to be warmer, and incidentally more polluted. The westerly winds tend to be cooler, and the air cleaner. Figure 2.7 illustrates the relationships between temperature and wind direction in winter and in summer at Hamburg from Reidat's work [31]. Such directional data is representative of areas of northern Europe close to the maritime influence. It demonstrates the need for winter protection against north easterly and easterly winds.

In preliminary site analysis it is useful to set such directional wind temperature data alongside the monthly mean and clear day solar radiation data for different seasons plotted in a similar directional way. Such diagrams may be easily constructed for vertical surfaces, using the various CEC Solar Radiation Atlas slope radiation tables discussed above. Figure 2.8 provides an example of such a directional diagram.

Wind also influences the diurnal range of temperature. In general the stronger the wind blows, the less impact the daily radiation energy cycle makes on the diurnal temperature range of the air layers close to the ground. Under windy conditions, the heat energy from the ground tends to be dissipated through a thicker layer of the atmosphere. As hilltop sites are windy, by day, the temperatures on hill tops in sunny weather tend not to rise as much as in sheltered sites like valleys. Hilltops tend to feel particularly cold because of the strong convective body cooling caused by the frequent strong winds. The long wave radiative exposure is also considerable. By night in still weather however, hilltops are free from the cold air drainage effects. These can cause low temperatures on still nights in nearby valleys. Even above level ground there are effects. The diurnal temperature range over level ground tends to decrease with height. Temperatures on still clear nights may rise as one moves upwards from the radiatively cooled ground surface layers.

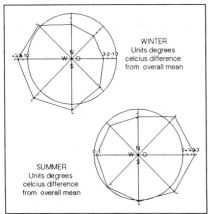

Figure 2.7. The relationship between wind direction and mean daily temperature in winter and in summer at Hamburg. The basic circle defines the mean temperature taken for all directions. Source Reidat [31].

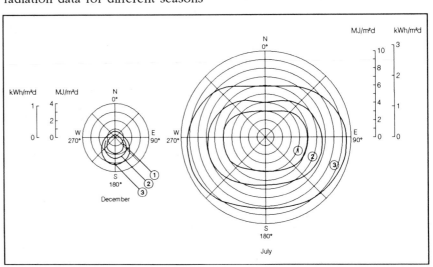

Figure 2.8. Sensitivity to orientation of mean monthly solar short wave radiation gains through vertical clear glass in December and July at Kew, London on an unobstructed site. 1: transmitted energy through double glazing: 2: transmitted energy through single glazing 3: incident daily solar radiation. (6mm clear float glass assumed).

2.12

WIND, POLLUTION AND SOLAR RADIATION

The wind is a very important dispersing agent for atmospheric pollutants. Still conditions in towns can lead on to considerable accumulations of pollutants in the urban atmosphere. Blocking of wind flows by mountains tends to increase pollution concentrations, as there is less dispersal to the wider atmosphere. The pollutants absorb and scatter the solar radiation, decreasing the strength of the direct beam, and increasing, under clear sky conditions the amount of diffuse radiation. The visibility range through the atmosphere gives important clues concerning the clarity of the atmosphere for the transmission of solar radiation. Visibility in polluted towns tends to decrease strongly in still anticyclonic weather and increase in windy cyclonic weather. Inversions are liable to develop in valleys in winter trapping pollutants within them.

Studies in Torino [32] have shown there is a very strong correlation between the pollution index and global radiation. Awareness of prevalent wind directions in relation to major pollutants sources is very helpful in site selection. Polluted sites not only provide less solar radiation, but also cause deterioration of the transmission characteristics of transparent collector surfaces and windows, as well as accelerating corrosion.

While urban variations in solar radiation do occur due to atmospheric pollution, with modern environmental control policies for atmospheric pollution, the rural/urban differential is far smaller than it used to be. High windiness promotes rapid dispersal of pollution, so the greatest urban/rural differences tend to occur with anticyclonic conditions with low wind speeds, especially when there are temperature inversions aloft. Much of the variation in beam strength day to day depends more on the origin of the meteorological air masses blowing over the site. Polar air masses tend to be dry, cold and clean. Their presence is often associated with high wind speeds. Maritime air masses from the North Atlantic contain more water vapour, but are basically very clean on

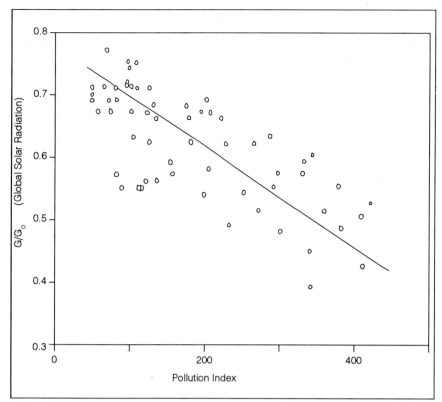

Figure 2.9: Correlation between the pollution index and global solar radiation for Torino. Source [32].

arrival. Tropical air masses from the mid Atlantic tend to be moist, and less clear, due to the high water content. They are often associated with relatively low wind speeds. Air masses travelling up from the Sahara to the south, may contain considerable amounts of dust particles from the deserts. Continental air masses too may accumulate heavy burdens of particles and pollutants, especially when wind speeds are low in anticyclonic weather.

The effects of pollution on solar beam strength are greatest in winter, when the path length is longest. In general the turbidity in winter is lower than in summer, but this basic clarity advantage is offset by the longer atmospheric path length.

Air drainage from sloping terrain and high moisture availability can produce low cloud and surface fog by humidification and cooling. This is, for example, very common in lakeside situations in Northern Italy. Often nearby mountain sites enjoy good sunshine, while the lake cities suffer continuous days of low cloud and fog.

Wind strength and direction, analysed in relation to terrain and pollution, therefore impact on the solar radiation assessment.

2.13

EUROPEAN SOLAR MICROCLIMATES

In order to improve the predictability of the technically usable heliothermal potential of a specific site, the CEC launched a research programme on solar microclimates [32]. The general aims of the study were:-

- to provide for a better understanding of the relationship between climatic factors and solar radiation;
- to determine the effect of anthropogenic pollution on the technically usable solar energy;
- to establish sound correlation models between weather data, pollution, regional orography and solar radiation patterns.

Eight characteristic sites were selected:-
- Brussels and Torino: industrial cities situated on plains;
- Grenoble and Saarbrucken: industrial towns in narrow valleys;
- Strasbourg: the microclimate of the upper Rhine valley;
- Vendee, France and Pentland Hills, Scotland: interaction of coastal marine and inland climate;

19

Sierra Nevada: strong climatic variations in mountain valleys.

At these sites, measuring networks for global solar radiation, meteorological data and pollutants were set up, or, where available, data records were collected from existing networks. The research project was organized as a concerted action of ten different research groups in seven member countries. One group acted as coordinator to assure compatability of measuring methods, data registration, correlation models, and the management of a central data bank. On the basis of the recorded data and the topographic and oreographic characteristics of the location, the site specific heliothermal potential is being correlated to site specific data and standard meteorological data using numerical models. The results are due for publication in 1992.

CONCLUSIONS ON THIS ASSEMBLY OF MACRO-CLIMATIC DATA

Very considerable progress has been made on the provision of macroclimatic data for solar energy applications in Europe, as a consequence of the various CEC programmes on solar radiation, wind and other climatic data. The impact of climatic factors on the radiation supply side is better understood on a common European basis than the impact of climate on the energy demand side. This is demonstrated by the CEC publications reviewed in this Chapter. Improved data are now becoming available for the proper analysis of summer and autumnal overheating risks. However the assessment of long wave radiation on obstructed building sites is not yet properly resolved, but progress has been achieved concerning the data needed to carry out overheating assessments on a sounder climatic basis. Appendix 8 adds new data to the existing publications, enabling some progress on natural cooling to be reached.

While consideration of the effects of obstruction on solar energy availability and on wind flows is set out later, designers must appreciate that it is the macroclimate that drives the microclimate and not vice versa. They must therefore start with a systematic analysis of the macroclimate. Such data help indicate how to structure initial design responses in intelligent ways to respond appropriately to climate. The emerging decisions can then be refined by subsequent microclimatic analyses following the procedures discussed later in this chapter.

THE WIND AND SITE PLANNING

2.14

INTRODUCTION

Shelter from wind is a key requirement in cold weather. Adequate indoor air movement is a key requirement in hot weather, especial in humid areas. It follows that designers need information about both wind speeds and wind directions at different seasons to handle properly design issues concerning wind shelter and summer ventilation. Knowledge of the wind macroclimate is needed first, in order, subsequently, to examine the detailed formation of the wind microclimate of the site.

This section outlines methodologies for estimating site wind conditions quantitatively. The contents are mainly based on the CEC European Wind Atlas (1). The European Wind Atlas is particularly helpful for assessing the impacts of the stronger winds that are typical of winter. It is of lesser value for estimating the lighter breezes that are typical of hot summer weather. Use is also made of WMO Technical Note No 175, "Meteorological Aspects of the Utilization of Wind as an energy source" [33].

The assessment of wind speeds for structural design purposes is not considered. This aspect of design is normally covered in the loading codes. These give prominence to the assessment of extreme winds, occurring only rarely.

The interactions between wind and temperature have been considered in the previous section.

The wind regime of Europe is influenced by three major factors, namely:-

1.   The large temperature difference between the Polar air to the north and the Sub-Tropical air to the south.

2.   The distribution of land and sea with the Atlantic Ocean to the west, Asia to the east, and the Mediterranean Sea and Africa to the south.

3.   The major orographic barriers such as the Alps, the Pyrenees and the Scandinavian mountain chain.

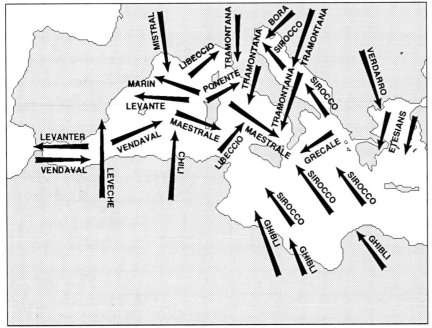

Figure 2.10: Principle winds of the Mediterranean region. Source - European Wind Atlas [1]. A brief description of each wind is given in Huschke (1959). (From: Series Grandi Progetti di Ricerca 3, ENEL, Italy)

The notable feature of the wind climate north of about 40°N is that the wind regime consists mainly of migratory cyclones and anticyclones moving in an eastward or northeastward direction. The wind impacts of these flows are greatest on the westerly Atlantic side, the winds becoming increasingly strong as one proceeds north west towards Scottish and Irish coasts.

In the Southern half of Europe, interactions between mountains, warmer seas, and hot deserts to the south of the Mediterranean, produce many well identified local winds: like the Mistral, a strong downslope northerly wind, which penetrates the Rhone valley, and spreads along the French Riviera to the Mediterranean sea for long periods in winter. Another such wind is the Sirocco, which brings air masses containing large amounts of dust from the Sahara into Southern Europe. Some of these special winds are plotted in figure 2.10.

An important special local wind phenomenon is the Föhn, which occurs in many mountainous regions of Europe. The best known Föhns are those which occur in and around the Alps. The North Föhn is a layer of cold air which overflows the chain of the Alps, resulting in a gusty flow of cold air down the southern slopes of the Alps. The South Föhn is the reverse of the North Föhn, a warm dry wind in the Alps, blowing down the northern slopes of the Alps.

A designer wishing to carry out a wind analysis should start by trying to reach a proper understanding of the basic forces driving the wind climate of the area in which he or she is to build at different seasons of the year.

2.15
DAILY AND YEARLY VARIATIONS IN WIND SPEED

The daily variation in wind speed varies very much with location. Oceanic and high mountain sites have low mean diurnal wind speed variations. In Northern Europe the diurnal variation in mean wind speed is relatively small in winter, but increases considerably in summer. South of the Alps and Pyrenees there is usually an appreciable increase of wind speed in daytime, followed by a night time drop throughout the year. The maximum mean wind speeds occur at different times of year in different regions. A winter maximum of mean wind speed is characteristic of Northern Europe. However there may be a summer maximum , for example, in the Canary Islands, which, in summer, come under the influence of the Trade Winds.

2.16
DAILY AND YEARLY VARIATIONS IN WIND SPEED

The design problem is how to get reliable quantitative wind data for assessing problems of shelter and air movement at site level. Figure 2.12 provides a schematic illustration of the local factors influencing wind. It is immediately evident that the quantitative prediction of wind climates is a complex task. One can state that all wind observations reflect some impact of the immediate local environment around the specific observing site. In some situations, these local impacts are very considerable; for example the

presence of a large building in a specific direction from the wind measurement anemometer. If observed wind data is to be generalized on a regional basis as opposed to a specific site observational basis, then a considerable amount of data processing has to be carried out to achieve satisfactory representations of regional wind macroclimates free from local disturbances, specific to that particular measurement site.

The CEC European Wind Atlas provides quantitative data on wind in a large number of sites, processed using a systematic methodology. The general methodology used to develop the CEC European Wind Atlas is illustrated in Figure 2.12. Three corrections are applied to obtain the Atlas regional data from the specific site observations: a correction for terrain cover, a correction for terrain shape, and a correction for sheltering obstacles. It follows that, in

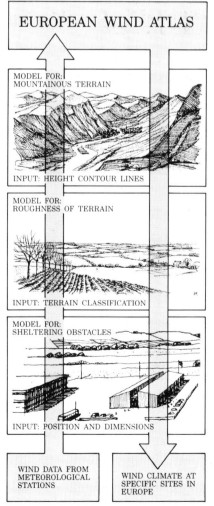

EUROPEAN WIND ATLAS

MODEL FOR:
MOUNTAINOUS TERRAIN

INPUT: HEIGHT CONTOUR LINES

MODEL FOR:
ROUGHNESS OF TERRAIN

INPUT: TERRAIN CLASSIFICATION

MODEL FOR:
SHELTERING OBSTACLES

INPUT: POSITION AND DIMENSIONS

| WIND DATA FROM METEOROLOGICAL STATIONS | WIND CLIMATE AT SPECIFIC SITES IN EUROPE |

Meteorological models are used to calculate the regional wind climatologies from the raw data. In the reverse process - the application of the Wind Atlas - the wind climate at any specific site may be calculated from the regional climatology.

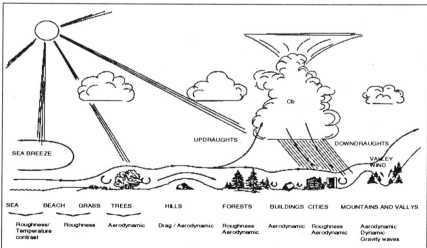

Figure 2.11: Schematic illustration of local factors influencing the wind. Source WMO Technical Note No 175 [33].

proceeding the other way, in order to estimate the wind quantitatively at any site from the regional data, there are three adjustments to be made to the regional data to obtain site data for a specific site. The correction processes have to be carried out for each wind direction. Often it will be sufficient to assess the corrections qualitatively, for example, in the design process when attempting to enhance wind shelter by site layout. In seeking shelter, it is very important to assess the wind as a directional phenomenon.

The CEC Wind Atlas classifies the wind direction in 30 degree sectors, making 12 sectors in all. Data from more than two hundred meteorological stations in the European Community Member States were selected for the calculation of regional wind climatologies. The detailed regional results for each site are presented in Chapter 7 of the Wind Atlas, Station Statistics and Climatologies. An example table is discussed later. Wind resource maps were also prepared, based on the wind energy available at 50 metres above ground level for five types of terrain, including hills and ridges. Radiosonde wind data was also used. The surface winds, in these cases were assessed by working downwards from the wind data aloft. The data were also entered on an IBM PC-compatible disc, which is available with the Atlas. A series of wind energy assessment maps were produced. Figure 2.13 provides the overall wind energy assessment map for the whole European community. The detailed locations of the stations used to develop regional data are marked on the national maps, which also shade the relative relief into four categories: none, light, 200-800m and above 800m. The relative relief is the difference between the highest and lowest terrain level within unit areas of 100 km .

The following comments amplify the basis of the Map in Figure 2.13:-

1. The resources refer to the power in the wind. A wind turbine can utilise between 20 and 30% of the available resource. The resources are calculated for an air density of 1.23 kg m^{-3}, corresponding to standard sea level pressure and a temperature of 15°C. Air density decreases with height but up to 1000 m above sea level the resulting reduction of the power densities is less than 10%.

2. Urban districts, forest and farm land with many windbreaks are sheltered terrain with roughness, Class 3.

3. Open landscape with few windbreaks, roughness Class 1. In general, the most favourable wind power inland sites on level land are found here.

4. The sea coast values pertain to a straight coastline, a uniform wind rose and a land surface with few windbreaks, roughness Class 1. Wind energy resources will be higher, and closer to open sea values, if winds from the sea occur more frequently, i.e. the wind rose is not in uniform and/or the land protrudes into the sea. Conversely, resources will generally be smaller, and closer to land values, if winds from land occur more frequently.

5. Open sea is considered to be more than 10 km offshore and is classed as roughness Class 0.

6. Hills and ridge data correspond to 50% overspeeding and were calculated for a site on the summit of a single axisymmetric hill with a height of 400 metres and a base diameter of 4 km. The overspeeding depends on the height, length and specific setting of the hill.

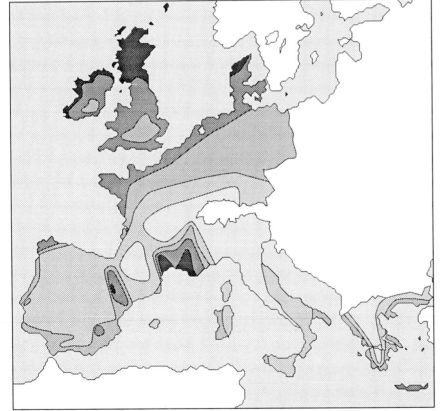

	Sheltered terrain		Open plain		At a sea coast		Open sea		Hills and ridges	
	ms⁻¹	Wm⁻²	ms⁻¹	Wm⁻²	ms⁻¹	Wm⁻²	ms⁻¹	Wm⁻²	ms⁻¹	Wm⁻²
	>6.0	>250	>7.5	>500	>8.5	>700	>9.0	>800	>11.5	>1800
	5.0-6.0	150-250	6.5-7.5	300-500	7.0-8.5	400-700	8.0-9.0	600-800	10-11	1200-1800
	4.5-5.0	100-150	5.5-6.5	200-300	6.0-7.0	250-400	7.0-8.0	400-600	8.5-10	700-1200
	3.5-4.5	50-100	4.5-5.5	100-200	5.0-6.0	150-250	5.5-7.0	200-400	7.0-8.5	400-700
	<3.5	<50	<4.5	<100	<5.0	<150	<5.5	<200	<7.0	<400

Figure 2.13: Wind resources at 50 metres above ground level for five different topographic conditions.

STATISTICAL DESCRIPTIONS OF WIND VELOCITY

The variability of wind speed

The CEC European Wind Atlas is based on statistical approaches to the analysis of wind. The wind velocity is a constantly varying vector quantity described by its speed and direction. The values obtained from the observations depend on the averaging period used. Structural wind loading design is normally based on gust velocities measured with anemometers with response periods of 2 seconds or less. The Wind Atlas is based on wind speeds averaged over longer periods of time. For each station processed in the CEC Wind Energy Atlas, data sets of wind observations were used, taken every three hours, in synoptic time (UTC), over a period of approximately ten years. The averaging periods used in the Atlas correspond mostly to averaging periods of 10 minutes about the observation hour, or the hourly average. Figure 2.14 plots an example of observed wind speeds over different periods of time. It starts with a 100 day record, then zooms in to progressively shorter periods of time, finally reaching 10 minutes. The greater variance in the longer time series is very evident in comparison with the short time series. Cycles of weather systems produce changes in wind speeds, anticyclonic weather tending to produce low wind speeds and cyclonic weather tending to produce windier conditions. Additionally there is the daily cycle which is linked to the convection caused by the heating of the ground by the sun, which is most pronounced in anticyclonic weather.

The data concerning variance can be converted to what is called a power spectrum. Figure 2.15 shows such a power spectrum. Time is plotted logarithmically on the bottom scale. The partitioning of the variance is plotted vertically. It will be noted first of all there is a strong daily peak. In Denmark there is also a high variance over periods of four to seven days associated with the passing of weather fronts. This wind pattern is characteristic of Northern Europe. The variance from month to month is much less. The European

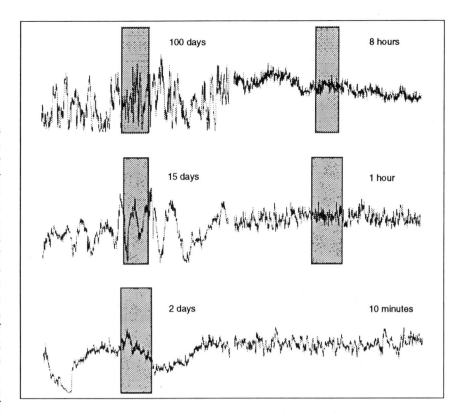

Figure 2.14. Wind speed measured 30m above flat homogeneous terrain in Denmark (Courtney, 1988). Each graph shows the measured wind speed over the time period indicated. The number of data points in each graph is 1200, each data point corresponding to the speed averaged over 1/1200 of the period. Vertical axis is wind speed, 0-20 m/s.

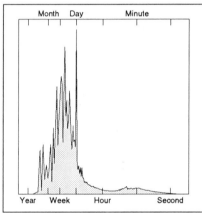

Figure 2.15. The power spectrum of wind speeds measured continuously over a flat homogeneous terrain in Denmark. The data were collected over one year with a sampling frequency of 8Hz. The spectrum is shown in a log-linear, area-true representation. Source CEC Wind Atlas.

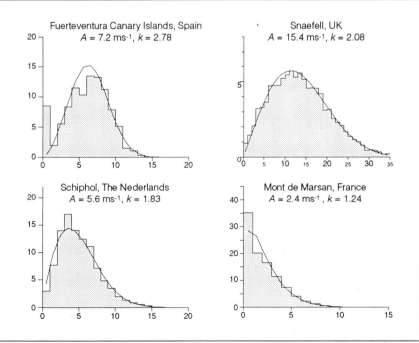

Figure 2.16: Histograms of measured wind speed data for four different stations used in the Atlas. The scale parameter A is related to the mean value of the wind speed. Horizontal axes: wind speed in metres per second. Vertical axes: frequency of occurrence in per cent per one metre/second band.

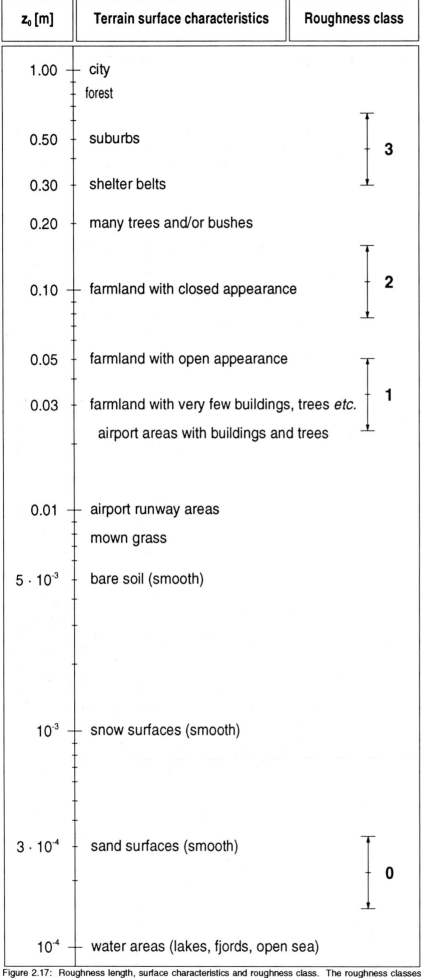

z_0 [m]	Terrain surface characteristics	Roughness class
1.00	city	
	forest	
0.50	suburbs	3
0.30	shelter belts	
0.20	many trees and/or bushes	
0.10	farmland with closed appearance	2
0.05	farmland with open appearance	
0.03	farmland with very few buildings, trees *etc.*	1
	airport areas with buildings and trees	
0.01	airport runway areas	
	mown grass	
$5 \cdot 10^{-3}$	bare soil (smooth)	
10^{-3}	snow surfaces (smooth)	
$3 \cdot 10^{-4}$	sand surfaces (smooth)	0
10^{-4}	water areas (lakes, fjords, open sea)	

Figure 2.17: Roughness length, surface characteristics and roughness class. The roughness classes are indicated by vertical bars. The central points give the reference values and the length of the bars indicates the typical range of uncertainty in roughness assessments.

Wind Atlas contains a power spectrum for each site processed, as discussed later.

The Statistics of Wind

The measured wind speeds can be classified statistically using bands of wind velocity to establish the relative frequency of occurrence of values lying in different wind speed bands. This may be done using winds observed from a specific sector, or the wind from all directions. The first method also establishes the frequency of occurrence of winds from different directions. Figure 2.16 illustrates frequency distributions for four sites. Snaefill is a mountain top site and very exposed. Mont de Marsan is a very sheltered site. Schipol Airport is fairly typical of an exposed level site in northern Europe. Fuerteventura in the Canaries comes under the influence of the Trade Winds in summer, but not in winter.

The Weibull Distribution

The wind statistical data of velocity are reasonably described by the two parameter Weibull distribution:-

$$f(u) = \frac{k}{A} \left(\frac{u}{A} \right)^{k-1} \exp\left(- \left(\frac{u}{A} \right)^k \right)$$

where:

f(u)   is the frequency of occurrence of wind speed, u, m/s.

A   is a fitting parameter known as the scale parameter, m/s.

k   is a fitting parameter known as the shape parameter, dimensionless.

A is fairly closely related to the mean wind speed, u. A reasonable approximation is:-

A = u x 1.125 m/s

Much of the wind data in the Atlas is presented in terms of the derived Weibull parameters A and k.

2.19

ADDITIONAL BASIC CONCEPTS

In order to understand fully the tabular presentations in the European Wind Atlas, it is necessary to master certain additional concepts.

The atmospheric boundary layer

The atmospheric boundary layer is the air layer close to the ground in which the wind flow is affected by the properties of the terrain ground cover below. It is the transition zone between the surface and the free atmosphere, where motion is unaffected by surface irregularities or turbulence introduced at different levels. The boundary layer depth may vary from less than 100m on clear nights over open country up to more than 2 km on a fine summer day, when vertical convection of warmed air influences the horizontal wind flows over considerable depths. The boundary layer increases in depth over cities.

The roughness length

The wind in the boundary layer is slowed down by the frictional resistance offered to the air flow by the terrain cover. The rougher the terrain cover, the greater is the slow down of the wind in the boundary layer above. The aerodynamic roughnesses of different types of terrain are usually parametrized by a length scale known as the roughness length, z_O

Figure 2.17 taken from the CEC European Wind Atlas shows the relationship between terrain surface cover characteristics and the roughness length. It also shows the categorization into the four basic roughness classes adopted for the regional assessment tables in European Wind Atlas, discussed in more detail later. A fifth category could be defined for a city with a roughness length of 1 metre. It will be noted that suburbs are categorized by a roughness class 3.

Figures 2.18 to Figures 2.21 characterize the terrain characteristics giving rise to the four basic roughness categories used in the European Wind Atlas.

When the flows are not disturbed by vertical convection, the wind speed u(z) at a height z m can be modelled as:-

$$u(z) = \frac{u*}{k} \log(z/z_O) \text{ m/s}$$

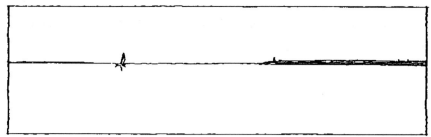

Figure 2.18: Example of terrain corresponding to roughness class 0: water areas (Z_0 = 0.0002m). This class comprises the sea, fjords and lakes. Source: European Wind Atlas [1].

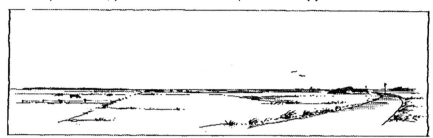

Figure 2.19: Example of terrain corresponding to roughness class 1: open areas with few windbreaks (Z_0 = 0.03m). The terrain appears to be very open and is flat or gently undulating. Single farms and stands of trees and bushes can be found. Source: European Wind Atlas.

Figure 2.20: Example of terrain corresponding to roughness class 2: farm land with windbreaks, the mean separation of which exceeds 1000m, and some scattered built-up areas (Z_0 = 0.1m). The terrain is characterized by large open areas between the many windbreaks, giving the landscape an open appearance. The terrain may be flat or undulating. There are many trees and buildings. Source: European Wind Atlas.

Figure 2.21: Example of terrain corresponding to roughness class 3: urban districts, forests, and farm land with many windbreaks (Z_0 = 0.40m). The farm land is characterized by the many closely spaced windbreaks, the average separation being a few hundred metres. Forest and urban areas also belong to this class. Source: European Wind Atlas.

HEIGHT	TERRAIN			
	Open flat Country	Country with scattered wind breaks	Urban	City
5.00	0.89	0.72	0.52	--
10.00	1.00	0.82	0.62	0.45
20.00	1.13	0.95	0.74	0.56
30.00	1.21	1.03	0.82	0.64
40.00	1.27	1.09	0.88	0.71
50.00	1.32	1.14	0.93	0.76
* Open flat country corresponds roughly to Category 1, refer Figure 2.18.				

Table 2.11: Corrections to apply to the 10m open country wind speed to allow for the effects of terrain, and height of roof top level to assess wind speed at roof top level.
Source CIBSE Guide A2, Weather and Solar Data.

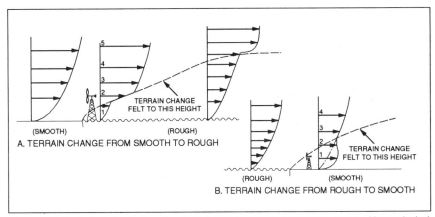

Figure 2.22 Vertical wind speed profiles near changes in terrain roughness. Source Meteorological Aspects of the Utilization of Wind as an Energy Source, WMO [33].

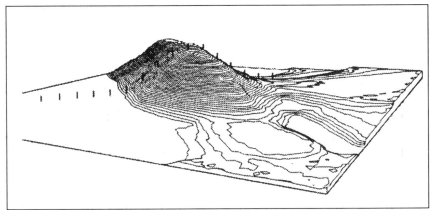

Figure 2.23: Perspective plot of the Askervein hill. The posts show the wind flow line studied. Source: CEC European Wind Atlas [1].

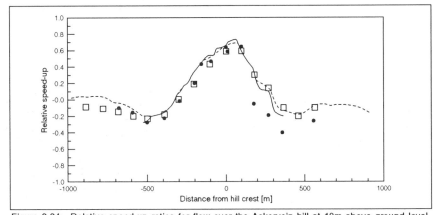

Figure 2.24: Relative speed-up ratios for flow over the Askervein hill at 10m above ground level. Measurements are indicated by dots and results from the CEC European Wind Atlas orographic model by squares. Results from two other numerical models are shown by a full and dashed line, respectively. Source: CEC European Wind Atlas[1].

Figure 2.25: European landscape of type 1: Plains, water areas and lowland regions far from mountains. Winds near the surface are modified by changing surface roughness and sheltering obstacles only.

where

z is the height above ground level, in metres.

z_O is the roughness length in metres

u_* is the friction velocity in metres/second

k is the Van Karman constant = 0.40

u_* depends on the density of the air r and the surface stress t.

These relationships were used in the development of the basic regional Tables presented in the European Wind Atlas at a range of standard heights, ranging from 10m to 200m for the four basic terrain roughness categories. These systematic tables are discussed later.

Flow over level ground without obstructions

For engineering and architectural design purposes, adjustments for the effects of height above ground level on the wind speed for flows above level terrain are usually made using a simple power law exponent. The wind speed, u, at a given level above ground of z metres is related to the 10 metre wind speed by the following expressions:-

$$u_z/u_{10} = (z/10)^a$$

where:

u_{10} = the wind speed at 10 metres

Typical values of a used are as follows:-

Open flat country	0.17
Country with scattered wind breaks	0.20
Urban	0.25
City	0.33

When the ground cover is continuously vertically high e.g. extensive buildings of roughly similar height, the wind tends to move above the obstructed level with a gradient corresponding to a ground plane displaced to mean obstruction level. If this ground plane displacement is d metres, then:-

$$u_z/u_{10} = \{(z-d)/10\}^a \text{ m/s for positive values of } (z-d)$$

Table 2.11 provides typical corrections estimated in this way. Table 2.11 makes clear the sheltering effects found in typical urban areas. The air movements below the displaced ground plane are mainly created by the shear forces above. Weak reverse flow circulation cells are set up below, as Figure 2.34 shows.

Transitions from one roughness type to another

When the roughness length changes, a new boundary layer begins to develop from the line of transition. It takes distance to establish the full transition to the new regime. Figure 2.22 illustrates the processes of transition from smooth terrain to rough terrain at the top, and from rough terrain to smooth terrain at the bottom. The urban boundary layer close to the ground builds up quickly, but a tall building located close to the windward urban boundary will be mainly immersed in an adverse flow regime, so when downward wind deflections occur, these will be far more adverse at ground level than in city centres, where the full urban boundary layer has developed. The European Wind Atlas contains quantitative procedures for assessing the impacts of such roughness changes. The Atlas should be consulted for details.

2.20

THE EFFECT OF HEIGHT VARIATIONS IN TERRAIN

Estimating the impacts of hills

The wind profile is considerably modified by the vertical variations in terrain. If the wind has to move over a hill, an acceleration of wind speed takes place over the brow. If one takes observations at a fixed height above ground level, as the wind traverses across a hill, one can calculate the relative speed-up factor as

$$S = (u_2 - u_1)/u_1$$

where:

u_2 is the wind at a selected height above ground level at some point on the hill, m/s

u_1 is the wind velocity at the same height above ground level over the terrain upstream of the hill, m/s

Figure 2.23 shows a cross section of a particular hill studied, while Figure 2.24 shows the relative speed up. Two noteworthy features are:

1. the relative speed up at the crest in this case is 80%, i.e. the undisturbed upstream 10m flow has to be multiplied by 1.80.

Figure 2.26: European landscape of type 2: Gently undulating and hilly regions far from mountains. Typical horizontal dimensions of the hills are less than a few kilometres. Winds near the surface are modified by changing surface roughness, sheltering obstacles and - most importantly - by the acceleration induced by the hills.

Figure 2.27: European landscape of type 3: Strongly undulating and highland regions ('Mittelgebirgs-relief'). Typical horizontal dimensions of the hills are several kilometres. Winds near the surface are modified by the topography as for landscape type 2. In addition, the larger scale orographic features may induce strong modifications of the entire atmospheric boundary layer.

Figure 2.28: European landscape of type 4: Foothill regions. In these broad sloping regions distinct and persistent flow systems occur, such as : Föhn, Bisa, Bora, Mistral, and Tramontana. These flows are caused by processes like channelling, deflection, leeside descent, and hydraulic intensification.

Figure 2.29: European landscape of type 5: High mountain massifs cut by deep valleys. The winds at the peaks may be representative of free atmospheric values depending on the specific conditions. In the valleys thermally-induced mountain valley winds dominate the wind climate. Except for leeside föhn, the winds in the valleys are decoupled from the free atmosphere winds.

2. the negative relative speed up is about -20% at the foot of the hill up wind and about -40% in the region just before the foot of the hill downwind.

The speed up depends on the shape of the hill. The CEC European Wind Atlas contains theoretical models for predicting the effects of wind flow over contoured terrains. The Atlas provides reasonably simple methods for estimating the speed up over isolated conical hills and infinitely long ridges. There are other more complex models for less simple terrain orography. The European Wind Atlas should be consulted for further details. The critical issue is to recognise the impacts of vertical terrain changes on the wind driven flows above them. Shelter is achieved upwind at the foot of the hill, as the flow starts to lift as well as downwind. There is value in keeping buildings down the slope if cold conditions prevail. However right at the bottom, radiatively cooled cold air draining down may pond on cold nights, so there are often advantages in moving slightly off valley bottoms. These radiatively driven flows were discussed earlier in this chapter.

Schiphol

| 52° 18' 00" N   04° 46' 00" E | UTM 31   E 620471 m   N 5795996 m | -4 m a.s.l. |

Location in the western part of the Netherlands at the national airport. In the northeast quadrant the Amsterdam urban area begins at a distance of approx. 3 km, while towards the west and south the region consists of flat polder land with occasional villages and tree-lined roads. The anemometer is placed between two runways. To the southwest there is a nearby dense patch of trees and small buildings, and to the northwest the large airport buildings lie at a distance of approx. 1 km.

Sect	z_{01}	x_1	x_{02}	x_2	z_{03}	x_3	z_{04}	x_4	z_{05}	x_5	z_{06}	Pct	Deg
0	0.01	4000	0.40										
30	0.01	3000	0.40										
60	0.01	2500	0.40										
90	0.01	1200	0.40										
120	0.01	900	0.40	1800	0.20								
150	0.01	1500	0.20										
180	0.01	100	0.20	200	0.01	2500	0.20						
210	0.01	80	0.20	150	0.01	3000	0.10	4000	0.20				
240	0.01	70	0.20	120	0.01	3500	0.10	5000	0.20				
270	0.01	100	0.20	150	0.01	1300	0.30	1800	0.03	21000	0.00		
300	0.01	1000	0.30	2200	0.01	4500	0.10	20000	0.00				
330	0.01	700	0.30	2000	0.01	4500	0.10						

Height of anemometer: 10.0 m a.g.l.   Period: 70010103 - 76123121

Sect	Freq	<1	2	3	4	5	6	7	8	9	11	13	15	17	>17	A	k
0	7.4	101	109	193	177	122	122	88	43	23	18	3	2	0	1	4.4	1.73
30	7.8	23	75	109	162	160	148	119	66	58	60	17	2	1	0	5.7	2.09
60	8.6	23	77	122	161	154	134	131	82	46	52	13	5	0	0	5.7	2.12
90	7.1	31	74	161	234	190	131	83	56	21	17	3	0	0	0	4.7	2.13
120	5.6	50	109	193	215	175	111	106	23	11	5	2	0	0	0	4.3	2.18
150	6.7	33	101	162	186	138	145	104	57	37	32	4	1	0	0	5.0	2.06
180	9.5	23	80	163	186	156	116	97	72	35	53	14	4	1	1	5.2	1.79
210	11.9	17	67	126	178	145	127	115	75	49	58	31	10	2	0	5.8	1.87
240	12.1	19	68	116	128	96	95	116	99	81	109	48	18	4	3	6.9	2.09
270	9.9	17	55	106	136	124	129	139	91	68	70	38	20	6	1	6.6	2.00
300	6.5	22	66	123	157	125	116	117	85	56	69	39	19	5	4	6.2	1.77
330	6.9	23	76	153	160	135	132	113	79	53	46	19	6	2	2	5.7	1.88
Total	100.0	30	77	140	170	141	124	112	72	48	54	22	8	2	1	5.6	1.83

UTC	Jan	Feb	Mar	Apr	May	Jun	Jul	Aug	Sep	Oct	Nov	Dec	Year
0	5.8	4.9	4.7	4.7	4.0	3.4	3.4	3.2	3.5	4.2	5.5	5.4	4.4
3	5.6	4.9	4.5	4.6	3.9	3.3	3.3	3.0	3.3	4.1	5.4	5.5	4.3
6	5.8	4.8	4.6	5.0	4.5	4.0	4.0	3.5	3.5	4.0	5.4	5.6	4.6
9	5.9	5.4	5.7	6.3	5.5	4.9	5.0	4.8	5.1	5.1	6.1	5.6	5.4
12	6.4	6.0	6.5	6.9	6.1	5.6	5.7	5.3	5.6	5.7	6.7	6.2	6.1
15	5.8	5.7	6.1	6.9	6.0	5.7	5.8	5.4	5.3	5.1	6.1	5.8	5.8
18	5.7	5.1	5.0	5.7	4.9	4.8	4.7	4.1	4.0	4.4	5.9	5.6	5.0
21	5.9	5.1	4.9	4.8	4.1	3.6	3.6	3.4	3.6	4.4	5.8	5.5	4.5
Day	5.9	5.3	5.3	5.6	4.9	4.4	4.4	4.1	4.2	4.6	5.8	5.6	5.0

Table 2.12: From European Wind Atlas

Types of European wind terrain

In the absence of nearby obstacles, the wind flows that occur are the result of the combination of terrain cover and terrain orography. The European Wind Atlas identifies five basic types of terrain orography. These 5 types are illustrated in Figures 2.25 to 2.29. As one moves from landscape type 1 through to type five, it becomes increasingly difficult to generalize the wind flows. Data from specific observing stations in category type 5 are extremely site specific. The valleys have their own wind climates, very different from the wind climates on the tops exposed the general atmospheric flows. At times, for example, there may be Föhn type winds. The analysis of air drainage patterns, up and down valley winds and so on, is of dominant importance in site wind analysis in such locations. Local knowledge becomes very important.

In terrain with landscape types 1 and 2, one can generalize the observational data by making adjustments for the terrain cover and obstructions in different directions to produce regionally representative data, which then can be assumed to apply over an area of 200 x 200 km as discussed earlier. In order to use this regional data quantitatively for specific sites, they have first to be adjusted, sector by sector, taking account of terrain cover, orography and local obstruction in each direction to get design data for a specific site. Refer figure 2.12.

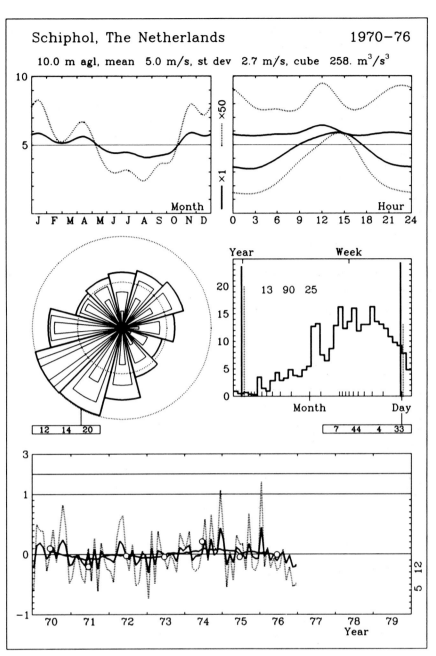

Figure 2.30: Wind climatological fingerprint for Schiphol.

2.21

THE CEC EUROPEAN WIND ATLAS TABLES

We are now in a position to comprehend the full scope of the CEC European Wind Atlas Tables. All the Tables have a similar form. Here the data for Schipol in the Netherlands will be discussed as an example.

Raw data tables

Table 2.12 provides the basic observed wind data for Schipol, and information about the obstructions in different directions, considered in 30 degree sectors. Such data always appear on the left hand page in the Atlas data series.

The contents of these tables are as follows:-

1. Details of the precise site location are given at the top.

2. A written description of the site follows.

3. A roughness rose is then provided, stated sectorally. The roughness length immediately adjacent to the anemometer in each sector is first stated, together with the termination point distance in metres. If there are changes in roughness in that direction, then the next roughness length is entered, together with its termination point distance, and so on.

4. A frequency classification of the raw observed anemometer data then follows using 1 m/s bands, arranged by sectors. The frequency of winds in each sector is recorded as a percentage x 10. The raw data Weibull parameters are also provided for each sector. Total percentage figures x 10 based on all sectors follow at the bottom of the middle part of the table, together with the Weibull parameters for the overall raw data. These are the roughness lengths used in the subsequent reduction to the regional level.

Frequencies of different wind speeds are given.

5. The monthly wind speed at 3 hour intervals expressed in Coordinated Universal Time, UTC (equivalent to GMT) are given. The conversion from Local Mean Time to GMT is discussed in Appendix 1 concerned with solar geometry. Basically the time zone correction is subtracted from local mean time. In summer, this has to include the summertime correction. Additionally the annual mean wind speeds at the same hours are recorded. The daily mean wind speeds are given at the bottom of the table.

The wind climatological fingerprints in the CEC European Wind Atlas.

Additionally a graphical statement is provided for each station. The graphical table for Schipol is copied in Figure 2.30.

Because the Atlas was primarily developed for wind power studies, where the power available is proportional to the cube of the wind speed, cubical relationships are developed as well as normal means and standard deviations. There are five graphs in each presentation:-

1. (Top left) The average seasonal variation of the measured wind speed, m/s (full line) and the cube of the wind speed (dashed line), $m^3 . s^{-3}$. The two vertical scale functions are given to the right.

2. (Top right) The monthly mean wind speed in m/s at different hours of the day in Universal Coordinated Time for the months of January and July (full line), also the corresponding cubed values for the same months (dashed line). The two vertical scales are read to the left. The summer wind speeds are normally the lowest, but this can be checked by reference to tables of type Table 2.12.

3. (Middle left) The relative frequencies of winds from each of the twelve sectors are shown in the left middle

graph as the radial extent of the twelve sectors, referenced to the frequency of the most prevalent direction, which is given a radius equivalent to unity. These are enclosed by broad lines. The relative contribution of each sector to the mean wind speed forms the next part of each segmental plot. The narrowest segmental plot gives the contribution of the cubes of the wind speed in each sector to the overall cube of the wind speed. In each case there are referenced to the most prevalent direction.

4. (Middle right) The wind power spectrum is given in the right middle graph. The description of this type of graph has been given in Section 2.18.1. Such graphs are of considerable significance in the assessment of wind energy applications, but do not contribute very much to building decision-making. Certain additional standard deviation data is superimposed. The Atlas should be consulted for details.

5. (Bottom) The year by year relative deviation from the long term monthly mean values is considered in the graphical time print at the bottom. The relative deviations of the means are plotted using the solid line. The relative differences in the values of the cubes are shown with the dotted line.

Standardized Regional Wind Tables

The second basic Tables for each site are the regionally representative Tables. The standardized regional wind tables were developed for each site using the procedures visually outlined in Figure 2.13. The processed data for each site provide regional data, obtained by correcting the observed sectorial data for the impacts of terrain cover, obstacles and orography. The derived data of regional applicability are presented for the four standard roughness classes identified in Figures 2.18 to

2.21. Table 2.13 provides the standardised regional wind data table from the Atlas for Schipol. These macroclimatically based Tables lie on the right hand side of the data page for each site.

The data in Table 2.13 are presented sector by sector in terms of the Weibull coefficients, A and k. The basic values were established at 10m, the recommended standard observing height. Then the other values for 25m, 50m, 100m and 200m above ground level were estimated using the roughness length and the logarithmic profile law discussed above in section 2.19. The percentage frequencies of occurance of the winds from each direction, after the corrections have been applied to the raw data, are given on the bottom line of each roughness section of the Table. On the right hand side of each roughness class section of the Table, the overall Weibull coefficients appear for winds from all directions.

The small table at the bottom provides the annual mean wind speed in m/s at different heights z and the total mean wind power in W/m^{-2} for the four roughness classes.

These basic regional tables provide the macroclimatic tables for design wind assessments. The main tables provide however only annual data.

Roughness Class 0

z	0	30	60	90	120	150	180	210	240	270	300	330	Total
10	6.8	7.6	8.2	7.5	6.8	7.0	7.7	8.0	9.0	9.0	8.6	8.3	8.0
	2.09	2.32	2.52	2.47	2.58	2.44	2.20	2.18	2.26	2.28	2.08	2.14	2.19
25	7.5	8.3	9.0	8.2	7.4	7.7	8.4	8.8	9.9	9.8	9.4	9.1	8.7
	2.16	2.39	2.60	2.55	2.66	2.51	2.26	2.25	2.31	2.33	2.12	2.20	2.25
50	8.1	8.9	9.6	8.8	7.9	8.2	9.0	9.4	10.5	10.5	10.1	9.7	9.4
	2.22	2.46	2.67	2.62	2.73	2.58	2.33	2.31	2.37	2.40	2.19	2.26	2.31
100	8.7	9.6	10.4	9.5	8.6	8.9	9.8	10.2	11.4	11.3	10.8	10.5	10.1
	2.15	2.38	2.58	2.53	2.65	2.50	2.25	2.24	2.31	2.34	2.13	2.20	2.26
200	9.6	10.7	11.6	10.6	9.5	9.9	10.8	11.3	12.4	12.4	11.8	11.5	11.2
	2.03	2.25	2.44	2.40	2.51	2.37	2.13	2.12	2.22	2.24	2.05	2.10	2.16
Freq	7.1	7.8	8.3	7.6	6.1	6.4	8.6	11.1	12.1	10.6	7.6	6.7	100.0

Roughness Class 1

z	0	30	60	90	120	150	180	210	240	270	300	330	Total
10	4.4	5.6	5.7	5.1	4.6	5.0	5.4	5.6	6.6	6.2	6.0	5.7	5.6
	1.72	2.05	2.11	2.10	2.14	2.05	1.79	1.85	2.03	1.96	1.74	1.86	1.87
25	5.3	6.7	6.8	6.1	5.5	6.0	6.4	6.8	7.8	7.4	7.2	6.8	6.7
	1.86	2.21	2.28	2.26	2.31	2.21	1.93	1.98	2.13	2.08	1.84	2.00	2.00
50	6.1	7.7	7.9	7.0	6.3	6.9	7.5	7.8	8.9	8.5	8.2	7.9	7.7
	2.09	2.48	2.57	2.55	2.60	2.48	2.17	2.21	2.31	2.28	1.99	2.23	2.21
100	7.3	9.2	9.3	8.3	7.5	8.2	8.9	9.2	10.3	9.9	9.5	9.3	9.0
	2.22	2.65	2.73	2.71	2.77	2.64	2.31	2.36	2.49	2.44	2.14	2.38	2.37
200	9.1	11.4	11.6	10.4	9.4	10.2	11.0	11.4	12.3	12.0	11.3	11.4	11.1
	2.12	2.53	2.61	2.59	2.64	2.53	2.20	2.26	2.39	2.34	2.06	2.28	2.29
Freq	7.3	7.8	8.5	7.2	5.7	6.6	9.2	11.7	12.1	10.1	6.8	6.8	100.0

Roughness Class 2

z	0	30	60	90	120	150	180	210	240	270	300	330	Total
10	3.8	5.0	5.0	4.4	4.0	4.4	4.7	4.9	5.8	5.4	5.3	4.9	4.9
	1.77	2.09	2.14	2.10	2.19	2.05	1.81	1.87	2.07	1.97	1.74	1.87	1.89
25	4.7	6.2	6.2	5.4	4.9	5.4	5.8	6.1	7.1	6.6	6.4	6.1	6.0
	1.89	2.24	2.29	2.25	2.34	2.20	1.93	1.99	2.17	2.08	1.82	1.99	2.00
50	5.5	7.2	7.2	6.3	5.8	6.4	6.9	7.2	8.3	7.7	7.4	7.1	7.0
	2.10	2.47	2.54	2.49	2.59	2.43	2.14	2.20	2.32	2.26	1.95	2.20	2.19
100	6.6	8.6	8.6	7.5	6.8	7.6	8.2	8.5	9.6	9.1	8.7	8.5	8.3
	2.30	2.72	2.79	2.73	2.85	2.67	2.35	2.41	2.55	2.48	2.13	2.42	2.40
200	8.1	10.6	10.6	9.3	8.5	9.4	10.1	10.4	11.5	11.0	10.4	10.4	10.2
	2.20	2.60	2.67	2.62	2.72	2.56	2.25	2.31	2.46	2.39	2.06	2.31	2.33
Freq	7.4	7.8	8.6	7.1	5.6	6.7	9.6	11.9	12.1	9.9	6.5	6.9	100.0

Roughness Class 3

z	0	30	60	90	120	150	180	210	240	270	300	330	Total
10	3.1	3.9	3.9	3.4	3.2	3.5	3.8	4.0	4.6	4.3	4.1	3.8	3.8
	1.81	2.10	2.13	2.11	2.15	1.99	1.83	1.88	2.07	1.96	1.75	1.83	1.89
25	4.1	5.2	5.1	4.5	4.2	4.6	4.9	5.2	6.0	5.6	5.4	4.9	5.1
	1.92	2.22	2.26	2.24	2.28	2.10	1.94	1.98	2.16	2.05	1.82	1.94	1.99
50	5.0	6.2	6.2	5.4	5.0	5.6	6.0	6.3	7.1	6.7	6.4	6.0	6.1
	2.09	2.42	2.45	2.43	2.48	2.29	2.11	2.13	2.29	2.19	1.94	2.10	2.14
100	6.0	7.5	7.4	6.5	6.1	6.7	7.2	7.5	8.4	7.9	7.6	7.2	7.3
	2.38	2.76	2.79	2.77	2.83	2.60	2.40	2.42	2.53	2.46	2.14	2.39	2.41
200	7.3	9.2	9.1	8.0	7.4	8.2	8.8	9.1	10.1	9.6	9.1	8.8	8.9
	2.30	2.65	2.69	2.67	2.72	2.51	2.32	2.34	2.50	2.40	2.11	2.31	2.36
Freq	7.4	7.9	8.4	6.9	5.8	7.1	9.9	11.9	11.8	9.4	6.6	7.0	100.0

z	Class 0		Class 1		Class 2		Class 3	
10	7.1	381	5.0	153	4.3	101	3.4	49
25	7.7	487	5.9	242	5.3	177	4.5	105
50	8.3	588	6.8	336	6.2	260	5.4	171
100	9.0	760	8.0	518	7.4	400	6.5	269
200	9.9	1052	9.8	981	9.0	748	7.9	488

Table 2.13: Standardised regional wind data table from the Atlas for Schipol. The data is divided into 30 degree sectors

USING THE EUROPEAN WIND ATLAS FOR SITE WIND ANALYSIS

The Atlas can be applied to design problems involving wind at differing levels of complexity, according to the designer's wishes.

Level 1 Basic relative windiness in different areas of Europe. Use country maps.

Level 2 Directional regional characteristics of wind, annual basis: use standardized regional table sectorial frequency distribution data to construct a wind rose for roughness class of area, using 10m wind speed values. In most suburban design situations roughness class 3 would be used. Then use wind rose to assess directional aspects of wind shelter provision.

Level 3 Assessment of monthly mean wind speeds at a fixed height and roughness class.

a. Select roughness class and height.

b. From the standardized regional wind table read off Weibull parameter A. Divide this value by 1.125 to get mean annual wind speed at that height in m/s.

c. Tabulate monthly mean wind speed, u_m, from wind climatological finger print at standard height, hence find annual mean wind speed u at standard height.

d. Multiply annual value obtained u_m/u to obtain the mean wind speed in the roughness class and height selected for month m.

The Atlas does not allow monthly directional estimates of wind to be prepared.

Level 4 Use full complexity of European Wind Atlas modelling methodology. Consult Wind Atlas for details.

In most cases, it will not be necessary to proceed beyond Level 3 in complexity.

REDUCING SITE WIND SPEEDS

Basic approaches

There are two basic approaches to the reduction of site wind speeds:-

1. The deflection of wind flows to provide shelter.
2. The dissipation of wind energy by frictional processes to reduce wind velocities.

In general the second approach, which converts the wind energy to heat energy, is more successful than the first, because, in the first approach, the energy in the deflected air stream is displaced somewhere else, often causing problems at another place. The role of the landscape in the dissipation of wind energy is discussed later. Deflected streams tend to induce turbulent eddies by the shear forces set up. These eddies are often of considerable size, and take time to pass a specific point. The recipient is first blown one way and then another way. Turbulent flows in the outdoor environment are less acceptable than smooth flows. As the direction of the turbulent flow usually cannot be anticipated, if the wind speeds are high this turbulence increases the difficulties of walking or moving about safely. Driving accuracy may be affected leading to accidents. The aim therefore is introduce shelter free from excessive turbulence.

Architecturally there tend to be three types of barriers against wind:-

1. Bluff bodies like buildings and external walls with sharp rectangular edges.
2. Porous screens, for example slatted wood ranch fences.
3. Trees and shrubs.

Typical flows round buildings are shown in Figures 2.35 and 2.36.

It will be noted in the case of a tall building in Figure 2.32 that the wind from a point about two thirds up the facade tends to move downwards forming a rotating turbulent vortex in front of the obstruction, blowing away from the obstruction at ground level.

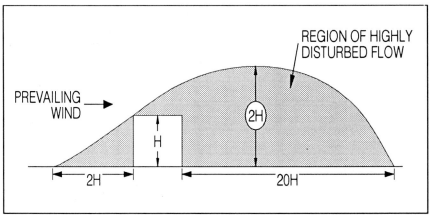

Figure 2.31: The general patern of disturbed flow around a small row of houses

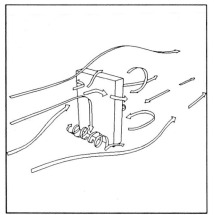

Figure 2.32: The general patern of flow around a tall isolated building.

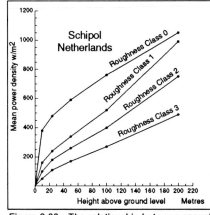

Figure 2.33. The relationship between annual mean wind power density and height above ground for four roughness classes at Schipol Airport. Basis: Table 2.13, bottom section, from CEC European Wind Atlas [1].

If there are gaps in the obstruction then the spiralling corkscrew vortex spirals around the edges of the obstruction producing very strong flows in the region of the corners, which form the vertical line where the flow separates from the facade. If the side facades are deep enough, the flow may re-attach to the facade further downstream. Behind the building, and to the side, there is a turbulent wake of considerable extent. The rotating vortex in front of a tall building is reinforced if a low building stands upstream from a high building.

Tall obstructions in the boundary layer basically intercept much more energy per unit facade area because of the vertical profile of the wind speed. Using the CEC European Wind Atlas Tables for different sites of the type illustrated in Table 2.13, bottom table, it is very easy to plot out a simple annual all-direction wind power density - height relationship graph for the different ground roughness classes. At the ground surface the flow rate is zero, so the associated power density is zero. Figure 2.33 plots the relationship for Schipol.

If we consider a vertical cylinder standing in the wind, one can calculate the annual mean power density arriving per unit width in the absence of the cylinder from Figure 2.33. The total power per linear width potentially intercepted by a vertical cylinder is given by the area under the curve, integrated to the height of the structure. Of course the actual flows go round, and only a small amount of the power would be removed by surface friction, but if any downward deflection occurs, as it often does, this mean power figure is representative of potential wind power in the environment likely to cause stress due to wind. The average power density is obtained by dividing the total power per linear width by the vertical height considered. The effects of structure height on the mean power density potentially incident is immediately evident. The mean potentially incident power density for a 100m building is about 6 times that of a 10m building. This is why regulation of vertical height is so important for wind control. What is even more evident, is the impact of roughness class on potentially incident power density. It is the overall landscape planning of the community that will determine the general roughness length. This factor is of great importance in windy areas. In potentially exposed areas, effective landscape development over a wide area is desirable to increase frictional drag at the surface, and so slow the wind.

Interactions between buildings

In considering the impact of wind on natural ventilation of buildings, Lee et al [34] have identified three basic flow regimes for the analysis of wind induced natural ventilation forces on low parallel spaced buildings located in line and also staggered. The regimes found are illustrated in Figure 2.34. The flow regimes are determined by the spacing height ratio S_c/H. Lee reached the following conclusions on the determination of the flow type regimes:-

1. The isolated roughness flow regime exists for values of $S_c/H > 2.4$ for both regular and staggered layouts.
2. The wake interference flow regime exists for values of $1.4 < S_c/H < 2.4$.
3. The skimming flow regime exists for values of $S_c/H < 1.4$ for both regular and staggered layouts.

Lee found for a space to height ratio of about 3 that the pressure difference across a building was roughly halved compared with an isolated building. The difference became virtually negligible with a ratio of 1. Summer requirements for air movement have to be balanced against winter needs for control of excessive infiltration. This points towards the advisability of using building spacing to height ratios in the region of 2.5 to 3.5. Such spacings can also give reasonable sunlight and daylight access.

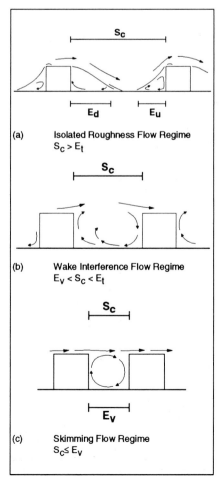

Figure 2.34. Three basic types of flow regimes identified by Lee et al [42]

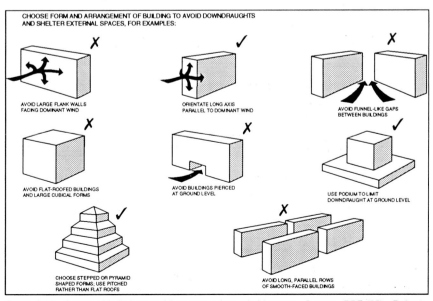

Figure 2.35: Choice of built form and layout to reduce wind impacts. Source: BRE (35). Refer to Appendix 2 for additional guidance.

Built form and shelter around buildings

The UK Building Research Establishment has attempted to provide practical advice about choice of built form and layouts in a recent digest [35]. Figure 2.35 summarizes their recommendations on the reduction of sensitivity of individual buildings to wind.

The issues of layout and shelter around buildings also require attention. It is best to keep all the buildings about the same height, so the main wind flow skims over the roofs. Buildings of five storeys or less usually produce acceptable conditions. Gandemer (36) has produced a useful set of guidelines for the assessment of layout wind risks. These are summarized in a visual form with notes in Appendix 2. The method makes use of a comfort factor that takes account of turbulence as well as changes in the mean speed ratios. The mean speed ratio sets the observed wind speed at the selected point on the site as a ratio to some reference windspeed in the undisturbed flow. The height of this reference point has to be defined. It is often sensible to use the undisturbed 10m wind speed as the reference value. A turbulent wind flow can be described in terms of its mean value, and the variations about that mean value. The root mean square average value of the instantaneous turbulence variations is a measure of the turbulence intensity. Gandemer defined a comfort parameter to provide information about the sheltering exposure effects of buildings to include account of turbulence measured with a hot wire anemometer as follows:-

$$\psi = \frac{u + \sqrt{u'^2}}{u_r + \sqrt{u_r^2}}$$

where: u and u' are the mean and turbulent components at the point considered.

u_r and u'_r are the corresponding undisturbed flowvalues at the reference point.

Good shelter design he indicated, will normally yield values of ψ less than 0.5. Values above 1 indicate a significant deterioration in conditions. Values above 1.5 represent a serious deterioration, and values above 2 will give rise to very serious public concern and protest on the inadequacies of the environmental design. Appendix 2 provides the detailed design guidance from Gandemer's work.

WIND FLOW THROUGH COURTYARD BUILDINGS

If the wind blows perpendicular to the walls of a building with a rectangular courtyard, the flow tends to lift over the first roof and pass right over the courtyard. The flow tends to that of a skimming flow regime. (refer to Figure 2.34c). Most of the pressure drop across the enclosing building then takes place across the windward section of the building. Bensalem and Sharples [37] [38] have studied, in wind tunnels, cross ventilation flows in courtyard buildings of various dimensions. They found the courtyard dimensions made relatively little impact on flow rates, when the wind was perpendicular to the wall. With cross ventilation, the flow rate through the front opening apertures was found to be about 0.38 to 0.50 times the undisturbed flow rate at roof level. The pressure drops across the side elements and leeward elements tended to be very small. The corresponding flow rates through the lateral cross ventilation openings were consequently small. Bensalem reported flow rate ratios of 0.05 to 0.18 of the roof level reference speed. The corresponding values for cross ventilation of the leeward rooms was 0.03 to 0.13 of the reference value. Averaging over the openings on all four walls gave an average flow rate ratio value of 0.18. On the other hand, when the wind struck at 45 degrees to the facade, the inflow ratios on the two windward facades fell to 0.24 from 0.36, but the airflow on the two leeward sides increased considerably. Averaging the flows over all four sides yielded an average flow rate ratio value of 0.28, with a good distribution between the openings in various rooms. At 30 degrees, similar phenomena were observed, but the pattern of flows was less uniform between the various apertures. When the flow is normal to the wall, a vertical rotating vortex fills the courtyard, i.e. the cross section looks like the flow illustrated in Figure 2.34c. When the wind blows corner on, rotating vortices run out along the edge of the roof from the corner on which the wind strikes. These vortices have the effect of deflecting the main displaced stream flowing between them downwards along the diagonal of the courtyard to strike the courtyard wall in the region of the inside courtyard corner. This stream then raises the pressure drops across the leeward walls set at an inclination to the main wind flow, which forces more air through the openings.

POROUS SCREENS

Artificial wind breaks can be used in the attempt to provide instant shelter, either as a permanent solution or as an expedient, until plants grow sufficiently to become effective. Solid walls and close fences are liable to generate excessive turbulence in their wakes. Refer to Figure 2.36. Permeable walls and fences are more effective. The optimum permeability is around 40% - 50%. If design enables permeability to vary, it should decrease from top to bottom. Sometimes gaps are left at the bottom of windbreaks to allow air drainage to reduce the risk of frost pockets being created.

WIND FLOW THROUGH LANDSCAPE

General principles

The landscape exerts a powerful influence on the wind microclimate. In the use of sites, it is always important to assess and place value on existing landscape features. It is better to use them positively rather than to remove them and start painfully and slowly again. Well treed areas enjoy a better controlled climate than landscapes carrying little such growth. Traditional European farming was often based on the extensive use of trees and hedgerows combined with small rectangular fields. Modern methods of mechanized farming have largely eliminated such an approach from economically viable farming areas. This has tended to increase wind exposure in rural areas dedicated to arable crops. However one does not need tall planting to secure considerable shelter from the wind. Basically one needs extensive areas of planting to offer a lot of frictional resistance at the surface to the wind flowing above. Enclosure of gardens by suitably designed porous fences helps reduce wind speeds and offers better gardening opportunities to further manipulate the wind microclimate by the growth of plants hedgerows and trees. When providing nearby shelter, it is important to achieve reasonably dense cover close to the ground. A lot of wind can penetrate below nearby trees. In some areas, winds from specific directions are very prevalent. In such cases, a policy of specific directional sheltering, for example, by shelter belts may prove valuable. In many cases the directions of adverse impact of the wind may vary widely, and such policies are then of limited benefit. There are also conflicts between the needs for winter wind shelter and the desire for good summer ventilation that have to be resolved. So in many cases widely scattered but carefully placed vegetation provides the best solution. In coastal areas, the impact of wind spray-carried sea salt may make it difficult to create effective shelter belts close to the coast.

2.27

SHELTER BELTS

Shelter belts are plantings of trees made more dense with suitable under canopy planting, established with the specific aim of controlling the wind. Such linear shelter belts are normally placed perpendicular to the direction of the primary wind which it is wished to control. This may be the prevailing wind, or an adverse wind from a particular quarter, for example cold north east winds. The European Atlas data makes it simple to identify the mean annual frequency of winds from different directions. It does not however provide a basis for seasonal directional analysis. Some temperature wind relationship information was provided in the previous Section. Shelter belts may be planted with conifers, or deciduous trees and plants. The resistance to flow will vary with the density of the foliage. Figure 2.36 shows a typical flow pattern. It will be noted there is some upwind impact, as well as considerable downwind impact. The actual performance depends on the shelter belt density. Figure 2.37 illustrates the wind speeds found at different levels around two shelter belts as a function of shelter belt height. The percentage reduction is greatest immediately behind the shelter belt. The wind speeds then gradually recover. There is still some impact for about 25 shelter belt heights behind the belt.

2.28

CONCLUSIONS AND WIND DESIGN

Successful achievement of good winter wind shelter combined with adequate summer air movement presents difficult design challenges. It requires the combination of a macroclimatic analysis and a microclimatic analysis. Wind exposure varies very much in different parts of Europe, so wind design priorities must vary. The balance between hot and cold weather also differs a lot in different regions. The strategies for wind shelter design therefore have to be set to meet the local challenges. Quantitative approaches are more likely to help the designer to make sounder decisions than simple qualitative approaches.

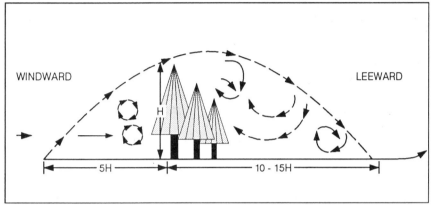

Figure 2.36, Basic flow characteristics of a shelter belt. (Based on van Eimern et al., 1964). Source reference [33].

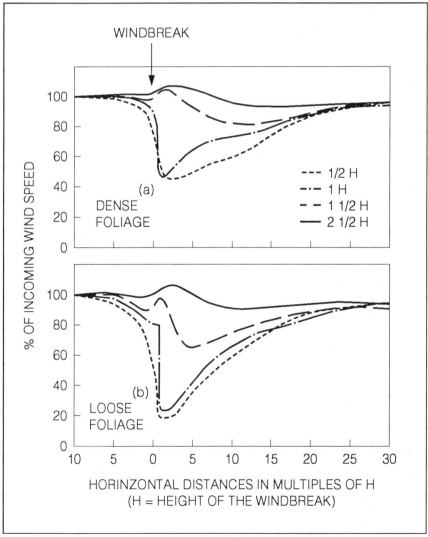

Figure 2.37. Percentage reduction in wind speed at different levels and distances expressed in multiples of shelter belt height. (Based on van Eimern et al. 1964). Source reference [33].

2.29

INTRODUCTION

Appropriate design tools are required to address the design challenges implicit in the consideration of the geometry of site and building layout for passive solar buildings. The resolution of the site and building layout problem can be aided through the application of geometrically based graphical design tools. The same design tools can assist in the consideration of the geometry of shading devices. In the underheated seasons the design aim is to promote effective insolation by avoiding overshading. In the overheated season, shading will be required. The design tools can thus be used to assess the two main facets of the overheated season shading design:-

1. The assessment of the degree of overshadowing of the various building facades by adjacent terrain, adjacent buildings and adjacent landscape, and also by facade projections.

2. The assessment of the performance of shading devices attached to the building itself to regulate overheating.

Additional graphical tools can be designed to facilitate the various consequent energy analyses.

A series of appendices accompanying this chapter deals with the geometry of the solar movements. The quantitative interpretation of the energy consequences of the geometrical facts of overshadowing and building shading requires a subsequent energy analysis. This energy analysis has to be split into two parts, a consideration of the impacts of overshadowing on direct beam energy availability, and, separately, consideration of the impacts on diffuse radiation availability. The monthly diffuse energy falling on vertical facades often exceeds the monthly mean solar beam energy. So the diffuse energy analysis is just as important as the direct beam analysis. The diffuse energy is also critical for successful daylighting.

2.30

TYPES OF GEOMETRIC DESIGN TOOL - DIRECT BEAM

The geometric analysis of direct beam availability can be conducted using one of five basic tools, namely:
- graphical plots
- manual trigonometric methods (rarely used)
- computer based trigonometric methods, often with graphical outputs
- scale models examined using a sundial device, and natural sunlight or an artificial light beam source
- scale models using a heliodon, a machine with a beam light source, which mechanically reproduces the geometric movements of the sun.

2.31

SUN ANGLES

In order to use the various graphic tools, it is first of all necessary to understand the meaning of the various angles used to describe the position of the sun.

Solar altitude and solar azimuth
The angular position of the sun as seen from a particular place on the surface of the earth varies from hour to hour and from season to season. The basic position of the sun at any instant can be described by two angles:-

1. The solar altitude (y)
2. The solar azimuth (a)$_s$

Figure 2.38 defines these angles. The reference plane for the solar altitude is the horizontal plane. The reference plane for the solar azimuth is the vertical plane running north south through the poles. The azimuth is referenced to due south in the Northern hemisphere. Figure 2.38 also defines the geometry of any inclined plane in terms of its surface tilt β expressed as inclination angle to the horizontal plane, and its wall azimuth or orientation, a$_w$, expressed as the angle between the projection of the normal to the surface on the horizontal plane, and the north south vertical plane. Finally figure 2.38 defines the angle of incidence of the direct beam on the inclined plane. This is simply the angle between the normal to the surface and the direction of the centre of the sun.

The variations in the solar altitude and solar azimuth are the consequence of the daily rotation of the earth about its polar axis, and the annual movement of the earth about the sun. The earth follows an elliptical path, with the poles inclined at an angle of 23.45 degrees to the plane of the ellipse traversed. The position of the sun may be predicted for any latitude, time of day, and any day of the year, using simple trigonometrical formulae, provided one works in solar time. Solar time at any place is referred to the precise time of day when the sun crosses the north south vertical plane. This moment is defined as solar noon. The sun has its greatest elevation at solar noon. For most circumstances it is adequate to work

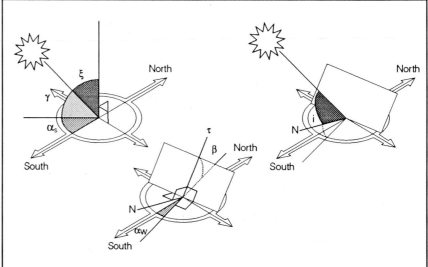

Figure 2.38: Diagrams illustrating solar altitude, solar azimuth, angle of incidence, surface tilt and wall azimuth.

in solar time, which fully defines the geometry of the problem. Methods for adjusting between local time and solar time in different times zones are included in Appendix 1, which also explains the detailed trigonometric calculation of the sun's position. The data from such calculations is used to construct the various graphical aids described below.

Cylindrical solar altitude and solar azimuth graphs

The data obtained by calculation can be plotted in a number of different ways. While horizontal sun dial type solar charts are useful for getting an overview of the solar geometry, other types of graphical plot are more effective for assessing overshading. A very useful plot is the vertical cylindrical solar altitude solar azimuth plot. Basically the altitude of the sun is plotted against the solar azimuth for specific days. The cylindrical plots can be cut along the north solar azimuth and laid flat. It should be noted that this is not the projection used for basic design using plan and elevation. The cylindrical chart has to be "interpreted" for orthogonal design purposes. Additional graphical aids are needed for this purpose.

Graphical charts of the solar altitudes plotted against solar azimuth for various dates in the year are given in Figures 2.39 a, b and c, covering latitudes 36°N to 56°N at 4 degree intervals of latitude.

The hour lines, in solar time, run up the chart, crossing the sun path lines for the various dates given. It should be noted that the charts cover a sector of 120° east to 120° west, which is only part of the full 360° cylinder. The truncation was done to produce more compact charts; however, as a consequence, the movements of the mid summer sun around sunrise and sunset lie off the charts at higher latitudes.

One of the dangers of this projection is that it is often interpreted erroneously as covering only 90 degrees each way from south. The northward solar penetration can thus get overlooked.

Obstruction data can be plotted onto the charts, but this must be done using the obstructing altitude appropriate to each wall solar azimuth angle. The angles cannot be read from drawings directly.. Interpretative difficulties often arise with this projection, because the plot of the roofline of a parallel obstructing building is a curved shape on this projection. The plotting of parallel obstructions can be aided using an auxiliary overlay aid as explained later.

Vertical and horizontal shadow angles

For design purposes it is useful to have information about the solar angles resolved on the planes of orthogonal projection normally used in design, namely plan and elevation. The angle between the direction of the sun, resolved in the plane of the elevation, and the horizontal plane is known as the vertical shadow angle (VSA). The angle between the direction of the sun resolved on the horizontal plane and wall azimuth angle (the direction

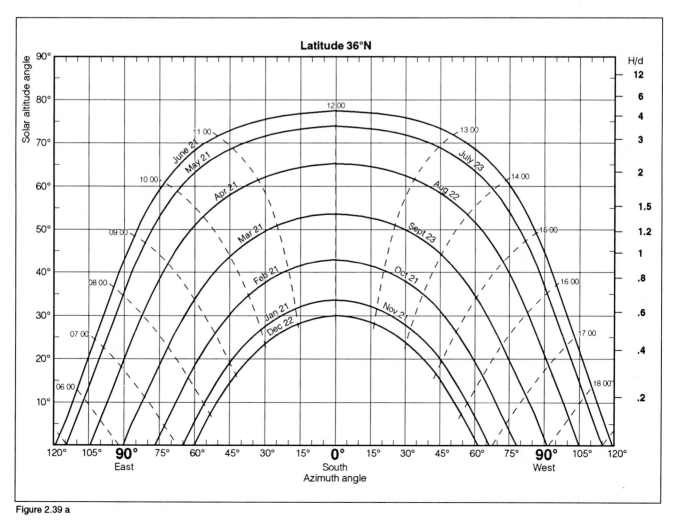

Figure 2.39 a

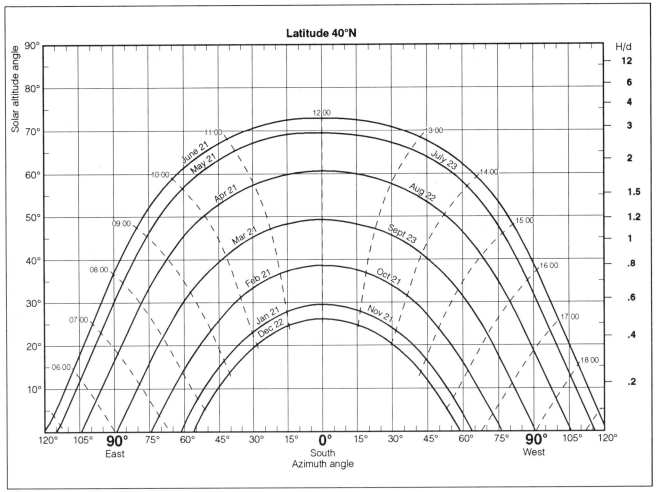

Figure 2.39 b

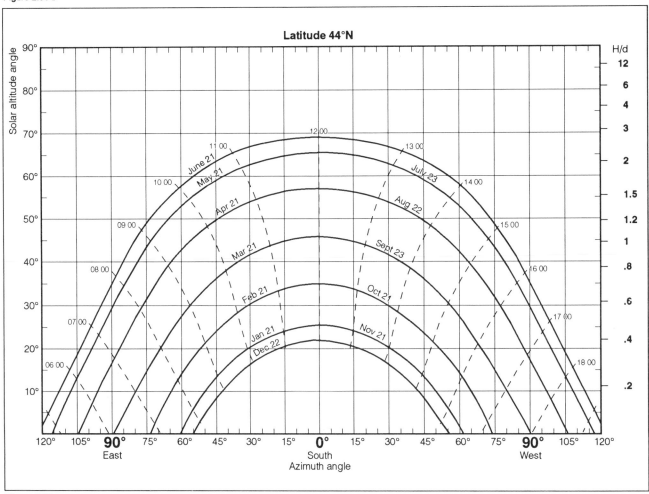

Figure 2.39 c

of the normal to the surface projected onto the horizontal plane) is known as the horizontal shadow angle (HSA). These angles are illustrated in Figure 2.40.

The vertical shadow angle in a given plane is obtained from the projection of the line to the centre of the sun onto the vertical plane containing the normal to that plane. The horizontal shadow angle is simply the angle between the wall azimuth, and the solar azimuth.

The vertical and horizontal shadow angles for a facade of given orientation may be calculated using trigonometry, as explained in Appendix 1. Alternatively they may be extracted graphically from a vertical cylindrical solar chart, using an appropriate overlay centred correctly on the direction of the required wall azimuth angle on the charts of Figure 2.39. This direction is defined by the wall azimuth angle of the plane under study. The overlay used is given in Figure 2.41 plotted to precisely the same vertical

scale as Figure 2.39. It will be noted the overlay diagram is only 180 degrees wide.

The transparent overlay may be used to extract the vertical and horizontal shadow angles from the solar charts (Figure 2.39) for any selected orientation and latitude. The overlay is first orientated as illustrated in Figure 2.42, so it is aligned in the required direction.

The date line under study is then identified on the solar chart below, and then the hour line intersections located. Dropping down vertically on the overlay for each point, one can read on the base line the required horizontal shadow angle for that date and hour. One can then read off the vertical shadow angle on the overlay by interpolating between the plotted curved lines on the

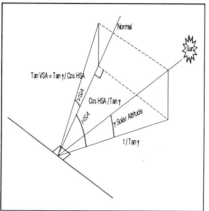

Figure 2.40. The relationship between the vertical shadow angle (VSA) and the solar altitude g depends on the horizontal shadow angle (HSA).

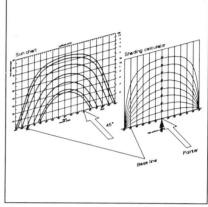

Figure 2.42. Procedure for aligning overlay on solar chart for a facade with a wall azimuth of +45 degrees, i.e. facing south west.

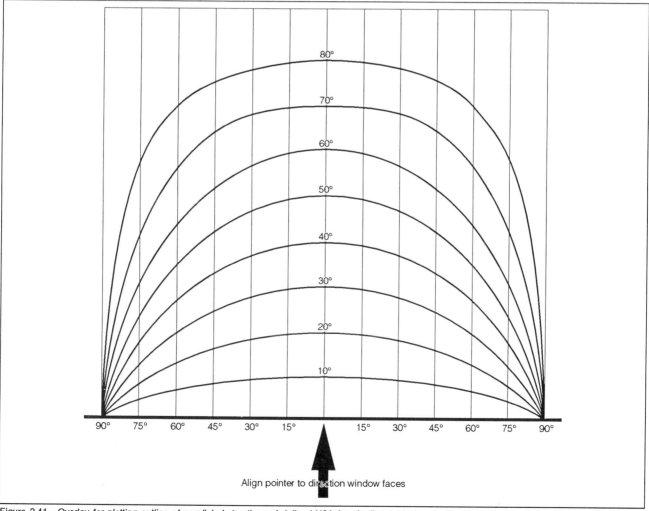

Figure 2.41. Overlay for plotting outline of parallel obstructions of defined VSA for shading analysis onto Figure 2.39 solar charts. This overlay must be precisely scaled to match the scales of the cylindrical charts.

overlay, which represent the various vertical shadow angles. The vertical and horizontal shadow angles for any location have to be extracted separately for each specific facade orientation under consideration.

Many designers, in fact, have a very poor perception of the movements of the sun expressed in terms of the projections they normally use for design. Figure 2.43 plots out for latitudes 35°N, 45°N and 55°N the vertical shadow angles on a north south section.

The following points are evident:

1. At the equinoxes, (21 Mar and 23 Sept), the vertical shadow angle on a north south section is constant throughout the day. The vertical shadow angle in degrees on these two days is actually given by 90-latitude.

2. In the winter half of the solar year, 23 Sept to 21 Mar, the vertical shadow angle on a north south section increases from sunrise to reach a maximum value at solar noon, and then decreases. In order to get reasonable winter insolation, it is not enough to design on the basis of noon time solar altitudes.

3. In the summer half of the solar year, 21 Mar and 23 Sept, the sun starts on the north side of the building, and moves over to the south, to reach a minimum vertical shadow angle on a south surface at solar noon. It then retreats back over the roof. An overhang on a south facing wall that provides complete shading at noon thus gives complete direct beam shading throughout the day in this part of the year.

It is immediately clear from Figure 2.43 for example, that nearby deciduous trees planted at acceptable distances to the south, will make little impact on direct beam insolation on a south facade at midsummer. The systematic extraction and graphical plotting of vertical and horizontal shadow angles for facade orientations under design consideration is very much recommended. This can be done in terms of simple angles in plan and section, as in Figure 2.43.

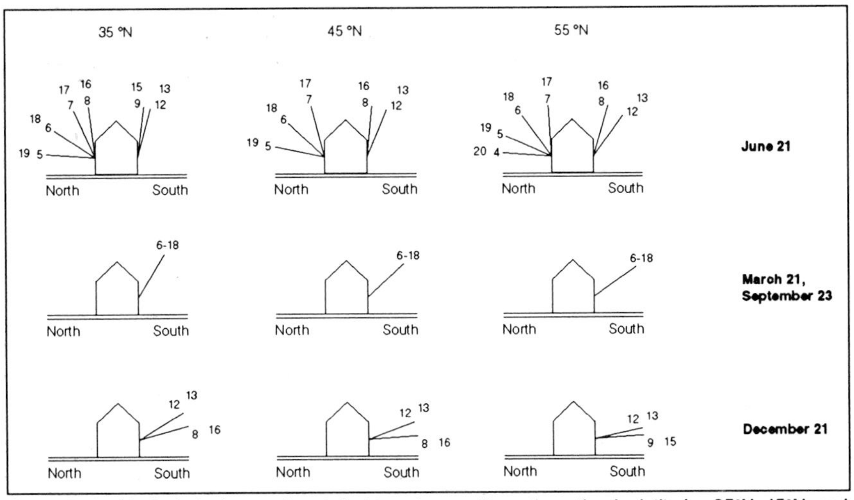

Let us now consider the plotting of a specific layout onto the solar chart overlay. Figure 2.44 shows the layout under study.

The horizontal shadow angles are read off from the normal to the facade. As the obstructing buildings are parallel, it is only necessary to extract the vertical shadow angles for the ridge and two verges. It is adequate to record the vertical shadow angle to the top of the tree. The outline can be transferred onto the overlay. The plotting of the parallel elements is achieved by using the vertical shadow angle lines for guidance. These are lines of constant VSA. The plotting points have to be established first using the HSAs measured from the drawing. Figure 2.45 shows the resultant plot on the overlay for use with the vertical cylindrical solar charts.

Alternatively it is also possible to plot design graphs for any orientation expressed in terms of the vertical shadow angle (y-axis) plotted against the horizontal shadow angle (x-axis). Figure 2.46 provides an example. The sector width is 180 degrees. The advantage of such a plot is that parallel obstructions plot as straight lines, as do horizontal shading overhangs of fixed width. Vertical obstructions plot as vertical lines. So it is easier to perceive what is happening during design, working with orthogonal drawings. When using a cylindrical projection, there is a considerable shape distortion of typical building profiles like parallel roof lines and shading devices like horizontal overhangs, which makes design interpretation more difficult.

Obstruction profile angles

Following the concept of using the vertical and horizontal shadow angle to describe the solar geometry in orthogonal terms, one can define two angles derived from plan and section to describe the geometry of obstructions. The horizontal obstruction profile angle is measured on plan from the normal to the obstructed surface resolved onto the horizontal plane. In section, the obstruction angle is resolved on the vertical plane containing the normal to the obstructed surface to obtain the vertical obstruction profile angle. These angles emerge naturally from the normal orthogonal design process.

Figure 2.43 Vertical shadow angles plotted for a true north south section for latitudes 35°N, 45°N, and 55°N for June 21st, 21 Mar/23 Sept, and 22 Dec. The numbers refer to specific hours in solar time.

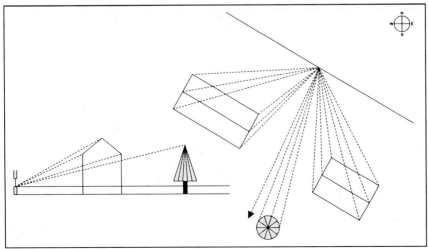

Figure 2.44. Layout plotted on diagrams that follow.

Angle of incidence

The angle of incidence at a given point on a plane is the angle between the line joining that point to the centre of the solar disc and the normal to the surface at that point. Refer to figure 2.38. If the beam is perpendicular to the surface, its angle of incidence is 0 degrees. At grazing incidence, it is approaching 90 degrees. Knowledge of the angle of incidence is very important in the analysis of direct solar radiation heat transfer for two reasons:-

1. The direct beam irradiance on any surface, I_s is related to direct beam intensity normal to the solar beam, I_n by the following expression: $I_s = I_n \cos i$ W/m^2 where i is the angle of incidence.

2. The transmittance of glazing materials is also a function of the angle of incidence.

2.32

ASSESSING THE OBSTRUCTION OF THE DIRECT SOLAR BEAM ON BUILDING SITES

It follows the more obliquely the sun strikes, the less energy that will penetrate. The calculation of the angle of incidence of the direct solar beam on any surface is explained in Appendix 1.

The problem of obtaining the site data to work with the various charts is now discussed:

Geometric classification of types of obstruction

The full assessment of direct beam obstruction of a site is a problem involving five dimensions, three dimensions of space, an hourly dimension, i.e. the passage of daily time, and a daily dimension, i.e. the passage of annual time through the year. A flat graphical device, such as Figure 2.39 overlaid with a web plot graph can present four dimensions, but never five. A sun's eye perspective on isometric uses up three dimensions, such as Figure 2.48. The geometric analysis of sun's eye views becomes essentially one of examining systematically sets of snap shots computed in terms of different times of day and times of year.

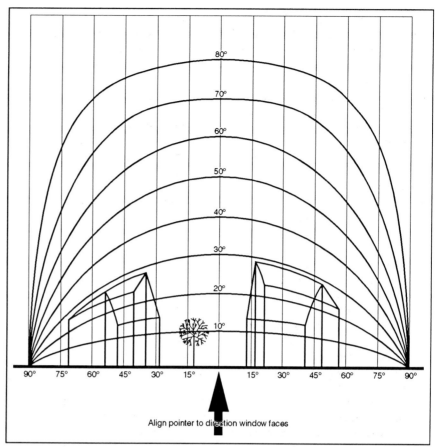

Figure 2.45. A plot of the obstructions shown in Figure 2.44 onto the solar chart overlay. The overlay can then be laid over the solar chart, but must be correctly orientated to match the direction of the normal to the facade. The overlay can then be used to assess the degree of insolation and of shading of the latitude in question.

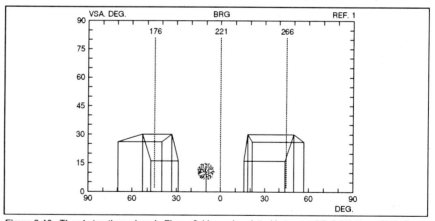

Figure 2.46. The obstructions given in Figure 2.44 may be plotted in terms of their VSA's and HSA's, as in this computer generated plot. a horizontal overhang plots on to this chart as a straight horizontal line of fixed VSA. The normal to the facade has a bearing of 221 degrees from true North.

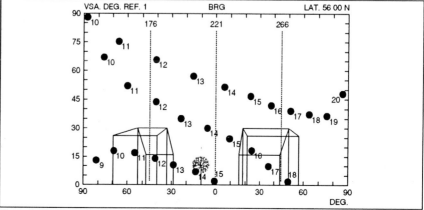

Figure 2.47. Computer generated hourly sun positions plotted onto the VSA/HSA diagram for Latitude 56°N for 22 December, 21 March/22 September, and 21 June superimposed on the building and tree outlines of Figure 2.46. It shows there is considerable winter sun obstruction due to the building on the left.

Figure 2.48. If the problem can be reduced to an analysis at a particular point in space, one spatial dimension can be eliminated. One can then examine the whole direct beam geometric problem visually on a single web chart. One has to examine under what conditions such a two dimensional spatial simplification is acceptable. The impact of changes in distance of obstructions on the changes of angular obstruction at different points on the site is the important factor in making this assessment.

Obstructions can be conveniently be classified into:

- distant

- middle field

- near field

Distant obstructions

Distant obstructions are obstructions that may be considered to exert the same solar geometric impact over a whole building site of limited area; for example, mountains and hills. Distant obstructions can be easily dealt with by one graphical obstruction plot for the whole site, using the same obstruction angles at all viewpoints.

Middle field obstructions

Middle field obstructions are obstructions that are far enough away to enable the impact of the sun over a limited area of the facade, say a specific window, to be assessed using a single geometric point study. Typical middle field obstructions are buildings on adjacent sites set at a reasonable distance, and trees not too close. The geometric impact of middle field obstructions can be assessed by a series of graphical obstruction studies at a range of representative points, say the mid points of relevant windows.

Near field obstructions

Near field obstructions are obstructions whose solar geometric impacts must be considered to vary continuously across the surface of a facade, for example shading devices attached to the building, passage walls, trees very close to buildings. Near field obstructions are best handled geometrically by techniques that present a sun's eye view perspective or isometric view point of the whole system, or by use of detailed plans and sections with shadow angles, using cross projection. Near field analysis essentially requires a series of "analytical snapshots" spaced in terms of day or time of year.

2.33

TECHNIQUES FOR THE ASSESSMENT OF DISTANT OBSTRUCTIONS

There are two main approaches, the use of contoured maps and the use of theodolite (or simplified theodolite type) surveys to assess the site obstruction profile by direct observation. In this case a single obstruction angle traverse from one point on the site is, by the definition of distant, adequate for the whole site. A true North South reference direction has to be established using a map, or a magnetic compass, applying the appropriate corrections for the magnetic variation at the site. Appendix 3 presents a method of analysis for mountainous areas using contoured maps. Once the angular profile is established in terms of obstruction angle, and associated obstruction azimuth, it can be plotted onto an latitude azimuth based solar chart. It then becomes simple to perceive the daily periods of obstruction due to distant objects at different types of year valid for all points on the site. Figure 2.48 shows a typical terrain plot superimposed on a computer generated cylindrical solar chart.

2.34

TECHNIQUES FOR THE ASSESSMENT OF MIDDLE FIELD OBSTRUCTIONS

The criterion for distinguishing between a distant obstruction and a middle field obstruction is the rate of change of the angular obstruction geometry across the site area under consideration. A change in the obstruction profile angle of less than 2 degrees across the facade area under consideration will not make a big impact on the geometric analysis. If the profile angle change over the area of analysis is greater than 2 degrees, then a more detailed analysis may need to be made at relevant points on the site. There are two approaches for obtaining the geometry`, use of site plans with measured building dimensions, and field survey of obstructions at a range of relevant points. The survey may be done using theodolites (or simplified theodolite) or photographic techniques. The field technique obviously cannot deal with the impacts of future development. There are often practical difficulties in achieving the appropriate survey points, if the areas under study are not to be at existing ground level. Photographic techniques enable the nature of complicated profiles

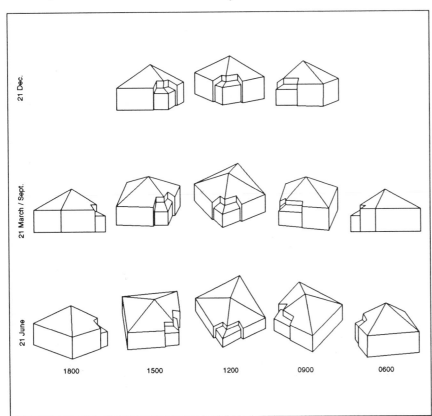

Figure 2.48: The sun's "eye" views of a building in the region of London for the winter solstice (top), the equinoxes (middle) and summer solstice (bottom). Ref [39]

to be captured quickly, and are particularly useful on urban sites, where the surroundings are geometrically complex. Appendix 4 provides a computer generated overlay for use with a 35 mm camera and a standard 21 mm wide angle lens. Two pictures enable a 180 sector to be covered. Obviously the camera must be level, for the horizon to be in the right place. Appendix 4 explains the details of the photographic method. A map and compass are needed to put on the correct reference direction. Additional simple survey techniques are also set out in Appendix 4.

Once the obstruction geometry is established, it can be plotted onto an appropriate solar geometry graph. For middle field obstructions, this has to be done separately for each region of the facade under study. Obviously an experienced designer will concentrate on a limited range of key areas on the facade.

The combined geometric effects of distant obstructions and middle field obstruction can be assessed by merging the two profiles, and overlaying the resulting profiles onto the vertical cylindrical sun chart.

TECHNIQUES FOR THE ASSESSMENT OF NEAR FIELD OBSTRUCTIONS

If the rate of change of the obstruction profile angle across a specific element (say a specific window) exceeds a few degrees, then the obstruction must be considered a near field obstruction. For example shading overhangs fall into this category. Near field obstructions on the building itself can easily be assessed geometrically from the drawings. The dimensions of other near field objects have to be determined by direct measurement, as three dimensional objects, using simple survey techniques. The near field impacts can only be considered with some fairly severe limitations using cylindrical projection solar charts, as the detailed three dimensional geometry becomes so important. The impacts of near field obstructions on the direct beam can be assessed using vertical and horizontal shadow angles on a drawing with a plan and section using cross projection. A graphical technique for producing isometric sun's eye views of buildings, given the solar altitude and azimuth is given in Appendix 5. Such views are normally formed using appropriate computer programmes.

ANALYSIS OF FACADE SHADING BY ATTACHED SHADING DEVICES

The geometric analysis of facade shading is a special assessment problem involving the analysis of near field obstructions. The analytical problem was clarified by Olgyay [40], who introduced some important simplifying procedures.

Horizontal overhang analysis

In this method, it is normally assumed any horizontal overhang is either continuous or at least runs to meet with vertical side fins. The vertical profile analysis is simplified to consider only two points, the base of the window, and the midpoint of the window. The profile angle from the base of the window gives the requirement for 100% beam shading. The profile angle from the middle of the window gives the 50% shading line.

The shading mask can then be prepared using the overlay Figure 2.41. This is then superimposed on the solar diagrams, Figure 2.39, with the centre line of the overlay lined up with the wall azimuth as

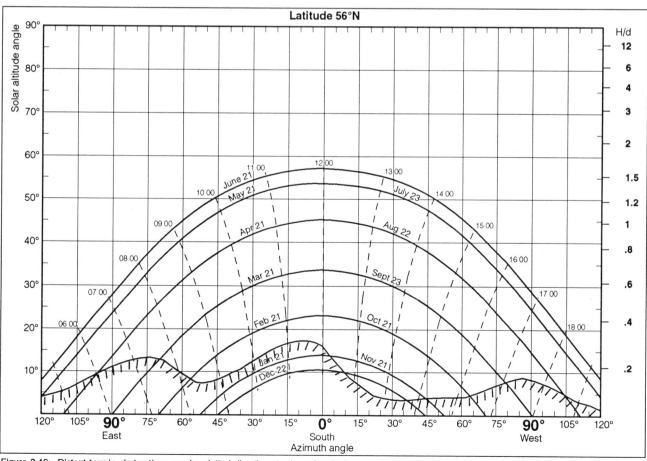

Figure 2.49: Distant terrain obstructions can be plotted directly onto the solar chart and are appropriate for the whole site. As obstructions become closer, different terrain outlines will be needed for different areas of the site.

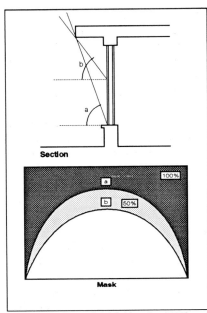

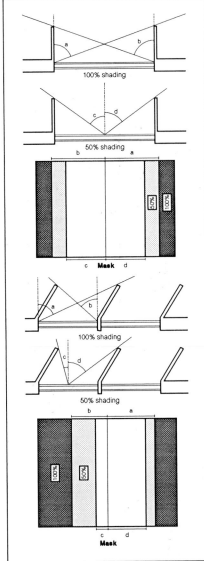

Figure 2.50 Top - Establishing the 100% and 50% vertical profile angles of an overhang. Bottom - Transferring these vertical profile angles, a and b, onto the Figure 2.41 overlay to produce a shading mask for the overhang.

Figure 2.51 The preparation of shading masks for vertical shading devices for 100% and 50% shading.

explained later. It is then possible to read off and record the periods of shading, month by month.

Vertical fin analysis

The procedures for making a shading overlay for vertical shading systems are very similar.

1. Draw an accurate plan of the window opening and any vertical shading devices.
2. Identify the conditions for 100% and 50% shading. Figure 2.51 illustrates the process.

Subsequent action

1. Using Figure 2.41 as an underlay, draw the shading mask to scale. Figure 2.51
2. Superimpose the shading mask on the solar chart for the appropriate latitude , and orientate the centre line to coincide with the wall azimuth under consideration. Read off periods of shading.

Combining horizontal overhangs and vertical fins

The combined effects are assessed by merging the two shading overlays, as in Figure 2.52. The process of overlaying the solar chart is also illustrated in Figure 2.52.

2.37

TOWN PLANNING ANALYSIS

Sometimes designers wish to examine the problem the other way round, by examining the impact of

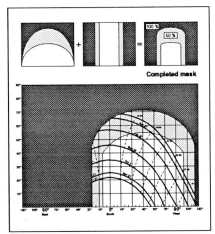

Figure 2.52: Combined vertical and horizontal shading mask, derived from figures 2.50 and 2.51 superimposed on solar chart with a wall solar azimuth of 45 degrees west, i.e. S.W. The periods of insolation at different dates can be read off for 100% shading and 50% shading.

their development on the insolation of adjacent sites. There are a number of techniques for doing this. One approach is to use a tool to help define a suitable solar envelope. The solar envelope may be defined as the greatest volume which a building can fill on a site without causing significant overshadowing of adjacent sites. This would obviously be of major importance in the development of a solar townscape, though it is likely to be more easily applicable in rural or suburban settings than on high density urban sites. One method for carrying out such an analysis is given in Appendix 6.

2.38

SHADING ANALYSIS THROUGH THE USE OF SCALE MODELS

The polar sundial

A sundial can be constructed in order to understand the impact of shading patterns cast by buildings, vegetation or shading devices. This will accurately predict shadows cast by an object at any given hour or month. This tool is to be used by designers with models constructed of the building and its surroundings.

The polar sundial in Figure 2.53 should be photocopied, stuck to a light card, cut out and folded as shown in Figure 2.54 with a 15 mm pin stuck in the middle of the dial.

The sundial may then be mounted on a tilting and pivoting table or on a fixed table with a nearby adjustable parallel light source used to indicate solar penetration and shading at the different hours of the day and months of the year. An alternative is to use a camera to record the view the sun would have of the building at different times of the day and year Figure 2.55.

It is not necessary to construct a new sundial for each latitude considered, as the polar sundial can be tilted to represent any latitude angle.

The heliodon

There are many types of heliodon. Figure 2.56 shows a typical simple version. Such devices have to make provision for mimicking the daily rotation of the earth about its polar axis, and the changes in the solar declination (the angle of the sun's rays make with the equatorial plane) from season to season. To be flexible they also need to allow the latitude to be altered. The advantage of the heliodon is that it allows mechanical reproduction of the geometric shadowing of the sun's rays on a small model for any latitude, time of day and year. The instrument has the added advantage that the model can be rotated on the inclined plane representing the surface of the earth to explore the impacts of a range of orientations. The results are best recorded photographically.

Figure 2.57 shows a heliodon photograph. In this case, the heliodon was been used to verify the output of a complex graphical computer programme for assessing the impacts of trees on sunlight.

ANALYSIS OF SOLAR AVAILABILITY

The detailed impacts of geometric layout and shading geometry can be analysed using the above techniques on a basis of the availability or non availability of sunlight at any specific time of day at any time of year at any analysed point on the facade. A monthly schedule of the hours of possible sunshine availability can be prepared. This provides the basis for any subsequent energy analysis.

In the case of vertical surfaces, one can, by identifying the 180° sector on the solar chart relevant to any particular facade orientation, establish the maximum possible hours of sunshine with no obstruction for that orientation (refer Figure 2.39). The actual hours of direct beam sunlight available after obstruction can then be expressed as a fraction of the maximum possible

hours of sunlight at that date at that location. This calculation gives some feeling for the relative performance achieved. This helps in reaching sound geometric decisions about layout and shading. In the winter period, the maximum possible sunshine duration for surfaces facing due south will be equal to the number of hours between sunrise and sunset. This is not so in the summer period, as the sun spends part of the day to the north of the facade. Refer to Figure 2.39. On a vertical east or west surface, the sun can only fall on the surface for half the day. Some design methods make use of the concept of a geometrically defined shading factor, as a correction factor to be applied to the incident diffuse radiation. However this approach produces a rather oversimplistic assessment of performance.

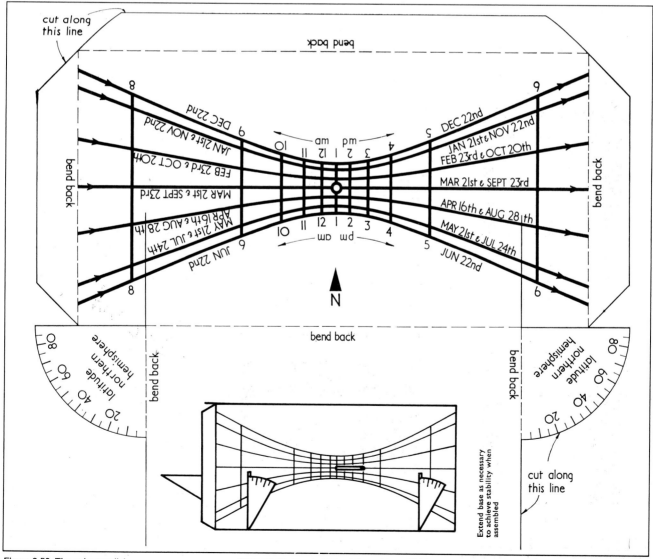

Figure 2.53: The polar sundial

Figure 2.54: Construction of the polar sundial

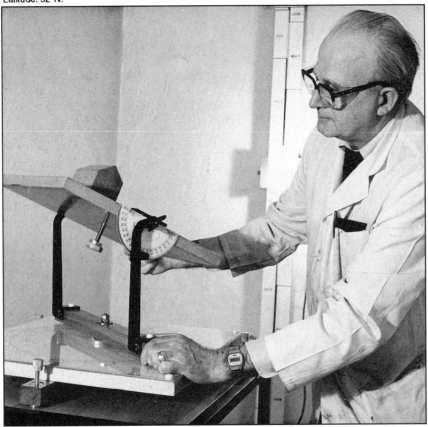

Figure 2.55. An example of the use of the polar sundial. Time of year: 09.20 on 21 May or 24 July. Latitude: 52°N.

Figure 2.56 Simple Heliodon with model on tilted board adjustable for latitude. The deliniation is allowed for by moving the lamp up and down the board. Time of day is achieved by rotation about the baseboard pivot.

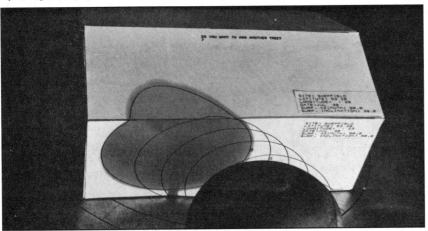

Figure 2.57. Photographic record of the impact of trees on the insolation of a building at a specific latitude, time of day and time of year as assessed by a heliodon. In this case, it was been used for checking the reliability of a computer programme for assessing landscape impacts.

REFERENCES

[1] CEC European Wind Atlas. Troen, I and Peterson, E.L. Risø National Laboratory, P.O. Box 49, DK-4000 Roskilde, Denmark. 1989

[2] Workshop on Passive Cooling, Aranovitch, E., de Oliveiro Fernandes, E., and Steemers, T.C., Ed., Ispra, 2-4 April, 1990, Commission of the European Communities. EUR . 1990

[3] The Thermal Radiance of Clear Skies, Berdahl, P. and Fromberg, R. Solar Energy, Vol. 29, pp 229-214. 1982

[4] Climate in the United Kingdom, a Handbook of Solar Radiation, Temperature, and other data for Thirteen Principal cities and towns, Section 4.3, Estimation of Long wave radiation exchanges between the external surfaces of buildings and the sky and ground, Page, J.K., and Lebens, R. Ed. HMSO, P.O. Box 276, London SW8 5DT, pp 335-367. 1986

[5] Test Reference Years TRYs: weather data sets for computer simulations of solar energy systems and energy consumption in buildings, Lund, H. Report No. EUR 9765, Commission of the European Communities, Brussels. 1985

[6] Short Reference Years and Test Reference Years for the EEC Countries: Technical University of Denmark, Thermal Insulation Laboratory, Final Report Contract ESF-029-DK, Lund, H. Report EUR 9402, published by the Thermal Insulation Laboratory, Building 118, Technical University of Denmark, DK-2800 Lyngby, Denmark. 1985

[7] Short Reference Years, Commission of the European Communities, CEC DGXII, EUR 10663, CEC, Brussels. 1986

[8] UNCHS (Habitrat) Manual on how to prepare locally based Handbooks for passive solar heating and natural cooling design, Page, J.K. and Balcomb, J.D. Eds., Chapter 11, Assessment of passive solar heating performance using the monthly SLR method, (Chapter written in association with Dr. J.D. Balcomb), UNCHS (Habitat), Nairobi. 1988

[9] Passive Solar Heating Analysis, a design manual, Douglas Balcomb, J. Jones, R.W., McFairland, R.D. and Wray, W.O., American Society of Heating, Refrigerating, and Air Conditioning Engineers. 1984

[10] Methode 5000, Claux, P., Franca, J.P., Gilles, R., Pesso, A., Pouget A. and Raoust, M. PYC Edition, 254, rue de Vaugirard, 75740 Paris Cedex 15. (in French). 1982

[11] The FIT method for estimating the heating requirements of buildings with direct gain passive solar systems, in Proc. 1989 2nd European Conference on Architecture, Paris, Dec. 1989, Oliveira, A.C. and de Oliveira Fernandes, E., Ed. T.C. Steemers and W. Palz, Kluwer Academic Publishers, P.O. Box 17, 3300 AA Dordrecht, The Netherlands, pp 441-443. 1990

[12] Solar Engineering of Thermal processes, Chapter 14, Design of solar heating systems, Duffie, J.A. and Beckman, W.A. , H. Wiley & Sons, New York, pp 485-511. 1980

[13] Atlas solaire francais, Claux, P., Gilles, R., Pesso, A., and Raoust, M., PYC Edition, 254, rue de Vaugirard, 75740 Paris cedex 15. (In French). 1982

[14] Climate in the United Kingdom, a Handbook of Solar Radiation, Temperature, and other data for Thirteen Principal cities and towns, Page, J.K., and Lebens, R. Ed. HMSO. 1986

[15] Commission of the European Communities' European Solar Radiation Atlas, Vol.I, (2nd edition) Kasten, F., Golchert, H.J., Dogniaux, R. & Lemoine, M., , Ed., W. Palz, Verlag, TUV, Rheinland, Cologne. 1984

[16] Commission of the European Communities' European Solar Radiation Atlas, Volume II, Inclined Surfaces, Page, J.K., Flynn, R.J., Dogniaux, R. and Preuveneers, G. Editor, W. Palz, Verlag, TUV, Rheinland. 1984

[17] Solar Energy R & D in the European Community, Project F, Solar radiation data, Volume 2, Palz, W. , D. Reidel, Dordrecht, Holland. 1986

[18] Prediction of solar radiation on inclined surfaces, Solar Energy R & D in the European Community, Project F, Solar radiation data. Volume 3. Page, J.K., D. Reidel, Dordrecht, Holland. 1986

[19] Climatic data for the design of solar buildings, in Proc 1989 2nd European Conference on Architecture, Paris, Dec. 1989, Bourges, B., Carvalho, M.J., Cherrey, M., Petrakis, M., Armenta, C., Ruiz, V., Frutos, F. and Caridad, J.M., Ed. T.C. Steemers and W. Palz, Kluwer Academic Publishers. 1990

[20] EUFRAT Climatic Data Handbook, Climatic data for the deisgn of renewable energy systems in Europe: frequency distribution of solar radiation and temperature, Bourges, B. Ed., in the course of publication by the CEC, Brussels. 1990

[21] Study for a Eurocode on the rational use of energy in buildings, Commission of the European Communities, Directorate-General, Internal Market and Industrial Affairs, Report No. 2973/III/86. 1986

[22] CIBSE Guide, Vol. A, Design data, CIBSE, Delta House, 222 Balham High Road, London SW12 9BS. 1986

[23] Climatic Atlas of Europe 1, Mean temperature and precipitation, World Meteorological Organization, 41 Ave. de Motta, Geneva. 1970

[24] Thermal insulation - calculation of space heating requirements for residential buildings, International Organization for Standardization, Geneva, ISO 9164. 1989

[25] Climate in the United Kingdom, Average monthly and yearly degree days to different base temperatures, Page, J.K. and Lebens, R., H.M.S.O., P.O.Box 276, London, SW8 5DT, pp 237-253. 1986

[26] ISO Preliminary title Standard for specifying accumulated temperature difference. ISO/DIS 6397

[27] Estimating monthly degree days, Building Services Engineering Research and Technology, Vol. 4, pp. 1159-1162. Hitchin, E.R., 1983

[28] UNCHS (Habitat) Manual on how to prepare locally based Handbooks for passive solar heating and natural cooling design, Appendix 5. The estimation of degree days for heating and cooling at any site, Page, J.K., UNCHS (Habitat), Nairobi, in course of publication. 1988

[29] CIBSE Guide, Vol. A, Design data, Section A8 Summertime temperatures in buildings, CIBSE, Delta House, 222 Balham High Road, London SW12 9BS, p. A.8.3. to A.8.15. CIBSE, 1986

[30] Climate and building in Britain, Lacy, R.E., HMSO, P.O. Box 276, London, SW8 5DT. 1977

[31] Weather data for the building industry, No. 1, Reidat, D., Hamburg, English translation, BRS Library Communication, No 933, Building Research Station, Garston, Watford, Herts., translated from Wetterrdaten fur das Bauwesen, Nr 1, Hamburg, Deutscher, Wetterdienst, Seewetteramt, Hamburg. 1957

[32] European solar microclimates - a research programme for improved solar energy harvest, in Proc. The European New Energies Conference, Vol. 3, 24-28 Oct., 1988, Scharmer, K., published by H.S. Stephens Associates, Agriculture House, 55, Goldington Road, Bedford, MK40 3LS, UK, pp- 396-399. 1990

[33] Meteorological Aspects of the utilization of wind as an energy source, World Meteorological Organisation, Technical Note No 175, WMO, Geneva, Switzerland. 1981

[34] A method for the assessment of wind induced natural ventilation forces acting on low rise building array, Lee, B.E., Hussain, M. and Soliman, F., Department of Building Science Note No. BS 50, Department of Building Science, Faculty of Architectural Studies, University of Sheffield, Western Bank, Sheffield, England. 1979

[35] Climate and site development; Part 3, Improving microclimate through design, BRE Digest Number 350, Building Research Establishment, Garston, Watford, WD2 7JR. (Contains an extensive bibliography). 1990

[36] Technische Universiteit Eindoven, Faculteit der Bouwkunde, Vakgrfoep Fago, Document 0469F/0072 in Dutch.

[37] Natural ventilation in courtyard and atrium buildings, in Proc. CEC 2nd European Conference on Architecture, Bersalem R, and Sharples, S., UNESCO, Paris, Ed. T.C Steemers, and W. Palz, Kluwer Academic Publishers. 1990

[38] Wind driven natural ventilation in courtyards and atria in urban settlements, Proc. 8th International PLEA Conference, Bersalem, R. and Sharples, S., Halifax, Nova Scotia, Canada, June 17-20th, 1990.

[39] Drawing a projected view of a building as 'seen' by the sun, Byrd, R,M., Lighting Research and Technology, Vol 22 (1), pp 53-54. 1990

[40] Solar Control and Shading Devices. Olgyay, A. & V., Princeton University Press. ISBN 0-691-02358-1. 1976

ADDITIONAL REFERENCES

Utilization of wind energy in urban areas - chance or utopian dream?, Grauthoff, M., Bochum, F.R.G.

Building sector energy conservation programme of Pakistan, Jamy, G.N., Islamabad, Pakistan

A linear goal programming model for urban energy-economy-environment interaction, Kambo, N.S., Handa, B.R., Bose, R.K., New Delhi, India

A simple computer model for estimating the energy consumption of residential buildings in different microclimatic conditions in cold regions, Rauhala, K. Espoo, Finland

Climate and building energy management, Taesler, R. Norrköping and Stockholm, Sweden

Conversion of exhaust heat to latent heat for the management of the thermal environment in urban areas, Tarumi, H, Fujii, S., Ito, N, Tokyo, Japan

The monitoring of air pollutants in Athens with particular reference to nitrogen dioxide, Boucher, K, Loughborough, UK

Climate and pollution in Paris, Escourrou, G. Meudon, France

Climate and air hygiene in the urban industrial area of Mainz-Wiesbaden, F.R.G.: results of a long-term study with various monitoring systems, Heidt, V. Mainz, F.R.G.

Nitrogen dioxide and oxidant in an urban region of Delhi, India, Kapoor, R.K., Singh, G., Tiwari, S. and Ali, K, New Delhi, India

Meteorological factos causing high dust concentration, Miyazaki, T., Yamaoka, S. Osaka, Japan

Wind circulation and air pollutant concentration in the coastal city of Ravenna, Tirabassi, T. (Bologna, Italy), Fortezza, F., Vandini, W. (Ravenna, Italy)

An urban air pollution model, Venegas, L.E., Mazzeo, N.A., Buenos Aires, Argentina

Source apportionment of aerosols in the Tokyo metropolitan area by chemical element balances, Yoshizumi, K., Tokyo, Japan

Size distribution of polycyclic aromatic hydrocarbons in Chinese cities, Shi-Wei Zhou, Gai-Yeng Ren, Shu-Min Li and Yi-Xin Wei, Beijing, China

Investigations of the microclimate in hospital wards, Czarniecki, W., Kopacz, M., Okolowicz, W., Gajweski, J., Grzedzinski, E., Warsaw, Poland

Climate and human health in the Parisian region, Escourrou, P., Meudon, France

Improving indoor thermal comfort by changing outdoor conditions, Höppe, P., Munich, F.R.G.

The effects of hypobaric conditions on man's thermal responses,, Ohno, H., Kuno, S., Saito, T., Kida, M., Nakahara, N., Nagoya, Japan

Human health and social factors in winter climates, Pressman, N., Waterloo, Ont., Canada

Tight or sick building syndrome, Thirumalaikolundusubramanian, P., Shanmuganandan, S., Uma, A., Madurai, India

The potential of land-use planning and development control to help achieve favourable microclimates around buildings: a European review, Keeble, E.J. (Watford, UK), Collins, M., Ryser, J. (London, UK)

An analysis of stochastic properties of room air temperature and the heating load during autumn, Matsumoto, M., Hokoi, S., Takamura, H., Kobe, Japan

Directional wind-chill data for planning sheltered microclimates around buildings, Prior, M.J. (Bracknell, UK), Keeble, E.J. (Watford, UK)

Built form of Shahjahanabad (Old Delhi): an evaluation from the climatic point of view, Saha, S.K., New Delhi, India

Buildings as climate modifiers, Torrance, V.R., Edinburgh, UK

The performance of external wall systems in tropical climates, Briffett, C., Singapore

A comparison of field surveys on the thermal environment in urban areas surrounding a large pond: when filled and when drained, Ishii, A., Iwamoto (Fukuoka, Japan), Katayama, T., Hayashi, T., Shiotsuki, Y., Kitavama, H. (Fukouka, Japan), Tsutsumi, J.-I., Nishida, M. (Fukuoka, Japan)

The value of long-period observatory records: an examination of the meteorological record for Durham University Observatory of the north-east of England, Kentworthy, J.M., Durham, UK

Mapping urban typologies, Paszynski, J., Warsaw, Poland

Gandemer, J. (1973), Inconfort du au vent aux abords des batiments: etude aerodynamique du champ de vitesse dans les ensembles batis: CSTB, Nantes, ADYM 12-73.

Gandemer, J. (1974), Etude de la simulation des structures gonfables, la simultude aerodynamique, CSTB, Nantes. ADYM 110-74.
Gandemer, J. & Barnaud, G., (1974), Simulation des proprietes dynamiques du vent en stabilite neutre dans la sofflerie a couche limite due C.S.T.B., Nantes. ADYM 1 -74.

CHAPTER THREE

Passive solar urban design

3.1

Introduction

Our main concern is with urban spaces, as affected by urban climate. With the objective of ensuring comfort for users of the open space and the optimal use of energy, we suggest ways of dealing with the urban fabric at different levels.

Urban spaces

We can deal with urban space either, (a), by considering the buildings, or, (b), by considering the space between buildings as the focus of interest.

Buildings are associated with the plot on which they stand. The plot is the ground surface manipulated by the architect. The building and its plot are an entity in the urban context and cannot be treated in isolation.

The remaining open space, not directly associated with a building, can be divided into different kinds of spaces, each having one or several definite functions. The urban designer works on that space, limited by the facades and by the roofs of the buildings. He considers the set of buildings as a whole built entity.

Methods of Investigation

Three levels of study, corresponding to three levels of action on urban fabric can be distinguished:

The first level is that of URBAN PLANNING, where action is performed at the scale of a whole city or more.

The second level is that of URBAN MORPHOLOGY, where the action is applied to a group of adjacent buildings and the spaces between.

The third level is that of BUILDING DESIGN, where the action is applied to an individual building.

Contents

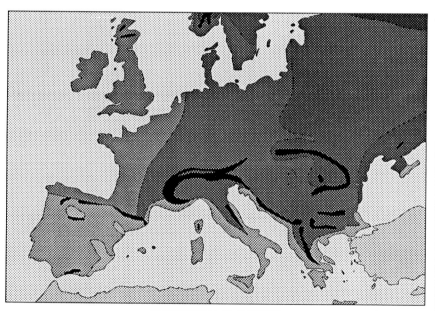

EUROPEAN CLIMATES

Semi-desertic climate with wet winters and arid summers

Mediterranean climate with wet winters and arid summers

Oceanic

Suboceanic

Subcontinental

Wet steppe climate with cold winters

Temperature cold boreal climate

High mountains alpine climate

Figure 3.1. European climates.

Space types

Our approach to the urban fabric may differ according to the particular climate and culture. We distinguish three sorts of European climate:

- suboceanic
- oceanic
- Mediterranean.

Those climates are, of course, locally modified by the presence of the urban forms. Five climatic factors are of particular interest:

- temperature (air temperature and radiant or surface temperature)
- solar radiation
- longwave radiation (especially by night)
- wind speed
- relative humidity (RH).

These factors have an effect on the urban space, but the space influences the local climate also.

Temperature

We have to consider both the air temperature in external spaces (as modified by the exchange of heat with buildings -through losses, combustion, etc), and the surface temperatures of the built forms and ground (affected by radiation and convection from surfaces heated during the day). This can affect comfort negatively (asymmetry in streets) or positively, (eg, by radiation from pavements during the night).

Solar radiation

This concerns the direct radiation coming from the sun, the diffused radiation from the sky and the reflected radiation from the environment.

Longwave radiation

Walls, ground and other surfaces have the capacity to re-emit the energy received from the sun. But the radiation emitted by the materials is long-wave radiation. Materials, normally transparent to direct solar radiation (glass) are opaque to infra-red radiation, which gives rise to the 'greenhouse effect'. Longwave radiation has an important impact on urban pollution because of CO_2 which, in the upper layers above cities is less transparent than the air. A second important consequence of longwave radiation is the cooling of building surfaces exposed to the clear winter sky. The effect is used for cooling in southern Mediterranean zones.

Wind speed

Because their energy is dissipated by friction, wind flows in towns tend to be less severe than meteorological winds occurring over open country. We note, on the other hand, the occurrence of wind corridors in cities.

Relative humidity

Air humidity tends to be higher in urban spaces than in open country, especially in hot arid climates. Towns may create more comfortable conditions by providing artificial water pools or fountains which increase the relative humidity by evaporation.

FIRST LEVEL: URBAN PLANNING

Ways of dealing with urban space

The goals are the same at each level of definition but the problems must be treated in different ways mainly because a higher level of precision is appropriate.

The problem is considered in its broadest form, which is: how to optimize the use of energy .

One solution is to use energy networks as for district heating, gas, and electricity . CREM's work in Martigny (Switzerland) has led to a computer management system for energy networks and pollution control (see Fig. 3.2) and the establishment of energy-use guidance plans.[13]

CREM's SYSURB project

The objective of this work is to develop computer-aided methods of management for urban systems. The SYSURB project is mainly aimed at the management of the distribution of energy in urban contexts, taking into account its various forms (water, gas, electricity, television, heating, etc). SYSURB aims to answer three fundamental questions:

- where and when is it necessary to build networks?
- how to maintain a network?
- how to evaluate the performance of a network?

The original feature of SYSURB is to consider all the networks of a town as a unique coordinated system. SYSURB can be run on personal computers and is available to every town however small. Development of the research may allow further adaptation of the package.

Intervention of architects and urban designers.

Architects do not participate directly in the creation of networks, but urban designers influence their extension and operation. They also have a part in promoting development, and controlling its impact.

To deal with urban organisation (functional zones and transport networks), some pointers are given by P.H. Steadman, P. Rickaby and T. de la Barra, in research they applied in England. [5]

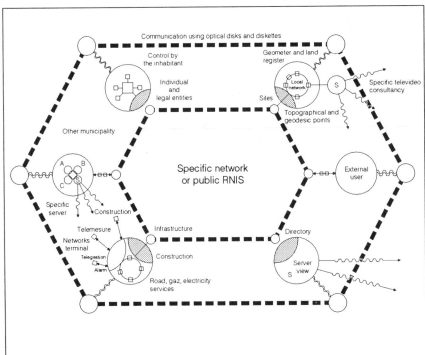

Figure 3.2: Computerised management system produced by CREM.

They considered five theoretical patterns for an urban region (Fig 3.4):

1. Concentration of new development into the central city,
2. Concentration along the main roads system,
3. Concentration into satellite towns,
4. Concentration along secondary roads,
5. Concentration into existing villages.

The more efficient arrangements seem to be those having one or several poles. Good results can be obtained even with a main roads system if appropriate transport networks are developed. For the city itself, the more efficient (and also probable) pattern is a radial one, involving the following areas in which intervention of urban planners would seem to be necessary.
(i) secondary urban centres (in which residential areas would be integrated with employment -which reduces daily commuting);
(ii) 'radial' areas of high density, developed along the main roads.

This does not of course claim to be a definitive treatment of the problem. Our purpose is to stress that good energy management may not be compatible with a random development of urban regions.

Pattern 1: the Concentrated-nucleated configuration:

The population removed from the rural hinterland is all located in the central urban area. The population density is increased and the pattern represents the possible result of an active, long-term policy of urban containment.

Comparison with the existing configuration leads to the conclusion that this pattern provides substantial fuel savings in domestic space-heating, though these savings have an associated cost in restricted dwelling sizes and locations. Amongst the five modified settlement patterns, this pattern ranks first for overall fuel-saving in two out of the three scenarios, and third for the cost of fuel savings in all three scenarios.

Pattern 2: the Concentrated-linear configuration:

In this pattern the redistributed population is located in eight spokes of linear 'ribbon development' radiating from the urban area along main roads.

There are fuelsavings in domestic space-heating but more fuel is used in passenger transport. There appear to be two reasons for this: increased trip lengths because a substantial portion of the redistributed population is located further from the central higher-level services than in pattern 0 (that existing); and congestion of the main roads within the development ribbons because of the concentration of trips on to them. The fuel savings cost more to achieve than in any of the five modified patterns. Thus of the five patterns, this is the least efficient option for fuel conservation. The result being largely attributable to traffic-congestion, it is possible that other transport systems operating along the development ribbons might produce fuel savings at acceptable cost. For example, cheap 'light-rail' rapid transit systems might well be appropriate to this concentrated-linear pattern of settlement.

Pattern 3: the Satellite towns configuration:

In this pattern the redistributed population is placed in eight satellite towns, all of which occur mid-way between two major centres. There are four 'primary satellites' and four

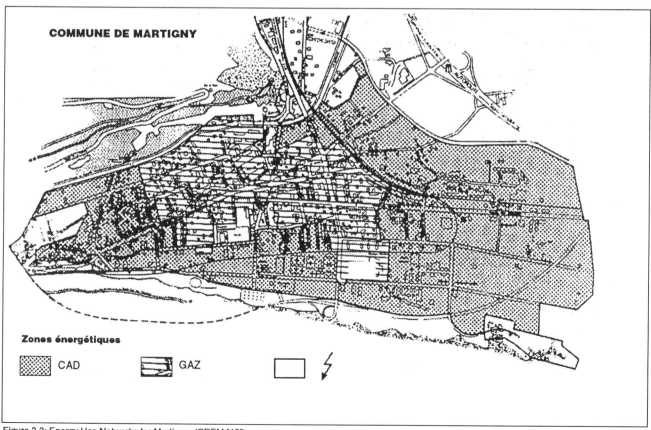

Figure 3.3: Energy Use Networks for Martigny. (CREM [13])

COMMUNE DE MARTIGNY

Zones énergétiques

CAD   GAZ

smaller 'secondary satellites'. Since all the satellite towns are 'shared' with the next adjacent major cities in the settlement pattern, only one quarter of the population of each primary satellite town, and one half of the population of each secondary satellite town, is included in the analysis.

There are fuel-savings in domestic space-heating but more fuel is used in passenger transport. Once again, this is accounted for by increased trip lengths: in this configuration the distributed population is located much further from the city centre than in the existing configuration, and journeys to work and to higher level services in the city centre are lengthened and subject to some congestion of the radial routes. Thus concentration of population into distant satellite towns appears less attractive - at least from the point of view of fuel-conservation - than the concentration of population into the existing city .

The savings to be made in domestic space heating are approximately the same, but pattern 1 provides additional fuel savings in passenger transport, with the associated positive benefit of increased accessibility.

Pattern 4: Dispersed linear configuration:

The population is located in linear development along minor rural roads. Main roads are not developed, so that high-speed inter-city travel is unrestricted.

There are fuel-savings in domestic space-heating but more fuel is used in passenger transport. Trip lengths in this pattern are shorter than in patterns 2 or 3 but still longer than in pattern 0, because some of the redistributed population is located further from the central city than in the existing configuration.

The fuel savings cost more than those of patterns 1 and 5, but are much less expensive than the overall savings in patterns 1 and 3.

Pattern 5: Dispersed nucleated villages configuration:

The redistributed population is located in twenty-four small villages, all of which are within the original rural hinterland of the existing city. Of the five modified settlement patterns, this one incorporates the distribution of population, employment and services which is closest to that of the existing configuration. It is also similar to the possible results of current settlement policies in some parts of England, where new development is directed into selected existing villages close to the urban area. The costs of local fuel savings are lower than in any of the other modified patterns. There are substantial positive benefits in the transport sector compared with pattern 0. Thus it appears that modest concentration of development into local centres within the hinterland of the existing city both saves fuel in transport and improves accessibility. Overall, the fuel savings in this pattern are not as

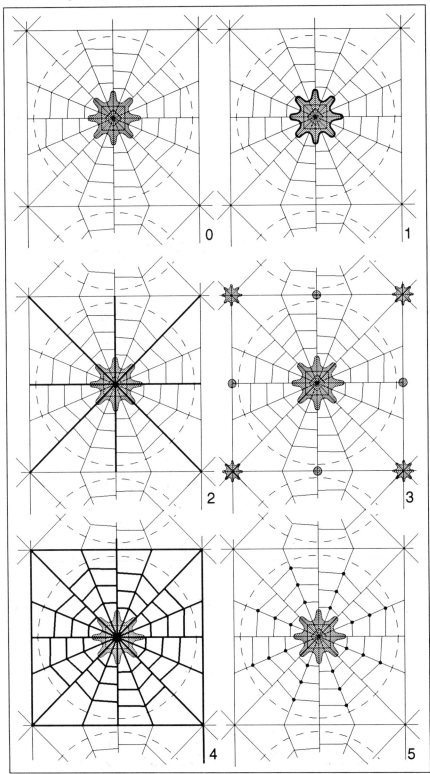

Figure 3.4: "Six regional settlement patterns, one as existing (pattern 0) and five variants: concentration of new development into the central city (pattern 1), along the main road system (pattern 2), into satellite towns (pattern 3), along secondary roads (pattern 4) and into existing villages (pattern 5)." [5]

great as those in patterns 1 and 3, but the costs of the savings are lower. These costs are about half of the equivalent costs in pattern 1 and about one quarter of those in pattern 3. Overall, in all scenarios, pattern 5 ranks third in overall fuel-use, better than patterns 2 and 4, but not as good as patterns 1 and 3. However, in pattern 5, the fuel savings are less costly than in any of the other modified patterns, and in the transport sector there are additional benefits from improved accessibility.

The study concerns only the problem of fuel and transport and the patterns are too general to indicate any intervention for passive solar design.

3.3

SECOND LEVEL: URBAN MORPHOLOGY

If we assume that the urban fabric is now fixed, the problem is then restricted to interactions between the city and the urban climate.

Means of intervention:
On built space, considered as a whole one means of control is density.

Increased building density implies an increase in the relative surface of the roofs. Radiative exchanges predominate at roof level (less near walls or near the ground). The effect is more significant when the buildings are high.

A favourable effect is particularly noticeable in hot climates (Mediterranean or semi-desert) characterised by high-density housing in narrow streets. Using appropriate materials, a large part of the radiation received by roofs can be reflected away to the sky and not absorbed by the building. This strategy can minimise the build-up of radiation and air temperature in external spaces especially near the ground.

In colder climates (or hot and wet climates), open space is often expanded: in the first place to collect the maximum solar radiation; in the second to allow adequate ventilation to dwellings. Note: With increasing density, the microclimatic contrasts in urban spaces also increase.

Preferred Orientation (with regard to sun):
Buildings are oriented according to the expected shading impact on the environment. Account must be taken of the variations due to sun movements according to latitude, time of day and time of year.

In warm southern European regions, the best orientation is north-south (with the best shading conditions in summer and the best lighting conditions in winter). The temperatures of north and south faces will be very different during daytime. Consequently, the architecture of the two façades should respond accordingly in terms of wall mass, insulation and the types and areas of glazing used. The arrangement of internal spaces should also take account of orientation. Walls facing private or a public spaces must be treated differently - see Figure 3.5.

The worst orientation is east-west: the highest intensity of solar radiation coincides with higher air temperatures. Using rooms (such as kitchens and bathrooms) as thermal barriers can help.

R. Knowles has defined the greatest volume (solar envelope) that can be built without shading the neighbouring sites [23]. In the east-west orientation, the facades receive solar radiation in the late afternoon when the temperature is highest.

The solar envelope defines the greatest volume of development that can be placed on a given site without shading neighbouring sites, and it must be calculated by convention at some appropriate time such as noon on the winter solstice. In an urban context the form of the solar envelope will be determined by the pattern and width of the surrounding streets, and by the orientation and

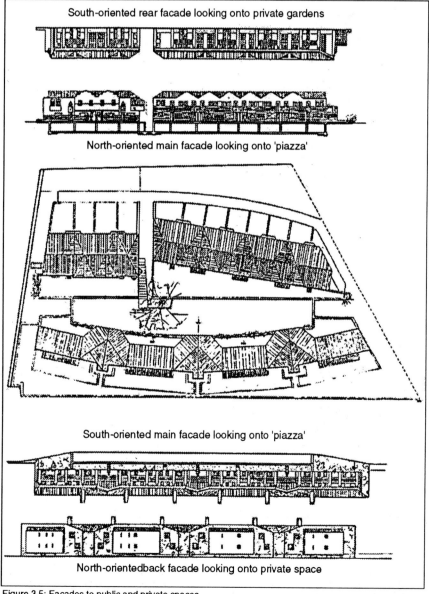

South-oriented rear facade looking onto private gardens

North-oriented main facade looking onto 'piazza'

South-oriented main facade looking onto 'piazza'

North-orientedback facade looking onto private space

Figure 3.5: Facades to public and private spaces

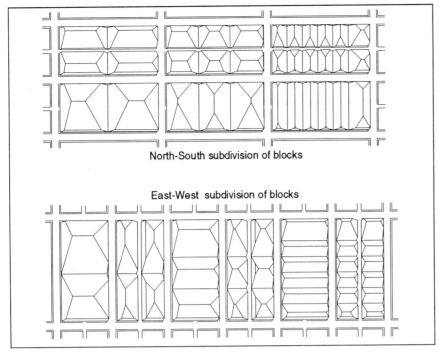

North-South subdivision of blocks

East-West subdivision of blocks

Figure 3.6: Solar envelope of city blocks

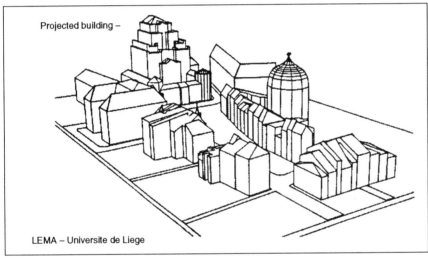

Projected building –

LEMA – Universite de Liege

Figure 3.7: Urban site with existing and proposed buildings [12]

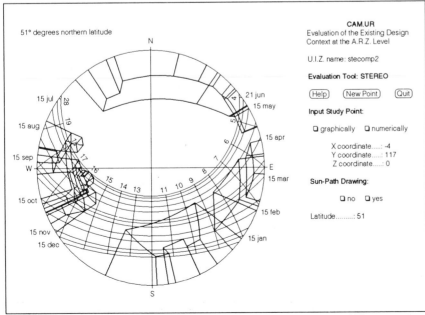

Figure 3.8. Overshadowing effect of the proposed building, using a sun-path diagram [12].

slope of the site. In many cases existing developments appear to break through winter solar envelopes, particularly on the northern sides of blocks, and so it is unlikely that idealized built form, wholly contained within the appropriate solar envelope, could be achieved for some considerable time, if ever. The solar envelope would not in any case be the sole determinant of block form, but would interact with other constraints, such as the requirements of access and density, to determine the overall form of the development.

The distance between buildings in southern European towns is often the result of a compromise between winter and summer conditions; greater distances allow for better lighting natural lighting in winter, smaller distances can help to prevent overheating in winter. The siting of buildings is largely irreversible, but shading and daylighting devices can often be used to improve poor thermal or visual conditions.

Solar protection can be increased by the mutual shading of buildings. Typically in Mediteranian towns, buildings are densely grouped and seperated only by narrow streets. Public or private courtyards are also common. Sloping ground will also affect solar exposure and shading (see chapter 5 – Heating).

In summer at southern latitudes, effective shading can be provided by buildings facing to the south where the seperation distance (i.e street width) on level ground is around 1/5 of the height of the building giving shade. When the sun moves to the west, its lower altitude allows buildings to be seperated by around 1.5 to 2 times the height of the building giving shade.

The microclimate in open spaces can be improved by the appropriate use of vegetation for shading and cooling.

In Mediterranean or semi-desert conditions especially, in large open spaces (such as squares), discomfort caused by exposure to the sun can quickly become unbearable. Vegetation can be a partial solution and may be preferred because:
- vegetation has lower thermal capacity and conductivity than construction materials
- vegetation absorbs solar radiation and reduces reflection

- evapo/transpiration increases relative humidity and thus decreases temperature
- vegetation helps to stabilise temperature, reducing extremes. (Man-made materials generally act otherwise)
- Vegetation can be used as a screen to reduce wind speed near the ground and thereby to reduce the movement of dust. (Wind is reduced but not totally blocked, which permits wind- cooling of spaces).
- Vegetation can be used as a shading device, to reduce noise,and to protect from light rain.

Note: In summer, leaves are a protection against the sun. In winter, when solar radiation is beneficial and contributes to heating, leaves of deciduous vegetation disappear.

Open spaces can be used to modify the microclimate around buildings, reducing solar radiation and temperature near the ground. Absorption, reflection, and heat exchanges in general, depend on the type and colour of materials used. Fountains and still or running water or ducted air can help to reduce the temperature of hard, paving materials.

Note: In hot climates, the presence of unbuilt, open areas of dry ground can be a source of dust. Possible solutions are: to cover the soil with an artificial covering or, better, with vegetation, and irrigation.

Water

Water is sometimes used in hot climates to lower the air temperature by humidification (using latent heat), or simply by heat exchange between hot air and cool water. Water can be used in different ways:

1. As artificial fog where very pure water at high pressure is transformed into droplets, evaporating quickly. This is applicable in small area such as walkways (especially if their axis corresponds with the usual direction of a light wind);

2. In a passive evaporative cooling tower where it is attached to a building, but may also be used for semi-confined external areas. A typical system consists of a down-draft tower which has at its top, vertical wetted cellulose pads. Water is supplied at the top of the pads and is collected at the bottom by a sump and recirculated by a pump. [16]

3. A fine spray of pure water falling down a vertical shaft entrains a large volume of air to produce a convective effect (heat exchange between ambient air and water droplets) and evaporative cooling.

Such a system has been prepared for the EXPO '92 in Seville [15]

More traditional devices can be used: fountains, and ponds (which are also sources of cool water for the previously mentioned techniques).

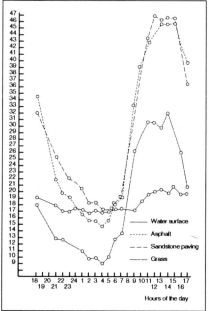

Figure 3.9: Surface temperature of various materials. [8]

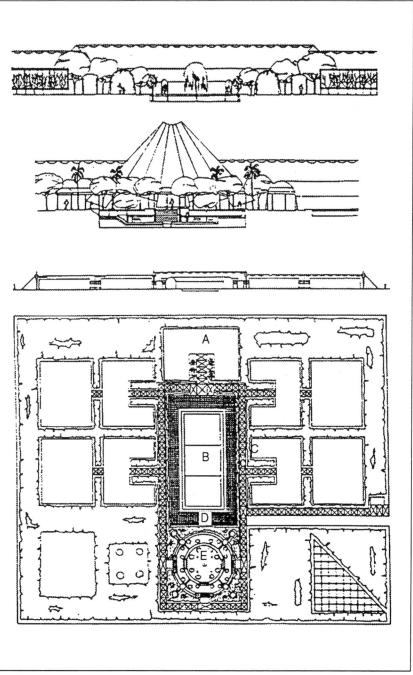

Figure 3.10: An example of passive cooling in open spaces: EXPO 1992 World Fair, Seville, Spain
A. Office, B. Central Pond, C. Pedestrian Pergola, D. Vehicle Pergola, E. Bioclimatic Rotunda [15]

55

Covered Way or Atrium

This is a circulation space (a street or a private space), covered with a transparent (or translucent) roof. Such a space can perform many functions and is more than a simple passage. See Chapter 9 - Atrium Design. Covered streets occur especially in the north of Europe (in oceanic climates) but the famous galleries in Milan and Naples can also be included. Advantages of an atrium.

- it acts as a solar collector;
- the internal air temperature may be lower than the external temperature in summer and higher in winter [4];
- it can be a shelter against wind, precipitation, urban pollution and noise;
- it is generally a very bright place (with natural lighting);
- it can be a means of renewing the city;
- it is a meeting place, with (generally) large dimensions (suitable for cultural / commercial functions). Extensive planting is often a feature.

In winter, atria can act as solar collectors. People do not have access to the upper zone where the temperature is high. The lower, occupied zone will have a lower, steadier temperature. Hot air may be taken from the upper zone and directed to the rooms adjacent to the atrium.

The renewal rate for the air volume is more or less the same for the atrium and for the adjacent buildings. To have an optimal efficiency, the volume of the atrium must be about the same as the volume of the buildings. As a result, the convective loss is minimal and the ventilation is sufficient.

In summer, the main objective is to cool the atrium and the adjacent or 'parent' buildings. Cool external air may typically be drawn through the parent building into the atrium at lower floor levels and expelled through vents at the top of the atrium.

Economy of the energy used in conditioning the air (heating in winter, ventilation in summer) is one of the main objectives. The bar chart (Figure 3.11) shows the energy consumption of the whole building for three situations. The base case is for the new building with no arcade, and shows a useful heat demand of 326 MWhr per annum. Simply covering over the open space would reduce this by 25% to 244 MWhr per annum, whilst the use of the arcade to provide pre-heated ventilation to the rooms facing the arcade reduces the heat demand further to 212 MWhr, a total reduction of 36%.

The Passage Choiseul in Paris, although not a perfect example, does effectively reduce summer temperatures in the atrium. It has, however, some notable deficiencies:

- Its slender outline increases the loss of heat which increases with the air speed.
- The walls have a low thermal inertia and a low absorption.
- Permanent ventilation is important.
- The glazing, due to pollution, has a low coefficient of transmission.

Note: The dimensions, shape and materials of an atrium are important. If the depth to height ratio increases, the convective heating loss also increases. Noise can be amplified by the atrium. Reflective materials must be used with caution. The optimum orientation will depend on available sunlight and prevailing winds.

THIRD LEVEL: BUILDING DESIGN

Intervention is restricted to individual sites. Architectural elements are considered only in the way they influence the character of the urban fabric. The principle design elements to consider are walls and roof, shading devices and courtyards.

Walls / Roofs

The envelope of the building (walls and roof) play two parts: Considered as an aspect of enclosed space the envelope is a filter controlling internal conditions of comfort. But considered as an aspect of open space, the envelope affects comfort in external spaces but cannot really control the climatic conditions.

In hot climates (or during the hot season), to minimise the heat gain in enclosed spaces in hot climates (or during the hot season) the envelope must have:

- light colour (to increase the reflection of radiation),
- significant thermal mass and capacity to absorb radiation during the day-time and release it during the night-time.

The effect is to keep the internal temperature as low and steady as possible.

In open spaces, these same characteristics can produce an increase in discomfort. This can be corrected by cooling the external space near the building. (This is beneficial for both internal and external spaces.) A first step is to act directly on walls by shading and, if possible, wetting them.

Shading Devices

Shading devices are useful, especially in Mediterranean and semi-desert climates but also in colder regions during the hot season They differ from one climatic region to another (they may be fixed in hot regions and movable in others).

Characteristics of shading

Shading depends on geometrical factors (varying with the time, the period of the year); and on the relative transparency of the materials. Usual materials are thin, light and relatively transparent (typical materials have a transparency factor $\geq 20\%$). Their

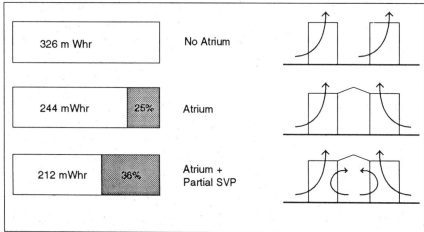

Figure 3.11. Annual useful demand (Casual gain: 10 (W/m^2)) [2].

surface temperature is typically higher than air temperature The re-emission depends on:
- colour and thermal conductivity.
Light colours and low conductances reduce the effect (vegetation is useful in this case too).

Shading is the first measure to use because it prevents the penetration of solar radiation thus reducing temperatures and the need for cooling.

Using ventilation for shading:

a) *Outdoor spaces in hot-dry climates:*
Planting is restricted by the availability of water and is more feasible in enclosed spaces than in the open. Evergreen trees, shrubs, cacti, creepers and a few types of grass can be used.

b) *Outdoor spaces in warm-humid climates:*
Shrubs or hedges must be used with care, so as not to reduce air flow near the ground, where it is most needed. Tall trees with clear trunks are preferable, as they cast a broad shadow whilst permitting penetration by breezes.

The first principle in passive cooling is to isolate the zone that is to be treated (white walls, screens of vegetation etc).

Courtyards
Mediterranean or semi-desert climate

Advantages
- During the day-time there is minimal solar penetration into the courtyard which limits the heating of internal walls of the courtyard. As a result, the temperature of the courtyard is lower than the external temperature.
- During the night-time the courtyard retains a pool of cool air, undisturbed by breezes.

Disadvantages
- A low height-to-width ratio reduces solar penetration during day-time but also reduces the longwave radiant cooling of the courtyard during the night.
(A possible solution is bringing cool air from the roofs to the courtyard during the night).
- The ventilation of the rooms opening onto the courtyard is sometimes difficult and therefore the internal temperature increases. (This problem might be alleviated by increasing the

dimensions of the openings.)
- The greater surface areas provided by the courtyard allow increased heat transfer by conduction.
- The temperature of the courtyard must be made as low as possible to avoid an excessive internal temperature. Vegetation (on pergolas, for example) or fountains are sometimes used to cool the air in the courtyard.

The internal air conditioning may be used to cool the courtyard

The system is beneficial for both internal and external spaces. These means are often used in individual houses (as in a patio). When the courtyard is not a private space but a hole in the heart of a (small) block of houses, it may be difficult to accommodate the varying needs of different occupants. The problem must be solved at an urban level: the block is considered as a whole and uniform rules applied.
- Cool air brought in courtyards
See figure 3.12

Guidelines
Courtyards can be designed to provide both sun and shade. Sunlight penetration must be considered in relation to climate and in relation to other priorities such as urban density. It may be preferable that some parts of a house are oriented towards a shaded space, filled with cool air and vegetation, rather than the harsh light of direct sun. Latitude and climate will influence form and in some regions it will be desirable to include direct sunlight, in others to exclude it. The courtyard can also be designed to satisfy different seasonal preferences with respect to direct sun. A whole urban structure may be formed with this type of element.

Tools / Software

* CAMUR, University of Liège, LEMA
* DIRAD, University of Liège, LEMA
* STEREO, University of Liège, LEMA
* SOLENE, Laboratoire CERMA, Ecole d'Architecture de Nantes (reference no. 17)
* GOSOL, Universität Stuttgart (reference no. 18)
* GABLE's 4D SERIES, C.A.D. Systems Ltd.

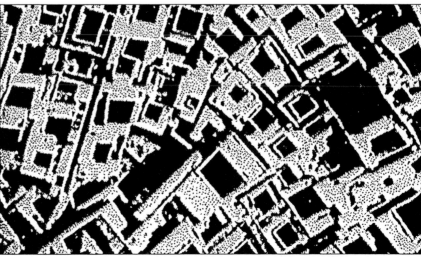
Figure 3.12: (Marrakesh / Morocco / Ariel view of an 'urban courtyard philosophy' / [10]

REFERENCES

[1]. Architecture, Urbanisme et Energie Bilan et perspectives Actes de colloques

[2]. Energy and Urban Built Form, Hawkes, Owers, Rockaby, Steadman, Butterworths

[3]. Metodi e Modelli al Calcolatore per il Controllo Energetico della Progettazione Edile-Impianistica,Synergia progetti srl

[4]. Problems Energetiques de la Ville, AFME - Direction de l'Architecture Plan-Construction

[5]. Rencontre Internationale - Gestion et Representation des "Transformations Urbaines", Ministère de l'Equipement, de l'Aménagement du Territoire et des Transports, GAMSAU / CERMA

[6]. Les villes Francaises, Barrère, P., Cassou, M., Mounat, MASSON-Collection géographie

[7]. Energy, Planning and Urbans Form, Owens, S., Pion

[8]. An Envronmental Analysis of a Medieval Urban Fabric and Dwelling in Ahmedabad India, Nanda, V., Thesis

[9]. An Introduction to Housing Layout, Architectural PRESS

[10]. Energy and Buildings for Temperate Climates - A Mediterranean Regional Approach, Fernandes and Yannas, Pergamon

[11]. Bioclimatic Design for Building Components and Systems, AgipNucleare

[12]. CAMUR research action, Dupagne, A., LEMA

[13]. Project SYSURB - La micro-informatique pour la gesion coordonée des réseaux, CREM Analyse de Systemes Energetiques Regionaux, IENER - CERS - Commune de Martigny

[14][. Passive and Hybrid Solar Low Houses - project of 12 dwellings in Osuna (Sevilla), Escuela Tecnica Superior de Arquitectura - Universidade de Sevilla

[15]. Climatic Contol for the Open Space of the 1992 World Fair, Seville, Spain, Lopez de Asiain, J., Cabeza Lainez, J.M., Ballestreros Rodriguez, A.L., Perez de Lama Halcon, J.

[16]. Cooling of Outdoor Spaces, Workshop interaction between physics and architecture in environmental conscious design, Givoni, B.

[17]. Soft 'SOLENE', Peneau, Jean-Pierre, Labroatoire CERMA - Ecole d'Architecture de Nantes, Rue Massenet à F-44300 Nantes, France. 'Imagerie numérique de l'ensoleillement en milieu urbain', in Actes du colloque MICAD 89, Ed. Hermes Ed., Paris (1989)

[18]. Soft 'GOSOL', Goretzki, Dipl. Ing. P., Universität Stuttgart, Institut für Bauökonomie, Keplerstrasse 11, D-7000 Stuttgart 10 - R.F.A.

[19]. Design Primer for Hot Climates, Konya, Allan.

[20]. Manual of Tropical Housing and Building, Part 1: Climatic Design, Koenigsberger, Ingersoll, Mayhew, Szokolay.

[21]. Housing in Arid Lands, Design and Planning, Gideon Golany, Gideon, Architectural Press, London 1980

[22]. Housing, Climate and Comfort, Evans, Martin.

[23] Sun-Rythm-Form, Knowles, R.L., Cambridge, Mass., MIT Press, ISBN 0-262-11078-4. 1981

58

THERMAL COMFORT

Contents

4.1

INTRODUCTION

The ambient environment has both a physical and emotional effect on man and is therefore of central importance in building design. One of the designer's main tasks is to create an environment inside and outside the building which is appropriate for all the human activities likely to take place there.

Comfort may be defined as the sensation of complete physical and mental well-being. Thus defined, it is only to a limited extent within the control of the designer. The occupants' biological, emotional and physical characteristics also come into play. Hence, if a group of people is subjected to the same climate, the individual members are unlikely to be satisfied simultaneously. The designer must aim to create optimal thermal comfort for the group as a whole, i.e. he or she must provide conditions under which the highest number of people in the group feel comfortable.

Thermal neutrality, where an individual desires neither a warmer nor a colder environment, is a necessary condition for thermal comfort. However, attainment of thermal neutrality does not necessarily ensure comfort. For example, a person who is exposed to an asymmetric radiant field may well be in thermal neutrality but is unlikely to be comfortable. In most situations encountered in buildings, however, the two conditions will coincide.

Achievement of human comfort is particularly important in passive solar buildings. The manner in which solar energy is collected, stored and distributed can have a profound effect on the comfort of the occupants.

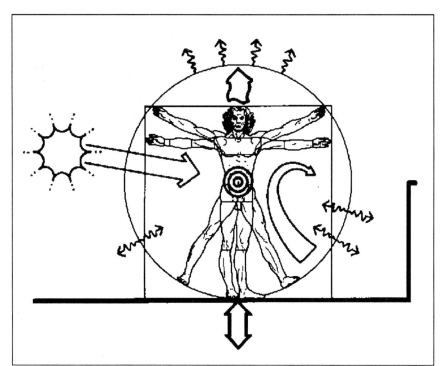

Figure 4.1. Heat exchange between the body and its thermal environment

ACTIVITY	W/m²	met
Resting		
Sleeping	40	0.7
Reclining	45	0.8
Seated, quiet	60	1.0
Standing, relaxed	70	1.2
Walking (on the level)		
0.89 m/s	115	2.0
1.34 m/s	150	2.6
1.79 m/s	220	3.8
Office Activities		
Reading, seated	55	1.0
Writing	60	1.0
Typing	65	1.1
Filing, seated	70	1.2
Filing, standing	80	1.4
Walking about	100	1.7
Lifting, packing	120	2.1
Driving / Flying		
Car	60-115	1.0-2.0
Aircraft, routine	70	1.2
Aircraft, instrument landing	105	1.8
Aircraft, combat	140	2.4
Heavy vehicle	185	3.2
Miscellaneous Occupational Activities		
Cooking	95-115	1.6-2.0
House cleaning	115-200	2.0-3.4
Seated, limb movement	130	2.2
Machine work		
sawing (light table)	105	1.8
light (electrical industry)	115-140	2.0-2.4
heavy	235	4.0
Handling 50kg bags	235	4.0
Pick and shovel work	235-280	4.0-4.8
Miscellaneous Leisure Activities		
Dancing, social	140-255	2.4-4.4
Calisthenics / exercise	175-235	3.0-4.0
Tennis, singles	210-270	3.6-4.0
Basketball	290-440	5.0-7.6
Wrestling, competitive	410-505	7.0-8.7

Table 4.1. Typical metabolic heat generation for various activities

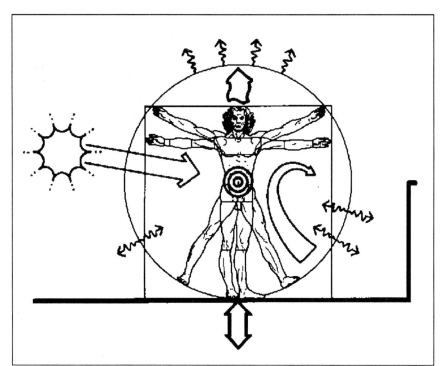

4.2
FACTORS INFLUENCING THERMAL COMFORT

The body converts food into energy. The rate at which this occurs depends largely on activity level. The energy produced in the conversion is dissipated by the body as heat or used for external work. The sensation of comfort depends to a great extent on the ease with which the body is able to achieve a balance between energy production and heat gain on the one hand and heat loss on the other so that the internal body temperature is maintained around 37° C.

The factors affecting comfort can be divided into personal variables (such as activity and clothing) and environmental variables (such as air temperature, mean radiant temperature, air velocity and air humidity). This second group is directly dependent on the design of the building and its heating and cooling systems.

The above factors affecting comfort (together with the operative temperature - a combination of air temperature and mean radiant temperature) are discussed in more detail in the following paragraphs.

4.3
ACTIVITY

The metabolic rate is the amount of energy produced per unit of time by the conversion of food. It is influenced by activity level and is expressed in mets, which are watts per square metre of body surface area. 1 met is the metabolic rate of a seated person when relaxing, i.e. 58 W/m². The average body surface area of an adult is around 1.8 m².

The metabolic rates for various activities are given in Table 4.1 and Figure 4.2.

4.4
CLOTHING

Clothing offers man thermal insulation against his environment. This thermal insulation can be expressed in terms of m²K/W or clo units. 1 clo corresponds roughly to the thermal resistance of a winter business suit, i.e. 0.155 m²K/W. The thermal insulation properties of various types of clothing are given in Table 4.2 and Figure 4.3.

4.5

AIR TEMPERATURE

The temperature of the air in the occupied zone of a space (t_a) is important for man's thermal balance and comfort. For people who spend most of their time sitting down it is the average air temperature from the floor to a height of 1.1 m which is significant. In such situations, it is recommended that measurements be taken at a height of 0.6 m above the floor.

4.6

MEAN RADIANT TEMPERATURE

The mean radiant temperature (t_r) is an average temperature of the surrounding surfaces (Figure 4.4). It includes the effect of incident solar radiation and has as great an impact on human comfort as air temperature.

Mean radiant temperatures are usually determined using a globe thermometer, i.e. a black spherical shell with a thermal sensor in the centre of the globe. The mean radiant temperature is computed from the globe temperature, the air temperature and the air velocity.

As the interior surfaces of the external walls of a poorly-insulated building are generally colder than those of a similar well-insulated building, the air temperatures of a well-insulated building can be maintained below those in a poorly-insulated building for the same comfort level.

The surfaces of windows experience large temperature fluctuations. The mean radiant temperature close to these surfaces may therefore be lower or higher than that in the rest of the space. Cool surfaces (such as the glazing of a large window in winter) can also cause discomfort due to radiant asymmetry.

A person who is directly exposed to solar radiation may experience a mean radiant temperature much higher than the air temperature. For instance, the mean radiant temperature can be as much as 25°C above air temperature in the case of a sedentary person whose entire body is exposed to maximum solar radiation. Therefore,

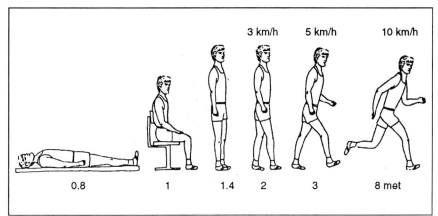

Figure 4.2. Metabolic rates of different activities. (1 met = 58 W/m^2)

CLOTHING	THERMAL RESISTANCE	
	m^2.K/W	clo
Nude	0	0
Shorts	0.015	0.1
Typical tropical clothing ensemble: briefs, shorts, open-neck shirt with short sleeves, light socks and sandals	0.045	0.3
Light summer ensemble: briefs, long light-weight trousers, open-neck shirt with short sleeves, light socks and shoes	0.08	0.5
Light working ensemble; Light underwear, cotton work shirt with long sleeves, work trousers, woollen socks and shoes	0.11	0.7
Typical indoor winter ensemble: Underwear, shirt with long sleeves, trousers, jacket or sweater with long sleeves, heavy socks and shoes	0.16	1.0
Heavy traditional European business suit: cotton underwear with long legs and sleeves, shirt, suit including trousers, jacket and waistcoat, woollen socks and heavy shoes	0.23	1.5

Table 4.2.: Thermal insulation provided by various combinations of clothing.

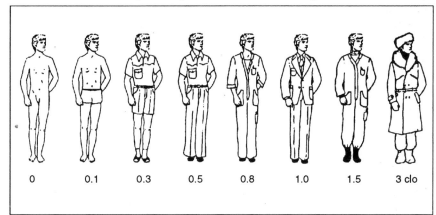

Figure 4.3. Thermal insulation properties of typical combinations of clothing. (1 clo = 0.155 m^{2}K/W)

exposure to solar radiation indoors can easily cause discomfort. This can be intensified by the asymmetry between the exposed side of the body and the side in the shade.

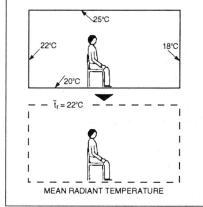

Figure 4.4.
The concept of mean radiant temperature.

4.7
AIR VELOCITY

Air velocity has an effect on convective heat loss from the body. Air at a greater velocity will seem cooler. Therefore it is essential that air velocities are kept low in winter so that thermal comfort can be experienced at the lowest temperature level. Sedentary people are particularly sensitive to draughts, i.e. unwanted local cooling.

Careful design of air handling equipment is necessary to prevent excessive air velocities. Particular attention should be paid to the placing and sizing of outlets.

Poorly weather-stripped buildings with large cold surfaces and high rooms tend to create undesirable patterns of air movement.

4.8
AIR HUMIDITY

At moderate air temperatures (say, 15-25°C) under steady state conditions (i.e. when a person stays in the same space for a long time), the humidity of the air has little effect on thermal sensations: an increase of 10% in relative humidity will have the same effect as a mere 0.3°C rise in air temperature.

Under transient conditions (i.e. when a person moves from indoors to outdoors or from one space to another with a different humidity), however, the thermal effect of the change in humidity can be 2-3 times greater [1].

In warm environments (i.e. >30°C), the effect of humidity changes can have a considerable effect on thermal comfort.

Although in most situations encountered in buildings air humidity has only a moderate thermal impact, there are a number of reasons why extreme humidity levels should be avoided. High levels can, for instance, cause problems of mould, mites, static electricity and dry mucous membranes. Maintenance of relative humidities between 30% and 60% will reduce such problems.

4.9
OPERATIVE TEMPERATURE

The air temperature and mean radiant temperature are often considered as one parameter, known as the perceived or operative temperature. For low air velocities (i.e. <0.2 m/s), the operative temperature is the average of the air and mean radiant temperatures.

4.10
PREDICTED MEAN VOTE (PMV) AND PREDICTED PERCENTAGE OF DISSATISFIED (PPD) INDICES

The combined effect of the above parameters on the comfort of human beings is described in the international standard for thermal comfort, ISO 7730 [2].

An estimate of how warm or cool a particular environment will feel is obtained by calculating the Predicted Mean Vote (PMV). This is an index which predicts the mean value of the votes of a large group of people on the following seven-point thermal sensation scale:

+3 hot
+2 warm
+1 slightly warm
 0 neutral
-1 slightly cool
-2 cool
-3 cold

The PMV can be expressed mathematically and can easily be calculated using a small computer program listed in the appendix on

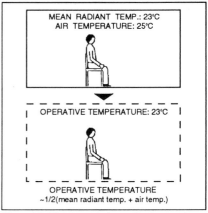

Figure 4.5: The concept of operative temperature.

Design Method. It can also be determined directly from tables in ISO 7730 [2].

As indicated above, the PMV index predicts the mean value of the thermal votes of a large number of people exposed to the same environment. There are, however, significant differences between people so that the individual votes are scattered around the mean value. It is useful, therefore, to know the percentage of people in a group likely to feel uncomfortably warm or cool in a particular environment. This can be established using the Predicted Percentage of Dissatisfied (PPD) index.

There is a direct relationship between the PPD and PMV (Figure 4.6). ISO 7730 [2] recommends that the PPD be <10%. This corresponds to - 0.5 < PMV < + 0.5. What this means in practice can be shown by means of the comfort diagrams discussed below.

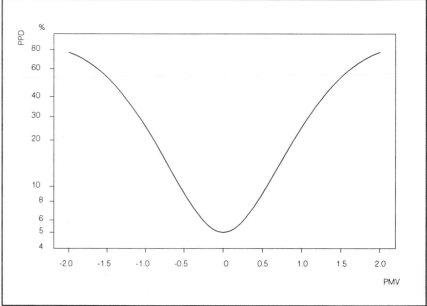

Figure 4.6: Predicted Percentage of Dissatisfied (PPD) as a function of Predicted Mean Vote (PMV)

4.11

COMFORT DIAGRAMS

Figure 4.7 shows the optimal operative temperature for different types of activity and clothing. In winter, for instance, a common situation found in dwellings and offices, etc., is one where a person undergoes mostly sedentary activities (1.2 met) and has a clothing level of 1 clo. From Figure 4.7 it can be seen that the required temperature for this situation is 22°C ± 2°C. For sedentary activities under typical warm summer conditions with light clothing (0.5 clo), the operative temperature is 24.5°C ± 1.5°C.

The comfort diagram shown in Figure 4.8 gives different combinations of air temperature and mean radiant temperature which will provide optimal conditions. It should be noted that the mean radiant temperature in a passive solar building may on occasion be higher than that in a similar conventional building. Thus passive solar buildings may require a lower air temperature to be comfortable.

4.12

LOCAL DISCOMFORT

Thermal comfort implies satisfaction with the thermal environment. It is not enough for the body as a whole to feel thermal neutrality as specified by the PMV and PPD indices. It is additionally necessary that no individual part of the body should be uncomfortably warm or cool. Local discomfort should be avoided. Sedentary people are particularly sensitive to local discomfort whereas those with a higher level of activity are less likely to complain.

Local discomfort can be caused by too warm or too cool a floor, by an excessive temperature difference between the head and ankles, by radiant asymmetry where one side of the body is too warm and the other too cool, or by draughts.

Radiant asymmetry can be caused by direct exposure to sunshine. It can also be experienced close to large window areas.

Draught (i.e. local cooling caused by air movement) is the most common form of local discomfort. Air velocities below 0.15 m/s can cause discomfort. It is therefore important to keep air velocities low.

Quantitative units for local discomfort are given in ISO 7730 [2].

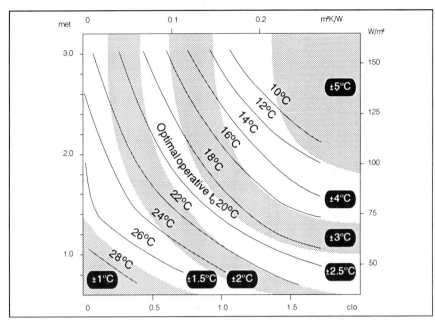

Figure 4.7: The optimum operative temperature as a function of activity and clothing. The shaded or unshaded bands indicate the comfort ranges (±Δt) around the optimal temperature within which 80% or more of the occupants are expected to find the thermal conditions acceptable. (The relative humidity is 50%)

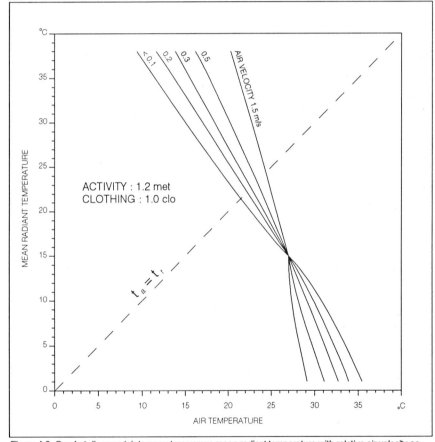

Figure 4.8: Comfort diagram (air temperature versus mean radiant temperature with relative air velocity as parameter) for persons wearing 1.0 clo at an activity of 1.2 met. (The relative humidity is 50%.)

4.13

THERMAL COMFORT IN PASSIVE SOLAR BUILDINGS

The same basic criteria for thermal comfort apply regardless of whether the predominant need is for heating or cooling. However, passive solar buildings have certain specific characteristics which distinguish them from conventional buildings. In 'free running' passive solar buildings the temperature can fluctuate during the day. Heat is stored in the building fabric during the day when solar radiation is present and the stored heat is given off at night when the temperature falls. In a passive solar building, too, there may be significant differences between, for example, the spaces facing north and those facing south, according to function.

ISO 7730 [2] is based on studies carried out under steady-state conditions.However, comprehensive research has recently been carried out to investigate human reactions to thermal transients as they occur in passive solar buildings [3]. The results show that during steadily increasing or decreasing temperatures with changes of up to 5°C an hour people sense the actual thermal environment in the same way that they do under steady-state conditions. During step changes of the operative temperature such as those which occur when people walk from north-facing spaces to south-facing spaces, they sense the change at once. For changes to higher temperatures, the steady-state sensation is felt immediately. For changes to lower temperatures, the new environment initially feels cooler than the steady-state situation; there is then a gradual return to the steady-state level.

For constant clothing levels, the temperature ranges for comfort are relatively small (see Figure 4.7). However, if people are willing to modify their clothing during the day, then a much wider temperature range is acceptable. Good use can be made of this in passive solar buildings. The differences between the thermal requirements of individuals should be kept in mind. The PMV predicts conditions which will satisfy the majority. However, in spaces which are occupied by only a few people, it is essential that the thermal conditions are easily modified by each occupant. Some people actually prefer rather low temperatures or are ready to accept some discomfort to save money and energy. In such cases a design with the flexibility to meet personal requirements is recommended.

Draughts can often cause problems in spaces with large windows. During the night, thermal convection downwards along the cold surfaces can cause discomfort from air movement. Double or triple glazing and moderate window heights will reduce these problems. Heat sources under the windows can also counteract draughts.

The same cold glass which causes draughts can also bring (to a lesser extent) some discomfort from asymmetric radiation. The worst cases of asymmetric radiation occur, however, when people are exposed to direct solar radiation inside a building. Such situations are usually only acceptable for short periods of time.

In passive solar buildings, the floor is often used for thermal storage and the floor temperature may therefore fluctuate considerably. Although in general this does not cause problems, temperatures above 29°C and below 19°C can cause complaints about feet which are too hot or too cold.

4.14
FIELD-BASED RESEARCH IN THERMAL COMFORT

The research on thermal comfort in passive solar buildings described in the preceding section has been backed up by additional field trials to investigate people's experiences of indoor climates, both transient and steady-state, in buildings they use in real life. The study has been carried out in houses, schools, offices and hospitals in Germany, France and the UK. Some are conventional buildings. Others contain passive solar features. [7]

The results show that generally the temperatures required for thermal comfort are significantly lower in all building types than those predicted from models established through laboratory-based research. They gave, for instance, the following operative temperatures for optimal conditions: 21°C for office workers in Germany and France; 19°C for schoolteachers in France; 19°C for hospital occupants in the UK; and still lower temperatures for houses in the UK and France. These figures compare with predicted operative temperatures for thermal neutrality in the range 23-25°C.

The field trials showed that there is no apparent difference between the temperatures required for thermal comfort in buildings containing passive solar systems and those without such features. Further, although changes in temperature over time are noticeable in both conventional and solar buildings, these are not rated by occupants as having a significant effect on thermal comfort. The same is true for temperature differences between different parts of the building.

REFERENCES

[1] Impact of air humidity on thermal comfort during step-changes. R.J. de Dear, H. N.Knudsen and P. O. Fanger. ASHRAE Trans. 1989, Vol. 95, Part 2.
[2] ISO 7730. Moderate thermal environments - determination of the PMV and PPD indices and specification of the conditions for thermal comfort. International Standards Organization, Geneva. 1984.
[3] Thermal comfort in passive solar buildings. Final report, CEC research project EN3S-0035-DK(B). H. N. Knudsen, R.J. de Dear, J.W. Ring, T. L. Li, T. W. Püntner and P. O. Fanger. Laboratory of Heating and Air Conditioning, Technical University of Denmark. MaY 1989.
[4] Thermal comfort analysis and applications in environmental engineering. P. O. Fanger. McGraw-Hill Book Company, New York. 1973.
[5] Indoor climate. D. A. McIntyre. Applied Science Publishers, London. 1980.
[6] Bioengineering, thermal physiology and comfort. K. Cena and J. A. Clark. Elsevier Scientific Publishing Company, Amsterdam-Oxford-New York. 1981.
[7] Thermal Comfort in Buildings with Passive Solar Features – Field Study. Final Report to C.E.C., Griffiths,J., University of Surrey,1990.

PASSIVE SOLAR HEATING

5.1

INTRODUCTION

Heat losses from buildings occur mainly by conduction through external surfaces and by infiltration and ventilation through cracks and openings in the building envelope. Significant reductions in these heat losses will have the effect of shortening the 'heating season', but will also decrease the efficiency of passive solar heating systems as illustrated in figure 5.1.

Reduced heat demand, brought about by improved insulation and draught-sealing, results in a restriction of the heating season to the time of year which is least favourable for solar gains. Nevertheless, passive solar heating can still provide a significant contribution to energy saving. Although the total amount of useful solar energy is decreased due to the shortening of the heating season, additional insulation increases the ratio of solar gains to the heating demand of the building. This

underlines the importance of appropriate sizing of passive solar elements for heat gain and storage.

The design of the microclimate around the building should be the primary consideration to minimise heat losses and maximise the opportunities for solar gain (see Chapter 2).

For the design of the building itself various passive solar configurations are first outlined in this chapter. Heat collection, storage and distribution are then discussed separately, for clarity, although in practice these processes will usually operate simultaneously, as for example in the Transwall system which can collect, store and distribute heat (see Figure 5.14). Finally a section on auxiliary heating offers some guidance on the selection and use of appropriate 'conventional' heating systems for use when additional heating is required.

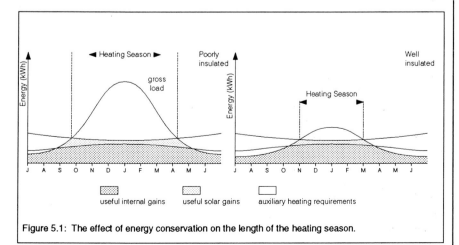

Figure 5.1: The effect of energy conservation on the length of the heating season.

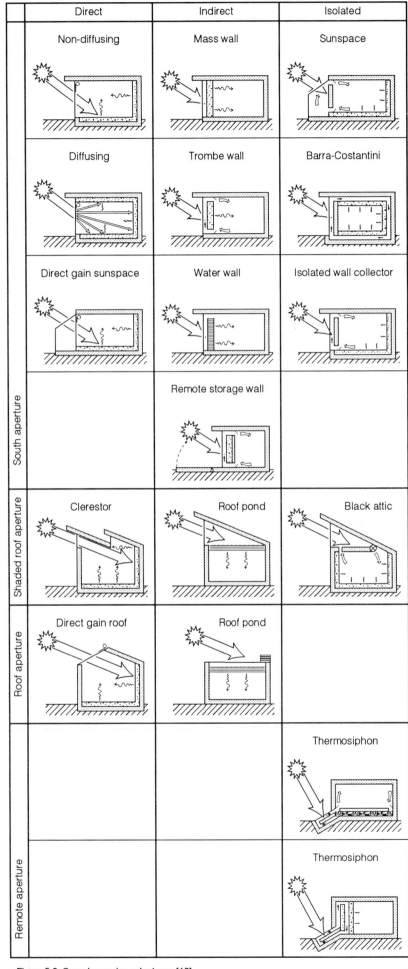

	Direct	Indirect	Isolated
South aperture	Non-diffusing	Mass wall	Sunspace
	Diffusing	Trombe wall	Barra-Costantini
	Direct gain sunspace	Water wall	Isolated wall collector
		Remote storage wall	
Shaded roof aperture	Clerestory	Roof pond	Black attic
Roof aperture	Direct gain roof	Roof pond	
Remote aperture			Thermosiphon
			Thermosiphon

Figure 5.2: Generic passive solar types [12].

PASSIVE SOLAR CONFIGURATIONS

Various passive systems for heat gain are outlined here, starting with the basic generic types and going on to examine in more detail the more common specific systems.

Generic categories can be defined by three factors: the characteristics of the collection aperture, the interaction of incoming solar radiation and heat storage and the method of delivering energy to the heated space [12]. Heat gain, storage and distribution are discussed later in this chapter.

Direct Gain

The simplest system is that of direct gain, consisting primarily of a well-insulated building with a relatively large expanse of south-facing glazing which admits the low-angle rays of the winter sun. Direct gain systems use the occupied spaces of the building to collect, store and distribute solar heat and, properly designed can be the most effective and practical option for European conditions. In summer, the high altitude of the sun reduces the insolation transmitted through this glazing and an overhang can exclude the sun completely. The building needs thermal mass to store heat during the day and to re-emit it at night. This thermal mass is usually in the form of externally insulated masonry walls and/or a solid floor with underfloor insulation. The sun shines directly on to the thermal mass, energy is stored and lower air temperature fluctuations are achieved.

Requirements:

The basic requirements for a direct gain system are: a large south-facing glazed aperture with the living space directly behind: exposed thermal mass in the ceiling and/or floor and/or walls, the area and capacity of which is appropriately sized and positioned for solar exposure and storage: and a means of isolating the thermal storage from exterior climatic conditions. For the first requirement, an appropriate amount of vertical glazing, often double-glazed to minimize heat loss, is oriented south to admit the maximum useful radiation, while limiting the amount of solar gain in summer.

In northern Europe, triple glazing, movable insulation applied at night to double glazing, or low-emissivity glazing are recommended for the solar aperture in order to avoid excessive heat losses.

Many modern buildings have large south-facing windows, but it is often the lack of appropriate thermal storage or occupants' behaviour (for example the use of blinds to reduce glare) which prevents them from benefiting fully from their solar gains. Conversely, commercial buildings with large areas of glazing can suffer from excessive solar gain and, if adequate shading is not provided, require additional cooling.

Also important is the choice of heating system and its controls, both of which can have a large influence on the performance of a direct gain system.

Variations:

Beyond these basic requirements there is a series of variations and controls which provides alternatives within direct gain systems. The most common variations are in the location of thermal mass. The best location of thermal mass is governed by the physical laws of heat flow by radiation and convection. Within these limitations primary storage can have various configurations: in the floor, in free standing mass within the room, in the ceiling, or in internal walls or insulated external walls.

The distribution or concentration of the thermal mass provides the first subdivision of direct gain passive types. Both subdivisions have south-facing apertures but differ in the way in which the sunlight is handled when it enters the building. One allows the sunlight to fall on a concentrated area of thermal mass (Figure 5.4) and the other diffuses or reflects the sunlight so that it is distributed over a large area of thermal mass (Figure 5.5). Care should be taken to avoid visual discomfort caused by glare.

The use of diffusing glass, blinds or reflection from a light-coloured surface behind clear glass, will all have the effect of spreading the incoming solar radiation evenly throughout the room. Such devices should however only be used above eye-level to avoid glare.

Thermal storage materials used vary from concrete, brick and ceramics, to water and other liquids, either singly or in various combinations.

Controls:

To add to the efficiency and the usefulness of direct gain and other passive systems, several controls should be considered. The large areas of glazing required by direct gain buildings can result in extreme temperature swings (in both directions) within the living space: sufficient thermal storage must be placed about the space to absorb and store excess energy and so moderate these fluctuations. To prevent overheating, shading is usually required for south-facing glass. The high altitude of the summer sun often allows overhangs to provide adequate shading to vertical south glazings. Exhausts and vents will also help cool interior spaces when summer temperatures are high. To prevent unwanted heat loss in winter, or at night, insulation is necessary to provide a low U-value for the glazed area. Insulation of the glazing in the form of movable panels, curtains and shutters can be effective in preventing unwanted heat loss. Movable insulation can also alleviate overheating at either end of the heating season. Without these control considerations, a passive system can cause considerable discomfort due to winter losses and summer, spring and autumn overheating.

Advantages:
- Direct gain is the simplest solar heating system and can be the easiest to build. In many cases it is achieved simply by the relocation of windows.
- The large areas of glazing not only admit solar radiation for heating but also allow high levels of natural daylighting and good visual connections to the outside.
- Glazing is well researched and cheap and materials are readily available.
- The overall system can be one of the least expensive methods of solar space heating.

Disadvantages:
- Large areas of glass can result in glare by day and loss of privacy at night.
- Ultraviolet radiation in the sunlight will degrade fabrics and photographs.
- If large areas of glazing are used, large amounts of thermal mass will usually be needed to

modulate temperature swings which can be expensive if the mass serves no structural purpose. Highly insulated buildings will require smaller glazed areas and less thermal mass.
- Even with thermal mass, diurnal temperature swings will occur.
- Night-time insulation of the solar aperture is normally a necessity in Northern European climates and can be costly. Specially treated glazing materials can reduce heat losses at all times.

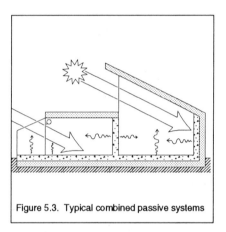

Figure 5.3. Typical combined passive systems

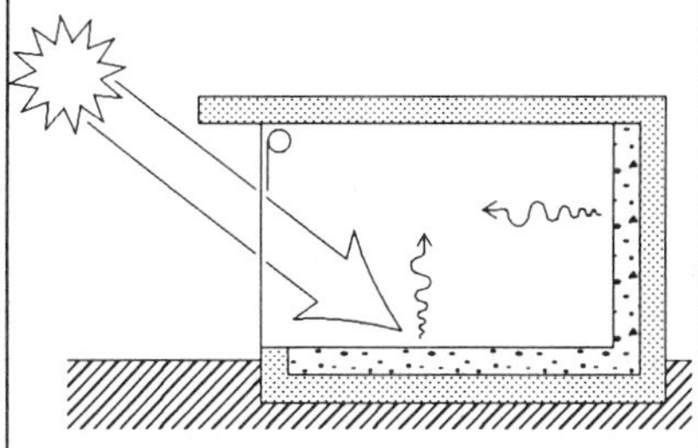

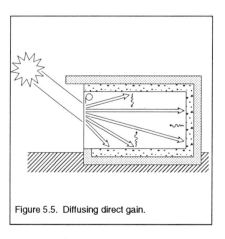

Figure 5.4. Non-diffusing direct gain

Figure 5.5. Diffusing direct gain.

Indirect Gain

The Trombe wall, mass wall, water wall and roof pond are all indirect gain systems, which combine the collecting, storage and distribution functions within some part of the building envelope which encloses the living spaces.

Mass and Trombe Wall

In the mass and Trombe wall systems, the thermal storage mass for the building is a south-facing wall of masonry or concrete construction with the external surface glazed to reduce heat losses to the outside. The difference

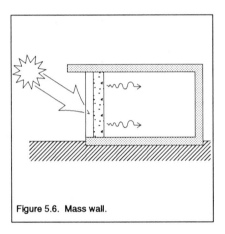

Figure 5.6. Mass wall.

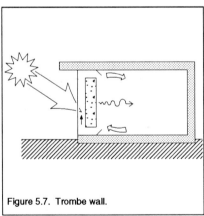

Figure 5.7. Trombe wall.

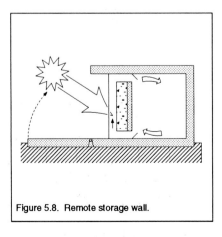

Figure 5.8. Remote storage wall.

between a mass (Figure 5.6) and Trombe wall (Figure 5.7) is that the latter has vents top and bottom to allow air to circulate through to the heated space. The Trombe wall system was named after the pioneering work by Felix Trombe and Jacques Michel at Odeillo in France.

The required elements of the mass and Trombe wall systems are a large south-facing glazed collector area, with thermal storage mass directly behind it. New materials such as transparent insulation can be particularly suitable for Trombe wall applications. The range of storage materials includes concrete, stone and composites of brick and block. Solar radiation falls on the mass wall and is absorbed by it, causing the surface of the masonry to warm up. This heat, in the form of a dampened temperature rise, is transferred through the wall to the inner surface by conduction, from where it radiates and is convected to the living space. The timelag and dampening of the temperature wave in this transfer depends on the type and thickness of the storage material chosen: the timelag is approximately 18 minutes per 10 mm for concrete. Wall thicknesses of greater than 100 mm do not significantly increase heat conduction to the living space. The Trombe wall also allows for the distribution of the collected heat by natural convection: the volume of air in the intervening space between glazing and storage mass can reach high temperatures of 60°C on clear days. By means of openings or vents at the top and bottom of the storage mass, hot air will rise and enter the living space drawing cooler room air through the lower vents back into the collector air space. The vents should be controllable by means of dampers to prevent reverse circulation at night which can reduce the effectiveness of the Trombe wall by about 10 per cent. The means of storage, distribution and the insulation of the wall from the external air all affect the operational efficiency of the Trombe and mass wall systems.

Controls:

Controls for the operation of the Trombe wall system are important.

For optimum efficiency in winter it is necessary to reduce wasteful heat loss to the sky at night or on an overcast day: this can be achieved by external insulated shutters, by improving the insulation value of the glazing (double-glazing or heat-reflecting glass or by the use of transparent insulation) and by applying a selective coating to the surface of the masonry, which has a high absorptivity to solar radiation but a low emissivity for thermal radiation. In summer, unwanted heating of the storage mass can be prevented by means of overhangs, closing external insulation or by the use of external opening vents. In some climates the Trombe wall may be used as a solar chimney in summer: in this way the continual air movement exhausts hot air from the house, drawing in cooler air from (say) the north side of the house for ventilation.

In Northern European climates in mid-winter, where there is insufficient solar energy during the day to heat the wall, the high U-value of the Trombe or mass wall can be a heating burden. It is sometimes possible with the Trombe wall to insulate the storage mass so that it is thermally isolated from the system, and then use the solar aperture and vent controls to form an isolated wall collector. Windows may be placed in Trombe walls to provide light and views.

Advantages:
- Glare, privacy and ultraviolet degradation of fabrics are not a problem.
- Temperature swings in the living space are lower than with direct gain systems.
- The time delay between absorption of the solar energy and delivery of the thermal energy to the living space can be an advantage for night time heating (rather than in the evening).

Disadvantages:
- The external surface of the mass wall is relatively hot as conduction of energy through the wall is slow and can lead to considerable loss of energy to the external environment thus reducing efficiency.
- The controls mentioned above can be expensive.

- Two south walls, one glazed and one massive are required, with obvious cost and space disadvantages.
- Discomfort can be caused at either end of the heating season by overheated air from the Trombe wall during the day or uncontrolled thermal radiation from the inside surface of either type on warm evenings. These effects can be reduced by venting.
- The need for sufficient thermal mass must be balanced with the requirements for views from the living space and daylighting.
- The Trombe wall must be designed for access to clean the glazed walls.
- Condensation on the glass can be a problem.
- In Northern European climates the use of a mass or Trombe wall can result in heating burden in mid-Winter.

Remote Storage Walls

The remote storage wall (Figure 5.8) is similar in form to the Trombe wall, but is insulated on the room side, to prevent energy transmission by conduction and radiation: all heat transfer is by convection, possibly fan-assisted.

The performance of such a system is questionable in Northern Europe, and would only work with night insulation.

An alternative form of this system has vents to the outside air at the bottom of the collector and to the heated space at the top, creating a siphonic "open loop" which supplies pre-warmed fresh air to the living zone. Screens or filters may be needed to prevent dust or insects entering the living space. This system is not as well researched as previous examples.

Water Wall

The water wall is similar to the mass and Trombe wall systems except that contained water replaces the solid wall (Figures 5.9, 5.10). Water walls may be an attractive system where low-mass construction is required. Because water has a greater heat capacity per unit volume than brick or concrete and because convection currents within the water cause it to

act as an almost isothermal heat store, the system can work more efficiently that the mass or Trombe wall.

Requirements and variations:

The water wall system must also have a large south-facing glazed area on the outside of the contained water storage. The water may be contained by various methods, the type of container affecting the heat storage capacity and the speed of distribution of the stored heat. Containers made of metal or glass in the form of tubes, bins or drums and water-filled concrete walls have all been used. The selection of the material and form of containment is an important factor in the operational efficiency as well as the economics of the water wall.

Controls:

Owing to the isothermal nature of water, the distribution of the collected solar energy within the storage is almost immediate. This is in contrast to the longer time lag of the mass and Trombe walls. If designed for a climate where indoor heating is undesirable until the cooler evening hours, the system may require distribution control: in such circumstances the addition of insulation between storage and the living space may be necessary.

Advantages:
- The isothermal nature of heat storage results in a reduced temperature of the external surface and so less energy is lost to the night-sky and to the outside air.
- Glare, privacy and ultraviolet degradation of fabrics are not a problem.
- Temperature swings in the living space are lower than with direct gain or convective loop systems.
- The storage can remain warm and continue to supply energy to the living space well into the evening.
- The performance of thermal storage walls is well researched.

Isolated Gain

In isolated gain systems, solar collection is thermally isolated from the living spaces of the building. In

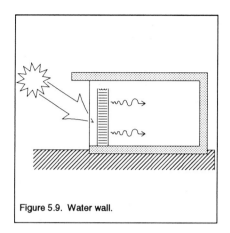

Figure 5.9. Water wall.

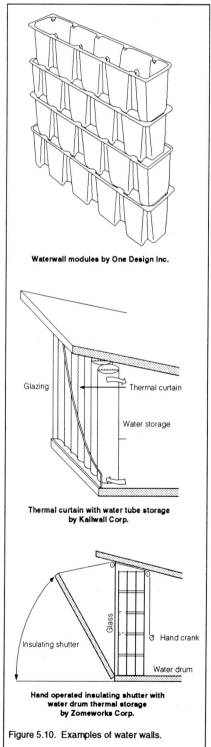

Waterwall modules by One Design Inc.

Glazing — Thermal curtain

Water storage

Thermal curtain with water tube storage by Kalwall Corp.

Insulating shutter — Glass — Hand crank — Water drum

Hand operated insulating shutter with water drum thermal storage by Zomeworks Corp.

Figure 5.10. Examples of water walls.

true passive systems, energy transfer from the collector to the living space or to the storage and from the storage to the living space will be by non-mechanical processes, such as convection or radiation. The most common of these processes for transferring energy from the collector is the particular form of convection known as a thermosiphonic loop: air is heated in the collector, becomes buoyant and rises, drawing in cooler air from below: the warmer air transfers its energy to remote storage or to the room and its occupants, becomes cooler and sinks to the bottom of the collector, from where the cycle continues, as long as the

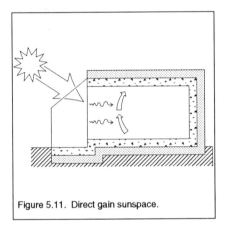

Figure 5.11. Direct gain sunspace.

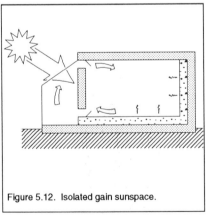

Figure 5.12. Isolated gain sunspace.

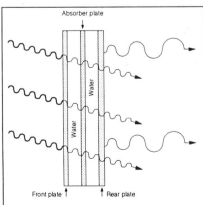

Figure 5.13. Schematic diagram of the Transwall system, illustrating transfer of solar energy.

collector is being sufficiently warmed. The thermosiphonic principle may also be used for transferring heat to the living space from some types of storage, such as underfloor rock-beds and isolated mass walls, though in many cases ordinary convection or radiation may be adequate. In 'hybrid' systems, a fan or fans will be used to move the warm air, or to reinforce the thermosiphonic loop.

Isolated gain systems may be appropriate as retrofit options depending on the circumstances, but in general it is preferable to integrate passive solar heating systems in the architecture of the building, especially where a new building is being designed. In addition, heat storage in specially constructed rock beds is difficult to justify for European buildings which typically use heavy mass construction.

Sunspace

Description:
The attached sunspace or conservatory (Figure 5.11, 5.12) consists of a glazed enclosure on the south side of a building. Depending on the climate and the way in which the sunspace is used, there may be a heat storage wall separating the sunspace from the building, or other storage within the sunspace: this serves to stabilize the temperature in both the sunspace and the house. Normally the minimum sunspace temperature is not regulated and it is not supplied with auxiliary heating. In many cases a sunspace is used as a preheat the ventilation air required by the house.

Requirements and variations:

It is possible to use a sunspace in two different ways for solar energy collection. The sunspace can act as an unheated direct gain space, in which case one introduces mass - which may be in the wall, or floor, in masonry or water - and movable insulation, so that the space is seen as a cheap extension to the house, habitable for much of the year. In principle, this is like a Trombe wall system where glazing and the wall has been increased. Alternatively it is possible to use the sunspace as a collector, in which case the emphasis is on lightweight surfaces, and the extraction of the warm air to remote storage, within or below the heated building.

Temperatures within this sunspace will vary greatly and so it may not be suitable for living or growing plants unless some solar control is used, and it is certainly not recommended in southern European climates. In Northern Europe it is not appropriate to use rock-beds as remote storage charged by a sunspace. In both types of sunspace it is essential that they are without auxiliary heating if a net solar gain is to be achieved without resort to sophisticated night insulation.

Sunspaces can take a wide variety of geometrical configurations; as simple add-ons to the south wall, semi-projecting or completely recessed into the building (i.e. surrounded on three sides by living space), covering part or the whole width of the house, and one, one-and-a-half, two or more storeys high. Even detached greenhouses may provide warm air to dwellings through fans and ducts.

Controls:

The method of distribution of the collected energy in a sunspace will be determined by the external climate, the use of the sunspace as collector or direct gain space, and the connections between the sunspace and the living space: fans will usually be needed if the sunspace is to be used mainly as a collector. Shading should be provided in mid- to southern Europe to prevent overheating in summer and some kind of venting capability must be seen as a minimum level of control throughout Europe. Vertical instead of inclined glazing can help to reduce the need for shading. Movable insulation would prevent unnecessary heat losses on winter nights or cloudy days but its cost-effectiveness needs to be examined: it is less likely to be cost-effective in southern Europe unless a combined shading and insulating device is designed. A minimum of auxiliary heating may be provided for frost prevention, if the sunspace is used for plants, and humidity control is an important consideration in dwellings with plant- or water-occupied sunspaces. In some situations it may be possible to arrange the heating schedule of the living space so that it works in phase with the sunspace temperature.

Advantages:
- The interior "climate" of a house can be greatly improved by the addition of a thermal "buffer" between the living space and the outside air. A sunspace can run the full width of the house - and the full height - reducing fabric and ventilation losses. Temperature swings in the living space are lower than with direct gain systems.
- Sunspaces also serve non-energetic purposes: for example as additional living space or as a greenhouse.
- Sunspaces are readily adaptable to existing houses.
- Sunspaces can be easily combined with other passive systems.

Disadvantages:
- In hot climates there are potential overheating problems in summer.
- Sunspaces can experience large temperature swings.
- The glazed roof of the sunspace can be sufficiently cool at night to cause condensation on its internal surface.
- Thermal energy is delivered to the house as warm air - it is less easy to store heat from air than from direct solar radiation.
- The increased humidity caused by growing plants may cause condensation and discomfort in the building.
- A sunspace, as an extension to the living space, cannot be used throughout the year.
- The sunspace can provide relatively small energy savings in comparison with its cost, although its amenity value should be taken into account.

Dual Gain

Several dual gain systems have been built, and research continues. These systems are designed to profit from the main advantages of each category involved. For instance a system combining direct and indirect gain would permit direct beam transmission as well as indirect heat recovery from storage within the system.

Such a combination can be illustrated by a water wall made of transparent containers filled with clear water. In the Transwall system (Figure 5.13, 5.14) the proportion of direct and indirect gain is permanently fixed by the geometry and the materials of the system. The absorber is a 3 mm thick grey-tinted glass (transmission = 31%).

Another example of the same combination is illustrated by a set of aluminium rotating blades filled with a phase change material (PCM) (Figure 5.15, 5.16).

Phase changing materials are still undergoing development. They are capable of storing large quantities of heat by comparison with conventional building materials (for example approximately six times as much as brick or concrete). The system illustrated in Figure 5.16 enables the user to control the proportion of the solar beam transmitted directly or intercepted by the PCM aluminium blades. At the same time, the users modulate light, sight and space perception. Such a system can be presented as a combination of the direct gain system with movable shading and insulating devices and mass or Trombe wall.

Figure 5.14. Transwall prototype installed in the Ames Laboratory Passive Technology Test Facility.

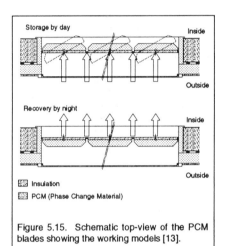

Figure 5.15. Schematic top-view of the PCM blades showing the working models [13].

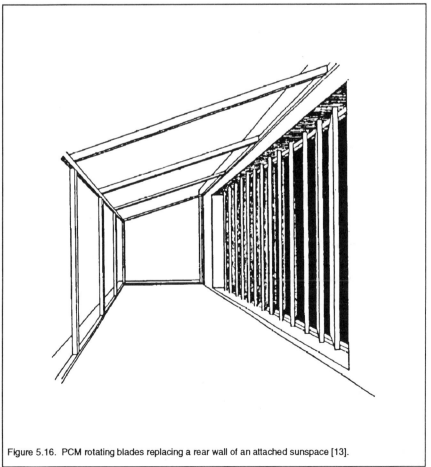

Figure 5.16. PCM rotating blades replacing a rear wall of an attached sunspace [13].

COLLECTION

Net solar gain is the amount of useful solar energy collected, less heat lost from the building through the collector apertures. In addition to careful selection and design of the storage and distribution systems to make optimum use of the available energy collected, the following factors can influence the net gain:

To maximise solar energy collected:
- The choice of a favourable orientation and collector slope or tilt
- The installation of reflectors
- The avoidance of shading (when appropriate)
- The choice of glazing with a high transmittance value for solar radiation
- The overall design of the system to achieve a high absorptance of solar radiation at appropriate times of day and year.

To minimise heat loss:
- The use of glazing with a low thermal transmittance
- The installation of movable insulation (used only at night in a direct system).

There will be an improved absorptance of solar radiation if the absorbing surface is black or if there is a high apparent absorptance as in a direct system (see Appendix 9 on radiation).

Orientation and Tilt
In general, the largest quantity of solar energy is received from a clear sky by a collector with a south-oriented sloped plane where the ideal angle of slope varies with latitude and the time of year. This may, of course be reduced by the presence of clouds or shading by nearby buildings or vegetation.

The amount of this energy which can be used will depend on the available heat storage within the building and the pattern of demand. This pattern can affect the choice of orientation where, for example, the greatest demand is in the afternoon and consequently an orientation west of south should be chosen.

If the collector cannot face true south the heat gain is only slightly reduced up to an angle of 30° east or west of south. However, the effectiveness of horizontal sunshades will be reduced significantly and this may have a greater bearing on the choice of orientation. Chapter 2 and Appendices 1, 4, and 6 provide design tools which give guidance on orientation.

Figure 5.17 shows that for a vertical surface in winter, most solar energy falls on a south-facing surface. Conversely, in summer more energy falls on the eastern, western, southeastern and southwestern vertical planes. This will have the effect of reducing summer overheating problems if south-oriented glazing is used.

The most frequently occurring tilt is 90° (vertical). This is the result of spatial considerations and practical matters such as cleaning, the discharge of condensed water, etc. Moreover, movable insulation and sunshades are usually easier to fix on vertical planes. Since a number of aspects have to be considered when determining the optimal tilt, one single general solution cannot be given. Reduction of the slope of a vertical south-oriented surface results in:

- a greater quantity of solar energy during the heating season (the tilt at which maximal energy is received increases with increasing latitude);
- More long-wave radiation loss to the atmosphere; the plane will 'see' a larger section of the cold sky vault;
- more overheating problems in the building (especially in summer);
- unfavourable conditions for the application of horizontal sunshades.

If the gradient is increased the reverse is true.

Reflection
The irradiation at an aperture can be increased by placing reflectors in front of the glazing system. The reflectors can be metallic surfaces sometimes protected by glazing. Water, snow or other light-coloured surfaces can have a similar effect.

The reflection can either be specular or diffuse. Polished metallic surfaces give a mainly specular reflectance while most other surfaces give a diffuse

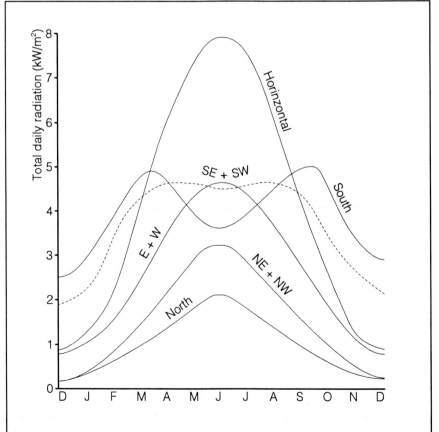

Figure 5.17. Total daily solar radiation on various surface orientations for a clear day in London (Kew): ground reflection included [1].

reflectance. It may be necessary to adjust the angle of a metallic reflector to maximise the useful gains throughout the heating season. While this is common for reflectors in other solar applications, in passive solar buildings they may cause glare problems, especially if placed in front of windows below eye-level.

The reflection from diffuse reflectors will not be as effective. Their reflectances are often lower than those from metallic surfaces and the reflected radiation per unit area which then travels in the direction towards the adjacent aperture is smaller. Diffuse reflectors have the advantage of being cheap and there is no need for periodic adjustment.

The reflectivity of water is not as great as is often thought. It also varies considerably with the angle of incidence. At low solar altitudes the surface reflectance is highest but even then it is only 35% and this will be reduced if there are waves. At high solar altitudes the surface reflectance goes down to 2% because most of the sunlight is transmitted through the water (see Table 5.1).

The specular reflectances of different materials are shown in Table 5.2. The effect on received solar energy of a horizontal specular reflector at the foot of the solar aperture for different reflectances of the reflector is shown in Figure 5.18.

Glazing

Glass is the most commonly used glazing material, although transparent or translucent plastic materials may be specified, depending on the circumstances. The transmittance of glazing as a function of the angle of incidence is shown in Figure 5.19.

Most glazing materials have poor thermal resistances, the main thermal barrier being the separation of internal air from external air. The challenge is to maintain efficient transmittance of solar energy and to reduce heat loss. The thermal and solar properties of the glazing can be improved by various means:

- Adding one or more layers of glazing improves the insulating properties by forming a layer of trapped air but it does slightly decrease the solar transmittance. In

addition, a heavy gas can be placed in the cavity between the layers of glazing and a selective surface added to the glass. The heavy gas (such as argon) reduces convective heat losses and the selective surface is transparent to solar (short-wave) radiation but reflects thermal (long-wave) radiation from the occupied space to the outside. By combining these two effects it is possible to reduce the heat transfer between the inner and outer panes of glass significantly, with only a small reduction in solar trans-

Angle of Incidence	Reflectance
0°	0.02
45°	0.03
60°	0.06
75°	0.21
80°	0.35

Table 5.1: Surface reflectances of water for different angles of incidence with an index of refraction of n = 1.333.

SURFACE	REFLECTANCE
Polished Aluminium	0.70
White Paint	0.6 - 0.9
Aluminium Paint	0.45

Table 5.2: Reflectances of different surfaces. See also Appendix 10.6

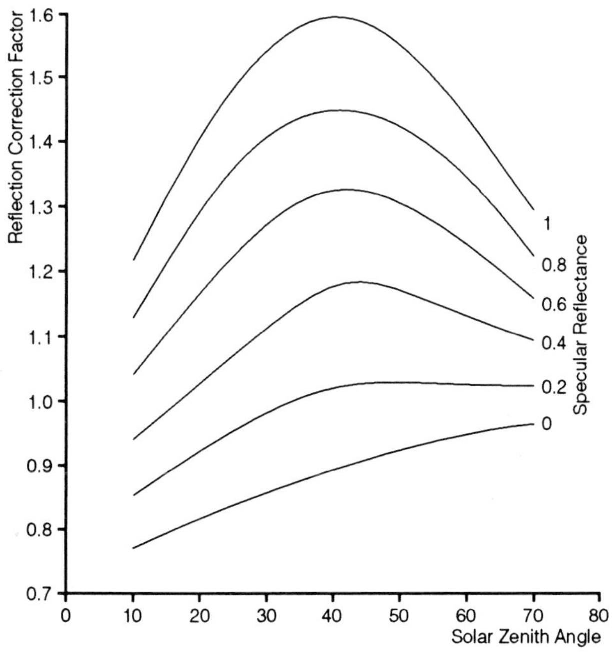

Figure 5.18. Total daily solar radiation on various surface orientations for a clear day in London (Kew): ground reflection included [1].Figure 5.18. The effect on insolation of a horizontal specular reflector, in front of the collector plane, as a function of the minimal solar zenith angle (= 90° - solar altitude at solar noon). These curves represent a south-facing aperture, five times as wide as it is high, with a reflector depth equal to the height of the aperture, and length equal to that of the aperture, under clear sky conditions [3].

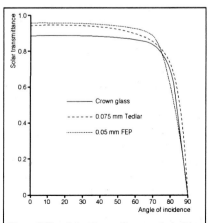

Figure 5.19. Solar transmittances of glazings as a function of angle of incidence [4].

mittance. Air or gas-filled space between glazing panes has a thermal resistance that is proportional to its width up to approximately 20 mm. It then remains constant to about 60 mm and decreases slightly after that.

- Low iron content glass can be used. This has a higher solar transmittance than ordinary glass and is only slightly more expensive to produce.

- Reflecting glazing which can help to avoid indoor overheating in summer also admits less sunshine during the heating season and may therefore not be compatible with maximising passive solar gain. Glazings are being developed with variable transmittances, so offering improved control of insolation.

- Other possible materials include various types of transparent polymer sheets. Some of these have very high solar transmittances and therefore can be used in multi-pane structures giving

good thermal and solar performances. But many are transparent to (long-wave) thermal radiation, not as durable as glass and their frames can be expensive to construct.

- Heat losses due to thermal radiation to the exterior may be reduced by using glazing which has been coated with a low emissivity layer.

- It may sometimes be necessary to use sunshades to maintain thermal and visual comfort in areas close to glazing. However, this will reduce the overall solar gain in the building, the effects of which may be significant during the heating season.

When considering an alternative to glass, points to consider are: How does its durability compare with that of glass? Are its fire resistance characteristics suitable? Other considerations will be cost, colour degradation, transparency and safety, especially for overhead glazing.

The window frame will also play a part in the thermal losses. There are other issues which will influence the choice of frame, such as stability, ease of cleaning, colour etc. Wood or PVC frames have good thermal properties. Aluminium frames with a thermal barrier often have slightly higher heat losses (see Figure 5.20), but less than aluminium frames without a thermal barrier or steel frames which may cause severe condensation problems.

Transparent Insulation Materials

Transparent insulation materials (TIM) are a new class of materials which combine the properties of good optical transmission (described by the solar transmittance t) and good thermal insulation (described by the thermal heat loss coefficient U or thermal conductivity k).

One of the most obvious applications for TIM is on the sunny facades of buildings, replacing conventional opaque insulating materials. Well-designed TIM facades can reduce the annual energy requirements for space heating in new and retrofitted houses to one quarter that of

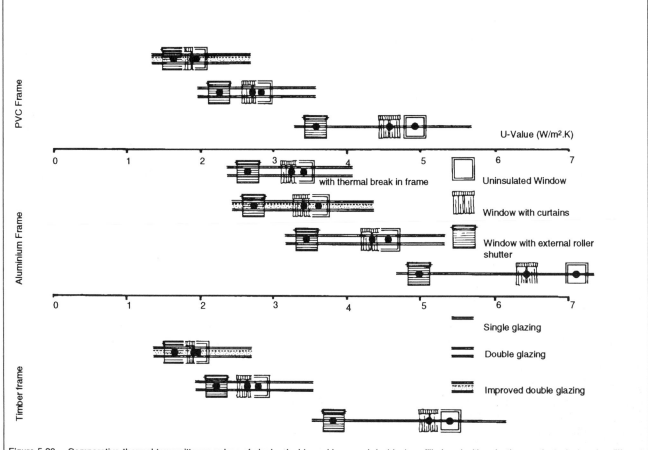

Figure 5.20. Comparative thermal transmittance values of single, double and improved double (gas filled and with selective coating) glazing, for different frame materials and with curtains or external roller shutters [5].

comparable buildings with conventional wall insulation. They can replace windows where light but not vision is required.

A variety of different materials can be characterised as TIM and they fall broadly into four categories:

Type A: with material structures parallel to the surface plane (multiple glazings, plastic films, IR-reflective glass)

Type B: with material structures perpendicular to the absorber (parallel slats, honeycombs, capillaries)

Type C: scattering structures (duct plates, foams, bubbles)

Type D: quasi-homogeneous materials (glass fibres, aerogels).

A number of materials made from glass or plastic are being produced now, but some are only available on a laboratory scale. Their optical and thermal properties listed below can be compared with those of 'conventional' glazing:

During the 1980s R+D in transparent insulation materials (TIM) and subsequently in transparent insulation systems (TIS) has been supported by the Commission of the European Communities as well as by national bodies, mainly in Denmark, Germany, Switzerland and Sweden.

The physical principle of TIM walls is similar to the well-known Trombe wall: the dark-coloured wall surface acts as an absorber, collecting solar energy and converting it into useful heat. Most of the solar heat can be stored in an adjacent high-mass wall, but some is still lost through the

Figure 5.21. Transparent insulation used in a house in Freiburg. Architect H. Novazov

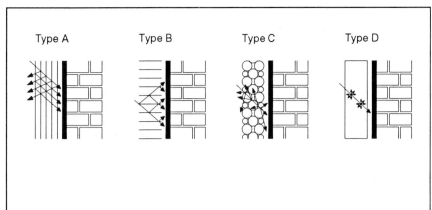

Figure 5.22. Classification of transparent insulation materials. Source BINE projekt Info-Service [6]

MATERIAL	THICKNESS (mm)	tdif	L10°C (W/ m²K)
Single glass			
normal float	4	0.78	
low iron	4	0.84	
Double glazed window			
float	16	0.67	5.20
low iron	16	0.76	5.20
PMMA-type foam			
type A1K	16	0.58	3.60
type A2K	16	0.48	3.10
type (C)	16	0.60	3.50
type (C) film cover	16	0.53	3.50
Capillary structure			
PC	60	0.63	1.45
PC	100	0.60	0.79
Honeycomb structures			
sinusoidal PS	108	0.66	1.28
hexagonal PS	43	0.75	2.21
hexagonal PA	98	0.61	0.97
sinusoidal PVC	108	0.45	1.12
hexagonal PVC	108	0.48	0.92
AREL/PC	100	0.71	1.10
Aerogel windows			
GRA 12968c	12	0.52	1.67
GRA 12968c	20	0.44	1.00

Remarks:

tdif : Hemispherical -hemispherical global transmittance
L10°C : Heat loss coefficient at 10°C (without heat transfer to air as included in k/U-value

PC : Polycarbonate
PS : Polystyrene
PA : Polyamid
PVC : Polyvinylchloride

Table 5.3. Optical and thermal properties of TIM

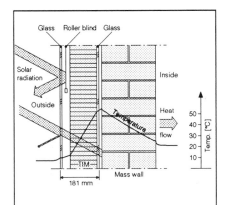

Figure 5.23. Principles of passive heating with transparent insulation. Source BINE Projekt Info-Service [7]

Figure 5.24. 'Aubea' direct gain house, Aubigny-sur-Nere, Cher, France. Insulating shutter. Architects: S.C.P.A. Chauz-Pesso-Raoust.

insulation. Depending on the wall material and the indoor and outdoor temperatures, heat moves inwards through the wall, with a time shift. The wall surface facing into the living space eventually heats up and acts like a large, low temperature wall heater which itself further contributes to energy saving: comfort can be achieved at reduced air temperatures. Because of the time delay of heat flow through the TIM wall, a very good complementarity with direct solar heat gains through windows is achieved. Sun shading devices are needed to avoid overheating on sunny days in spring, summer and autumn.

Heat Retention

Instead of conceiving of the glazing as an element with fixed thermal characteristics, it is possible to use a movable insulation system during times when there is no sunshine or at night.

The simplest of these movable systems is a curtain. It is possible to improve the thermal resistance of curtains by adding low emissivity layers, by incorporating an insulating lining and by careful attention to the edge details where the curtain meets the wall, floor, windowboard or ceiling. Other interior night insulation systems include insulated shutters. Condensation on the glass is a potential problem which can arise with internal movable insulation. This can be a serious problem and may affect the building fabric.

If insulation is applied behind glazing and not removed during sunshine, thermal stresses can build up in the glass causing it to break. This will depend on how well the edge fixing details compensate for thermal movement within the glass and/or on the presence of partial shading of the glass, causing differential expansion.

The cavity between the panes of glass in a double glazed unit can also accommodate movable insulation. Different options include drapes, low emissivity foils, blinds and Beadwall [8]. The latter is a system where polystyrene beads are pumped into the cavity between double glazing at night and removed when the sun is shining.

External movable insulation panels are another possibility, but they must be robust enough to withstand rain, wind, ice and ultraviolet radiation.

The main principles governing movable insulation are that the insulation layer should as far as practicable be placed on the exterior surface of the building element (for both reduction of condensation and use under summer conditions) and that one should attempt to create a layer of still, trapped air with seals between the glazing and the movable insulation system that are as tight as possible (especially for winter conditions).

It is difficult to achieve airtight seals with movable insulating systems. Without good seals the insulation effect will be reduced drastically. The time commitment of the user in opening and closing most of the systems, other than those with mechanical controls, may be considerable and the practicability of such interventions should be realistically assessed.

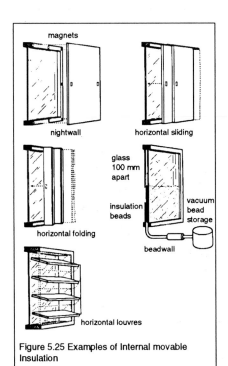

Figure 5.25 Examples of Internal movable Insulation

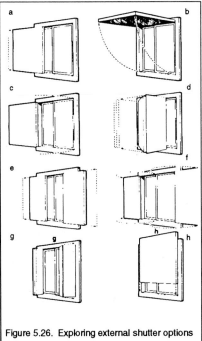

Figure 5.26. Exploring external shutter options

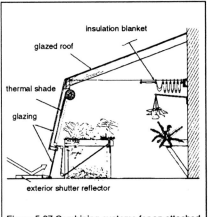

Figure 5.27 Combining syatems for an attached sunspace [10]

Movable insulation can fulfil more than one function: in winter the heat loss can be reduced and in summer it can be used as a shading device. Moreover, in winter, night insulation greatly increases comfort near the windows in the evening. For this reason night insulation or glazing with a low U-value is particularly useful where there are large windows.

There are many solutions for insulating building elements, ranging from various methods of window insulation to insulating Trombe Walls. Some of these are illustrated here (Figures 5.24 to 5.28). In an indirect system, the increase in thermal resistance between the absorber surface and the exterior environment has two favourable effects: the total U-value decreases, which results in less conductive heat loss, and the fraction of incoming solar gain increases.

Figure 5.28 shows a comparison of the performance of Trombe and water walls with different external thermal resistances. For a wall with a matt black absorber and single glazing, one or more of the following measures have been taken: double glazing night insulation (U = 0.5 W/m2K), spectrally selective absorber (low emissivity and high absorptance for solar radiation). The importance of the reduction of heat loss is clearly visible. Although the figures are relative, and depend largely on climate, the trend will be generally valid.

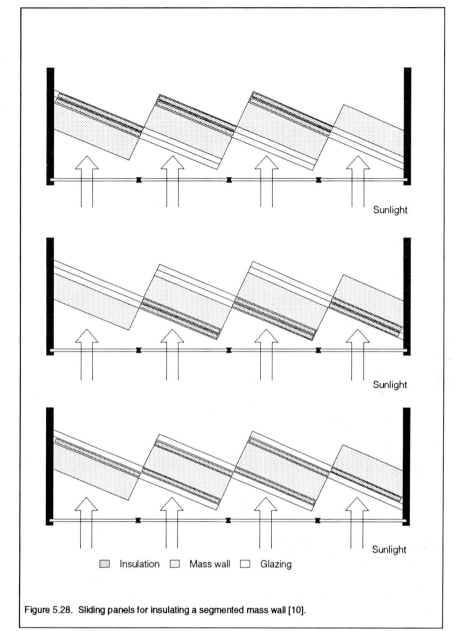

Figure 5.28. Sliding panels for insulating a segmented mass wall [10].

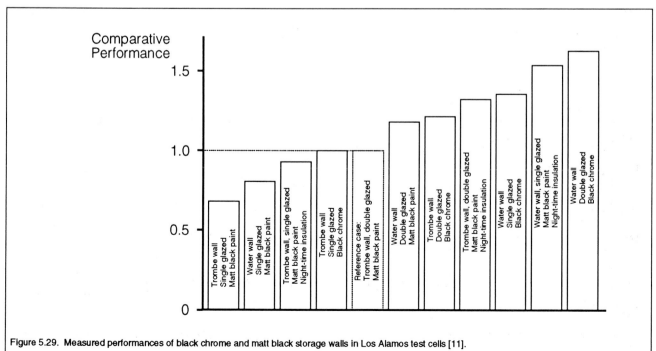

Figure 5.29. Measured performances of black chrome and matt black storage walls in Los Alamos test cells [11].

STORAGE

On a clear day the solar energy gained by a passive solar system may considerably exceed the heat demand. Energy savings are achieved by storing the surplus energy and by using it at a later time, when there is a need for it. If too much solar heat is released in a room, overheating will result. To avoid this, the occupant will typically lower a shading device or use extra ventilation. This implies that part of the potentially useful energy is wasted. Thus, thermal storage has a dual purpose: saving energy by storing the surplus until a later time, and avoiding overheating. The latter is particularly important in summer. Thermal storage can in certain circumstances be utilized for the heat released from lighting, electrical equipment and occupants.

There are three types of thermal storage:

Primary thermal storage is defined as that area of storage which receives direct solar radiation. It is usually defined as being the area illuminated by the sun (beam radiation) at noon on either of the Equinoxes.

Secondary thermal storage is that area of storage material which is located outside the sunlit area but is within thermal radiative contact with it. Secondary storage is supplementary to primary storage.

Remote thermal storage is hidden from the view of both primary and secondary storage and

therefore not in radiative contact with them. Transfer of heat to remote storage is by convection, either natural or fan-induced.

The suitability of the storage depends on a number of factors which may be divided into two categories:

- the size and material of the storage.
- the means whereby solar heat is charged and released.

Material and Thickness

The specific heat storage capacity of a material is the amount of heat (in kJ) which has to be added to a quantity of material (1 kg) to raise its temperature by 1K. The heat storage capacity of ordinary building materials and water (between freezing point and boiling point) is practically independent of temperature. Phase change materials (PCM) absorb heat when melting and release heat when re-solidifying. This could be said of many materials (such as water/ice), but the materials which are of interest here are those whose melting and solidifying temperature render them suitable for heat storage in buildings. Absorption and release occur over a small temperature interval: the melting range. In this interval these materials have a larger heat storage capacity.

The storage capacity of ordinary building materials is mainly dependent on the density of the material (Table 5.4). The mass (in this context also called thermal

mass) is a good measure of thermal capacity. However, the mass available for storage is generally less than the total mass.

The mass available for heat storage is largely determined by the frequency with which the storage is charged and discharged. Temperature variation in the material is reduced as the distance to the heated surfaces increases: the material will gradually participate less in storage. (See Appendix 9.1 on Radiation). The thickness playing a role is the 'effective thickness'. For the most important charge and discharge rhythm (once every 24 hours) and with the heat supply on only one side, the effective thickness of masonry building materials amounts to approximately 60 mm to 120 mm. As a result of this phenomenon it is of little use for thermal storage purposes to construct walls and floors which are thicker than 80 mm to 160 mm when heat is supplied on only one side. Water has a very large storage capacity per unit of volume, (see Table 5.4). In addition to this benefit, water may be used in thick layers, since, as a result of circulating flows, a practically isothermal storage exists. Figure 5.25 illustrates that under similar conditions a water wall has a higher output than a Trombe wall.

Phase change materials have advantages with respect to water only if the storage temperature is above melting point. For an equal

Material	Density kg/m³ (r)	Specific admittance kJ/m²h1/2K	Thermal conductivity W/mK (k)	Specific heat capacity kJ/kgK (Cp)	Thermal capacitance kJ/m²K	mm
Normal concrete	2400	142	2.10	1.0	576	240
Lightweight concrete	1000	39	0.38	1.0	240	240
Gas entrained concrete	400	15	0.14	1.0	96	240
Fired clay brick	1400	58	0.60	1.0	336	240
Hollow concrete block	1400	63	0.70	1.0	336	240
Asphalt	2300	91	0.90	1.0	138	60
Sand and cement screed	2000	106	1.40	1.0	120	60
Marble bedded in mortar	2800/2000	188/106	3.5/1.4	1.0	136	20+40
Tiles/mortar	2000/2000	90/106	1.0/1.4	1.0	120	20+40
Aluminium	2700	1310	200	0.8	130	60
Steel	7800	860	60	0.4	187	60
Wood	600	26	1.40	2.1	78	60
Rigid foam insulation	20	30-45	0.03	1.5	1.80	60
Horizontal layer of still air	1.25	-	0.30	1.0	0.08	60
For comparison:						
Water	1000	98	0.58	4.2	1008	240

Table 5.4: Specific heats and densities of different materials [32].

quantity of stored heat, a PCM storage has a lower temperature than an equally large storage of water. This property is a considerable advantage for the charging and the heat loss of the storage. If a PCM in a room has a melting point temperature slightly higher than the required room temperature, the material will act as a kind of thermostat: surplus heat in the room is stored without a significant rise in air temperature: when heat is required, it is released from the storage, the temperature remaining almost constant.

The above considerations underline the importance of the melting range, when a PCM is chosen. As these materials are poor thermal conductors (and also to avoid salt segregation) they are only used in thin layers. Additional problems are supercooling and containment (Figure 5.30).

Charge and Discharge

The most efficient means of transferring solar energy to storage is by placing the storage material in direct sunlight (primary storage). Another possibility is thermal radiative contact with a sunlit area (secondary storage). The mean radiant temperature "seen" by the storage must then be higher than the surface temperature of the storage. The third and least efficient means is by convection. Air must first be heated by a sunlit area (absorber) and then transferred to the (remote) storage. The air temperature must then be higher than the storage temperature. The discharge of storage occurs through thermal radiation and/or by convection.

A dark colour such as black is most effective for the absorption of a large quantity of solar radiation by primary storage (see Appendix 9.1 on Radiation). A low emissivity and small air movement over the surface effectively reduces the heat loss from the surface when temperature rises. These measures are important in mass and Trombe walls (see Figure 5.29). In a direct system, storage generally occurs in the building construction itself (walls, floor, ceiling): the above mentioned measures are then difficult to realize. A dark floor may be used but is not absolutely necessary due to multiple reflections within the room. They

are also not critical since solar heat which is not stored "primarily" may still be transferred to secondary storage, the remainder being released in the room by convection. A diffusing system "spreads" the solar radiation "over" a much larger surface area, which makes a considerable primary storage possible - but may create unacceptable glare problems.

In case of secondary storage it is important that it should be "seen" by the primary storage. In addition, emissivity should be high, but this is always the case when ordinary surfaces are involved, whatever their colour may be. The surface with the best "view" of primary storage will receive most solar radiation.

Heat transfer to remote storage occurs exclusively by convection: forced convection, with the aid of a fan, or free convection. Remote storage may consist of a rock-bed, a sub-floor, a storage volume placed away from the building or the mass of the remainder of the building. The heat release can be controlled by dampers and/or fans. The heat may also be transferred to a room by conduction through a wall or floor.

Effect of Mass

Research has shown that thermal mass has a considerable effect on room air temperature fluctuations. The results of a computer simulation of a direct gain system with different thicknesses of primary and secondary mass are shown in Figure 5.31.

The area of the secondary thermal mass is maintained equal to the area of primary thermal mass (target). When only the thickness of the secondary thermal mass layer is changed, it can be seen that a thin layer will cause large daily air temperature fluctuations within the room. As only the thickness of the secondary thermal mass layer is increased, the air temperature fluctuation within the room reduces until a point is reached where an increase in thickness will no longer reduce the air temperature fluctuation: this point is termed the turning point thickness. Each of the subsidiary curves has been drawn using a different thickness of primary thermal mass and again

changing only the thickness of the secondary thermal mass. It can be seen from these curves that the turning point thickness of the secondary thermal mass is not greatly affected by a change in the thickness of the target slab. It was found that a change in the proportional area of the secondary thermal mass, with all other parameters unaltered will cause a change in the turning point thickness of the secondary thermal mass.

In a heavy-weight building with night set-back, the temperature will decrease less during the night than in a light-weight building. Therefore energy savings from night set-back are smaller in heavy-weight buildings. But generally this disadvantage does not outweigh the benefits of the mass (constant air temperature, better use of passive solar energy). It may be more sensible to use a light-weight construction only in rooms which are occupied intermittently.

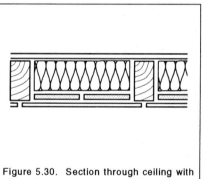

Figure 5.30. Section through ceiling with phase-change material in bags [16].

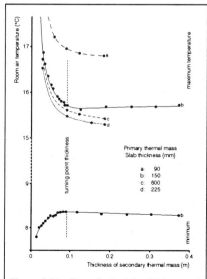

Figure 5.31. The effect of the thickness of secondary thermal mass on daily room air temperature fluctuation. Both the primary and the secondary storage are of concrete [17].

5.5

DISTRIBUTION

Heat distribution normally is required after collection and storage of solar energy. The aim of distribution is to have solar heat reach the locations where it may be of use. Heat distribution is directly dependent on building design and the heating system.

In many ways the challenge to the designer is to reduce the need for distribution as much as possible. The most efficient mode of distributing solar energy is to design the layout of the rooms in such a way that the solar energy is collected and stored in or adjacent to the room where it will be used: this is known as thermal zoning. If this is not possible, the energy will have to be transported to other rooms in the building.

The distribution of solar energy in a room must be designed to prevent large differences between surface temperatures and the air temperature near the ceiling and floor. If sufficient primary and secondary mass is available in a direct system, the distribution by the exchange of heat between the walls (radiation) and from the walls to the air (convection) will be adequate. Temperature differences are then a result of the heating system, or poor heat resistance of the windows. Distribution in a direct gain system is improved by using diffusing glass. This can however cause glare problems and should therefore be placed above eye-level. In a thermosiphon system (e.g. Trombe wall) the risk exists of a large vertical temperature gradient in the room. This will depend on the mixing of the indoor air with the outflowing air, on the heating system and air infiltration. High surface temperatures may arise for example with mass walls. Secondary storage within good 'view' of the wall (unobstructed by furniture), as well as sufficient convection over the surface are then important. A better distribution is obtained if the heat enters the room evenly over the entire floor surface (e.g. by remote storage). The only restriction is the maximum temperature of the comfort range.

If too much solar heat is released in the room, it may be partly transferred to the adjacent room by simply opening a door. Air circulation from one room to the other is enhanced if the door height is extended up to ceiling level: this avoids stagnant pockets of hot air near the ceiling (Figure 5.32). The amount of thermal radiation transferred to the adjacent room will generally be small (Figure 5.33).

A remote storage system, or a large surplus of solar heat on the south side will necessitate heat distribution to all locations requiring heat. A central air heating system may already regulate the distribution by means of the re-circulation of air. Designing a building where distribution takes place exclusively by means of natural air flows (free convection and flow as a result of wind pressure) is a hazardous task. Figure 5.34 shows an example of such an attempt in a multi-zone passive solar dwelling.

The flow resistances (e.g. in the rock-bed) may result in a velocity which is lower that velocities which result from infiltration and ventilation. The system would no longer work in that case. Moreover, a serious acoustic problem would arise in the dwelling. It is safer to rely upon ducts and fans for the transfer of heat. Figure 5.35 illustrates the use of a ventilator for the distribution of incoming solar energy.

Interzonal Airflow
Design Guide for Buoyancy Flows
Where a temperature difference occurs between adjacent rooms, large heat flows will result if doors (or other apertures) between them are opened, for example over 1 kW for a standard door at 7K temperature difference. This is known as natural convection, or buoyancy flow, because it is promoted by the density difference between the two bodies of air at different temperature levels.

A recirculatory flow pattern will be established with warmer air flowing out of the top half of the opening to be replaced with cooler air through the bottom half.

In passive solar buildings this mode of heat transfer may be used to great effect to distribute heat from areas with high solar gain. For example heat from an attached

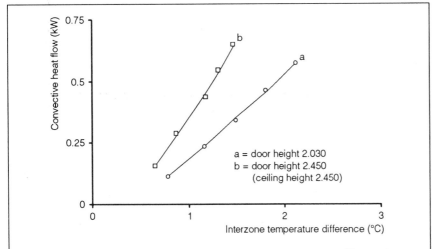

Figure 5.32: Convective heat flow through aperture versus interzone temperature difference for standard and full ceiling height doorways. [17]

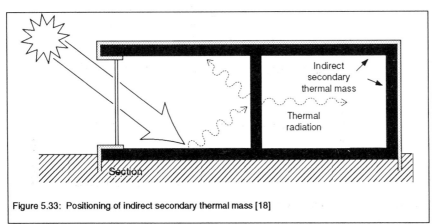

Figure 5.33: Positioning of indirect secondary thermal mass [18]

conservatory or highly glazed room may be transferred into the body of the house via doorways or vents. The same principle also applies to cooling overheated areas by natural ventilation through openings to the outside.

A recent review of the subject [19] has found simple expressions which provide good agreement with experimental work for natural convection flows. The equation for air flowrate written as:

$$q = \frac{CW}{3}\left[g\left(\frac{\Delta T}{\bar{T}}\right)\right]^{1/2} H^{3/2}$$

where q is flowrate (m³/s)
C is an empirically derived coefficient of discharge between 0 and 1
W is the width of the opening (m)
g is gravity (m³/s)
ΔT is the temperature difference (K)
T is the mean temperature (K)
H is the opening height (m)

The heat transfer rate q Watts will then be given by:

$$q = Q\rho C_P \Delta T$$

where C_P is the specific heat and ρ the average density.

Although this simple theory assumes isothermal zones (when in practice stratification is usually present) it has been found to be as accurate as stratified theory as long as appropriate values of the empirical coefficient C are chosen. For the air discharge coefficient a value of C=0.65 gives good agreement with data. Experimental data indicates that heat transfer in the stratified case is typically increased by 20% compared to the isothermal case, because the air being transferred is at a greater than average temperature difference. This can be allowed for by using a coefficient of discharge of 0.8 for heat transfer. Experimental results correspond to horizontally adjacent openings such as windows and doors for temperature differences 0k to 10, and relate to most building situations. Limited data for natural convection flow up and down stairwells indicates that coefficients approximately half these values are appropriate. Assuming typical values for horizontally related openings, heat transfer rates (q) may

be estimated by hand calculation or spreadsheet from:

$$q = 57.72W[H\Delta T]^{3/2} \quad W$$

or even more simply for a 2 m high, 0.7 m wide standard doorway:

$$q = 114.28[\Delta T]^{3/2}$$

Flowrate and heat transfer rate may be scaled linearly with aperture width. Identical side by side openings are additive. However, the variation of flowrate and heat transfer is proportional to the aperture height to the power of 1.5. The important repercussion from the design point of view is that tall narrow openings are more effective than square openings of the same area. A doorway on its side will only transfer 60% as much heat as in its upright position. Flowrate and heat transfer through two identical vertically displaced openings may be calculated using the same equations.

The net flow is that of an opening of the outside dimensions less that of an opening the size of the solid middle section. In this way it may be shown that two 1 m high openings at 2 m centres would improve heat transfer by 50% compared to a single 2 m high opening.

Heat transfer for a range of aperture heights, in 0.5 m increments, is presented in Figure 5.36. Heat transfer rates under steady temperature differences may simply be read off for the particular aperture height and proportioned to the width of the opening.

To further simplify calculation of heat transfer for any given temperature difference the data has been linearised for two ranges of temperatures. The two temperature ranges considered are 0°C to 5°C

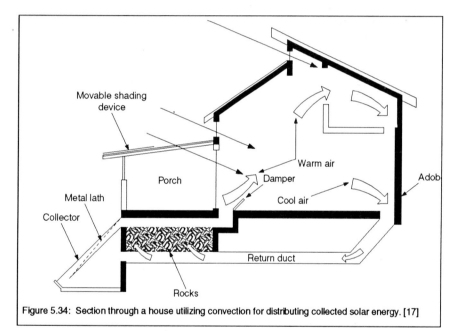

Figure 5.34: Section through a house utilizing convection for distributing collected solar energy. [17]

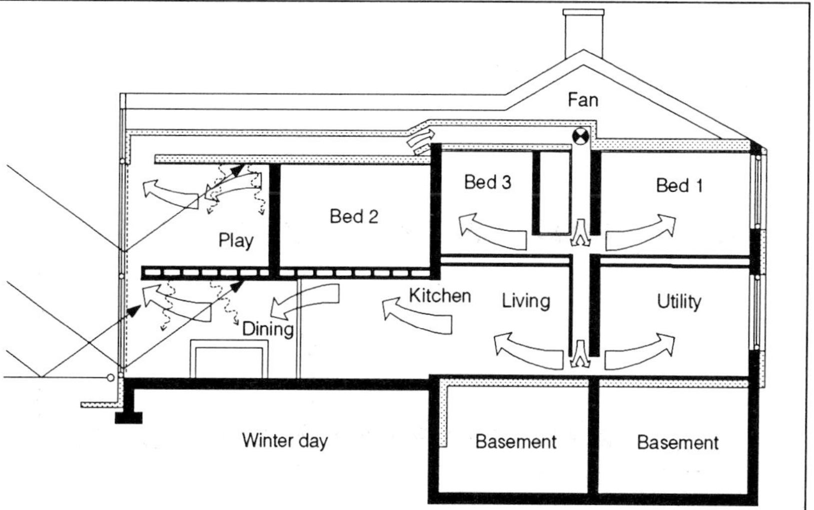

Figure 5.35: Section through a design in which a fan is used to enhance convective heat distribution. [17]

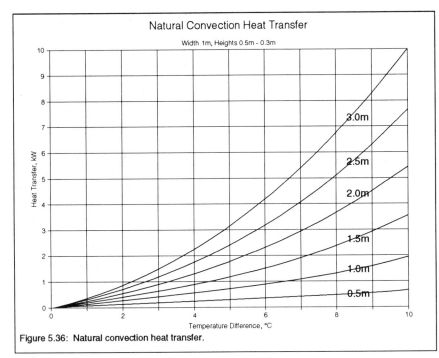

Figure 5.36: Natural convection heat transfer.

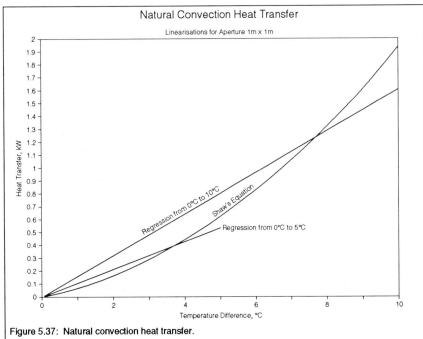

Figure 5.37: Natural convection heat transfer.

Height, m	0°C-5°C	0°C-10°C
0.25	53.6	79.4
0.5	75.7	112.3
0.75	92.8	137.5
1.0	107.1	158.8
1.25	119.8	177.5
1.5	131.2	194.5
1.75	141.7	210.0
2.0	151.5	224.5
2.25	160.7	238.2
2.5	169.4	251.1
2.75	177.6	263.3
3.0	185.5	275.0

Table 5.5 : Equivalent U-values for Various Aperture Heights

and 0°C to 10°C and the relative accuracy of the two linearisations is illustrated in Figure 5.34. For most purposes the 0°C to 5°C range will be applicable, particularly as the convection flow lowers the temperature difference. However where large solar gains are expected, in southern Europe, the 0°C to 10°C correlation will be more appropriate.

Because the data has been linearised it is possible to recommend equivalent "U" values for openings where natural convection is predominant. This will be very useful for hand calculations and in thermal simulation programs which do not have an interzonal airflow model incorporated. However because of the non-linear dependence on aperture height it is still necessary to provide different U values for different aperture heights. Table 5.5 presents the data in tabular form, whereas Figure 5.35 and Figure 5.36 present the data in graphical format. It may be seen for instance that, based on the 0°C to 5°C correlation, the equivalent U value for a 2 m high door is 150 W/m2K.

Data is not available for a range of aperture heights in the case of flow through a doorway and up and down a stairwell. However, the data from measurements in one house indicate that, based on the 0°C to 5°C correlation, the equivalent U value for flow between floors of a house via a 2 m high door is 75 W/m2K.

Both of these figures are approximations but represent an improvement upon the practice of ignoring buoyancy-driven heat transfer through open doors. The correlation for horizontally adjacent zones is in keeping with data from a number of projects and should provide accuracy within ±50%. The correlation for flow up and down stairs is based upon original data, which therefore cannot be compared to previous work. Given the greater dependence of this type of flow to boundary conditions, of the stairwell and of infiltration, a higher degree of variability will be expected [19]. Note that pressure differences between rooms will have a

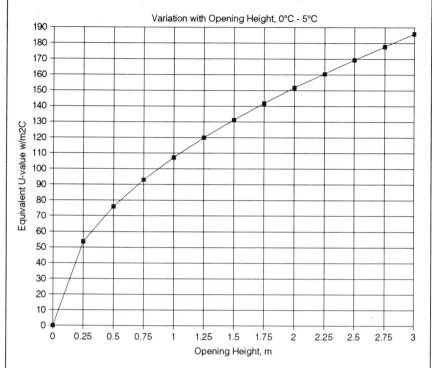

Figure 5.38(a)

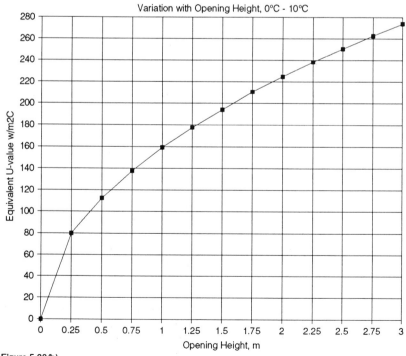

Figure 5.38(b)

Figure 5.38(a) and 5.38(b): Equivalent U-values of an aperture

significant effect on these airflow rates.

In nearly all locations there is an uncertainty in the weather pattern. Without interseasonal storage (whose cost-effectiveness for single projects is dubious at present) it is not possible to dispense entirely with an auxiliary heating system in northern and central Europe The auxiliary heating system will have to provide supplementary heat in certain parts of the building, at certain times, for longer or shorter periods and with variable intensities in order to obtain the required thermal comfort level.

This already suggests some of the criteria influencing the choice, design and sizing of the auxiliary heating system.

System Selection

The first criterion influencing the choice of an auxiliary heating system is the design of the passive building itself and its architectural and thermal properties: the choice of passive solar components, the method of heat distribution (water or air), the storage provided (capacity and location), partitions within the internal volume, and the use of certain architectural or passive control devices, etc.

The second criterion is more psychological in nature: local user habits and traditions of heating modes and systems used. For example, some regions prefer the use of water distribution systems with room control systems and others prefer air systems. Also of great importance are the life-style and behavioural aspects such as patterns and intermittancy of occupation and the comfort levels expected.

Local technical constraints will also influence the choice: for example, the lack of availability of one or more of the energy sources on the building site.

The final criterion when choosing an auxiliary heating system is cost:

- investment in the heating system in relation to other building costs.
- operating costs of the system (maintenance, energy

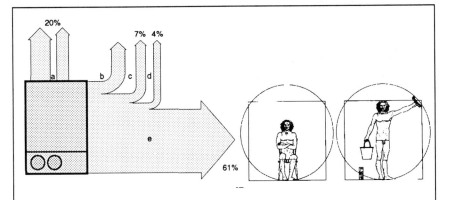

a. Boiler and chimney losses

b. Distribution losses

c. regulation losses

d. emission losses

e. effective heat emmission

Figure 5.39: A diagrammatic representation of the paths of inefficiency of a conventional gas or oil heating system. From left to right: boiler and chimney losses (20%), distribution losses (8%), regulation losses (7%), emission losses (4%), effective heat emission: 61% [27]. The use of high efficiency boilers (eg. condensing gas boilers) can significantly improve these figures.

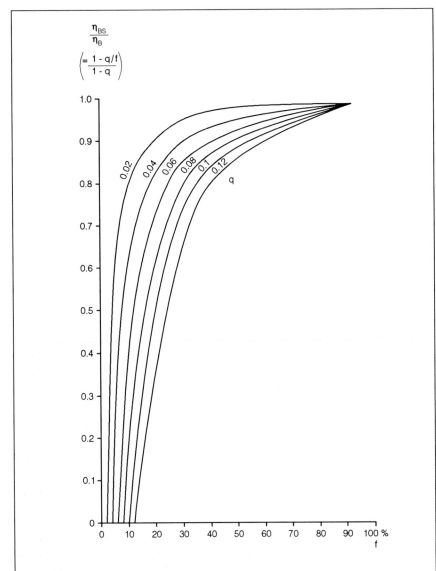

Figure 5.40: The relationship between nominal boiler production efficiency η_B and the actual seasonal efficiency η_{BS}, as a function of load factor (f) and boiler heat loss factor (q). [28]

consumption, energy price) in relation to the lifetime of the building and heating plant.

System Design

The heating system consists of four parts:

- production unit;
- distribution system;
- heat emitters;
- controls.

These four parts may be either distributed throughout the building or concentrated in one appliance, and may range in degree of sophistication from very complex to very simple. Examples of these are a water-to-water heat pump with a piped distribution system, floor heating and micro-processor control system, and a gas heater in each room with a manual control valve.

Each part of the heating system has its own nominal (full operating condition) and seasonal (average) efficiencies and associated factors affecting its performance and therefore affecting the final energy consumption:

η_{BS} The production unit or boiler's conversion efficiency of fuel to useful heat. This depends on the design of the production unit, chimney losses, and unutilised losses of the boiler to the room air.

η_d The efficiency of distribution. This relates to the length of the distribution system, the temperature of transported fluids and insulation employed to reduce pipe losses.

η_e The efficiency of the heat emitters. This is related to the way heat is to be emitted (by radiation or by convection) and any possible obstructions to efficient emission (e.g. curtains hanging in front of radiators).

η_c The efficiency of control of the distributed heat in rooms, either occupied or unoccupied.

The overall efficiency is then:

$$\eta = \eta_{BS} \times \eta_d \times \eta_e \times \eta_c$$

A typical profile for a conventional house is shown in Figure 5.39 (Note that the overall seasonal efficiency is about 60%).

Values for the various efficiencies are to be found in Table 5.5. The seasonal efficiency can be estimated with the aid of Figure 5.40 or the formula:

$$\eta_{BS} = \eta_B \times \frac{1 - q/f}{1 - q}$$

where:

η_{BS} = seasonal efficiency of the boiler
η_B = nominal boiler efficiency
q = heat loss factor of the boiler
f = load factor (working time as a fraction of total time)

Example: a boiler with $\eta_B = 0.85$ (commercially indicated value)

f = 0.3
q = 0.06
η_{BS} = 0.85 x η_B = 0.85 = 72%

When lower temperatures are used q decreases, load factor f increases, hBS will therefore increase.

The seasonal efficiency of the auxiliary heating system can be crucial in a passive solar house. A low seasonal efficiency will result from an oversized system (low f value), though different fuels are associated with more or less serious efficiency penalties.

The auxiliary heating system of a passive solar building is sized conventionally, using winter design conditions without solar gains.

It is important to note that oversizing the production unit has practically always a negative influence on the total energy consumption.

Note that while electric heating may appear to be highly efficient in terms of 'delivered' energy, this does not take account of generating and transmission losses which can reduce its true efficiency to around 35%. More rigorous comparisons are therefore often made on the basis of 'primary' energy use.

Guidelines:

Particular attention should be given to the heat losses of every part of the auxiliary heating system in order to increase seasonal efficiency.

The production unit::
- place the chimney centrally within the building if possible, and where practicable and safe, use flue gas heat recovery
- insulate the boiler well or

η_{BS} production:

Values depend on the load:
- boiler for central heating - see Figure 7.54
- electric heating - 0.90 to 0.95.

η_d distribution:

Values of 0.68 to 0.98 can be found.

η_e emission:

- radiators	0.9 to 0.95	well insulated room
- warm air or ventilo	0.9 to 0.95	well insulated room
		well insulated room
- convectors	0.8	poorly insulated room
- floor heating }		depending on emitter
ceiling heating }	0.75 to 1	and insulation level of room

η_c control (optimistic values):

a) centralized controls referring to external temperature to with thermostatic valves and radiators to rooms: ~1
b) as a) but with floor heating: ~0.93
c) as a) but without thermostatic valves: ~0.93
d) centralized but manually operated controls with thermostatic valves and radiators: ~0.91
e) as d) but with floor heating: ~0.89
f) as d) without thermostatic valves but with 1 central room thermostat: ~0.87
g) as d) but with floor heating: ~0.87
h) as f) but with floor heating: ~0.83

Table 5.5: Typical values of seasonal efficiencies for various systems [26].

make direct use of the heat
- work at as low a temperature as possible

The distribution system:
- the shortest possible distribution system
- fast-reacting system with local controls in each room if possible.

Heat Emission and Comfort

Heat is emitted by either convection or radiation from terminal units.

The proportional split between these two modes is different from one heating system to another. The systems may be classified as follows (going from the most convective to the least convective): forced air-convector; radiator; floor heating; ceiling heating.

The heat emission system is installed in a room in order to provide the required thermal comfort by combating the convective or radiative heat losses of the building envelope (draughts, cold surfaces). Different systems will create different environments even if the thermostats are set at the same value and placed at the same position within the room. The performance of a heating system in terms of comfort conditions is influenced by the level of insulation of the room.

Research has shown that for a poorly insulated external wall (single glazed for example), only radiators below the windows of that wall, or air heating systems with outlets under the windows can compensate for the discomfort of such walls. The better insulated the room, the less important it is which system is chosen for the provision of given comfort conditions. This means, for example, that for well-insulated rooms with a low infiltration rate, either a radiator on a side wall or a hot air system with outlet on the opposite wall, with an appropriate air injection rate (greater than 5 air changes per hour), will provide adequate comfort. But the nature of the conditions provided by these two systems can vary widely in terms of both air and environmental temperatures.

Figure 5.41 shows an example of the vertical temperature distribution in a room with different heating systems. Figure 5.42 shows the horizontal distribution of environmental temperatures in the same room taken at a level of 0.7 m above the floor. The thermal characteristics of the room are significant.

It is important to note that the characteristics of the emitters can easily be changed by simple errors. Figure 5.43 shows how curtains hung in the wrong position can drastically affect the energy consumption of a heating system.

Integration of Building and Heating System

There seems to be a tendency to return to separate room heaters with local controls for passive solar buildings. The decision to use a central production unit with a complex distribution system is often related to its possible integration with hybrid or active solar systems, but the associated installation costs are frequently prohibitive except when retrofitting a building which has an existing system that can be easily incorporated. When re-using an existing system, oversizing problems may arise as a result of thermal modification to the building.

The possibility of coupling the house design, the auxiliary heating system, and eventually heat recovery systems, must be an important consideration in passive solar designs.

Dynamic Behaviour and Control of Auxiliary Heating Systems

The dynamic behaviour of the different parts of the heating system is of particular importance. The response time of the control system must be less than the response time of the emitters, which in turn must be less than the response time of the room. This dynamic relationship allows efficient response of the heating system to both solar gains and intermittent occupancy of the rooms.

Table 5.7 lists some currently-encountered time constant of rooms, heating systems and controls.

A good control system will react quickly to variations of internal loads or to the size of solar radiation gains in passive solar houses and so preserve thermal comfort. Position

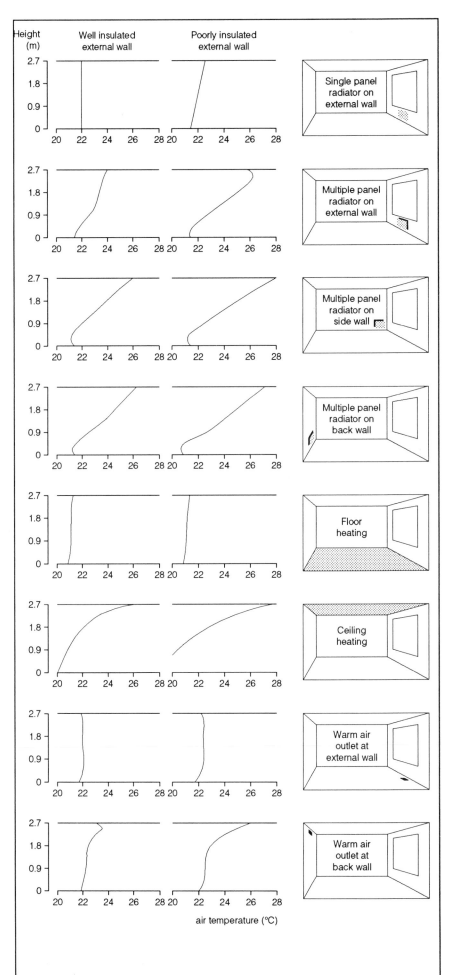

Figure 5.41. Vertical air temperature profiles at the centre of a room with different types of emission systems [29].

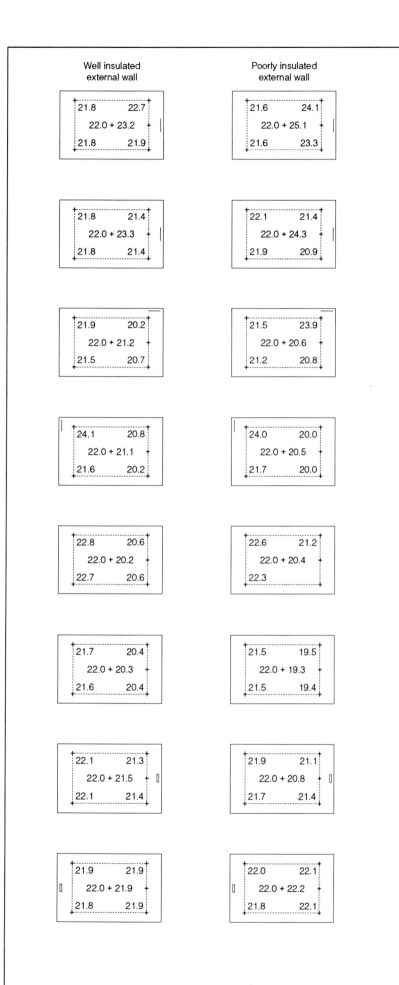

Well insulated external wall	Poorly insulated external wall
21.8 22.7 22.0 + 23.2 21.8 21.9	21.6 24.1 22.0 + 25.1 21.6 23.3
21.8 21.4 22.0 + 23.3 21.8 21.4	22.1 21.4 22.0 + 24.3 21.9 20.9
21.9 20.2 22.0 + 21.2 21.5 20.7	21.5 23.9 22.0 + 20.6 21.2 20.8
24.1 20.8 22.0 + 21.1 21.6 20.2	24.0 20.0 22.0 + 20.5 21.7 20.0
22.8 20.6 22.0 + 20.2 22.7 20.6	22.6 21.2 22.0 + 20.4 22.3
21.7 20.4 22.0 + 20.3 21.6 20.4	21.5 19.5 22.0 + 19.3 21.5 19.4
22.1 21.3 22.0 + 21.5 22.1 21.4	21.9 21.1 22.0 + 20.8 21.7 21.4
21.9 21.9 22.0 + 21.9 21.8 21.9	22.0 22.1 22.0 + 22.2 21.8 22.1

Figure 5.42. Resultant environmental temperature distribution within the same room as in Figure 5.41, with different forms of emission systems. Temperatures are measured at a height of 0.7m above floor level. [30]

and construction of control units must relate to these problems.

In general, the control system must be able to detect and react to at least the room air temperature. For systems with longer reaction times and/or those which are more centralized (central heating systems), a monitoring capability of both outside air temperature and solar energy would be advantageous. Modern technology, in the form of sensors and micro-processors ,offers greater possibilities in sensing and reacting to external climatic conditions. A classic example of a system which does not interact well with passive solar such as night storage radiators, could well be more interesting if controlled by more accurate meteorological forecasting (one or more days in advance).

User Education

Heating systems can be complex to operate and the entire benefit of a low energy consumption system design can be lost by improper use. Even the relatively simple controls available with commonly-used central heating systems can lead to enormous variations in energy consumed, when combined with differing user requirements (use of bedrooms etc. for additional living space), and available day-to-day finance. Surveys of identical houses with identical heating systems have shown variations by as much as four to one in fuel consumption (UK Building Research Establishment). These variations are strongly influenced by differences in ventilation rates. It is apparent that user education and follow-up are therefore very important.

Combining Systems

Often the most appropriate strategy for solar heating is to use within the building a combination of the systems previously discussed. This approach enables the various components to complement each other and improve the overall performance of the buildings.

For example, a direct gain system typically responds very quickly to solar input; that is, a direct gain building will heat up rapidly when solar energy enters the space. In contrast, a mass wall absorbs solar radiation for several hours before it begins to heat the living space, and a well designed mass wall will continue to supply usable heat long after sunset. A combination of both a direct gain system and a mass wall, therefore, in one building will, in certain climates, supply relatively constant heat throughout the day. The direct gain space will supply the building with heat during daylight hours, while the mass wall will reach its peak output during the evening. Again, both systems should be properly sized to complement one another according to the climate zone and heating requirement. In this way heat is supplied to the space during those hours when it is needed and does not overheat the space during the day.

Heating and cooling systems can also be integrated in climates where this is necessary, although some types can be made to perform either way: typically, for example,

roof ponds can be used for radiative heating or night-sky cooling, while Trombe walls can produce convective or radiative heat and induced ventilation. It seems likely that any heating system using thermosiphonic circulation to charge a thermal mass could be used to some degree to provide cooling, if the collector is covered by day and the night-time dampers left open, to allow a reverse cycle.

Comparative monitoring of the performance of combined active and passive systems suggests that in certain climates these can work well together, with the active system able to pick up a significant fraction of the space heating load 'left over' after the passive system. The effect, however, of the passive system is to shorten the heating season to just the mid-winter months when it is difficult for the active system to contribute significantly.

REFERENCES

[1] Solar Energy - A UK assessment, Page, J.K., UK-ISES, 1976.

[2] Documents of Second European Passive Solar Competition, 1982, Lebens, R., published in "Passive Solar Architecture in Europe 2, The Architectural Press, London, UK, 1983.

[3] Passive Solar Design Handbook (Volumes I, II, and III), Balcomb, J.D., National Technical Information Service, 5285 Port Royal Road, Springfield, VA, 22161, USA. 1980 and 1982.

[4] Solar Energy Technology Handbook Part A, Engineering Fundamentals, Dickinson, W., Cheremisinoff, Marcel Dekker, 1980.

[5] B.B.R.I. Measurements of the Thermal Performance of Windows, Caluwaerts, P., International Energy Agency Conference, Berlin, FRG, April 1981.

[6] BINE Projekt Info-Service, Leaflet No. 2 März 1990, BINE, Mechenstrasse 57, 5300 Bonn 1, FRG, ISSN 0937-8367

[7] BINE Projekt Info-Service, Leaflet No. 3 April 1990, BINE, Mechenstrasse 57, 5300 Bonn 1, FRG, ISSN 0937-8367

[8] Patented System by Zomeworks Corp., Beadwall (registered trademark), P.O. Box 712, Albuquerque, New Mexico 87103, USA.

[9] Natural Solar Architecture, Wright, D., van Nostrand Reinhold Company, 1978.

[10] Movable Insulation, Langdon, W.K., Rodale Press, 1980.

[11] Performance of Night Insulation and Selective Absorber Coatings in LASL Test Cells, Hyde, J.C., Proceedings of the 5th National Passive Solar Conference, Amherst, UISA, 1980.

[12] A Classification Scheme for the Common Passive and Hybrid Heating and Cooling Systems, Holtz, M.J., Place, W., Proceedings of the 3rd National Passive Solar Conference, San Jose, California, USA, 1979.

[13] Passive Solar Component using Rotating Blades with Latent Heat Storage, Achard, P., Mayer, D., White, M.D., Proceedings of the IEA Workshop on Latent Heat Stores - Technology and Applications, Stuttgart, 7-9 March 1984, V. Lottner editor, Projektleitung Energieforschung, 1985.

[14] Buildings, Climate and Energy, Markus, T.A., Morris, E.N., Pitman, 1980.

[15] Langtidstest genom cykling av ett saltsmaltlager med glaubersalt som energilagringsmaterial, Hetenyi, J., Lagerkvist, K.O., Teknisk rapport SP-RAPP, 1981:05, Statens Provningsanstalt, VVS-teknik, Boras, Sweden, 1981.

[16] Heat-Pac Brochure, Climator ab, Box 34, S-54500 Toreboda, Sweden.

[17] LASL Similarity Studies: Part 1, Hotzone/Coldzone: A Quantitative Study of Natural Heat Distribution Mechanisms in Passive Solar Buildings, Wray, W.O., Weber, D.D., Proceedings of the 4th National Passive Solar Energy Conference, Kansas City, USA, 1979.

Troom		
- Light construction	:	4 hrs
- Normal construction	:	9 to 12 hrs
- Heavy construction	:	16 to 26 hrs
Theating system		
- Radiator	:	2 to 10 minutes
- Electric convector	:	1 to 5 minutes
- Floor heating	:	2.5 h
- with carpet	:	3.3 h
Tcontrol		
- Thermostatic valves	:	10 to 20 minutes

N.B. Total reaction time ~ time constant x 3
1 h is the time taken for the air to change temperature for all building weights.

Table 5.7: Typical time constants (time after which 63% of the final value of an energy step is obtained) of rooms, heating systems and controls. See also the section on passive solar control systems.

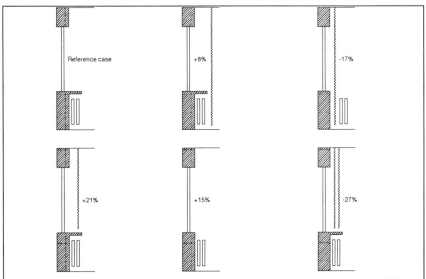

Figure 5.43: Influence of the window sill and curtains on the energy losses through a window. [31]

[18] Passive Solar Heating Design, Lebens, R., Applied Science Publishers, 1980.

[19] Measurement of Long Term Interzone Airflows, Walker, J.H., Littler, J.G.F., CEC Final Report for DGXII, Contract No: EN3S.0037.UK(H).

[20] Natural Convection Research and Solar Building Applications, Anderson, R., Passive Solar Journal 3(1), 1986.

[21] Inter-zone Convective Heat Transfer in Buildings: A Review, Bakarat, S.A., ASME Journal of Solar Energy 109, 1987.

[22] Natural Convection Through Rectangular Openings in Partitions, Part 1 - Vertical Partitions, Brown, W.G., Solvason, K.R., Int J. Heat and Mass Transfer 5, 1962.

[23] Heat and Mass Transfer by Natural Convection and Combined Natural Convection and Forced Air Flow Through Large Rectangular Openings in a Vertical Partition, Shaw, B.H., Instn. Mechanical Engineers, 1971.

[24] Natural Convection Airflow and Heat Transport in Buildings: Experimental Results, Balcomb, J.D., Jones, G.F., 10th National Passive Solar Conference, Raleigh, 1985.

[25] Analysis and Measurements of Interzonal Natural Convection Heat Transfer in Buildings, Hill, D., Kirkpatrick, A., Burns, P., ASME Journal of Solar Energy Engineering 108, 1986.

[26] Rendements Energétiques des Systèmes de Chauffage Brochure technique - excercices d'intégration. Hannay, J., Dols, J.M., SPPS, Bruxelles, Belgium. 1982.

[27] L'Economie d'Energie dans les Habitations, Uyttenbroek, J. et al., C.S.T.C., revue N° 4, Bruxelles, Belgium, 1979.

[28] Rendement d'Exploitation des Chaudières de Chauffage Central pour Maisons Individuelles, Guillaume, M., B.B.R.I., October, 1982.

[29] Influence of Heating System on Thermal Comfort and Energy Consumption in Rooms, Caluwaerts, Marret, B.B.R.I., XXI International Congress for Building Services Engineering, Berlin, FRG, 1980.

[30] Heat Losses of Buildings with Different Heating Systems, Lebrun, J., Marret, D., University of Liege, Belgium, ASHRAE Journal 1979.

[31] Energiebesparing dor gebruik van gordijnen, Bossers, P.A., Dubbeld, M., Maaskant, P., TNO-IMG, Delft, Netherlands, Report N° C.442, 1978.

[32] BINE Construction Information Leaflet, Verlag TÜV Rheinland GmbH, Köln, ISBN 3-88585-799-5, 1991.

NATURAL COOLING

6.1

INTRODUCTION

Strictly defined, the term passive cooling applies only to those processes of heat dissipation that will occur naturally, that is without the mediation of mechanical components or energy inputs. The definition encompasses situations where the coupling of spaces and building elements to ambient heat sinks (air, sky, earth, water) by means of natural modes of heat transfer leads to an appreciable cooling effect indoors. However, before taking measures to dissipate unwanted heat, it is prudent to consider how the build-up of unwanted heat can be prevented or limited in the first place. In this context we have chosen to consider natural cooling in a somewhat wider sense than the strict definition above suggests, to include preventive measures for controlling cooling loads as well as the possibility of mechanically assisted (hybrid) heat transfer to enhance the natural process of passive cooling. [1]

As with the design of passive solar buildings where heating is the main concern, the implementation of passive cooling techniques is a multi-layered process inextricably linked with the architectural design of the building in its environment and the anticipated patterns of use. The processes involved in passive cooling are fundamentally linked to those of the climate and the earth's daily energy exchange. Similarly, the body's comfort tolerances will influence the choice of cooling techniques used in different circumstances. Modification of the microclimate around a building can help to improve comfort conditions in and around the building, while reducing cooling loads. This can be achieved by lowering outdoor temperatures through solar protection, wind channelling, evapotranspiration of plants and the evaporation of water.

In the last three years the use of mechanical air conditioners in Greece for example, has increased by 900 per cent, and there have been similar increases in other southern European countries [2]. Extensive use of such air conditioners can cause serious problems in electricity supply and atmospheric pollution. Recent comparative studies of indoor air quality in air-conditioned and naturally ventilated office buildings have shown that illness indices are higher in air-conditioned buildings [68] [69]. In addition, greater use of air conditioners using CFC-based refrigerant fluids can increase damage to the environment by depleting the ozone layer if these fluids are lost to the atmosphere during manufacture, repair or through mechanical failure of the air conditioner. There has been a trend in recent years to use refrigerant fluids which have a less damaging effect on the environment, and development of improved fluids continues.

There are many options available to the designer for cooling domestic and non-domestic buildings which can help to avoid the use of mechanical air conditioning while achieving comparable comfort levels with much lower energy use and consequent savings in atmospheric pollution. Some of these options are totally passive, requiring no mechanical assistance, while others may need a relatively small input from mechanical devices such as fans or pumps. A considerable amount of development in such hybrid systems in Europe continues.

The range of climates with which we are mainly concerned in southern Europe (roughly between latitudes 35°N and 45°N) includes those of Portugal, Spain, southern France, Greece and much of Italy. The Mediterranean Sea has a major influence on most of these climates, while the Atlantic Ocean influences the climates of Portugal and parts of Spain.

Energy needs for cooling can be two or three times those for heating on an annual basis, although this position tends to reverse towards the north.

However we should note that natural cooling is of relevance in all our regions, given that overglazed offices for instance can result in overheating even the most northerly locations.

A useful design strategy for the overheating season is to first control the amount of heat from solar radiation and heated air reaching the building, then to minimise the effect of unwanted solar heat within the building skin or at openings, next to reduce internal or casual heat gains from appliances and occupants and finally where necessary to use environmental heat sinks to absorb any remaining unwanted heat. Examples of the latter include natural ventilation, ground cooling, evaporative cooling and radiative cooling. This chapter is structured accordingly, although in practice a combination of these cooling techniques is almost invariably in operation.

6.2

CONCEPTS

Overheating

The definition of overheating depends on a number of factors: but for the purpose of this Chapter we can assume overheating to be in excess of 27°C resultant temperature under still air conditions with a relative humidity of around 50% for lightly clothed, moderately-active occupants. The effect of a variation in one or more of these conditions can be assessed using a bioclimatic chart or other appropriate comfort scale. (see Chapter 4 - 'Thermal Comfort')

The effect of direct exposure of the body to solar radiation can be assessed in terms of an equivalent rise in temperature as a function of the level of irradiance and the absorbance of the body's skin and clothing. Variations in clothing and physical activity and the degree of acclimatisation will also have an effect on thermal comfort.

The effect of heat gains on the mean indoor temperature can be roughly assessed by:

$$T_{mean} = T_o + G/HLC$$

where:

T_{mean}	=	mean indoor temperature (°C)
T_o	=	mean outdoor temperature (°C)
G	=	heat gain in watts
HLC	=	global heat loss coefficient in watts/K

As T_{mean} approaches the threshold conditions the expression suggests that heat gain should be minimised and/or that heat losses should be maximised. This provides the basis of the design strategies outlined later in this chapter.

Swings in T_{mean} and the timing of temperature peaks are strongly affected by the thermal storage capacity of the building. Depending on the thermal inertia of the building, heat discharge from storage within the building fabric can add substantial cooling loads in the latter part of the day. However a range of measures, shown later in this chapter, can be used to reduce the extent and to control the timing of these peaks.

Generally, the need for cooling depends less on climate and more on building type, occupancy patterns and especially, on building design. In the south of Europe, where mean outdoor temperatures will tend to approach or exceed the comfort temperature threshold from June to August, cooling loads as a function of exposure to solar gain and the amount of internal gain can be particularly high. Outside the Mediterranean region, mean outdoor temperatures are very rarely close to the overheating threshold and the nature and length of the cooling season is almost entirely dependent on building type and design.

Solar Control

Direct solar radiation can be prevented from reaching all or part of the walls, roof or windows of a building by the use of shading. Shading can be provided by natural vegetation, neighbouring buildings or the surrounding landscape - for example on the north facing slopes of hills and valleys. Shading devices on the building (fixed or movable, the latter being manually or automatically controlled) can prevent direct radiation reaching critical parts such as windows, doors and even roofs. Indirect solar gain from the sky, or reflected from surrounding buildings or the ground and air heated by irradiated surfaces such as roads and pavements can also contribute significantly to the cooling load. The 'heat-island effect' which occurs in cities where a large concentration of heat is emitted from buildings can also cause appreciable cooling loads.

External Gain Control

Solar radiation reaching the building will cause a gain to be transferred through the building skin by conduction and may lead to an unwanted increase in internal temperatures unless measures to prevent this are adopted. Under these conditions, heat will be radiated to cooler objects within the building, transferred to the internal air by convection, and to other internal elements such as floors and ceilings by conduction. (See Chapter 4 - 'Thermal Comfort' and Appendix 9 'Fundamentals of Heat Transfer')

Internal Gain Control

A significant amount of heat is produced by appliances, electric lighting and occupants which, during the overheating season, can lead to uncomfortably high internal temperatures unless action is taken. The use of natural daylight (see Chapter 7) to replace artificial light where appropriate and the use of high efficiency artificial lighting where necessary can reduce cooling loads appreciably, especially in commercial buildings (which tend to have a low surface-to-volume ratio and thus retain more heat). High efficiency domestic appliances and office equipment and their appropriate location within the building to allow the heat given off to be expelled easily can also reduce cooling loads. The heat produced by occupants can also be significant in crowded spaces. When seated, an average adult produces about 110 watts, but this can rise to about 800 watts when playing basketball for example. Occupation densities and types of activities therefore need to be considered. (see Chapter 4 on Thermal Comfort).

Thermal Inertia

The thermal inertia of the building fabric may be used to reduce heat flow to the interior of the building. This is especially useful where heavyweight construction is used, as is usually the case in Europe. Materials with a high heat storage capacity, such as concrete and brick, heat up and cool down relatively slowly. When solar radiation strikes an opaque or solid surface such as a wall or roof the exterior surface absorbs part of the radiation and converts it to heat. Part of the heat is directly re-emitted to the outside. The remainder is conducted through the wall or roof at a rate which depends on the thermal diffusion characteristics of the material(s). When the temperature of the exterior surface drops due to a fall in ambient temperature, part of the stored heat is emitted outside. At night, the air temperature inside the building is usually higher than the temperature outside. The heat flow is therefore to the outside and the temperature of the wall or roof continues to decrease, thus eventually cooling the interior. The contribution of thermal inertia to natural cooling is particularly useful where there are significant daily variations in external temperatures - in hot, dry climates or mountainous regions of southern Europe, for instance.

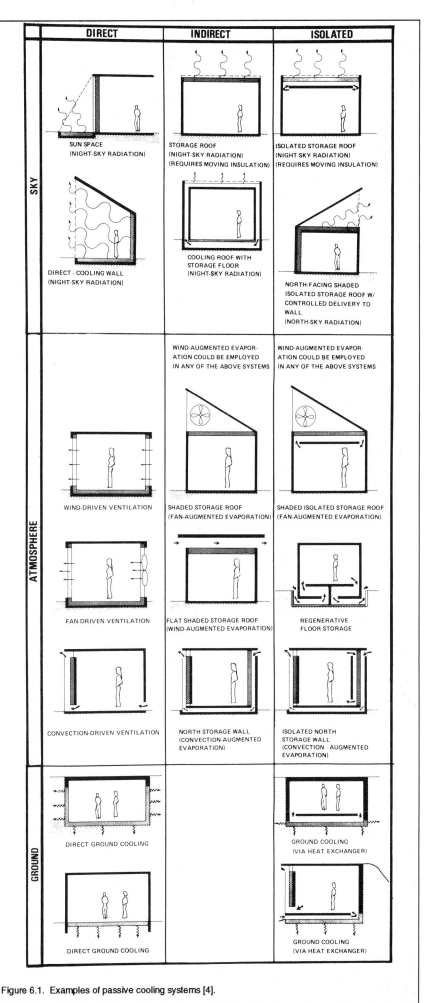

Figure 6.1. Examples of passive cooling systems [4].

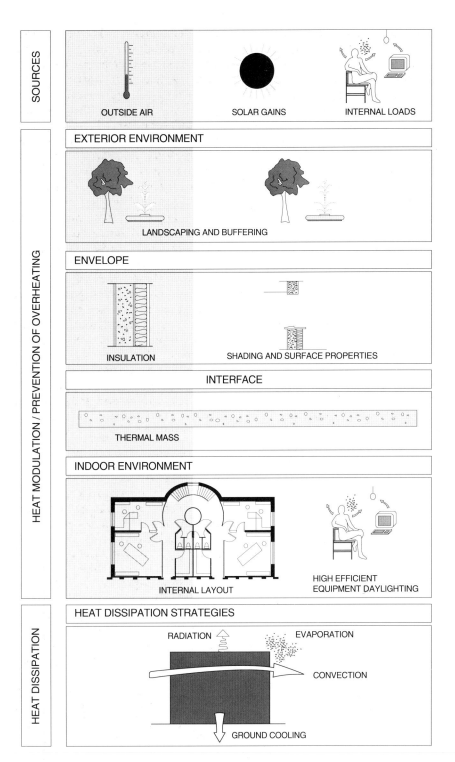

Figure 6.2. Heat sources and methods of preventing overheating [1]

Environmental Heat Sinks

If the methods of avoiding unwanted solar or casual heat gains have been applied and temperatures within the building are still predicted to be at times too high, then the surrounding air and ground can often be used to provide cooling.

Ventilation

Natural ventilation can produce a significant cooling effect, depending on the configuration of the building on the site and the surrounding spaces, the direction and strength of wind flows and the time of day. The layout of internal spaces in plan and section according to function is important, particularly for air movement indoors and the potential for cross ventilation. To be effective, ventilation air should be cooler than internal air, but air movement past an object also creates a chilling effect. Ventilation air passed through tunnels or pipes below ground can be appreciably cooler than the air surrounding buildings.

Ground Cooling

Increasing the building's contact with the ground can provide additional cooling. Throughout the year temperatures below the surface are more constant than air temperatures, varying negligibly several metres deep where they are significantly cooler than surface or air temperatures in summer. The ground can therefore be considered as an almost infinite heat sink. Examples are 'earth sheltered' buildings (built partially underground) and the use of 'berms', or built up earth, in contact with perimeter walls. These techniques need to be considered carefully to avoid problems, especially in winter, including damp penetration, condensation and insufficient daylight.

Evaporative Cooling

Evaporation occurs whenever the vapour pressure of water, in the form of droplets or a wetted surface, is higher than the partial pressure of the water vapour in the adjacent atmosphere. The phase change of water from liquid to vapour is accompanied by the release of a quantity of sensible heat from the surrounding air; in direct evaporative cooling this lowers the dry bulb temperature of the air while increasing its moisture content.

When evaporation takes place on the internal surface of a sealed container such as a tube the surface temperature decreases. Adjacent air outside the container is also cooled but without any increase in this moisture content. This process is called indirect evaporative cooling.

Both direct and indirect processes can take place passively, using only elements of the building envelope; or they may be mechanically-assisted 'hybrid' systems.

Radiative Cooling

Heat will be lost by electromagnetic radiation from a warmer body to a cooler one. This can be of practical use by allowing heat built-up in the building fabric during the day to be lost by radiation from the external surfaces of the building to the night sky.

If the building envelope (walls, roof, etc.) is of sufficient mass, it will absorb solar heat during the day without causing any significant temperature increase on the internal surfaces of the envelope or in the internal spaces. Heat accumulated during the day, from solar radiation , can be lost to the night sky by radiation and to the usually cooler night air by convection. Normally the required mass is provided by heavyweight construction materials such as concrete or masonry, by water contained in roof ponds, or in 'water walls'. Few examples of roof ponds exist in Europe. There are more in the US, where climatic conditions are more suitable.

6.3

HEAT SOURCES

External

Short Wave Radiation
Uncontrolled solar radiation transmitted indoors through glazing is a major source of heat gain, particularly from May to September in all parts of Europe. Its transmission through opaque elements is mainly of concern in the case of uninsulated buildings in the Mediterranean region.

Outdoor Air Temperature

Conduction of heat through the building fabric and convective transfers due to ventilation and air infiltration, while of negligible importance in terms of cooling loads in northern Europe, can be a major source of heat gain in the south. Thermal insulation and weatherstripping can substantially decrease these inputs.

Internal

There are three main categories:

Metabolic heat produced by occupants which can be a major heat source in densely occupied buildings.

Artificial lighting, especially if it is of low efficiency, can also contribute to heat gains which can be considerable in commercial buildings

Appliances and equipment; more energy-efficient domestic appliances and commercial equipment are now becoming available, but heat gain can still be significant especially in commercial buildings (computer rooms, large kitchens, etc.).

Occupancy Patterns

Under free-running conditions, the elevation of indoor temperature due to occupancy and appliances may represent a threshold value for overheating and is therefore an important criterion in the consideration of heat gain prevention and heat dissipation strategies.

The magnitude and composition of internal loads tends to vary considerably with occupancy density and the type of heat-generating appliances used. With domestic buildings, internal sensible gains tend to be a problem mainly in kitchens, while in commercial buildings such gains are more widely distributed within a building.
Detailed occupancy patterns and hourly internal gain profiles are routinely used as inputs to simulation studies. These are commonly based on rough assumptions of occupancy and appliance use rather than on detailed modelling of these parameters.

Behavioural Aspects

Field studies and occupant surveys have provided information on
- occupant expectations with respect to thermal and visual

environmental conditions and the kind of daily or seasonal adjustments people tend to make;
- the adjustments that occupants are prepared to make in response to particular design strategies that demand some form of user response or participation; and the reaction of occupants to specific design measures that are incorporated for the purpose of energy conservation. (see Chapter 4: Thermal Comfort).

Such studies have been mainly in northern regions, and there is less information on variations across Europe. For example, the tradition of outdoor sleeping in some regions; the influence of the midday siesta where it is still practiced; changes in house usage between winter and summer; migration to a summer house; use of separate daytime and nighttime spaces and acclimatisation between winter and summer are behavioural aspects which can affect comfort.

Management and Control of Internal Loads

Two aspects merit consideration:
- maximising daylighting (in conjunction with protection from solar heat gains) as a means of reducing heat gains from artificial lighting;
- development and use of high efficiency appliances. There is a trend toward more energy-efficient appliances. However, this also parallels a trend toward greater ownership of household appliances, as well as an increasing use of heat-emitting equipment in offices.

Field studies have shown that heat losses through boiler casings and hot water cylinders can represent a considerable proportion of the internal heat gains in domestic buildings. Additional insulation on these appliances can help improve their efficiency as well as reduce internal heat gains in summer.

The concentration of sources of internal loads in particular spaces in a building and the relationship of such spaces to potential heat sinks are important considerations which have a bearing both on the management of internal loads and on the application of heat dissipation techniques.

Figure 6.3. Hedges used as a major element in landscape design - source Alex Lohr, Köln.

CONTROL OF HEAT GAIN

Solar control involves the prevention of unwanted solar gain taking into consideration the following:

- Microclimate and site design
- building form and external finishes
- building envelope
- shading
- thermal insulation
- control of internal gains.

Microclimate and Site Design

Siting and Site Layout

Siting and site layout, followed by landscaping can improve the microclimate around a building, taking advantage of existing topographical features, adjacent buildings and vegetation for solar protection and local breezes or the presence of water or vegetation for natural cooling.

Good site layout can reduce cooling loads appreciably by optimizing natural solar protection and local breezes. For example, streets with an east-west orientation can help to reduce summer solar gain on south facades. Staggered layouts can enhance natural ventilation. Many traditional site layouts in southern Europe exemplify these concerns, often providing good comfort conditions within buildings as well as in courtyards, streets, parks and gardens.

Landscaping

Landscaping can improve the microclimate both in winter and summer, providing shading, evaporative cooling and wind channelling in summer, or shelter in winter. Vegetation absorbs large amounts of solar radiation helping to keep the air and ground beneath cool, while evapotranspiration can further reduce temperatures.

Trees, shrubs and vines can provide protection from solar radiation in summer and deciduous trees can allow greater solar access in winter. Solar control and wind channelling may help to create attractive spaces for outdoor activities and fast growing varieties can often be used to provide these conditions quickly in new developments. Some care in the choice and placement of vegetation

on or near buildings should, however, be taken to avoid structural damage. [5]

Grass and other ground-cover planting can also influence the microclimate, keeping ground temperatures lower than most hard surfaces as a result of evapo-transpiration and their ability to reduce the impact of solar radiation. Robinette [6] has reported surface temperatures of 38°C for grass, compared to 61°C for asphalt and 73°C for artificial turf. Rizvi and Talib [7] have presented a review of selected plant species for energy conservation. A detailed sample of landscaping materials and their characteristics is given in [8].

Windbreaks can enhance air-pressure differences around buildings and can improve cross ventilation. Hedging, for example, can allow a gentle breeze to filter through the foliage, while a masonry windbreak can create a calm, sheltered zone behind it. Gaps in windbreaks, openings between buildings or openings between the ground and a canopy of trees can create wind channels, increasing wind speeds by about 20%.

Other landscape techniques include the use of ponds, streams, fountains, sprays and cascades for evaporation where water is available in summer. These are particularly effective in dry conditions where relative humidities are low. According to Yellot [9], under mean conditions of wind, dry-bulb and wet-bulb temperatures, the energy released by 1m² of open water surface is almost 200 watts.

Performance data

Bowen [10] has reported reductions of 2°C to 3°C due to plant evapotranspiration. Duckworth and Sandbery [11] found temperatures in the heavily vegetated San Fransisco Golden Gate park to be around 8°C cooler than nearby, less vegetated areas. Gold [12] has reported that almost half of the total available spring and summer solar radiation can be dissipated by natural evaporative cooling. Theoretical analyses suggest that evapo-transpiration from one tree can save between 1 and 2.4 MJ of electricity in air conditioning per year [13]. Moffat [14] has reported that an average full size tree evaporates 1460 kg of water on a sunny day, which is equivalent to 870 MJ of cooling capacity. He also stated that the latent heat transfer from wet grass can result in temperatures which are 6°C to 8°C cooler than exposed soil and that one hectare of grass can transfer more than 120 GJ per day. Finally, Labs [15] has reported that shading by vegetation and other surface materials can provide 1 to 2 months timelag between surface temperatures and the earth in contact with underground structures.

Landscaping should be integrated in the design and construction phases of the building project for maximum effectiveness. Consideration should be given to maintenance costs, the time taken for plant growth, the seasonal availability of water, planning constraints and the appropriate selection of plant species and other landscape features for the site in question.

Building Form and External Finishes

The configuration of the building and the arrangement of internal spaces according to function can help to influence the exposure to incident solar radiation, the availability of natural daylight, and air flows within and around the building.

Building Shape

In general a compact building will have a relatively small exposed surface, or in other words a low 'surface-to-volume ratio'. This can offer advantages for the control of both heat losses and heat gains through the building skin without any conflict between design priorities for winter or summer conditions.

In practice a whole range of matters in addition to thermal performance will influence the designer's determination of building form. Multi-storey office or apartment blocks, for example, arranged in rows of varying length along main roads, are common in southern Europe where aspects such as orientation, siting and building form are largely outside the control of the designer. Where the designer can determine the basic building form, then a range of design options to improve thermal performance becomes available including compact design, courtyards, construction on pilotis, the use of wing walls and maximization of roof spaces for radiative cooling.

The relationship between building form and solar/thermal transmission is not however critical, as there are many strategies for counteracting the negative effects of form in the design of the building skin. More important are the effects of building form on wind channelling and air flow patterns, and the opportunities for enhancing the use of natural daylight.

Internal Layout

Thermal zoning may be used as a buffering strategy in the arrangement and use of internal spaces and to improve cross ventilation [16]. Deep plan arrangements and subdivisions into many layers of spaces require careful design as they can inhibit air movement within the building.

Intermediate Spaces

Intermediate spaces such as conservatories which may be incorporated as part of a heating strategy can add to the cooling loads of the parent building and may require effective heat dissipation techniques in summer. The interface between such intermediate spaces and the parent building is usually of critical importance.

Unglazed intermediate spaces such as verandas, arcades or courtyards can create their own microclimates and have a role in wind channelling and solar protection for adjacent openings and surfaces [17].

Building Envelope

The building envelope must respond to a range of climatic conditions including extremes of temperature, solar radiation, precipitation and wind. The envelope may at different times be required to act as a buffer, a filter or a concentrator. Thermal and solar transmission can be controlled by appropriate selection and manipulation of surface materials.

Compromises may be necessary between the needs for daylighting, air flow, solar protection and of building operation, with the overall objective of minimizing unwanted heat gains through the envelope, particularly through glazing.

Design strategies for the building envelope include:

- Design of openings
- Solar control and shading systems
- Thermal insulation
- Thermal mass
- Airtightness

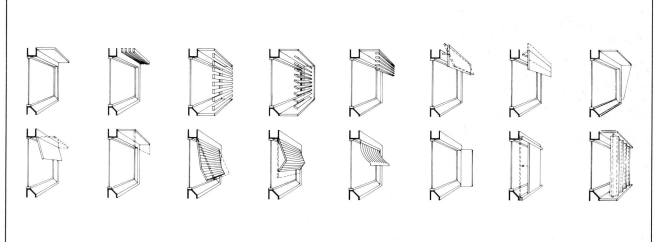

Figure 6.4: Typical external shading devices.

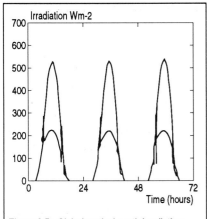

Figure 6.5: Global vertical south irradiation measured under and above the awning.

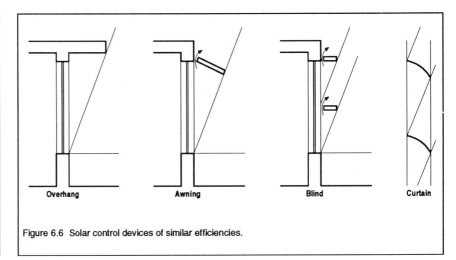

Overhang   Awning   Blind   Curtain

Figure 6.6 Solar control devices of similar efficiencies.

Position of shading and type of sun protection		Solar gain factors (*)	
		Types of glazing	
Shading	Type of sun protection	Single	Double
	None	0.76	0.64
	Lightly heat absorbing glass	0.51	0.38
None	Densely heat absorbing glass	0.39	0.25
	Lacquer coated glass, grey	0.56	-
	Heat reflecting glass, gold		
	(sealed unit when double)	0.26	0.25
	Dark green open weave plastic blind	0.62	0.56
Internal	White venetian blind	0.46	0.46
	White cotton curtain	0.41	0.40
	Cream holland linen blind	0.30	0.33
Mid-pane	White venetian blind	-	0.28
	Dark green open weave plastic blind	0.22	0.17
External	Canvas roller blind	0.14	0.11
	White louvered sunbreaker, blades at 45°	0.14	0.11

(*)   The solar gain factor of a transparent material is the fraction of incident solar energy passing through the material.

Table 6.1: Solar gain factors for various types of glazings and shading (strictly for UK only, approximately correct correct world-wide)

Material	% Direct Transmittance	% Diffuse Transmittance
Canvas	0	0
Plastic	25	15
Aluminium (separated slats)	0	20

Table 6.2: Solar Transmittance of awning materials

Colour	% Transmitted	% Reflected	% Absorbed
Light-colour, translucent	25	60	15
White, opaque	0	80	20
Dark, opaque	0	12	88

Table 6.3: Solar radiation transmission through external shading.

Design of Openings

The balance between heating, cooling and daylighting is a critical consideration for the choice of orientation and sizing of openings. The design of openings will usually depend on the building type and may be influenced by building regulations, particularly with regard to maximum or minimum sizes for glazed areas. The use of additional devices, such as overhangs and shutters, may allow the designer some scope to correct or limit unfavourable orientations or large glazed areas.

The building envelope design strategy must encompass winter and summer conditions so that, for example, excessive solar gains in summer can be controlled while adequate daylight is available throughout the year, thus avoiding the need for artificial lighting during the day and the consequent cooling loads.

The sizing of north facing openings is less affected by seasonal variations and may be determined largely by daylight and cross-ventilation requirements. North facing openings can provide an almost uniform daylight source.

Effective cross-ventilation

Colour	Type	% Transmission	% Reflection	% Absorption
Light-coloured	horizontal	5	55	40
Medium-coloured	horizontal	5	35	60
White	vertical (closed)	0	77	23

Table 6.4: Transmission, reflection and absorption characteristics of Venetian blinds (for blinds at 45° tilt with sunlight perpendicular to the slats)

Description	Glass Thickness (mm)	U-value (winter)	U-value (summer)	Visible	Solar	S.C.
Single, clear	3	1.16	1.04	0.90	0.84	1.00
	6	1.13	1.04	0.89	0.78	0.95
	10	1.11	1.03	0.88	0.72	0.90
	13	1.09	1.03	0.86	0.67	0.86
Single, heat absorbing	6	1.13	1.10	0.52	0.96	0.71
Air space						
Double	5	0.62	0.65		0.71	0.88
	6	0.58	0.61			
	13	0.49	0.57	0.80		0.82
	13					
Low ε Coating						
	ε = 0.20	0.32	0.38			
	ε = 0.40	0.42	0.49	0.14		0.25
	ε = 0.60	0.43	0.51			
Triple	6	0.39	0.44			0.71
Acrylic, single glazed	6	0.96	0.89	0.92	0.85	0.98
Acrylic, reflective coating	6	0.88	0.83	0.14	0.12	0.21

Table 6.5: Table of shading coefficients and U-values for various types of glazing. Source - Generalitat de Catalunya

TREE SPECIES	COMMON NAME	DENSITY
Acer griseum	Paperbark Maple	76-85%
Acer macimowixzianum	Nikko Maple	90-93%
Acer platanoides	Norway Maple	90-96%
Acer rubrum	Red Maple	78-84%
Fagus sylvatica	European Beech	85-93%
Fraxinus pennsylvan	Red Ash, Green Ash	87-89%
Ginkgo biloba	Ginkgo, Maidenhair	74-82%
Gleditsia triacanthos	Honey Locust	49-50%
Quercus bicolor	Swamp White Oak	81-84%
Quercus macrocarpa	Burr Oak	87%
Quercus palustris	Pin Oak	64%
Quercus robur	English Oak	72-88%
Quercus rubra	Northern Red Oak	76-80%
Quercus velutina	Black Oak	74-88%

Table 6.6: Typical tree canopy densities

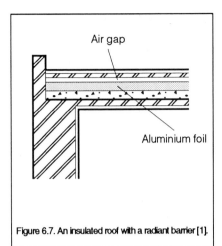

Figure 6.7. An insulated roof with a radiant barrier [1].

typically requires large openings distributed across opposing facades, with minimal internal barriers to impede the air flow needed for cooling [18,19]. For single sided ventilation, the shape of the opening becomes important, horizontal formats being more efficient in stimulating internal air velocities [20]. The design of openings should be undertaken in conjunction with the overall solar control strategy.

Solar Control and Shading Systems

Blocking the sun before it reaches the building, particularly the glazed, but also the opaque surfaces (including the roof) and reflecting the solar radiation is fundamental to the prevention of heat gain.

The appropriate choice from a wide range of fixed and moveable shading systems will depend on location, orientation, building type and the overall cooling, heating and daylighting strategies adopted in the design phase of the building. The architectural potential of shading is, however, frequently not exploited fully, especially if it is not an early consideration.

While shading systems must provide good solar protection in summer, they should not reduce solar gains in winter, obstruct natural lighting or impede natural ventilation. Well designed shading systems can actually enhance natural ventilation and daylighting.

Shading systems can block the direct component of solar radiation but are usually not as effective in reducing the diffuse and reflected components. New components and materials such as electrochromic, thermochromic, holographic glasses will soon start reaching the market.

The efficiency of shading systems may be expressed in terms of the shading coefficient (SC) which is the amount of solar radiation on the shaded surface, expressed as a fraction of the radiation falling on the shading device.

Shading systems may be fixed or movable and may be positioned externally, internally or within double glazed panels. Vegetation may also provide significant shading.

Fixed Shading Systems

Fixed shading systems include structural elements such as balconies and projecting fins or shelves and non-structural elements such as canopies, blinds, louvres and screens.

The orientation and shape of the opening to be shaded, relative to the position of the sun at different times of day and year, is critical to the design of fixed systems. Each orientation will need to be examined separately, taking account of direct and diffuse or reflected components of the overall solar radiation throughout the day and year. Typically horizontal shading is used for south facades, whereas vertical or diagonal fins or louvres are often more efficient on the east or west facades.

Fixed shading systems are most commonly used on the external facades where they can prevent direct radiation from reaching glazing or other openings and where heat absorbed by the shading system can be dissipated to the outside air. If fitted internally, heat will build up between the shading system and the glazing thus reducing the efficiency of the system, typically by around 30%.

The following table shows the amounts of solar energy passing through a range of typical fixed shading devices and glazing materials, measured in the UK. The 'solar gain factor' is the amount of energy admitted, expressed as a fraction of the total direct solar radiation.

Table 6.1 shows the greater effectiveness of external shading, resulting in the elimination of up to 90% of the direct solar gain.

The efficiency of internal blinds and curtains depends on the amount of energy they reflect back toward the glazing and the amount

of this energy transferred to the outside. Absorbing and low emissivity glasses should therefore not be used with internal blinds.

Movable Shading Systems

Movable shading is used externally or internally. Control can be either manual or power assisted and may be automated to respond to changing conditions such as current radiation levels and daylighting or thermal requirements.

Awnings can reduce heat gain by up to 65% in summer on south facades and by up to 80% on east or west facades. The geometry of awnings is similar to that for horizontal overhangs but efficiencies will also depend on how opaque the material is to both direct and indirect radiation as well as the presence of dust or dirt which may change the absorbtion or radiation characteristics of the awning. Normally, an air gap between the awning and the building facade should be provided to permit air circulation. The efficiency of fabric awnings may deteriorate with age or weather damage.

Venetian blinds can permit simultaneous ventilation and shading which is controllable and may allow daylight to be reflected, onto the ceiling for example.

With the exception of reflectant blinds, curtains and blinds fitted internally are less satisfactory as they provide shade only after the solar radiation has passed through glazing. Typical summer U-values are 1.06 for an uncovered window, 0.81 for a single glazed window with tightly-fitted curtains, and around 0.65 where double blinds or curtains are fitted. The use of curtains or internal blinds for shading may often conflict with daylighting or ventilation needs.

Glazing

Glazing may be clear or may have special coatings or treatments which enhance its reflective or heat absorbing characteristics. Electrochromic glasses allow the radiation transmission properties to be altered by varying an electric current which is passed through the glazing panel.

Vegetation

The position and density of foliage are the main considerations in using vegetation for shading. Deciduous vegetation can be helpful in providing greater daylight access in winter. Other factors include the rate of growth, size when mature, suitability for climate and soil, root development and maintenance.

The actual shading cast by trees and other vegetation will of course depend on the specific characteristics of the building, its immediate environment and the time of day and year. Several computer programmes can simulate such shading patterns in detail, but as a general rule, for southern European locations the best positions for tree planting are to the east and west of the building.

Thermal Insulation

Thermal insulation may combine two physical processes: reducing thermal transmittance of the envelope and maximising longwave radiation. Usually only the first principle is taken into account. Both these processes can be incorporated in the concept of radiant barriers [21,22]. In a multilayer element, a low emissivity material (such as an aluminium foil) next to an air gap will impede radiation, thus reducing the temperature of the inner layer and also room radiant temperature. At night, the foil will block radiant exchange, reducing night cooling. Reduction in cooling load of some 10% has been reported.

Thermal Mass

Floors, internal walls, roofs, partitions, and furniture, can provide thermal storage capacity, usually referred to as thermal mass. Thermal mass may have a significant effect on comfort, energy consumption and peak cooling load. It can store both heat and "coolth" and if well designed and positioned, can act as a regulator, smoothing temperature swings, delaying peak temperature, decreasing mean radiant temperature, and providing improved comfort conditions. The need for appropriately-sized and placed thermal mass should be a major consideration in the design of the building envelope and internal structure.

Thermal mass moderates temperature swing by absorbing heat directly from sunpatches or from the air. The stored heat can be dissipated in the evening or at night,

via nocturnal ventilation cooling. In some cases, the stored heat provides useful warmth in the early morning.

The key issue is to maximise the convective heat transfer between the thermal mass and the air. Unfortunately, this parameter is usually accounted for by an empirical value. This difficulty becomes critical when we have to deal with a dual winter/summer optimization approach. Another difficulty is due to the thermal inertia of the mass compared to the relatively fast processes at the boundary layers. For optimum design, thermal mass should be considered in conjunction with the heating and cooling controls to be used. Much R & D remains to be done in this area.

The role of thermal mass is particularly important for temperature regulation under conditions of continuous building occupation. Shaviv's numerical approach has given some recommendations for buildings in Israel [23] and a study by INSA [24] has been carried out under European conditions.

Airtightness

Design for winter conditions tends to promote airtightness in buildings, and this is entirely compatible with the requirement for heat gain minimisation in hot weather. In summer, when the outside air temperature is higher than the inside air temperature, any air infiltration will represent a cooling load for the building. When ventilation is to be used as a cooling technique, infiltration may negatively affect the flow circulation induced by purpose-designed openings.

NATURAL COOLING

Ventilation

Ventilation provides cooling by using air to carry heat away from the building and from the human body. Air movement may be induced either by natural forces (wind and stack effect) or mechanical power. Air flow patterns are the result of differences in the pressure distribution around and within the building. Air moves from high pressure regions to low pressure ones.

When the outside air temperature is lower than the inside air temperature, building ventilation may exhaust internal heat gains or solar gains during the day and may flush the building with cool air during the night if required. Indoor air movement enhances the convective exchange at the skin surface and increases the rate of evaporation of moisture from the skin. Evaporation is a very powerful cooling mechanism which may bring a feeling of comfort to occupants under hot conditions. However, in order to be effective, the surrounding air should not be too humid (relative humidity less than 85%). Turbulent air movement favours both of these mechanisms for heat removal.

Both the design of the building itself and its surrounding spaces can have a major impact on the effectiveness of natural cooling.

Hourly air change rates (ACH) may vary considerably, depending on the circumstances. For ventilative cooling in domestic or office situations, ASHRAE standards recommend a maximum of 0.75 to 1 ACH, whereas in densely occupied theatres or bars for example, controlled ACH rates of 30 to 50 are sometimes required. High ACH rates may affect comfort conditions and cause a nuisance; for example a candle flickers at about 0.5 m/sec and loose papers can be blown about if the air speed reaches 1.5 m/sec.

The quantity of air passing through an opening can be expressed as:

$$Q = C_v A V$$

where

Q = airflow (m³/sec)

A = area of opening (m²)

V = wind velocity (m/sec)

C_v = opening effectiveness coefficient (usually 0.3 to 0.6)

The capacity of the ventilation air for heat removal can be expressed as:

$$Q = rVC_p (T_{exhaust} - T_{supply})$$
$$= 0.35 \, V\Delta T \text{ (Watts)}$$

where:

r = air density (kg/m³)
V = volume flow rate (m³/sec)
C_p = specific heat capacity (J/kg °C)
$T_{exhaust}$ = Temperature of exhausting air (°C)
T_{supply} = Temperature of incoming air (°C)

As a rule of thumb, increasing the air velocity by 0.15 m/sec can compensate for an increase of 1K felt by occupants at moderate humidity levels (less than 70% RH).

The rate of airflow through the building will be affected by the location, sizing and flow characteristics of openings, the effects of indoor obstacles on air movement and the effects of the external shape of the building in relation to wind direction, for example wing walls. Airflow through buildings should be considered in three dimensions.

For the forces of buoyancy to act, there must be a significant temperature gradient between the internal and external air, and little resistance to air flow. The total airflow normally results from a combination of buoyancy and wind pressure differences, and is affected by the size and location of openings. The design of window systems for ventilation should also take account of daylighting, solar gain, security and acoustic considerations.

Ventilation Strategies

Wind Towers

Wind towers draw upon the force of the wind to generate air movement within the building [25,26]. There are various systems based on this principle. The wind-scoop inlets of the tower, oriented toward the windward side capture the wind and drive the air down the chimney. The air exits through a leeward opening of the building. The air flow is enhanced by cold night air. Alternatively, the chimney cap is designed to create a low pressure region at the top of the tower, and the resultant drop in air pressure causes air to flow up the chimney. A windward opening should be associated with the system for air inlet. The anabatic process benefits in this case from the buoyancy of the warm inside air.

Both these principles may be combined in a single tower providing both admittance and exhaust of air. A self-contained system is thus created.

Solar Chimney

Solar chimneys use the sun to warm-up the internal surface of the chimney. Buoyancy forces due to temperature difference help induce an upward flow along the plate. The chimney width should be close to the boundary layer width in order to avoid potential backward flow [27].

Limits

- Wind-induced ventilation would be an ideal strategy if winds were steady in direction and intensity (greater than 3 m/s). In reality of course winds are extremely variable, and detailed (microclimatic) weather data are not readily available at most sites.
- If the night-time temperature remains above the interior operative temperature then night-time ventilation is undesirable.
- An air change rate of 20 to 40 ACH can maximise the cooling benefits of ventilation, but lower rates may be sufficient.
- Taking advantage of prevailing winds in a complex changing environment is a difficult exercise for

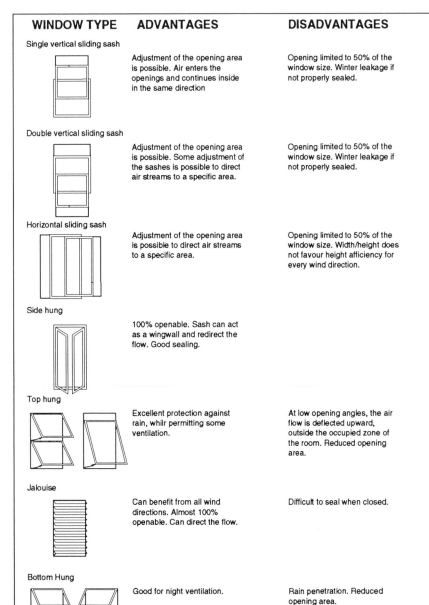

WINDOW TYPE	ADVANTAGES	DISADVANTAGES
Single vertical sliding sash	Adjustment of the opening area is possible. Air enters the openings and continues inside in the same direction	Opening limited to 50% of the window size. Winter leakage if not properly sealed.
Double vertical sliding sash	Adjustment of the opening area is possible. Some adjustment of the sashes is possible to direct air streams to a specific area.	Opening limited to 50% of the window size. Winter leakage if not properly sealed.
Horizontal sliding sash	Adjustment of the opening area is possible to direct air streams to a specific area.	Opening limited to 50% of the window size. Width/height does not favour height afficiency for every wind direction.
Side hung	100% openable. Sash can act as a wingwall and redirect the flow. Good sealing.	
Top hung	Excellent protection against rain, whilr permitting some ventilation.	At low opening angles, the air flow is deflected upward, outside the occupied zone of the room. Reduced opening area.
Jalouise	Can benefit from all wind directions. Almost 100% openable. Can direct the flow.	Difficult to seal when closed.
Bottom Hung	Good for night ventilation.	Rain penetration. Reduced opening area.

Figure 6.8. Advantages and disadvantages of common window types.

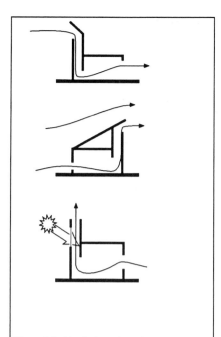

Figure 6.9. Ventilation using wind towers and a solar chimney [1].

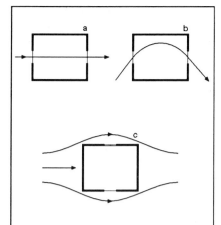

Figure 6.10 a-c When inlet and outlet are aligned with the wind, the air flow is short-circuited and poor secondary flow is generated next to the main stream.
If the wind is oblique to the openings, the air flow circulates in the entire building.
If the wind is parallel to the opening, no significant movement occurs within the occupied space.

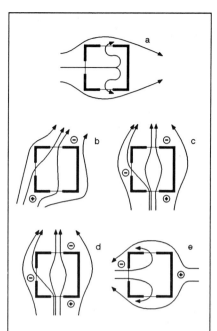

Figure 6.11 a-e Cross ventilation can be further enhanced by placing two outlets on the building sidewalls. This design option also addresses the frequent shifts in wind direction.
The distribution of the openings on the building facade is a key issue for efficient natural ventilation. Ventilation is a three dimensional function and correct location of openings including windows is essential. The position of the inlet dominates the air flow pattern within the space. The position of the outlet is of secondary importance.

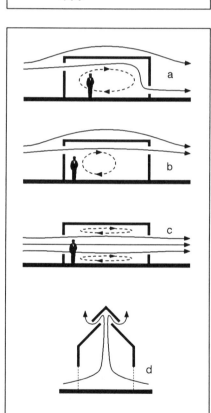

Figure 6.12 a-d High inlets do not generate a strong air velocity in the occupied zone and are thus less suitable for occupant cooling. However, this configuration is often interesting for night ventilation because the air stream may be directed toward the storage element, for example a massive ceiling. Moreover, the high position offers better security against intruders. Openings at body height generally offer good cross ventilation.When the building is too deep to offer cross ventilation, or when opposing openings are not possible, roof openings may be used to encourage an anabatic flow. The roof opening should be designed to create a low pressure region next to the opening to enhance the natural stack effect.

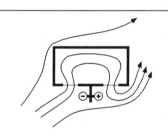

Figure 6.13 Cross ventilation is often optimal if a room has three openings on different facades. Unfortunately, this configuration is rare as most rooms have only one external wall. With one open window, the ventilation is mainly due to the turbulent fluctuations of the wind, and internal air movement is not significant. Ventilation can be improved if two windows can be placed on the facade, as far apart as possible. Wind fluctuations generate pressure differences between the two windows and induce air circulation within the space.
Wingwalls for windward windows can enhance the pressure differences between two windows and induce air circulation within the space. Wingwalls for windward windows considerably increase cross ventilation in the room. Unfortunately, they are ineffective for leeward facades.

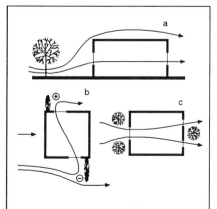

Figure 6.15 Influence of vegetation on airflows, also providing air filtration, noise reduction and shading

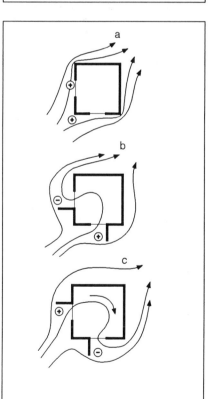

Figure 6.14 a-c If the room has apertures on adjacent walls, wingwalls can considerably enhance cross-ventilation [31]. On the other hand, wingwalls modify the initial flow within the space.
Vegetation can affect the air flow pattern in the same ways as neighbouring buildings or wingwalls [32]. Moreover, it also offer air filtering, noise reduction and shading

which there are as yet no comprehensive design tools. Forced ventilation greatly extends the opportunities for using ventilation cooling. A box fan, an oscillating fan or a ceiling fan can supplement natural ventilation by increasing air velocities and convection exchange.

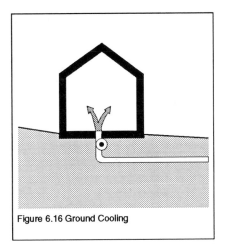

Figure 6.16 Ground Cooling

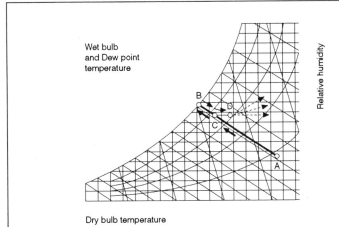

Wet bulb and Dew point temperature

Relative humidity

Dry bulb temperature

Figure 6.17: Point A represents the indoor air temperature. Evaporative cooling takes place along the line AB and point C represents the temperature of the air after evaporative cooling has occurred. The line CD represents a certain level of heating from the remaining water. Finally, when the cooled air mixes with warmer indoor air, the temperature and humidity will follow one of the paths from point D.

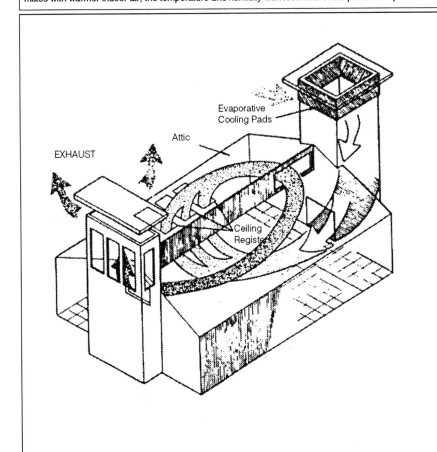

Evaporative Cooling Pads

Attic

EXHAUST

Ceiling Registers

Figure 6.18: Wind towers using evaporative pads can enhance evaporative cooling.

Ground cooling

Physical Processes

During the summer, soil temperature at certain depths are considerably lower than the ambient air temperature, thus offering an important source for the dissipation of a building's excess heat. The time lag between the annual cycle in temperature for the earth and the air increases with depth. Seasonal variation in ground temperature decreases with depth, with moisture content, and with soil conductivity.

Heat dissipation to the ground can be achieved by conduction or by convection. For the former, part of the building envelope must be in direct contact with the soil, while in the case of the latter, air from the building or from outside is circulated through buried pipes where it is precooled before it enters the building.

Strategies

Direct Contact

Heat dissipation to the ground by conduction via direct contact of the building envelope with the soil is a well known technique. Many traditional earth-coupled dwellings have been reported [70]. An extensive review and performance data of earth sheltered buildings are given in [71,72]. An analysis of analytical and numerical methods for estimating the performance of earth-sheltered buildings is presented in [73].

Underground buildings offer various advantages, including protection from noise, dust, radiation and storms, limited air infiltration and potentially increased fire security. They provide benefits under both cooling and heating conditions and they represent the ultimate application of thermal mass to the building. The potential for large-scale construction of such buildings is fairly limited. High cost and poor daylighting conditions are frequent problems.

On the other hand, buildings in partial contact with the soil offer interesting cooling possibilities. Thousands of such buildings have been constructed in Europe, especially in hilly areas. These buildings are characterised by reduced heat losses and an increase in comfort resulting from their proximity to the ground. However, there is little information on the performance of this type of building and there is uncertainty on the calculation of heat transfer from or to the ground.

Buried Pipes

The use of buried pipes for precooling of air is a technique developed recently. An early application of a similar concept dating from the 16th century and making use of natural cavities ("covoli"), in the hills of Vicenza, Italy has been reported by Fanchiotti and Scudo [74].

Recent applications have been based on the use of a series of plastic or metal underground pipes [36]. Air from the building, or external fresh air, is passed through the pipes and then introduced into the building. The temperature drop of the circulated air is a function of the inlet air dry bulk temperature, the ground temperature, the thermal characteristics of the pipes and soil, as well as the air velocity, and pipe dimensions. A review of the existing applications of buried pipes, as well as methods for calculating the efficiency of such systems is given in [37]. A parametric study is required to optimise the system [38].

As a threshold value for the application of this system, the ground temperature around the tube should be at least 5K - 6K lower than the air temperature. The performance of buried pipes systems has been studied by various authors. Experimental data reported in [39], show that the drop in air temperature inside a 30m pipe buried at 1.5m, is close to 10K- 12K. Data reported in [40] show that a temperature drop of 9K - 10K is achieved in a similar tube using lower inlet temperatures.

Special problems related to the use of buried pipes are : possible condensation of water inside the tubes, or evaporation of accumulated water; and control of the system. Existing knowledge on the topic is mainly on the thermal characteristics and the performance of single earth tubes used mainly for heating purposes or a system of tubes in conjunction with heat pumps. There is a lack of information on the practical problems related to the operation of real scale earth cooling, the thermal impacts for the building, and on the control problems that may arise from interaction with conventional systems. Such lack of information limits effective design and implementation of the system and the wider adoption of earth cooling.

Evaporative cooling
Physical Principles

Evaporation occurs whenever the vapour pressure of water (in the form of droplets or a wetted surface) is higher than the partial pressure of the water vapour in the adjacent atmosphere. The phase change of water from liquid to vapour is accompanied by the release of a large quantity of sensible heat from the air that lowers the dry bulb temperature of the air while the moisture content of the air is increased. The efficiency of the evaporation process depends on the temperatures of the air and water, the vapour content of the air and the rate of airflow past the water surface. The provision of shading and the supply of cool, dry air will enhance the evaporative process. Evaporation is characterized by a displacement along a constant wet bulb line, AB (Figure 6.17). Where the decrease in the dry bulb temperature is accompanied by an increase in the moisture content of the air, the process is commonly referred to as "direct evaporative cooling". When the evaporation of water takes place on a surface, or inside a tube, resulting in a decrease of surface temperatures, it is possible to cool air adjacent to these surfaces without increasing its moisture content. In this case, the process is referred to as "indirect evaporative cooling" and is characterized by a displacement along a constant moisture content line CD.

Systems

In direct evaporative cooling systems the moisture content of the cooled air is increased, raising the relative humidity of the indoor space. This may be acceptable, especially if the

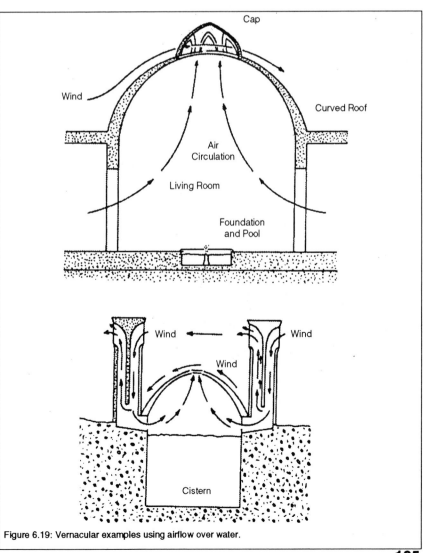

Figure 6.19: Vernacular examples using airflow over water.

air change rate is sufficient. Otherwise comfort may be adversely affected and condensation or mould growth may occur. The system should be capable of isolation or control when cooling is not required, for example in winter.

Typically, acceptable performance figures for direct evaporative systems are:

1. A saturation efficiency of the cooling process of 70% or better
2. A maximum indoor air velocity of 1 metre per second
3. The air temperature of the indoor space should be around 2K higher than the discharge air temperature and its RH should be below 70%.
4. The resulting temperature of the indoor space should be about 4K below the outdoor dry bulb temperature.

Many examples of direct evaporative systems exist in vernacular architecture, particularly that of hot, arid regions, where ponds, cisterns or wetted surfaces are typically placed in the incoming ventilation air stream.

Such direct systems typically use little or no auxiliary power and simple but robust technologies and can avoid the need for large surfaces of water and the movement of large volumes of air, and are thus particularly suited to arid regions. Their main disadvantage is in the increased moisture content in the ventilation air supplied to the indoor spaces.

Indirect evaporative systems avoid problems associated with increased humidity levels and are particularly suited to tropical or subtropical regions where RH values of 70% or more are common. Air change rates can be lower than with direct systems and dessicants or other means of de-humidification are not normally necessary. Indirect systems are, however more complex and often more costly and may be difficult to integrate in the building, particularly if retrofitted.

Evaporative cooling techniques can be grouped into two major categories: passive and hybrid. Hybrid systems are those which rely on equipment installed in the building to supply cooling, while passive

techniques rely on elements of the building envelope and/or the adjacent landscape.

With this distinction, we obtain four major categories of systems and techniques:

- Passive direct systems and techniques
- Passive indirect systems and techniques
- Hybrid direct systems
- Hybrid indirect systems

Passive direct systems and techniques include the use of vegetation for evapotranspiration, as well as the use of fountains, pools and ponds. An important technique known as the 'volume cooler' is used in traditional architecture [41]. The system is based on the use of a tower where water contained in a jar or sprayed is precipitated. External air introduced into the tower is cooled by evaporation and then transferred into the building.

A contemporary version of this technique was presented by Cumming and Thomson [42]. Here a wet cellulose pad is installed at the top of a downdraft tower below the roof, where the air is humidified. Measurements have shown that with a dry bulb temperature of the incoming air of 35.6°C and wet bulb of 22.2°C, the exit air temperature was close to 24°C. A similar system has been installed at the site of EXPO '92 at Seville [43].

Passive indirect evaporative techniques include mainly the roof spray and open water pond.

Roof Spray

The exterior surface of the roof is kept wet using sprayers. The sensible heat of the roof surface is converted into latent heat of vaporization and the water evaporates. A temperature gradient is created between the inside and outside surfaces causing cooling of the building. A threshold condition for the operation of the technique is that the temperature of the roof should be higher than the wet bulb temperature of the air.

Experience with this type of technique has been acquired through many commercial applications in USA, [44], and from experimental work by Yellott, [9] and by Jain and Rao [45]. The observed reduction in cooling load is close to 25% [46]. However,

there is no information on the performance of the system in Europe. There may be large variations from one installation to another. A method to assess the suitability of this system in a given climate has been presented by Kishore [47].

There are a number of problems associated with this type of technique. Many studies have shown that it is not cost effective and that improving the thermal insulation of a roof may be a better alternative. Also there are problems associated with the appearance of the piping and possible damage to the roof with the freezing of piping, etc.

Roof pond

The roof pond consists of a shaded water pond over an uninsulated concrete roof. Evaporation of the water to the dry atmosphere occurs during day and night time. The temperature of the roof follows closely the ambient wet bulb temperature, while the ceiling acts as a radiant convective cooling panel for the space. Thus the indoor air and radiant temperatures can be lowered without elevating indoor humidity levels.

A theoretical threshold condition for the application of the system is that the temperature of the roof should be higher than the wet bulb temperature of the air. Givoni, [48], has suggested that the wet-bulb should be lower than 20°C.

Knowledge on this technique has developed through experiments made by Yellott and Hay, [9], Jain and Rao , [45], Pittinger White and Yellot, [49]. Important information is also given by Givoni [48] and Sodha [50].

The performance of this system was measured extensively by Givoni [48] and Jain, Rao [45]. According to Givoni a decrease of 2K - 3K in the ceiling temperature was achieved while Jain and Rao reported a decrease of 13K. However there are no evaluations of the system for European conditions. Problems or limitations with this technique are that it is confined to single-storey buildings with flat concrete roofs or to the upper storey of multi-storey buildings; and capital costs can be high. There is again a question of whether a well insulated roof of conventional construction may be more appropriate.

Hybrid Indirect Evaporative Cooling Systems

These are based on the use of a heat exchanger where the indoor ventilated air passes through the primary circuit where evaporation occurs while the outdoor air passes through the secondary circuit. This decreases the temperature of the air without any moisture increase. There is significant industrial development of this type of system with more than 10 manufacturers worldwide. There are three main types of indirect coolers: the plate, the tubular or the rotary type coolers. Detailed description is given in [51].

The threshold value for the operation of the system is that the indoor wet bulb temperature should be lower than the outdoor dry bulb. In practice, the indoor wet bulb should be lower than 21°C. According to reported data [52], the performance of the system is quite satisfactory.

The threshold value for the use of such a system is that the ambient wet bulb temperature should be lower than 24°C. The performance of this type of system is strongly related to its saturation efficiency which is defined as:

$$\text{Saturation Efficiency} = \frac{Tdb_{in} - Tdb_{out}}{Tdb_{in} - Twb_{in}}$$

where Tdb_{in}, Tdb_{out} are the dry bulb temperature of the air at the inlet and outlet of the system respectively and Twb_{in} is the wet bulb temperature of the incoming air. Typical saturation efficiencies for those systems range between 60 and 80 per cent.

The main problem associated with direct evaporative coolers is the increase in the moisture content of the air. Hence, their use should be always combined with a humidity control system. Other problems may arise due to the porosity and the capillarity of the cooler, the filter capacity, etc.

Energy savings of up to 60% compared to compression refrigeration systems may be achieved in hot dry regions. However, the efficiency of the system is strongly influenced by the wet bulb temperature of the outside air.

As indirect evaporative systems do not add moisture to the building, no humidity control is required for their operation. Corrosive components should be avoided for maintenance reasons. Effective filtering is necessary to stop accumulation of dust particles.

Where the ambient air temperature is too high, a two stage evaporative system can be used. This consists of an indirect cooler coupled with a direct and/or an indirect cooler. These may be coupled with a refrigerative air conditioning unit.

There are many applications of these systems, especially in California, and a number of established manufacturers, also in the USA. Energy savings for such a system have been reported at close to 50% compared to an equivalent air conditioning system [51]. Problems associated with these systems are the same as those of direct and indirect systems.

Radiative cooling
Principles

Any object emits energy by electromagnetic radiation. If two elements at different temperatures are facing one another, a net radiant heat loss from the hotter element will occur. If the coldest element is kept at a fixed temperature, the other element will cool down to reach equilibrium with the colder element. This physical principle forms the basis of radiative cooling. If there was no atmosphere, the external envelope of the building would 'see' the deep space - an extremely cold source, and radiative cooling would be ideal. An intermediate heat sink is the sky. Any building element which 'sees' the sky exchanges heat with it. In order to have an appreciable net heat flux between two bodies, the temperature differences should be significant. Low sky temperatures are associated with clear skies. Opaque surfaces should have a maximum reflectivity in the short wave region of the spectrum to reflect solar radiation and a maximum emissivity in the long wave region to favour radiation from the building to the night sky. High reflectance of vertical surfaces in urban sites may contribute to an increase in solar input for adjacent buildings. However, this drawback may sometimes present an advantage by enhancing daylight for other buildings. Night radiation from vertical surfaces is limited, whereas roof surfaces radiate intensively and interact less with other buildings. (see Appendix 9 - Fundamentals)

Although deep space can be considered an infinite heat sink for radiant emission from the earth, atmospheric carbon dioxide, water vapour and pollutants intervene, absorbing much of the earth's longwave radiation. As a result, the atmosphere takes on a fictitious or 'effective' temperature which the earth 'sees' and which is closer to the earth's temperature. Consequently, the sky radiates longwave heat energy back to the earth at a rate similar in magnitude, but slightly less than the rate of the earth's radiation to the sky. This difference in radiation rates determines the maximum potential of radiative cooling systems.

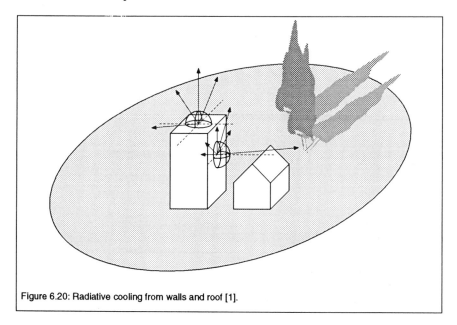

Figure 6.20: Radiative cooling from walls and roof [1].

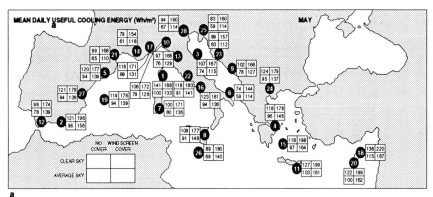

a

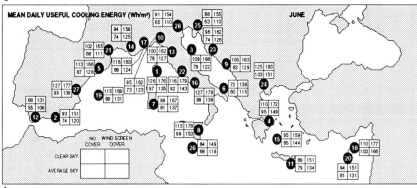

b

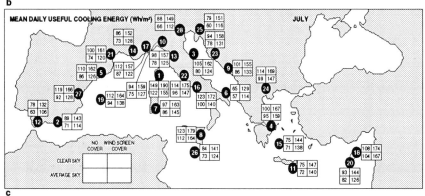

c

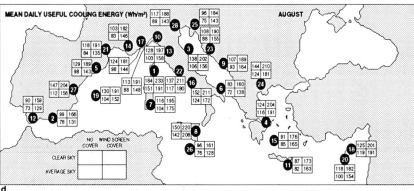

d

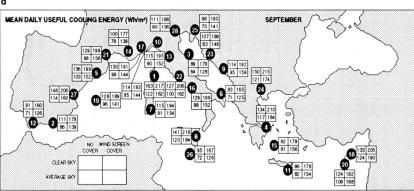

e

Figure 6.21: Mean useful cooling energy (Wh/m²)

To evaluate this cooling potential it is necessary to know the night sky effective temperature. In the absence of data it can be calculated as follows:

For a clear sky:-
$$T_s = (0.787 + 0.0028\ T_{dp})\ 0.25\ T_{air}$$

For a cloudy sky:-
$$T_s = (1 - F_w) - T_{air} + F_w\ T_c$$
where:

T_s is the sky temperature

T_{dp} is the dewpoint temperature

T_{air} is the air temperature

F_w is the fraction of blackbody radiation in the 8 to 13 micron wavelength region. (F_w normally varies from 0.2 to 0.4 depending on the location)

T_c is the temperature of the cloud base

As the longwave absorbtion of the atmosphere is directly related to the water vapour content of the air, the effective temperature of the sky is a function of the humidity and the dry bulb temperature of the air near the ground. The effectiveness of radiant cooling can be significant, particularly in hot, dry conditions. For example, under hot, dry conditions, with a clear sky, a roof temperature of 27°C and a maximum value of emissivity for the roof materials, the net radiant heat loss would be around 160 W/m².

Radiative cooling can be adversely affected by the convection of heat from ambient air to the radiating surface. This effect can significantly reduce the effectiveness of radiant cooling and may require the use of windscreens which are transparent to longwave radiation. Any dust or condensation which forms on these screens will impede the radiative cooling process.

Strategies

Radiative Roofs

The roof surface is the element of the envelope that has the best view of the sky dome and represents the most appropriate surface for radiative cooling. The roof also receives a large amount of solar gains during the day. In order to limit this impact, the roof reflectivity in the short wavelength spectrum should be high. Maximizing night radiation requires a high emissivity in the longwave range. A white paint coating or an aluminium sheet has good spectral properties for radiation. Highly selective surfaces have been proposed but no distinct advantages have been observed over more conventional radiators.

The performance of radiators is limited by two main phenomena: the temperature drop of the radiating surface is limited to the outside temperature due to convection, particularly in the presence of wind, and, the cold temperature of the radiators increases the likelihood of condensation on the radiating surfaces thus reducing their efficiency. In order to limit the wind effect, windscreens may be used.

The efficiency of the system in meeting building cooling loads and reducing the internal radiant temperature depends largely on the thermal coupling between interior and exterior environment through the roof. Different systems are proposed: the radiators may directly cool the roof mass, or may cool air that is then flushed into the building via a mechanical fan.

Air Cooling

In order to transfer the cooling effect of the nocturnal radiation from the external surface of the roof to the building interior, air can be circulated under the cold radiator and then introduced into the building, preferably next to the storage mass.

A painted metallic sheet with an air gap of 50-100 cm beneath is a typical radiator. Fins can be added to maximise the rate of heat transfer from the internal air to the cooling panel at the expense of a higher power consumption for the fan. The air can also circulate in tubes attached to the metal sheet. Insulation should be placed under the air gap to maximise the heat transfer to the air.

Movable Insulation

Exposing the thermal storage to the sky during the night while protecting it during the day by movable insulation optimises the potential of radiant cooling. Operable insulation requires storage spaces at night and manual or mechanical operation. The storage mass can be represented by a massive roof or by water bags. Direct thermal contact with the ceiling is required.

Roofponds have demonstrated high efficiency in hot, dry climates but they cannot provide latent cooling. Moreover, they represent a cooling technique which is suitable only for the upper storey. They induce an extra financial cost because of increased weight and the risk of potential leakages if not properly constructed.

The system can be reversed in winter to benefit from daytime solar gains and to limit night radiation and transmission losses.

Movable Thermal Mass

At night, a pond situated above an insulated layer is filled with water and is cooled by night-time radiation. In the morning, the cool water is drained below the insulated layer and can absorb a large quantity of heat and contribute to a decrease in the mean radiant temperature of the upper floor.

Traditional solutions

Many examples of cooling by natural means are to be found in the vernacular architecture of the various regions of southern Europe. The more successful examples have stood the test of time and offer useful guidance for buildings of a similar scale and function in similar surroundings. However, while many of the techniques employed in these vernacular buildings are the result of empirical development, our current understanding of building science, together with developments in construction materials, procedures and control allow us to use a wider range of techniques which may be more effective. In addition, the scale of buildings and our expectations of building performance have changed, rendering some vernacular techniques inappropriate particularly those in small-scale buildings.

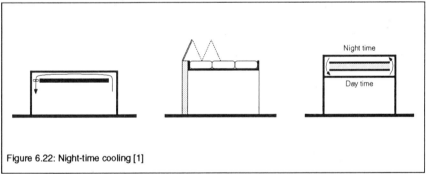

Figure 6.22: Night-time cooling [1]

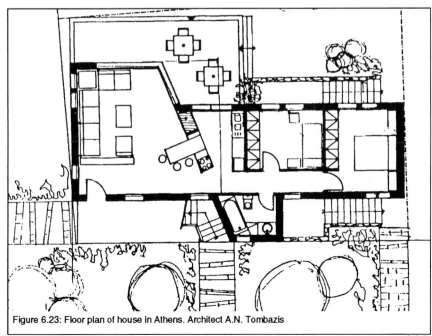

Figure 6.23: Floor plan of house in Athens. Architect A.N. Tombazis

STUDY OF THE COMPARATIVE PERFORMANCES OF FOUR PASSIVE COOLING SYSTEMS

Very little information is currently available on the comparative performance of the various natural cooling techniques in Europe, or on their relative advantages or disadvantages.

The following study aims to provide comparative information on the effects and performance of the most significant natural cooling techniques in a house in Athens. Night ventilation techniques, ground cooling via earth to air heat exchangers and direct and indirect evaporative cooling systems are examined. For each technique, the relative influence on the thermal behaviour on the building was investigated by means of a sensitivity analysis of the main parameters determining their performance.

The reference building
The building used in this study is a typical 80 m2 one storey house dwelling [54] situated in the Athens region. It was designed by A. Tombazis. The floor plan of this building is illustrated in Figure 6.23 and consists of two bedrooms, a living room, a kitchen and a bathroom.

The building is well insulated. The U-values of the walls, floor and ceiling are 0.53, 0.31 and 0.89 W.m-2°C respectively. All windows are single glazed and well shaded during the summer period. More details on the reference building are given in [54].

Climatic data
The climate in Athens (Latitude = 37.58 N) is characterized as "warm Mediterranean", with mild, relatively wet winters and warm dry summers. An analysis of the summer climatic data of Athens for cooling purposes is given in [55] and [56].

During the simulation procedure, Test Reference Summer weather data were used [67]. Test Reference Summer weather data were created in order to provide hourly values of all the necessary climatic data for cooling purposes, i.e., ground temperature at various depths, sky temperature, wet bulb temperature etc. Primary data are taken from the National Observatory of Athens [57].

The simulation procedure
The thermal behaviour of the building was simulated using CASAMO - CLIM. developed by Ecole National des Mines de Paris, France [58]. This programme has been developed especially for cooling purposes and validated against real buildings.

The programme allows a detailed description of the building to be simulated, and provides dynamic calculation of the shading levels. The programme calculates the hourly variation of the indoor air temperature and of relative humidity in the building. The necessary cooling load in order to obtain a defined set point temperature can also be calculated.

In order to simulate performance of the passive and hybrid cooling components integrated into the building, special algorithms describing the performance of these systems, have been developed and integrated into the basic code.

Four different types of passive and hybrid cooling systems and techniques have been simulated:

1. Earth to air Heat Exchangers (buried pipes)
2. Direct Evaporative Cooling components
3. Indirect Evaporative Cooling components
4. Night ventilation techniques.

Ground Cooling
The simulated system consisted of an underground horizontal P.V.C. pipe, whose inlet sucks air from the environment using an electric fan. The air was cooled by circulation underground and is then introduced into the building.

The system consisted of a 50 m long pipe, having an internal diameter of 0.2 m, placed at a depth of 4 m. The thickness of the pipe wall was 0.005 m. The air velocity inside the pipe was 5 m/sec.

In order to calculate the impact of the ground cooling system on the building, a subroutine for the thermal performance of the buried pipes has been developed and integrated into the code. For this purpose the algorithms proposed in [59] and [60] have been used.

A major question associated with the use of an earth to air heat exchanger, is the choice of the optimal depth at which the exchanger is to be buried.

In order to investigate the influence of this parameter on the thermal behaviour of the building, a series of simulations have been performed for depths of 1.5 m to 6.5 m. The choice of this range was based on the ground temperature variation curves as a function of the depth. The ground temperature at depths of up to 1.5 m is not low enough to permit efficient performance of a heat exchanger. At depths greater than 6.5 m the ground temperature remains practically steady. In the study it was found than a 2.5°C reduction of the peak indoor temperature could be obtained as the depth of earth to air heat exchanger ranged between 1.5 and 6.5 m respectively.

The most important air temperature decrease was obtained for depths ranging between 3 m and 6.5 m. More precisely, the maximum indoor temperature decrease observed in June, occurred when the exchanger was placed at a depth of 4 m. For July and August this maximum value occurs for a depth of 5 m. Consequently, the optimal contribution of the heat exchangers to the indoor temperation reduction in the building occurs at depths between 4 and 5 m.

The results are consistent with ground temperature variation as a function of depth, where it can be observed that the minimum values occur within the range of 3.5 m to 5 m. Temperatures below 6.5 m can in some cases increase slightly.

At night the outdoor air temperature may be lower than below-ground temperatures resulting in an increase in the temperature of ventilation air after it has passed through the heat exchanger. In this case a control system may be required.

The effect on the indoor temperature of the building of changing the length, the internal diameter and the air velocity inside the exchanger, has also been investigated. During these simulations the depth of the exchanger was fixed at 4 m. The length was increased from 50 m to 70 m, the diameter from 200mm to 220mm and the air velocity was decreased from 5 m/sec to 3 m/sec.

It was observed that when the length of the exchanger was increased from to m to 70 m, the corresponding indoor temperature drop was about 0.5 °K. This is because the temperature decrease in the final 20 m of the exchanger is not as significant as the temperature decrease in the first 50 m.

The increase of the diameter of the exchanger from 200mm to 220mm lead to a more important decrease in the indoor temperature of the building of about 1.5°C. This is because increasing the diameter and maintaining a constant air velocity in the exchanger, increases the cooled air flow rate and consequently the cooling energy 'injected' to the building.

On the contrary, when the air velocity is reduced, the cooling energy supplied to the building is also reduced, resulting in higher indoor temperatures.

Direct evaporative cooling

The direct evaporative cooling device considered in this study was a parallel plate pad evaporative cooler. This cooler consisted of a centifugal fan (diameter 0.25 m). The parallel matrix using a 50 m^2 wetted area was composed of 38 plates. The dimensions of each plate were 1.20 m x 600mm x 3.56mm and the gap between each of the plates in the matrix was 4.4mm

The water distribution system pump had a capacity of 250 m^3/h under a head of 3 m.

Using the ambient dry and wet bulb temperatures as input data, the fan rate (in r.p.m.) and the water mass flow rate, the temperature and the relative humidity at the outlet of the cooler were calculated.

In order to analyse the impact of the parameters regulating the performance of the system to the building, a sensitivity analysis was performed. Thus the influence of the fan speed as well as the flow rate of the water humidifying the parallel plate matrix was investigated.

The results showed that a maximum reduction in the peak indoor air temperature of about 4-6°K is possible. An increase in the water flow rate has an almost negligible variation on the indoor temperature of the building. However, the effects of changes in fan speed were significant giving, for

example, a mean temperature reduction of 1.5 K when the fan speed was increased from 800 rpm to 1500 rpm.

Indirect evaporative cooling

Indirect evaporative cooling systems can cool buildings efficiently without increasing the moisture content of the indoor air. Various types of indirect evaporative cooling systems have been proposed [61]; however plate type indirect evaporative coolers have given very encouraging results and have already seriously penetrated the market [62]. The type of cooler considered in this study research consisted of a plastic heat exchanger with dimpled sheets of a hydraulic polymer, two fans, a water pump and simple water sprays [63].

In order to calculate the efficiency of the system, a saturation efficiency algorithm proposed in [64] was used. The air temperature at the outlet of the cooler was then calculated using the outdoor dry and wet bulb temperatures as inputs. An analysis of the impact on the building of the velocity with which the cooled air circulated through the primary circuit of the heat exchanger of the cooler was performed (the evaporation takes place in the secondary circuit). Two air velocities were considered, 0.3 m/sec and 0.1 m/sec. The results show that an increase in air velocity is inversely proportional to the indoor temperature. In all cases, a temperature decrease of at least 1.5K was obtained, compared with the indoor temperature levels when no cooling load was provided. During June, acceptable indoor temperature levels were obtained even with the lower air velocity. However, in order to create comfortable indoor conditions during the day on July and August the higher air velocity has to be chosen.

Light ventilation

Night ventilation can provide effective cooling for buildings while contributing to increased indoor comfort during the day. A range of night cooling applications for cooling buildings is described in [66] and [67].

The building in the study was ventilated from 21:00 until 07:00. Various ventilation rates were considered, ranging from 2 ACH and

8 ACH with steps of 2 ACH. The resulting indoor air temperature profiles were compared with a reference case in which a ventilation rate of 1 ACH has been assumed during all day.

As expected, the indoor temperature decreased by increasing the ventilation rate. However the maximum reduction in peak indoor temperatures did not exceed 1 K.

The decrease in indoor temperature during the day is more significant in June and August than in July due to the high night ambient temperatures which occur during this month. However, simulations have shown that while the use of night ventilation can contribute to the cooling of the building the contribution is not sufficient to produce to acceptable temperature levels during the day and for this reason a complementary cooling system is normally required.

Summary:

The use of night ventilation techniques can provide part of the cooling load required; however the use of additional cooling systems is often necessary.

When earth-to-air heat exchangers are used, the pipe has to be buried at a depth of between 3.5 m and 5 m. It was found that the diameter and air velocity variations significantly affect the indoor temperature of the building. Increasing of the length of the exchanger over a certain value does not have a significant impact on temperature within the building.

The most important parameter affecting the indoor temperature of the building using the direct evaporative cooling system chosen, was the fan rate of the cooler, rather than the flow rate of the water humidifying the cooler pad. The indoor temperature decrease was satisfactory for a fan rate higher than 1500 rpm.

The use of indirect evaporative coolers in June and August produced acceptable temperature levels for an air speed though the cooler of 0.1 m/sec, but during July an air speed of 0.3 m/sec was required.

REFERENCES

[1] Horizontal study on Passive cooling, Santamouris, M. (Ed), Antinucci. M,Fleury,B.,Lopez d'Asian,J., Maldonado,E., Tombazis,A., Yannas, S. Commission of the European Communities DGX11, Building 200 Action, 1990.

[2] Workshop on Passive Cooling, Santamouris, M, Proceedings of Passive Cooling Workshop, Ispra, 1990. EUR 13078 EN, ISBN 92-826-1690-8

[3] Analysis of the Summer Ambient Temperature for Cooling Purposes, Tselepikaki, I., Santamouris, M., Melitisiotis, D., University of Athens & Protechna Ltd., Themistokleous 87, 10683 Athens.

[4] "A Classification Scheme for the Common Passive and Hybrid Heating and Cooling Systems". Holtz, M.J., Place W. Proceedings of the 3rd National Passive Solar Conference, San Jose, California, USA. 1979.

[5] Landscaping as a Passive Solar Strategy, Montgomery, D.A., Passive Solar Journal 4(1), 79, 1987

[6] Landscaping Planning for Energy Conservation, Robinette, G., Reston, VA, Environmental design Ress, 1977

[7] Landscape as Energy and Environment Conservation in the Arid Regions Saudi Arabi, Risvi, S., Talib, K., International Conference on Passive and Hybrid Cooling, Miami Beach, 1981

[8] Earth Sheltered Community Design, Energy Efficient Residential Development, Stirling, Carmody, R.J., Elnicky, G., N.Y., Var Nostrand Reinheld Company, 1981

[9] Passive and Hybrid Cooling Research, Yellot, J.Y., Advances in Solar Energy, 1983

[10] Heating and Cooling of Building Sites Through Landscape Planning, Bowen, A., Passive Cooling Handbook, Newark; DE: AS / ISES, 1980

[11] The effect of Horizontal and Vertical Temperature Gradients, Duckworth, E., Sandberg, J., Bulletin of Meteorological Society, 1954

[12] Influence of Surface Conditions on Ground Temperatures, Gold, L., Canadian Journal of Earth Science 4, 199, 1967

[13] Residential Cooling Loads and the Urban Heat Island: the Effect of Albedo, Taha, H., Akbari, H., Rosenfeld, A., Huang, J., LBL Report 24008, 1988

[14] Landscape Design Hot Save Energy, Moffat, A., Schiller, M., New York: William Norrow and Company, 1981

[15] The Underground Advantage: Climate of Soils Passive Solar Subdivision, Window and Underground, Labs, K., Miniapolis, MN, MASEC: MASEC

[16] Studies of Air Motion in a Room Having a Door Opening into a Lounge, Chand, I., Krishak, N., Civil Engineering Construction and Public Works Journal, July 1971

[17] Effect of Verandah on Room Air Motion, Chand, I., Civil Engineering Construction and Public Works Journal, November 1973

[18] A Design Procedure to size Windows for Naturally Ventilated Rooms, Chandra, S., Florida Solar Energy Center, PF-46-43

[19] Effect of Multiple Windows on Indoor Air Motion, Chand, I., The Indian Engineer, Oct. 1969

[20] Effect of the Distribution of Fenestration Area on the Quantum of National Ventilation in Building, Chand, I., Architectural Science Review, pp 130-133, Dec. 1970

[21] Radiant Energy Transfer and Radiant Barrier Systems in Building, Fairey, P., Design Note, Florida Solar Energy Center, DN-6-86

[22] Designing and Installing Radiant Barrier Systems, Fairey, P., Florida Solar Energy Center, DN-7-84

[23] On the Optimum Design of SHULDING: Devices for Windows, Shaviv, E., Plea Conference, Porto 1988

[24] l'Inertie Thermique par le Béton, Economie d'Energie et Confort d'Eté, Depecher, P., Brau, J, Ronssean, S., (INSA de Lyon), Centre d'Information de l'Industrie Cimentiére (CIC).

[25] A Passive Cooling Heating System for Hot Arid Regions, Bahadori, M., 13th National Passive Solar Conference, Cambridge, USA, pp 364-367, June 1988

[26] Evaluation of Pressure Coefficients and Estimation of Air Flow Rates in Buildings Employing Wind Towers, Karakjatsanis, C., Bahadori, N., Vickery,B., Solar Energy, Vol 37, n°5, pp 363-374, 1986

[27] Moving Air, Using Stored Solar Energy, Bouchair, A., 13th National Passive Solar Conference, Cambridge, June 1988

[28] Controlling Air Movement, Boutet, T., McGraw Hill Book Company, 1987

[29] Cooling with Ventilation, Chandra, S., Fairey, P., Houston, M., Florida Solar Energy Center, SERI Report n°Sp 273-2966, Dec. 1986

[30] Low Energy Cooling, Abrams, D., Van Nostrand Reun Hold Company, New York, 1986

[31] Wingwalls to Improve Natural Ventilation: Full-Scale Results and Design Strategies, Chandra, S., Houston, M., Fairey, P., Kerestecioglu, A. Florida Solar Energy Center, PF-47-83

[32] Influence of Landscape Elements on Wind Induced Air Motion in Wide Span Buildings, Chand, I., Sharma, V., Krishak, N., Indian Journal of Technology, Vol 15, pp 369-374, Sept. 1977

[33] Studies of Air Motion Produced by Ceiling Fans, Chand, I., Research and Industry, Vol 18, n°3, pp 50-53, June 1973

[34] Air Conditioning Fan Speed Controller for Comfort and Dehumidification, Khattar, M., Florida Solar Energy Center, FS-31-85

[35] Fans to Reduce Cooling Costs in the Southeast, Chandra, S., Florida Solar Energy Center, EN-13-85

[36] On the use of Atmospheric Heat Sinks for Heat Dissipation, Energy in Buildings Journal (in print), Agas, G., Matsaggos, T., Santamouris, M., Argyriou, A.

[37] Performance Evolution of Passive and Hybrid Cooling for a Hotel Complex, Tombazis, A., Argiriou, A., Santamouris, M., International Journal of Solar Energy, in Press, 1990

[38] Earth Tubes for Passive Cooling, Schiller, G., Master Thesis, University of California, Berkeley, USA, June 1982

[39] Construction and Operation of a Hybrid Low Energy Green House, Santamouris, M., Report to the EEC, DG/17, 12/86 Demonstration Project, 1989

[40] Use of Heat Surplus from a Greenhouse for Soil Heating, Santamouris, M., Yianoulis, P., Rigopoulos, R., Argiriou, A., Kesaridis, S., Proc. Conf. ENERGEX 82, Canada, 1982

[41] Natural Cooling, Canha de Peidade, Proceedings of the Summer School on Passive Application in the Mediterranean, Cephalonia, JRC Ispra 1988

[42] Passive Cooling with Natural Draft Cooling Towers in Combination with Solar Chimneys, Cunmningham, W.A., Thompson, T.L., Proceedings PLEA Conference, 1986

[43] Comfort in Urban Spaces of Southern Europe, Lopez de Asiain, J., Proceedings 2nd European Conference on Architecture, Paris, 1989

[44] Low Energy Cooling, Abrahms, Van Nostrand Reinhold Company, New York

[45] Experimental Studies on the Effect of Roof System Cooling on Unconditioned Building, Jain, S.P., Rao, K.R., Building Science, 9, 91974

[46] Automatic Roof Cooling, Holden, L.H., Aril Showers Company, Washington DC, 2, 1957

[47] Assessment of Natural Cooling Potential for Buildings in Different Climatic Conditions, Kishore, V.N., Building and Environment, Vol 23, n°3, pp 215-223, 1986

[48] Models for Passive Cooling, Givoni, B., Plea Conference, Porto 1978

[49] The Energy Roof, a New Approach to Solar Heating and Cooling, Pittinger, A.L., White, W.R., Yellot, J.C., Proceedings of the 2nd National Passive Solar Conference ASES, 1978

[50] Reduction of Heat Flux by a Flowing Water Layer of an Insulated Roof, Sodha, M.S., Kumar, A., Singh, A., Tiwari, G.N., Building Environment 15, 133, 1980

[51] Evaporative Cooling. A Nationwide Low Energy Alternance, Watt, J., Passive Solar Journal, 1987

[52] Passive and Hybrid Cooling in Greece, Santamouris, M., Proceeding of the B2000 Meeting, Barcelona, 1988

[53] Potential of Radiative Cooling in Southern Europe, Argirou, A., Santamouris, M., Balaras, C., Jeter, S., University of Athens, Physics Department, Laboratory of Meteorology, Ippocratous 33, GR-10680 Athens. The George W. Woodruff School of Mechaical Engineering, Georgia Institute of Technology, Atlanta, Georgia, U.S.A.

[54] PROTECHNA and MELETITIKI: Potential of Passive and Hybrid Cooling of buildings in Greece. CRES, EEC-Valoren Program. 1991.

[55] Statistical and Persistence Analysis of high ambient temperatures in Athens for cooling purposes. Tselepidaki, I., Santamouris, M., Energy in Buildings. 1991.

[56] Analysis of the Summer ambient temperatures for cooling purposes.

Tselepidaki, I., Santamouris, M., Melitsiotis, D., Submitted to Solar Energy. 1990.

[57] Monthly Climatological Bulletin. National Observatory of Athens. Published Annually.

[58] Manual Description of CASAMO-CLIM. Dialogic, 1988.

[59] Earth Tubes for Passive Cooling. The Development of a Transient Numerical Model for Predicting the Performance of Earth/Air Heat Exchangers. Schiller, G., MSc. Report, MIT, 1982.

[60] Comparative Study of 13 different algorithms predicting the performance of earth to air heat exchangers against experimental data. Tzaferis, A., Liparakis, D., Dissertation Thesis. TEI Pirea, Greece, 1990.

[61] Evaporative Cooling. A Nationwide low Energy Alternative. Watt, J.R., Passive Solar Journal, 4, 3, 293-311, 1987.

[62] Passive and Hybrid Cooling Research. Yellot, J.I., Advances in Solar Energy, 241-263, 1983.

[63] Application of the CSIRO plate heat exchangers for low energy cooling of Telecom buildings. Pescod, D., Rudhoe, R.K., Proc. Institution of Engineers Conf. 80, 2, Adelaide, Australia, 1990.

[64] An evaporative air cooler using a plate heat exchanger, Pescod, D., CSIRO - Division of Mechanical Engineering, Technical Report No. TR2, Higget, Victoria, 1974.

[65] Convective cooling. Fleury, B., Ibid 5, 1990

[66] Passove and Hybrid cooling projects in Greece. Santamouris, M., In proc. Building 2000 Workshop, 155-193, Barcelona, Spain 1988.

[67] Window design practical directions for passive heating and cooling in heavy and light buildings. Hoffman, M.E., Gideon, M., Proc. Windows in building Design and Maintenance, p.p. 277-286, Sweden, 1984.

[68] Building Indices Based on Questionnaire Responses, Hedge, A., Sterling, E.M., Sterling, T.D., Proc. IAQ 86, ASHRAE, Atlanta, USA, pp 31-43, 1986.

[69] Energy Audits in Public and Commercial Buildings in Greece. Goudelas, G., Moustris, K., Santamouris, M., Proc. Conference on Renewable Energy Sources for Local Development, Chios, Greece, 1991.

[70] Building Underground: A Tempered Climate, Earth as Insulation, and the Surface Undersurface, Intersurface, Labs, K., Energy Efficient Buildings, McGraw Hill, 1980.

[71] The Climate of Earth Sheltered Buildings, Andreadaki, E., Ph.D. Thesis, University of alonika, 1986.

[72] Mulligan, H., Ph.D. Thesis, University of Cambridge, 1986.

[73] Earth Contact Buildings: Application Thermal Analysis and Energy Benefits, Carmody, J.C., Meixel, G.D., Labs, K.B., Shenm, L.S., Advances in Solar Energy, Vol. 2, p 297, 1985.

[74] Large Scale Underground Cooling System in Italian 16th Century Palladian Villa, Fanchiotti, A., Scudo, G., Passive Cooling Conference, Miami, 1981.

DAYLIGHTING

Contents

7.1

INTRODUCTION

This chapter provides a practical approach to the design of daylighting in buildings.

Why should we use natural instead of artificial light? Alternatively, why should we rely on electric light when natural light is so abundant? The use of natural daylight can bring significant advantages in cost savings, reducing cooling loads created by artificial lighting and the consequent atmospheric pollution as well as contributing to a healthier living and working environment. Natural daylight in buildings is undoubtedly agreeable, but clearly it is not 'free'; there are costs involved in bringing natural light into a building and these will depend on the types of fenestration systems used, the size and configuration of the building and any obstacles to light surrounding and within the building in question. More subtly, however, it is important to ensure an appropriate lighting and thermal environment free of discomfort from glare, overheating or excessive cooling which can occur due to incorrect window sizing.

Good daylighting design is inseparable from good architectural design and should be considered from the earliest stages of the design process. This approach is much more effective than applying daylighting techniques to a largely completed design.

Some requirements concerning good quality natural lighting within buildings are given in the Daylighting and Visual Comfort sections of the companion volume to this publication entitled 'Energy Conscious Design - a Primer for European Architects'. They cover three aspects:
- the amount of light needed (illuminance);
- visual comfort (luminance contrast and glare index);
- psychological considerations (external view, perception of time of day, agreeableness of the internal space).

This chapter considers ways of evaluating how far these requirements have been achieved. They involve establishing the following:
- critical indoor illuminance;
- daylight factors at various points in the building;
- critical outdoor illuminance;
- natural light availability.

Figure 7.1(a) Critical indoor illuminance concept

Figure 7.1(b) Critical indoor illuminance concept

Figure 7.1(c) Critical indoor illuminance concept

CRITICAL INDOOR ILLUMINANCE

The requirements for electrical lighting are often expressed in terms of the illumination level on a reference plane. In an office building this is typically the workplane. In a sports facility it might be the floor.

The section on visual comfort in 'Energy Conscious Design - a Primer for European Architects' lists optimal illuminance values for various types of task as given in the Building Energy Code published by the Chartered Institute of Building Services Engineers (UK). These values were chosen as presenting a reasonable compromise between the lighting requirements of occupants on the one hand and installation and running costs on the other hand. Given the choice, most people would opt for higher illuminance levels and one advantage of daylighting techniques is that at certain times of day they offer large amounts of light at low cost.

Although most people prefer more, rather than less, light at a task point, their attitude towards artificial and natural light is very different. They tend to accept a range of lighting conditions as long as the light source is natural. The amount of light they need is to some extent a function of the mood of the moment, the type of task being performed and, above all, the way in which natural light penetrates the room.

An individual's decision to turn the lights on is based on a perception that the space is becoming darker. This often takes place when the illuminance level is below the required value for electric lighting. Where natural light is abundant on the other hand, the occupant might not notice that the lights are on - and will often forget to switch them off.

In designing daylighting systems, therefore it is necessary to establish for a particular point the illuminance level when artificial lighting will be necessary if it is not reached by the natural lighting. This is known as the critical illuminance level. The general concept is illustrated in Figure 7.1.

In Figure 7.1, occupant A will tend to switch on the lights earlier than occupant B because:

(a) the illuminance I_A due to natural light alone will always be lower than I_B;

(b) light will come in at a more oblique angle to A's workplane than it will to B's.

As indicated above, it is frequently found that the critical indoor illuminance at A's workplane is lower than the recommended illuminance for artificial lighting. Near windows, for instance, it is often found that people do not turn on lights until the illuminance falls below 200 lux whereas electric lights usually provide at least 400 lux.

Determining the value of the critical illuminance level can be a difficult task due to varying circumstances and subjective factors. Designers are recommended to carry out their own tests by, say, evaluating their offices or other familiar spaces and noticing when they appear to be dark. Measurement of the illuminance at this moment will provide the designer with useful information. The author, for instance, found that he tends to turn the lights on when the illuminance on his desk falls below about 200 lux whereas at night electric lamps provide only a little less than 400 lux on the same desk.

There are no absolute guidelines for determining the difference between the critical illuminance level and the requirements for electric lighting because it is largely a function of human behaviour. The designer has to set the general objective, therefore, of bringing a reasonable amount of light (say, a daylight factor of 3-5% in northern Europe and 1-3% in southern Europe) to the workplane while maintaining an agreeable quality of light.

7.3

DAYLIGHT FACTOR DISTRIBUTION

To characterize the way in which natural light penetrates a building, it is useful to examine the distribution of indoor illuminance as a function of the external luminous conditions. This is known as the daylight factor distribution and, typically, is carried out with reference to an overcast sky. Figure 7.2 defines the daylight factor at point P on the workplane of a desk.

In Figure 7.2, light reaches P in a number of ways - from the sky vault, from the external environment and from reflections on the internal surfaces of the room. For an unobstructed location, the illuminance at P has been shown to be proportional to the external horizontal illuminance for the given type of sky (i.e. luminous distribution). As indicated above, it is usual for the reference type of sky to be overcast. This can be characterized either by a simple model (say, uniform sky, constant luminance) or by a standard overcast sky as specified by the Commission Internationale de l'Eclairage (CIE) with a luminous distribution which varies from 1-3 going from the horizon to the zenith [1].

Since the illuminance at P is due to light from the sky and light from the external and internal environments, the daylight factor (DF) at P can be said to be the sum of three parameters each of which relates to one of these contributions. Therefore, for any point P indoors:

$$DF(P) = SC(P) + ERC(P) + IRC(P)$$

where SC is the sky component (i.e. light from the sky vault), ERC is the externally-reflected component (i.e. light from external environment) and IRC is the internally-reflected component (i.e. light from internal environment).

It follows, therefore, that if no sky is visible from P, SC(P) = 0. Further, if no window can be seen from P, SC = ERC = 0 and DF(P) = IRC(P).

For over 80% of the floor area in most buildings, daylight factors fall in the range 0 to 5%. Close to the windows, however, they can reach 10-15%. The values of the daylight factors, therefore, provide a guide to the brightness achieved in a room (or part of a room) from daylight. A classification of the brightness of building zones on the basis of daylight factors is given in Table 7.1.

Familiarity with daylight factors enables the contribution which natural light will make to the occupants' lighting requirement to be evaluated. The set of curves in Figure 7.3, which have been extracted from a publication of the CIE [2], can be used to determine the availability of outdoor light as a function of the site latitude.

In Figure 7.3, for a given latitude selected on the horizontal axis, the fractions of time are given between 09.00 hours and 17.00 hours when the values of the outdoor illuminance on the horizontal plane (shown on the vertical axis) are exceeded. To illustrate: for a site at a latitude of 40°N it is necessary to determine how often a value of 400 lux is exceeded at locations in the building where daylight factors are 4.5%. An illuminance of 400 lux will be obtained in these locations for an outdoor illuminance level of 400 divided by 4.5%, i.e. 8,888 lux. From Figure 7.3, it can be seen that the indoor level of 400 lux will be exceeded for 87% of the time between 09.00 hours and 17.00 hours.

zone	Daylight factor (DF)	Daylight contribution
bright	>6%	very large
average	3-6%	good
dark	1-3%	fair
very dark	0-1%	poor

Table 7.1: Correspondence of daylight factors and brightness of building zone.

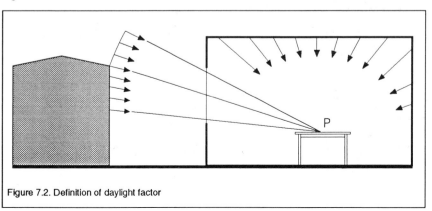

Figure 7.2. Definition of daylight factor

Limitations of the Daylight Factor Concept

The daylight factor can be a significant parameter which describes the amount of light entering a room under overcast conditions. Under certain conditions, however, the concept has its limitations. For instance, for partly cloudy skies, the daylight factor can be 0.2 to 5 times the value for overcast conditions. The maximum differences are found with side lighting when the natural light comes in from a facade. For roof apertures with horizontal glazing the daylight factor values tend to be stable provided direct sunlight does not enter the zone.

The daylight factor is a characteristic of the geometry of a space and is independant of the site and climate. It can be used to describe the performance of a natural lighting system at a particular indoor point. It does not assess the quality of the indoor luminous environment.

At points of equivalent daylight factor value, a space may appear as bright or dark according to the way the light has penetrated and the amount of contrast in the field of vision. This leads to very different luminous environments.

A daylight factor value of 1% on a desk, for instance, will make the desk look dark. The same value in a hallway will seem pleasant. A 3% value on a desk may feel more comfortable than 4% if the former case gives less glare. For this reason, it may be preferable to place a desk at right angles to a window than facing it.

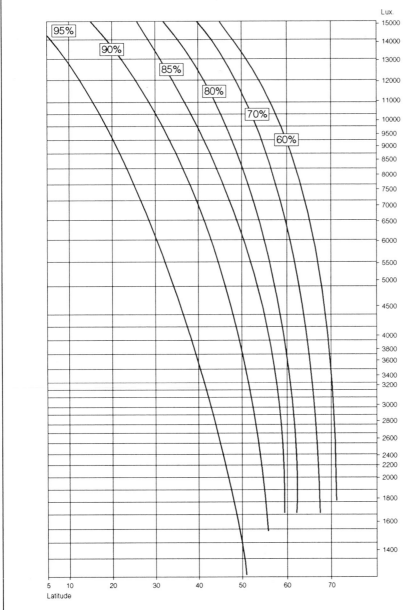

Figure 7.3. Availability of outdoor light as a function of site latitude.
From publication C.I.E. no. 16 (E 3.2.), 1970

7.4

CRITICAL OUTDOOR ILLUMINANCE

Limitations to the Use of Natural Light

A number of limitations prevent natural lighting techniques being the panacea for the lighting requirements of buildings. The first is related to the duration of day, which itself depends on latitude and season. Figure 7.4 shows duration of day as a function of latitude for certain specific days. Length of day is the same throughout the world at the equinox. It differs considerably according to location near the solstices. Therefore, daylighting techniques have a restricted use above 55°N because of the lack of natural light for one half of the year. The closer we move towards the equator, the more consistent the day length becomes.

The second limitation concerns the amount of light available. Around noon, a typical overcast sky in summer is often more than twice as bright as an overcast sky in winter. This is because in summer the sun is in a higher position above the cloud cover. Light availability can also be restricted by the presence of neighbouring buildings or vegetation. Further, outdoor light levels at the beginnings and ends of days are often too low to permit useful daylighting indoors.

Figure 7.5 shows the change with the hour of the day of illuminance levels due to the sky vault (direct sunlight is not included) for each month of the year at Kew in England. It should be noted that for the three months of November, December and January the average illuminances throughout the day are below 10,000 lux. Around noon during April to September, the illuminances exceed 20,000 lux.

From the above, it follows that for any building there is a critical outdoor lighting level (the critical outdoor illuminance) which needs to be exceeded if the indoor lighting requirements are to be met with natural light. We can define this illuminance level for a particular building as for example,10,000 lux or 20,000 lux - above which the building can be considered to be daylit and the maximum number of electrical lamps turned off.

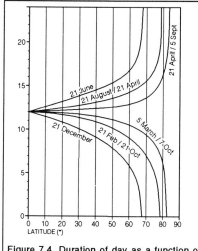

Figure 7.4. Duration of day as a function of latitude

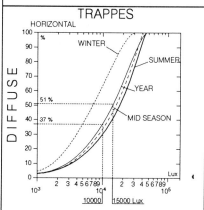

TRAPPES

Figure 7.7 Duration of day as a function of latitude

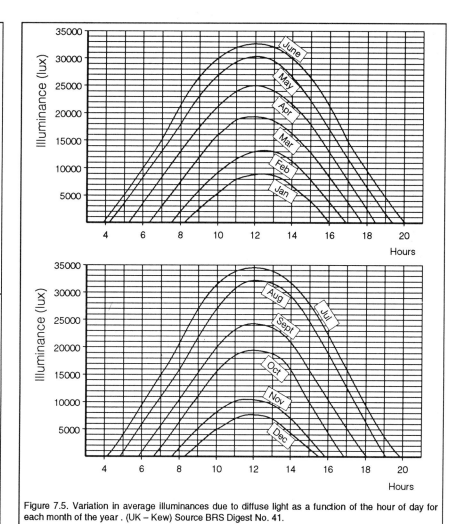

Figure 7.5. Variation in average illuminances due to diffuse light as a function of the hour of day for each month of the year . (UK – Kew) Source BRS Digest No. 41.

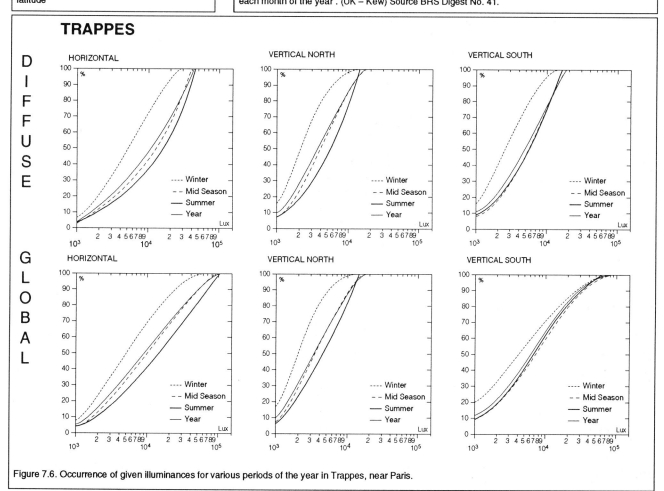

TRAPPES

Figure 7.6. Occurrence of given illuminances for various periods of the year in Trappes, near Paris.

119

Occurrence of Critical Outdoor Illuminance

Daylight availability differs from one site to another. A good way to describe a luminous climate is in terms of how often a given outdoor illuminance level is exceeded between specified hours - say, from 08.00 hours to 19.00 hours or from sunrise to sunset. This requires a knowledge of continuous measurements of the horizontal diffuse illuminance. Unfortunately, such measurements have only been made at a few locations in Europe. Radiation measurements, however, are more common. Correlations between the two parameters have been made and the distribution of diffuse illuminance deduced for about 30 European cities. The complete set of data can be found in the European Reference Book on Daylighting [3].

Figure 7.6 presents copies of graphs in the European Reference Book on Daylighting showing the occurrence of given illuminance levels in Trappes (in the suburbs of Paris) in winter, summer, between seasons and the year as a whole.

Figure 7.7 indicates how the graphs in Figure 7.6 can be used. Let us assume, for instance, that we have a building which is satisfactorily daylit when at least 300 lux is available. We want to know how often this level is exceeded between the hours of sunrise and sunset if the direct sunlight is not taken into account. With a daylight factor of 2%, the critical outdoor illuminance would be 300 divided by 0.02 = 15,000 lux. We can see from the graph that this value is not reached 51% of the time - and therefore it is exceeded 49% of the time. Thus, an indoor value of 300 lux will be exceeded approximately 49% of the time.

If the daylight factor is increased to 3%, then the critical outdoor illuminance becomes 300 divided by 0.03 = 10,000 lux. This is exceeded 100 - 37 = 63% of the time between sunrise and sunset.

The total number of hours between sunrise and sunset is 12 x 365 = 4380. Therefore, achieving a daylight factor of 3% instead of 2% should theoretically give (0.63 - 0.49) x 4380 = 641 more hours a year of satisfactory daylighting, which is slightly below 2 hours a day on average.

120

7.5
ON-SITE EVALUATION USING CRITICAL OUTDOOR ILLUMINANCE

The above parameters (i.e. daylight factor and critical indoor and critical outdoor illuminances) can be used to evaluate the performance of particular configurations of windows and other apertures.

Before using such methods in design work, one should become familiar with them by carrying out on-site evaluations of some existing buildings. This will allow results to be compared with observations.

A suggested procedure for on-site evaluation which involves the use of one or two luxmeters (see Figures 7.8 and 7.9) is outlined below:

1. Visit the building on a day when the sky is overcast.
2. Measure the outdoor horizontal illuminance on the roof - if possible, without obstruction.
3. Measure the indoor illuminances simultaneously - use a radio or have a colleague write down the time of the measurements;
4. Map the internal space in terms of daylight factor values - for example, <1%, 1<DF<3%, 3<DF<6%, >6%.

Figure 7.8. Measurement of indoor illuminance on a desk. The body of the operator should be positioned away from the window.

Figure 7.9. Measurement of outdoor illuminance. The luxmeters should be put in as horizontal a position as possible

7.6
DAYLIGHT FACTORS AND SCALE MODELS

The use of a scale model is probably the most common technique for evaluating the daylighting performance of a building during the design phase. This is because light propagation is independent of scale provided that:

- the source (i.e. the sky) is identical;
- the geometry of the model retains the same ratios as the original;
- the reflective characteristics of the surface finishes are identical;
- furniture is simulated.

Daylight factor values in a scale model can be determined with luxmeters in the same way as in a real building. The steps are as follows:

1. Measure the outdoor horizontal illuminance.
2. Measure the indoor illuminance at one position.
3. Compute the ratio of indoor to outdoor illuminance (daylight factor).
4. Move to other indoor locations.
5. Map the light distribution within the interior.

The difficulty in carrying out this process is to obtain a stable source of light. Use of scale models under a real sky is most satisfactory when the sky is totally overcast or there is a thick and steady haze. Some laboratories and schools of architecture have an artificial sky (i.e. an area where lamps are used to simulate a real sky) to which models can be brought for measurement.

The Daylight Factor Meter

Designers who do not wish to purchase a portable luxmeter (prices range from 150 to 1000 ECUs) for determining daylight factors with a scale model can make use of a tool developed for designers participating in the European architecture competition "Working in the City" in 1988 / 1989. This tool, called a Daylight Factor Meter (DF Meter), is described in Appendix 13.3.

The principle of the DF Meter is to allow the operator to compare indoor and outdoor illuminances simultaneously, on the basis of the comparison of the brightness of the two surfaces - see Figure 7.10.

The aperture related to external illuminance at the top of the light tower can be continuously reduced or enlarged. The operator varies the aperture of the meter until the brightness of the two surfaces seen in the meter are identical. At this point, a scale by the aperture displays the corresponding value of the daylight factor.

At the top of the light tower the diffuser is tilted. The diffuser should be oriented so that it faces in the same direction as the apertures which bring most of the light indoors. The model should be rotated and readings should be taken with the model facing different directions, in order to compensate for any irregularities in the sky.

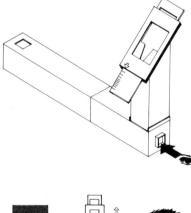

Figure 7.10: The DF Meter and its methos of operation

DAYLIGHT FACTOR SOFTWARE

The daylighting performance of a building can also be evaluated using various computational techniques which involve simulation of multiple reflections of light. There are various methods of doing this, some of which have been in use by lighting and thermal engineers for a long time.

The various steps of a computer program for evaluating daylighting performance can be defined as follows:

1. Simulation of the light source according to type of sky (overcast, clear, intermediate, average, etc.) and location of sun.
2. Simulation of the outdoor environment taking into account obstructions from neighbouring buildings and light reflection from the ground and the facades and other external elements of the building being simulated.

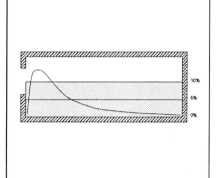

Figure 7.11. One-dimensional display of daylight factor distributions on the floor in a plane perpendicular to the window

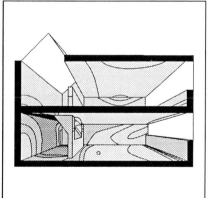

Figure 7.13. Three-dimensional display of illuminances on all the internal building surfaces, prepared using the GENELUX computer program.

3. Simulation of light propagation through the building's daylighting devices - glazing, louvres, translucent materials, etc.
4. Simulation of light propagation through the various indoor spaces, including simulation of the light reflection or transmission process at each surface.
5. Presentation of the final light distribution in terms of either illuminances or luminances as viewed by one observer..

The above specification is rather ambitious. Most computer programs take a more simplified approach by making a number of assumptions, typically as follows:

- Number of rooms. Often, one room only (sometimes with one borrowed light source) is studied.
- Geometry. Often, it is assumed that the space has a simple geometry and, say, less than twenty flat surfaces or a rectangular shape.
- Surfaces. Almost always, these

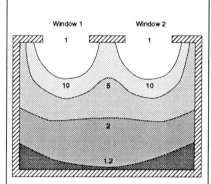

Figure 7.12. Two-dimensional display of isolux (or iso-daylight factor) curves on workplane or floor. This gives information on the spatial penetration of light in a room.

Figure 7.14. Three-dimensional display of luminances, prepared using the GENELUX computer program with a colour terminal. This technique attempts to show how an occupant

121

are assumed to be perfectly diffusing (like a carpet) although some programs simulate mirrors and a very few simulate shiny surfaces. Reflections between diffuse surfaces are simulated using various numerical techniques - some of which are sufficiently simple for the program to be used on a microcomputer.

- Number of light reflections. Programs using ray-tracing techniques rarely simulate many successive light reflections, except for mirror surfaces. They simulate diffuse light according to the theory of multiple reflections in a diffuse cavity. Some programs generating rays or photons from light sources, however, can simulate the process of successive light reflections in complex cavities with specific surfaces.

- Output. The most common output is in the form of isolux or daylight factor curves. Some graphical enhancement is often added. More sophisticated programs present pictures of rooms using colour graphics.

Examples of some output are given in Figures 7.11-7.14. They range from a very simple type of output (Figure 7.11) to a very detailed one (Figure 7.14). It is not always necessary for the designer to carry out a detailed simulation to make the right decisions on daylighting design.

A general review of computer programs for lighting design has been conducted as part of a European Commission sponsored research programme. Details are available in the European Reference Book on Daylighting [3].

It should be remembered that these programs are simulation programs and only simulate (with greater or lesser accuracy) the situation submitted to them. Their aim is to help decision-making. They do not propose solutions. It is therefore necessary for the designer to specify clearly the objectives of his or her design strategy.

All recommendations arising from these studies regarding daylight factor values in buildings should be looked at prudently. They are useful hints - but they only present one aspect of the overall lighting strategy which needs to be established by the designer.

The luminous climate, for instance, needs to be taken into account in developing an overall strategy since the availability of natural light is an important factor in determining the solutions finally adopted.

Furthermore, the energy-saving potential is not touched on in the available programs simulating light propagation. Some computer programs for determining the thermal performance of buildings, however, are equipped with a daylighting algorithm which can simulate a lighting installation operated in response to the insufficient availability of natural light. Examples of such programs are DOE2 [5] and BLAST [6]

Figure 7.15. Results of simulation of a museum with a roof aperture system in Grenoble

EXAMPLES OF DAYLIGHTING SYSTEMS

This section presents the results of some computer simulations. They show indoor daylight factor distributions obtained using various techniques for a number of cases under an overcast sky and uniform luminous distribution.

The daylighting systems in these examples should not be seen as necessarily the most appropriate solutions for other, similar, cases since the solution for a particular building will depend on its specific requirements. Rather, the light distribution pattern and the values of the daylight factors should be noted. In practice, the latter will not be attained in a real building because of:

- specific external obstructions;
- dirt deposits on the glazing;
- fenestration components which block additional light;
- absorption of light by furniture.

The daylighting factor values obtained by simulation must therefore be adjusted to take these into account. Typically, the weighting factor falls in the range 0.5 to 0.7. The value of 0.5 would correspond to a furnished space. The value of 0.7 is more suitable for a space, such as that in a museum, which is rather more empty.

The output in Figure 7.14 results from simulations carried out using the GENELUX computer program for the architect Camilio Cortesão, as part of a design study for a business innovation centre in Porto. The work was conducted as part of the Commission of the European Communities' Building 2000 programme. It contains light distribution patterns from two types of borrowed light system. The upper output presents a classical solution where the room on the lower floor receives light from the window on the upper part of the right-hand facade but where light cannot enter from the corridor on the upper floor. The second output shows that, with a tilted slope to the corridor roof, light collection in the lower room is much more satisfactory. Assuming the slope is a good reflector, the daylight factor in the darkest part of the room increases from 1% to more than 5%. In practice, of course, these conditions may be difficult to achieve.

For the museum in Grenoble designed by Group 6 Architects (see Figure 7.15), the aim was to design the roof apertures so that optimal natural light fell on the paintings, i.e. the daylight factors in the area of the paintings were in the range 0.7% to 1.5% so that degradation was reduced but there was sufficient illumination for display.

The school illustrated in Figure 7.16 is located near the Spanish border. It is an example of 'bioclimatic' architecture designed with a concern for daylighting. The south side of the building has two daylighting components - a light shelf and a clerestory. The clerestory contains reflective blinds fixed between two layers of glazing to prevent direct sunlight from reaching the students but which maintains a reasonable level of diffuse lighting inside the building.

SUNLIGHTING

In areas such as corridors or entrance halls where there are few visual constraints, direct sunlight can contribute to the quality of the space. In offices and similar spaces, the entry of beams of direct sunlight can be a source of discomfort and glare. If, however, the sunlight is diffused or directed onto the ceiling or walls then it can make a worthwhile contribution to lighting requirements.

The control of sunlight can be carried out by means of fixed or movable architectural components. These either diffuse the direct sunlight or guide it by means of directing element such as reflectors. Some examples of methods or components which allow sunlight to contribute without the risk of glare to the lighting requirements of the building are shown in Figure 7.17. They are: (a) the light well; (b) roof monitors; (c) the light shelf; (d) external reflectors; (e) the atrium; (f) light pipes; (g) the clerestory; (h) reflective blinds; (i) prismatic components; (j) tilted/reflective components; (k) claustras/cloisters; (l) external/internal shades. Thought should also be given to building orientation.

Care should be taken over introducing these components in northern Europe. Light shelves, for instance, will cut down on glare but they will also reduce the daylight admitting area and are not therefore in general suitable for regions where the skies are mainly overcast. There is, however, scope for increased use of well-designed roof apertures in cloudy climates.

In atria (see Chapter Nine), particular attention must be paid to controlling glare caused by direct sunlight or sunlight reflected from the atrium walls. Shading is a necessity. The shading devices can be located at the atrium wall facades and/or the atrium roof. Shading devices at the atrium roof can be fixed or movable. Movable devices offer greater flexibility to daily and seasonal variations in daylight availability; fixed shading devices are simpler. Diffusion of the sunlight is often achieved using fabrics. White fabrics have less impact on the colour spectrum of the light source. They are usually used for interior shading and can be designed as horizontal shades, as shades which follow the shape of the roof or as vertical banners. Light directing elements such as mirrors or reflectors are fixed shades and reduce the light-admitting area of the atrium.

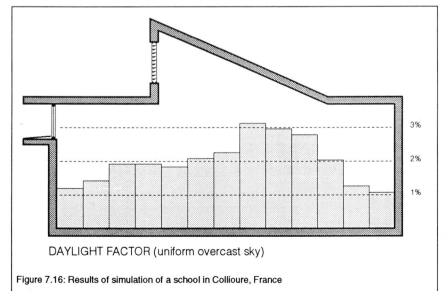

DAYLIGHT FACTOR (uniform overcast sky)

Figure 7.16: Results of simulation of a school in Collioure, France

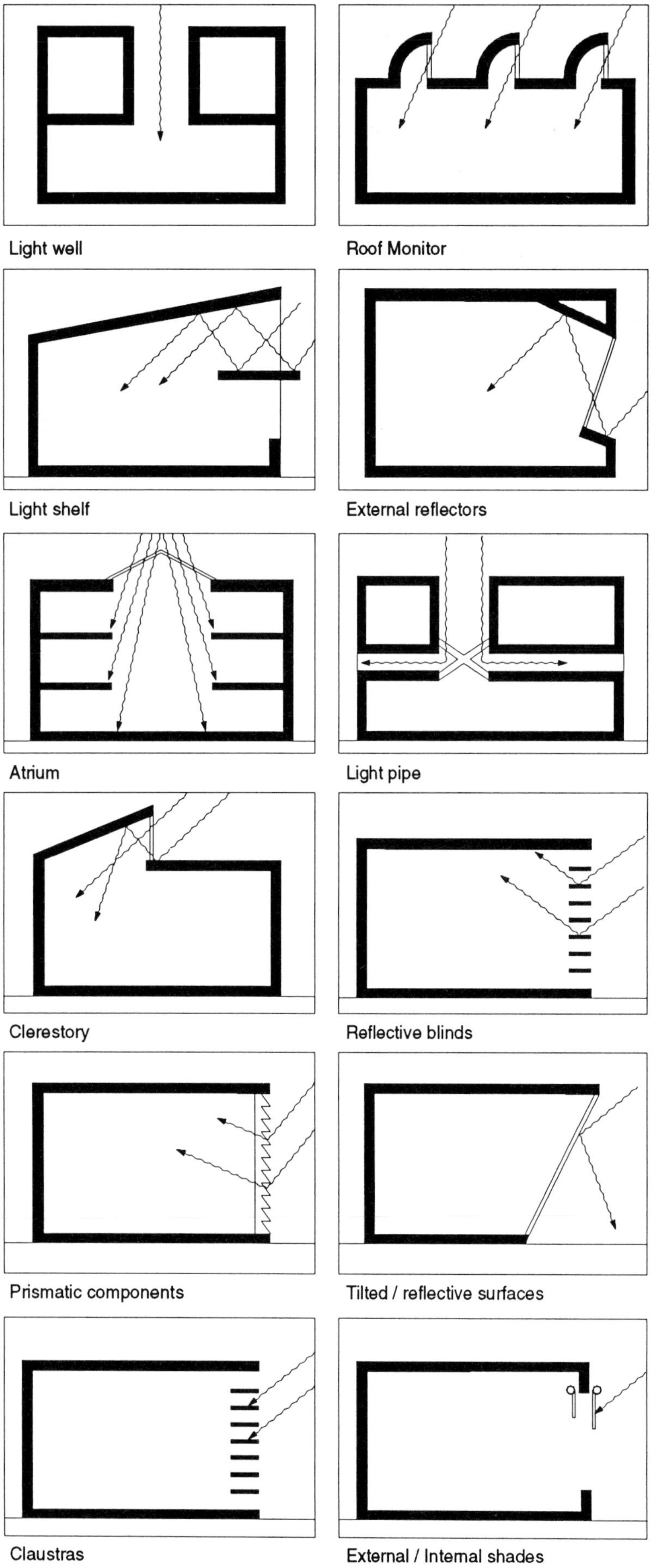

Light well

Roof Monitor

Light shelf

External reflectors

Atrium

Light pipe

Clerestory

Reflective blinds

Prismatic components

Tilted / reflective surfaces

Claustras

External / Internal shades

Figure 7.17: Components for the control of sunlight

124

CHECKLIST

In daylighting design, many choices have to be made in the proper order, otherwise the designer runs the risk of taking a step backwards and being caught in a diverging process. The decision-making process, therefore, must be properly organized and the following checklist has been included to help the designer do just this.

1. Site analysis:

Identify the directions with the most interesting and pleasant views. Consider sunlight availability on the site. Select appropriate orientation for the building and its facades. Remember that it can be more difficult to achieve good shading for west and east facades than it is for south-facing orientations.

2. Organization of indoor spaces:

Organize the indoor spaces with respect to daylighting requirements. Establish a hierarchy among the spaces, giving priority to those needing most daylight. Bring daylight into the building wherever it is necessary. Do not hesitate to create apertures in the roof. With conventional building technology, you can expect to bring light into a space up to a distance of 6 metres from the building envelope.

3. Aperture distribution:

Distribute and size apertures with respect to the volume of the indoor spaces to be lit. When possible, use borrowed light systems to bring light to deep interiors.

4. Sunshading/sunlighting:

Design solar shading devices with precision. Keep in mind the associated reduction of transmitted light under overcast conditions. When appropriate, consider solutions which allow sunlight to enter the building in a way which improves the quality of the indoor environment.

5. Verification of performance:

Designers need to evaluate how well light penetrates the proposed building. This is a real challenge. There are three techniques for doing this:

A.   If possible, compare the design with an existing building which has a similar configuration.

B. Build a scale model and evaluate light penetration under real sky conditions using a luxmeter or a Daylight Factor Meter.

C. Use a computer program. Very few of them, however, will be able to simulate the design with precision. A review of available programs is given in the European Reference Book on Daylighting [3].

6. *Detailed performance analysis:*
Adjust the configuration to improve the performance. This can be done by changing aperture sizes and other construction details. Particular attention should be given to ensuring low or easy maintenance. Daylighting systems which are difficult to maintain are unlikely to perform as the designer intended.

7. *Material selection:*
Select materials with respect to performance and the quality of the indoor environment. Remember that rooms which receive daylight in small quantities often need warmer colours to improve the luminous environment.

8. *Electric lighting:*
Develop a sound electric lighting strategy which takes full account of daylit zones. This involves taking care over the location of luminaires and the number of lamps connected to each light switch. Evaluate the possible benefits of task lighting as well as the automatic control of overall electric lighting with respect to daylight availability.

REFERENCES

[1] Commission Internationale de l'Eclairage (C.I.E.), Central Bureau, Kegelgasse 27, A-1030 Vienna, Austria.

[2] C.I.E. Publication No. 40

[3] European Reference Book on Daylighting (in print)

[4] Working in the City, O'Toole, S. & Lewis, J.O. (Eds.). Results of CEC European Architectural Ideas competition, including guidance notes supplied to competitors. Energy Research Group, School of Architecture, University College Dublin, Richview, Clonskeagh, IRL-Dublin 14. EUR 12919 EN / ISBN 0946 846 022 . IR£14.00.

[5] DOE2 : U.S. Department of Energy Simulation Program, Applied Sceince Decision, Simulation Research Group, Lawrence Berkely Laboratory, Berkely, U.S.A.

[6] BLAST is trademarked by the Construction Engineering Research Laboratory, U.S. Department of the Army, Champaign Illinois U.S.A.

CHAPTER EIGHT

CONTROL OF PASSIVE SOLAR SYSTEMS

8.1

INTRODUCTION

The dynamics of passive heating systems mean that maximum values may occur at times which do not always coincide with the need for heating. A successful application requires that the complementary contributions of solar and auxiliary systems are synchronized by passive controls and as far as necessary by a control system. For a conventional control system this is a difficult task, because of the complex nature and interactions of the passive solar systems. Therefore preference should be given to computer-based control systems with more advanced control algorithms. In this chapter conventional as well as computer-controlled systems are described.

Overheating is a problem encountered in poorly-designed passive solar buildings with inadequate control, particularly in summer during a succession of hot days. Apart from shading and other strategies described in Chapter 6 which seek to minimise unwanted gains, the problem can be diminished or completely eliminated by harnessing the cooling potential of night air without resorting to mechanical ventilation. This can be achieved by using a control system to operate motorized windows or vents. In this chapter special attention is paid to natural ventilation and its control.

Natural daylighting can also make a significant contribution to energy savings, but lighting and thermal needs can sometimes be in conflict. The use of an intelligent control system can reduce these conflicts and lead to improved thermal and visual comfort for the occupants of the building while improving energy efficiency.

Passive solar buildings should be designed to be self-regulating as far as possible, requiring minimal intervention by the occupant to operate heating, cooling or daylighting devices. However, this is not always feasible, and particularly in larger buildings or with a greater number of occupants or with more elaborate passive systems, it may be unrealistic to expect the occupants to operate manual systems to maximum advantage. Under these cir-cumstances, automatic control can help to optimize comfort and energy efficiency at all times. The performance of many passive buildings can be improved by automatic control, particularly in the case of conventional buildings retrofitted with passive systems.

In summary, the control system should manage passive solar gains, ventilation, cooling, auxiliary heating and daylighting in such a way that it minimizes energy consumption and provides good thermal and daylighting conditions in the building throughout the year.

This chapter begins with a review of the basic elements of a control system which include sensors, actuators, and controllers and continues with the control of individual systems for shading, ventilation, solar heating, and their integration. The need for appropriate control to maintain comfortable conditions and the efficient use of passive solar devices is demonstrated, as is the need for ventilation to minimise overheating.

Both the control strategy and the basic design of the building affect comfort and auxiliary energy consumption, and because of the complexity of the interactions it is difficult to give general rules. In practice one needs to simultaneously optimize the control strategy and the building design [1].

8.2

CONTROL SYSTEMS

Effective automatic control depends on the quality and correct interpretation of information provided by the sensors. Once this information has been processed, appropriate commands can then be sent to the actuators.

Sensors

In passive solar control systems the following sensors are of vital importance:

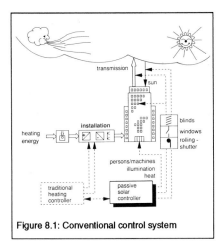

Figure 8.1: Conventional control system

Figure 8.2: Pyranometer

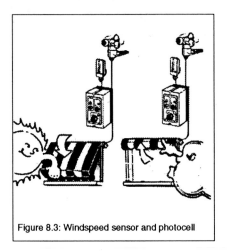

Figure 8.3: Windspeed sensor and photocell

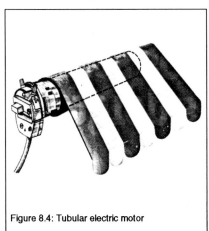

Figure 8.4: Tubular electric motor

- indoor and outdoor temperature sensors
- solar flux sensors

Sensors measuring energy consumption and windspeed are very useful for the optimization of passive gains. For more detailed information the reader should refer to the cataloques produced by manufacturers of control equipment for heating, ventilation and air conditioning (HVAC). Manufacturers of control systems for greenhouses can supply low cost sensors which are often suitable for passive control. The following paragraphs describe the generic types of sensor.

Temperature sensors

There are various types. but platinum resistors (PT100) are the most commonly used. Electronic sensors which give an instantaneous reading are now available, as are two-terminal integrated circuit temperature transducers that produce an output current which is proportional to the absolute temperature.

Solar flux sensor

The familiar pyranometer which is based on the principle of thermal conversion, is a highly accurate sensor most commonly used in meteorological data logging. The cost of pyranometers is probably too high for control purposes, but lower cost sensors are available:

The liquid crystal display (LCD) solar meter is a light sensor based on the principle of a specific electronic cell. It has an accuracy of about 98%, and can be set to provide control signals at different light intensities. The electronic transducer is a sensor which can be used for a range of control purposes including light sensing. It produces a linear output which can be readily interpreted by an electronic control system.

Wind sensors

If the windspeed exceeds the design limit, the controller must take action to avoid damage, for example to external shutters or awnings. Windspeed sensors are also necessary for the automatic control of natural ventilation.

Controllers

Many standard HVAC components can be used in control systems. Controllers based on temperature difference are useful for solar applications where for example, two temperatures are compared so that the collection of solar gains or the discharge of stored heat are optimized.

Optimal control systems

Optimal integration of passive gains requires an intelligent control system. Some exist, and the development of more advanced systems continues. Section 8.6 describes three such computer-based systems which have been developed within a concerted action of the PASTOR Project of the Commission of the European Communities.

Actuators

Control of passive solar gains essentially involves the following measures:

- Adjusting the position of insulating shutters, blinds, awnings etc., typically using a tube motor.
- Adjusting the position of air dampers with an electric or a pneumatic motor.

Examples include the control of convective airflow in Trombe walls or the control of a window vent. Automatic or manual switching determines the operation of the tube motor actuating these devices. Instead of electric motors, thermal motors in combination with thermostatic valves can be used. However, it is more difficult for these to be controlled digitally and digital control is normally essential for system optimization.

Hydraulic actuators which use freon vapour are another possibility. As the freon heats up, it evaporates and expands in the tube containing it. The resulting pressure increase can drive a piston connected to a mechanical linkage to open a window vent, for example. As with thermal motors, hydraulic actuators cannot easily be used with digital control systems.

Control variables

As comfort is one of the main objectives, its control and measurement are important. However, many of the available comfort sensors are too expensive for control applications. An economical solution is the use of 'effective temperature', where both air and radiant temperatures are measured. Measurement of these temperatures can allow the replacement of more complicated sensors, provided a lower threshold is used [2].

8.3

CONTROL OF INDIVIDUAL PASSIVE SOLAR DEVICES

Shading devices can be used for the following purposes:
- control of direct solar gain
- control of daylighting
- prevention of glare

Closed loop and feed forward

Shading devices can be controlled in a closed feedback configuration or by feedforward. In the first case a temperature controller adjusts the position of the shading device in order to keep the indoor temperature just below a specific threshold or a lighting controller adjusts the position to keep the illumination at a comfortable level. Feedforward is applied when it is desired to adjust the position only as a function of the sun and the amount of solar radiation, to avoid glare for instance. Both types of control can overrule each other; for example in the morning shading may be controlled by feedforward to avoid glare while later on in the day the temperature controller may take over by feedback control action in order to avoid overheating.

Psychological comfort

The ability to open windows at will is an important factor in the psychological sensation of comfort. Fully automated control may not always be accepted and provision should normally be made within automatic ventilation systems for the manual opening or closing of windows by occupants.

Active control of glare

Glare caused by direct solar radiation (or reflection) is one of the main reasons why people will use shading devices. Glare can be controlled by the manual or automatic operation of internal or external blinds or by the use of awnings. Thermally, in winter internal blinds are best, while in summer, external blinds are more appropriate. Control of glare is typically achieved by closing blinds or by adjusting the position of slats in venetian blinds. The control strategy is based on the penetration of direct solar radiation. This is the maximum horizontal distance of a sunlit floor surface from the window (glazing plane). Threshold values for radiation are in the range of 10 to 50 W/m² and it has been determined empirically

that this distance should be approximately 2 to 3 metres. At a 3 m distance the frequency of the use of blinds was 50% [3]. Glare, as previously mentioned, can also be reduced by adjusting the angle of slats in venetian blinds, but as it is important to maintain visual continuity, the best policy is to adjust the slats so that no direct sunlight reaches the occupants. Correct slat angles can be calculated from the profile angle of the sun. The appropriate slat angle at any given moment can be achieved by means of a control loop. The slat angle is measured with a potentiometer and the tube motor actuating the blind is advanced in measured steps to the

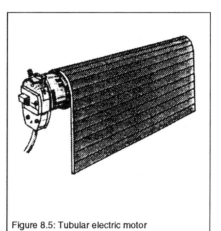

Figure 8.5: Tubular electric motor

correct position. This arrangement can also be used for daylighting control.

Optimal control

For optimal control, it is essential that the system 'knows' the current effects of shading devices and that it can predict future effects. Such prediction depends on an inbuilt model of the system which simulates the characteristics of the shading devices used. Simple models are given here for internal venetian blinds and overhangs or awnings.

Venetian blind model

The mathematical model below defines expressions for calculating the radiation and light transmitted.

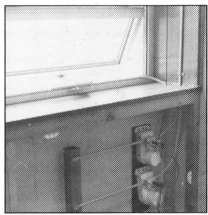

Figure 8.6: Motorized window

$$ZTA = \frac{\text{Solar energy transmitted by the window system}}{\text{Total solar energy falling on the window system}}$$

$$CF = \frac{\text{Solar energy transmitted as convective heat energy}}{\text{Total solar energy transmitted by the window system}}$$

$$LTA = \frac{\text{Visible solar energy entering through the window system LTA}}{\text{Visible solar energy falling on the window system}}$$

Euser [4] has determined values experimentally for these expressions for internal venetian blinds. These values can be approximated by polynomials as functions of the slat angle (α) of the venetian blinds and the incident angle (Φ) of the solar radiation:

$$L = X_1 \times \alpha^2 + X_2 \times \alpha + X_3$$

with:
$$X_1 = A_0 + (A_{45} - A_0) \times \Phi/45$$
$$X_2 = B_0 + (B_{45} - B_0) \times \Phi/45$$
$$X_3 = C_0 + (C_{45} - C_0) \times \Phi/45$$

By using the appropriate values for A_0 and A_{45}, B_0 and B_{45}, and C_0 and C_{45} from next Table, L can be replaced by respectively ZTA, LTA or CF.

	ZTA	LTA	CF
A_0	$+1.22449^{-05}$	$+5.71429^{-05}$	-1.22450^{-05}
B_0	-4.38780^{-03}	-1.02860^{-03}	$+4.67247^{-03}$
C_0	$+6.52653^{-01}$	$+5.57143^{-01}$	$+7.44898^{-02}$
A_{45}	$+1.21642^{-11}$	$+4.08163^{-06}$	-8.16330^{-05}
B_{45}	-1.42860^{-03}	-1.36730^{-03}	$+1.38612^{-03}$
C_{45}	$+4.74286^{-01}$	$+1.73265^{-01}$	$+3.07755^{-01}$

With the values of ZTA and CF as determined by these expressions, the various heat flows can be calculated [9].

$$Q_s = ZTA.Q_{sv} \quad \text{total incoming solar energy [W].}$$
$$Q_c = CF.ZTA.Q_{sv} \quad \text{convective part of } Q_s \text{ [W].}$$
$$Q_r = (1-CF).ZTA.Q_{sv} \quad \text{radiation part of } Q_s \text{ [W].}$$

with Q_{sv} total radiation on the window system.

Venetian Blind Model

$$\dot{Q}_r = f_D\dot{Q}_{svD} + f_d\dot{Q}_{svd} + \dot{Q}_{svr}$$

with

$$f_D = 1 - \frac{x_h\tan\beta + x_v\cos\gamma}{\sin\sigma_w\cos\gamma + \cos\sigma_w\tan\beta} \qquad \gamma \pm \frac{\pi}{2}$$

$$f_D = 1 - \frac{x_h}{\cos\sigma_w} \qquad \gamma = \pm\frac{\pi}{2} \wedge \sigma_w \pm \frac{\pi}{2}$$

$$f_D = 1 \qquad \gamma = \pm\frac{\pi}{2} \wedge \sigma_w = \frac{\pi}{2}$$

and

and

$$f_d = \sin\left(\arctan\frac{1 - x_v}{x_h}\right) \qquad \sigma_w = \frac{\pi}{2} \wedge x_h \pm 0$$

$$f_d = 1 \qquad \sigma_w = \frac{\pi}{2} \wedge x_h = 0$$

$$f_d = \frac{\sin\left(\arctan\dfrac{(1 - x_v)\tan\sigma_w}{x\tan\sigma_w - 1}\right)}{1 + \cos\sigma_w} \qquad \sigma_w \pm \frac{\pi}{2}$$

where

x_h = Horizontal swing of awning (relative to height of the window)
x_v = Vertical swing of awning (relative to height of the window)
β = Solar height
γ = Solar-surface azimuth
σ_w= Tilt angle of facade
Q_{svD} = Direct component of vertical solar radiation [W]
Q_{svd} = Diffuse component of vertical solar radiation [W]
Q_{svr} = Reflection component of vertical solar radiation [W]

Mathematical model for awnings

$$Q_{vent} = c.r.F.\Delta T \text{ [W]}$$

where:

ΔT = $T_{out} - T_{air}$ [K]
F = $S_{eff} * A_{eff}$ [m³/s]
S_{eff} = $\sqrt{C_1vw^2 + C_2h|\Delta T| + C_3}$ [m/s]
A_{eff} = $0.5 * A_1 * A_2 / \sqrt{(A_1{}^2 + A_2{}^2)}$ [m²]
A_1 = w*h [m²]
A_2 = 2*h*sin(0.5*a)
     *(h*cos(0.5*a)+w) [m²]
A_{eff} = effective window
     opening [m²]
c = heat capacity of air
     [1000 J/kgK]
C_1 = 0.001
C_2 = 0.0035
C_3 = 0.01
h = window height [m]
Q_{vent} = ventilation heat flow [W]
S_{eff} = effective air velocity [m/s]
T_{air} = Indoor air temperature [K]
T_{out} = Outdoor air temperature [K]
vw = wind speed [m/s]
w = window width [m]
a = window opening angle [deg]
r = density of air [kg/m³]
F = ventilation air flow [m³/s]

Characteristics of natural ventilation

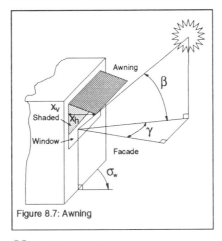

Figure 8.7: Awning

Model for awnings

The awning partially or totally prevents the admission of direct solar radiation into the room and partially the admittance of diffuse radiation. Figure 8.7 shows the effect of an awning.

The ratio of the solar radiation that reaches the facade of the building and the solar radiation that is not shaded by the awning depends on the position of the sun in relation to the room facade and the geometry of the awning.

Control of ventilaton

Ventilation is necessary for fresh air supply and for cooling in hot periods to avoid overheating. Mechanical ventilation is well understood and can be controlled by adjusting the speed of the fan motor or by on/off operation. Control of natural ventilation by adjusting window openings is much more complicated. Two problems arise when one attempts to use natural ventilation. The first is how the air flow can be measured in order to control the minimal fresh air required. The second is to determine the cooling effect of the ventilation air flow needed for night air cooling. In the latter case, windows should be opened as soon as the outdoor temperature in the evening falls to a level where it can be used to cause a decrease in indoor temperature. The exact moment when this should occur can be predicted (as can the extent to which the building should be cooled) by a control algorithm which will be discussed later in this chapter.

A simpler control strategy for cooling uses start and stop ventilation and is based on threshold indoor temperature values (24°C and 20°C), but this arrangement is less accurate and can waste heat unnecessarily.

Characteristics of natural ventilation

Both problems described above can be solved with the equations in the table to the left [5] :

These equations give quite a good prediction of the amount of ventilation which can be provided if the doors to the vented room are closed. However, the constants C_1, C_2 and C_3 are strongly dependent on the construction and orientation of the window. Open doors increase ventilation enormously but the actual rate is difficult to determine without measuring the pressure drops across the facade. Control of natural ventilation is a relatively new field and there is still much to be learned.

CONTROL OF DIRECT GAIN AND TROMBE WALLS

Control of Trombe walls must take into account the influence of direct gain. The same is true for rock storage. Heat delivery from the Trombe wall can be controlled by an air damper or vent and/or by the use of an electric fan when forced ventilation is required. Direct gain can be controlled by shading devices such as overhangs, awnings or external blinds.

Control strategy:
1. Night setback 11 pm to 7 am at 20°C, otherwise night insulation on.

Too hot :
- if T_{air} > 24°C, vent to the outside air
- if still > 24.5°C cancel vent and shade instead

Too cold :
- if T_{air} < T_{set} , turn on Trombe wall fan or open air dampers, if the space air temperature of the Trombe wall exeeds T_{air} by 10°C
- if still T_{air} < T_{set} , use auxiliary energy.

Sebold tested this strategy by computer simulation [1]. From his results the following general conclusions could be drawn:
- The control system uses much less auxiliary heat and provides a slightly more comfortable building.
- Night insulation on both the direct gain windows and on Trombe walls is the most useful control measure which can be taken and gives a 40-60% reduction in auxiliary heating.
- The Trombe wall fan control gives only a 3 - 8% reduction due to the presence of competing direct gain and is therefore of marginal benefit.
- Shading does not appear to be essential except during a few days of the year. However, night insulation can be provided by the same device, making this a very attractive option
- In hot climate zones, controls can be very useful in providing much improved comfort.
- Night air cooling was not considered. However, other research results show that it is a very useful means of improving comfort, the number of hours during which internal temperatures are too high being reduced substantially.

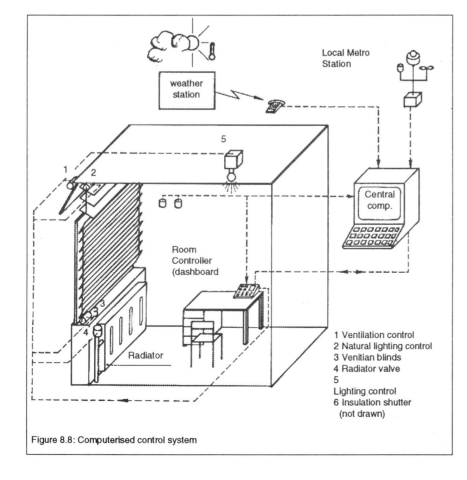

Figure 8.8: Computerised control system

Local Metro Station

weather station

Central comp.

Room Controller (dashboard

Radiator

1 Ventilation control
2 Natural lighting control
3 Venitian blinds
4 Radiator valve
5 Lighting control
6 Insulation shutter (not drawn)

ROCK STORES

Passive rock store control

Since heat delivery from the passive rock store is naturally controlled by an overlying slab floor, only the charging control needs to be considered. The rock store is charged when:

- The Trombe wall/glazing air space temperature exceeds the threshold (range 10° to 38°C)
- The temperature difference between the collector air and rock store outlet exceed a threshold (range 6 to 20K).

Sebold's simulation studies show that a passive rock store does not significantly reduce backup or auxiliary requirements and that these are of the same magnitude as the auxiliary energy consumed by the charging fan. A lower charge threshold caused a two-fold increase in the charge fan energy consumption, a 25% increase in the heat delivered by the rock store and a small reduction in auxiliary heating.

Active rock store control strategy

In tests it was found that the gross heat supplied by an active rock store (where air is moved past the rocks storing heat using a fan), although high, was typically only half of that of the passive rock store (where air moves past the rocks by natural convention). Nevertheless the auxiliary heating energy consumption was more efficiently reduced (twice as great as the passive rock store which showed a small reduction). An active rock store delivers a relatively large amount of heat during the night (1 - 6 am). Active rock store charge strategies were identical to the passive case except that different thresholds were allowed. Active discharge should be aimed at minimizing auxiliary energy use. The bin was discharged only when:
- the room temperature could not be maintained above the thermostat temperature by other means which did not involve auxiliary energy use.
- the hot side of the bin exceeds the room temperature by some threshold (typically 10K)
- its discharge provided the minimum auxiliary energy.

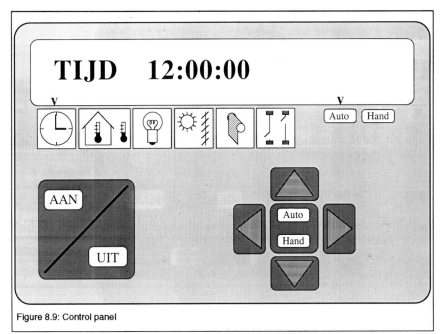

Figure 8.9: Control panel

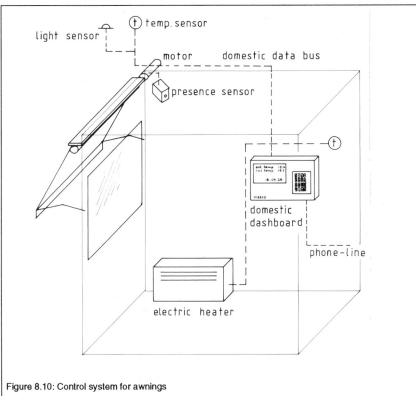

Figure 8.10: Control system for awnings

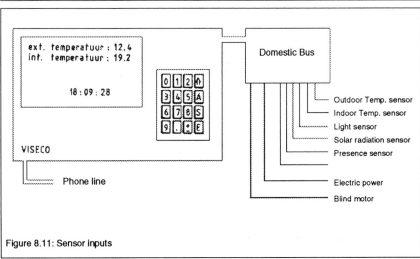

Figure 8.11: Sensor inputs

COMPUTERISED CONTROL

The following control systems have been designed as part of a concerted action of the Project PASTOR of the Commission of European Communities [11]. The control systems are designed for office buildings and dwellings to control the passive solar gains, natural ventilation, heating, and lighting in such a way that they minimize energy consumption and provide good thermal and daylighting conditions throughout the year. This is achieved by an integrated control system for window devices (blinds, vent windows, etc.) and the traditional lighting and heating system. Detailed information can be found in a series of publications [7],[8],[9].

New techniques are applied in the control systems such as adaptive modelling, adaptive control, weather prediction with local weather models and optimizing algorithms. These techniques allow optimal control strategies to be determined. The control system for dwellings has a control panel with limited control and it functions like an advisor. The control system for office buildings permits fully automatic control. In each room a control panel is provided which can communicate with a central computer and which takes care of the local control inputs. The central computer functions as an optimizer. The three control systems will be discussed briefly in the next paragraphs.

Passive office building control system

This system designed by the Technical University of Delft [11] takes the changing needs for heating, ventilation and lighting into account. It goes one step further than passive solar systems by using natural ventilation for cooling purposes. This system is therefore called the passive climate system.

Auxiliary heating is provided by radiators. Each radiator has a control valve in the case of hot water heating or an on/off control in case of electric heating. The window has a sun protection system (indoor venetian blinds), an outside insulating shutter and window openings for ventilation. In each room there is one controller, with a panel giving information to the occupants about the room's thermal state. In addition, it controls the indoor temperature and/or illumination automatically and can accommodate manual override by the occupant.

Control system design

The control system consists of a central computer and room controllers (figure 8.8). The central computer is the optimizer which provides the best setpoints for each room controller. These setpoints will be calculated by means of a thermal model of the building and a weather predictor. The thermal model can be identified from the outputs of sensors. The room controller sets the various components in the desired positions and controls the indoor temperature in a closed loop configuration. As a first step a prototype that controls a test-cell has been developed and will be tested. The room controller is designed as a control panel. It can operate in automatic or manual mode (figure 8.9). The ON/OFF switch activates the room controller and optimal conditions are achieved automatically. With the mode switched to the AUTO/MANUAL position the system works as a remote controller. A specific item can be selected and changed by remote control using the display cursor.

Control panel system for domestic use

This control system from NAPAC/ARMINES [11] is a commercial low-cost system for passive solar houses.

The proposed control system corresponds to the actual conditions in a room in a passive solar house with electric heating (which is typical for France). A control panel governs the operation of blinds according to information provided by the sensors and the user based on two criteria: comfort and energy saving. An important feature is the fact that total control of the system is not given to the control panel. Ventilation should not be controlled by the panel and regulation of the blind is limited by the user. The proposed passive system consists of a living room with solar gains that can be modulated by the use of blinds. An electric heater ensures the proper indoor temperature. Five sensors provide the control panel with the necessary information to enable the control of two actuators: the motor for the blinds and the electric heater. See figure 8.10.

Control system design

The control strategy implemented in the main program of the system uses the physical measurements and the setpoints defined by the occupants simultaneously, and controls both the heating system and the lighting. The problem is quite complex because of conflicting objectives: energy saving, and comfort.

To achieve realistic characteristics, the control operates on the following basis:
- Desired thermal and visual comfort levels are provided by the user.
- A thermal model of the building is automatically determined by

the system during the first few days of operation.

Direct gain control system

The aim with this system from CSTB/ARMINES [11] is to develop controllers using adaptive strategies for the combined control of direct gains and auxiliary heating in housing (see figure 8.12).

The system to be controlled is a one room building based on a test cell. It has a large window facing south and it has medium range thermal inertia. The room has an adjustable awning for sun shading and an electrical convector for heating.

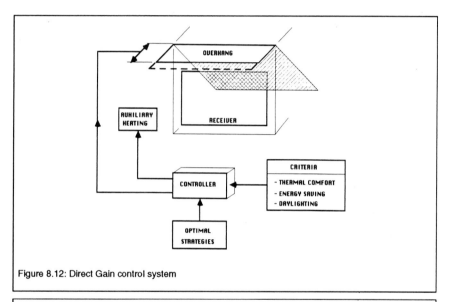

Figure 8.12: Direct Gain control system

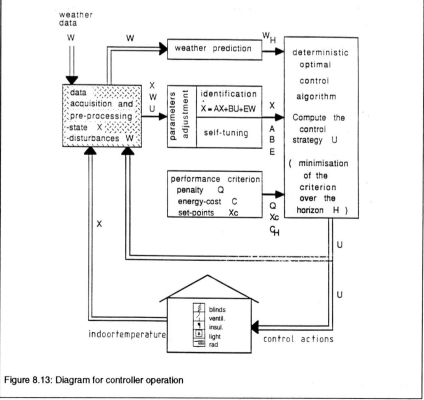

Figure 8.13: Diagram for controller operation

Control system design

The control strategies that were studied range from the very simple to the very sophisticated, using optimal control theory. An example of a simple strategy is based on a linear control law that governs the depth of an overhang and the heating power in accordance with the inside temperature or the available solar radiation. The output is a proportional, differential and integral function of the input. A more sophisticated one is an optimal control strategy, based on the minimum principle. It is able to adjust the equipment, taking into account economic criteria and comfort requirements. It uses a simplified thermal model to represent the dynamic behaviour of the building.

Internal structure

The internal structure of the controllers is represented schematically in figure 8.13. It functions in the following way. The outdoor climate variables W (horizontal global radiation, outdoor temperature, windspeed and wind direction) and the indoor temperatures (X) are measured. Then some calculations are carried out (preprocessing); for instance the solar radiation upon a vertical or inclined surface is calculated from the measured horizontal solar radiation. The data after preprocessing is used to identify the parameters of a model relating the temperatures X (air temperature and surface temperature) with the outdoor climate variables W and the control actions U (for example the position of the blinds, etc.). The parameters are found by means of a recursive least square parameter estimator [10]. The weather data W are also used to identify a weather model, which can predict the weather for the next 24 hours. A weather report can be added in order to improve prediction. This weather predictor WH and the model of the building are then used to determine the optimal control strategy U (desired values for room temperatures and the positions of the actuators). With feedback control these desired values are achieved. In the office building version these feedback loops are implemented in the room controllers, while in the dwelling version all functions are in one computer.

134

8.7
RESULTS OF SIMULATIONS OF CONTROLLED PASSIVE SOLAR SYSTEMS

Computer simulations have been carried out to answer the following questions:
- Is it possible to achieve a comfortable indoor climate in summer with a passive solar system?
- How much energy is required for heating?
- What are the influences of window size and the mass of the building?
- What is the influence of direct control strategies?

Simulations have been carried out with four types of buildings and two control strategies. these buildings have a window area of 40% or 80% of the facade area and light or heavy interior walls. A simple control system consists of feedback loops for radiator, window and awning control with fixed set-points and a simple start mechanism. An optimal control system uses prediction and optimizing techniques to calculate an optimal control strategy. A detailed description is given by Lute [12].

Figure 8.8 shows a summary of these results. The conclusion is that an acceptable indoor climate is possible in a building with heavy interior walls and a relatively small (40%) window area, when the internal heat load is lower than $10W/m^2$. The number of overheating incidences is within the comfort limit that is required by the Dutch National Building Service. Figure 8.8 also shows the effect of the optimal control system. The savings on heating vary from 14 to 40% and the decrease in the number of overheating incidences varies from 35 to 50%. When an internal heat load of $20W/m^2$ is present, the indoor climate in summer is unacceptable with the simple control system, while the optimal control system just exceeds the comfort limit. The conclusion is that a good control system has a beneficial effect on the indoor climate and energy consumption of a passive solar building.

To cater for higher internal heat loads than the previously-mentioned $20W/m^2$, it will be necessary to install a mechanical

cooling system. However the combination of a conventional HVAC system and a passive solar system can lead to a considerable amount of energy saving on cooling and yet an acceptable indoor climate. A condition for such a combination is that the control system can control both the cooling device and passive solar components such as awnings and vent windows. The following strategy can be considered:
- Use the vent windows for cooling during the night with cold outdoor air.
- Use the mechanical cooling only during the day, when cooling by ventilation is not sufficient.
- If a user wants to open the vent window anyway, even if this is not an advantage, then turn off the cooling device or when heating is required, decrease the set-point of the heater. The user should be notified that the control system will respond in this way.

The possibilities of this combined system of passive solar components and mechanical cooling are investigated for a building with a window size of 40% and heavy interior walls. Figures 8.9 and 8.10 show the results of different strategies for an internal heat load of $40W/m^2$.

The capacity of the cooling system chosen is $50W/m^2$, which would be sufficient in a conventional building to keep the indoor temperature within the comfort limit in summer. Figure 8.9 shows the energy consumption of the cooling system and the number of incidences of overheating for the following 3 situations.

A - Only mechanical cooling.
B - Mechanical cooling during the day and natural ventilation through the windows during the night.
C - As B, but also cooling by natural ventilation during the day, where possible.

The capacity of the cooling system has been selected to deal with situation A. Overheating for this situation is within comfort limits. Night ventilation reduces the energy consumption considerably (57%). When ventilation is also applied during the day, the energy consumption reduction approaches 84%. Comfort conditions have improved as the number of overheating incidences is reduced to zero.

Another advantage of this system is the possibility of reducing the capacity of the mechanical cooling device. Figure 8.10 shows the results using the same combined system, but with a cooling capacity of $25W/m^2$ (50% reduction). Situation A shows a large number of temperature excesses. Night ventilation reduces overheating considerably, however the number of incidences is still too high. When daytime cooling by natural ventilation is also in operation (situation C), the number of incidences of overheating drops below the comfort limit and again a large reduction in energy consumption is achieved. The conclusion is that the capacity can be lowered by 50% and that energy consumption can be reduced by over 85% with an intelligent ventilation control strategy. This strategy in essence says: "Cool the building by means of the vent windows until this is not possible anymore, then close the windows and use the cooling system".

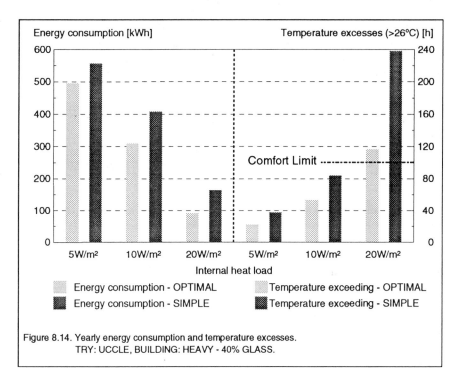

Figure 8.14. Yearly energy consumption and temperature excesses.
TRY: UCCLE, BUILDING: HEAVY - 40% GLASS.

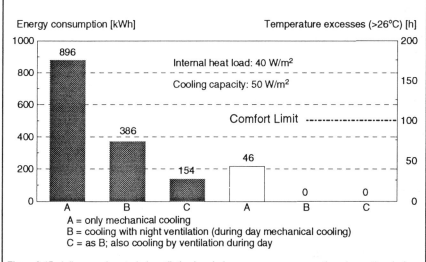

Figure 8.15: Influence of controled ventilation by windows on energy consumption of a cooling device and temperature excesses. (Cooling capacity chosen for system without vent windows)

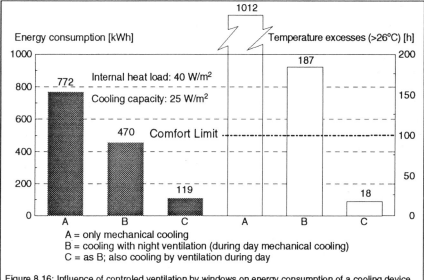

Figure 8.16: Influence of controled ventilation by windows on energy consumption of a cooling device and temperature excesses. (Cooling capacity 50% of required capacity)

135

REFERENCES

[1] Sebold, A. , Clinton, J.R. , Impact of Controls in Passive Solar Heating and Cooling of Buildings, Report Contract nr. DE AC04 - 79Al10891 US Department of Energy, 1980.

[2] Noll, S.A. , Wray, W.O. , A Microeconomic Approach to Passive Solar Design; Performance, Cost, Optimal Sizing and Comfort Analysis. Energy, the International Journal, vol 4, no. 4 , August 1979, page 575.

[3] Inoue, T. et. al. The Development of an Optimal Control System for Window Shading Devices based on investigations in office buildings, ASHRAE Transactions, 1988, V 92 Pt 2.(OT-88-03-2).

[4] Euser, P. and Knorr, K.T., Zonwering bij gebouwen. Jalouzieen bij enkele en dubbele beglazing, TPD Report 003.112-4-BIII-37, 1972.

[5] Phaff, J.C. Ventilation in buildings (Ventilatie van gebouwen, onderzoek naar de gevolgen van het openen van een raam op het binnenklimaat van een kamer). Report C448, IMAG-TNO, March 1980 (in Dutch).

[6] De Gids, W. and J.C Phaff, Ventilation rates and energy consumption due to open windows, Air Infiltration Review 4, nr.1 , November 1982

[7] Paassen, A.H.C. van , Digital Control of Passive Solar Systems; European Conference on Architecture, Munich , 6-10 April 1987.

[8] Paassen, A.H.C. van , Passive Building Control, CLIMA 2000, August 1989, Serajewo.

[9] Paassen, A.H.C. van , Final Report of Concerted Action on Passive Building Control, CEC Contract EN35-0036-NL, 1989.

[10] Iserman R. Digitale Regelsysteme; Springer Verlag, Heidelberg, 1977.

[11] Participants in this Action included: Technical University of Delft, Ecole des Mines de Paris, Centre Scientifique et Technique des Bâtiments, and the French company NAPAC, Paris.

[12] Lute P.J., Liem S.H., Passen van A.H.C. 1990. Control of passive climate systems with adjustable shutters, shading devices and vent windows Report MEMT 11, Delft University of Technology, The Netherlands, Novem-contract 11.26.112.10.

[13] Lute P.J., Liem S.H., Passen van A.H.C. 1990. Benefits of adjustable shutters, shading devices and vent windows in passive solar buildings Proceedings of International Symposium on Energy, Moisture and Climate in Buildings, CIB, Rotterdam, The Netherlands.

CHAPTER NINE

ATRIUM DESIGN

9.1

INTRODUCTION

Historically, atria were uncovered, internal patios within dwellings in southern Europe. They provided a tempered climate and were valued as protected, private outdoor spaces.

Public buildings adopted uncovered atria at a larger scale. With the emergence of new technologies for the production of metal and glass in the 19th century, glass covered atria, conservatories and arcades became popular, their use spreading to northern Europe and beyond. The indoor climate in these spaces was originally maintained throughout the year by passive means, but with developments in air-conditioning and an increased use of glass on building facades, passive atria, conservatories and arcades became less common and much experience in their use was lost.

The 1950s brought a revival of the atrium as a commercial amenity in offices, shopping malls and hotels, initially in North America and then in Europe. Most of these atria were mechanically air-conditioned and consequently wasted large quantities of energy.

In recent years the energy-saving potential of atria relying on passive solar principles has been rediscovered and applied to advantage, particularly at northern latitudes, typically in offices, buildings, hotels and shopping centres.

Apart from energy considerations, atria are commercially popular and this fact frequently dominates other considerations of energy use and location with their architectural potential often ill considered. It is of course an incorrect assumption that an atrium building which works well at one latitude will always work well at others.

It is important to be clear about the purpose of designing an atrium building from an energy standpoint. An unheated, glazed atrium allows us to increase the sizes of glazed areas to spaces 'looking in' to the atrium, thus improving the potential for natural daylight to replace electric light in those spaces, while greatly reducing heating losses from the whole building. As such the atrium may become a useful intermediate space for climate tolerant functions such as circulation, shopping and exhibitions. Used in this way an atrium can perform as a covered pedestrian street or square

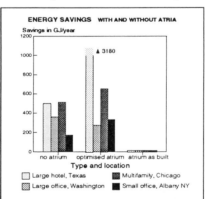

Figure 9.1: Energy Performance of Commercial Buildings with and without atria.(Redrawn from ASHRAE Paper [3]).

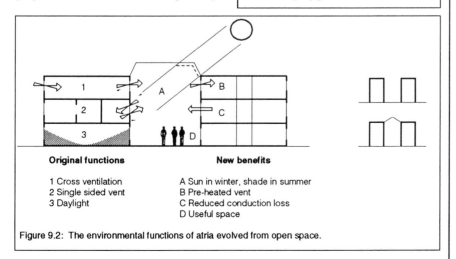

Original functions

1 Cross ventilation
2 Single sided vent
3 Daylight

New benefits

A Sun in winter, shade in summer
B Pre-heated vent
C Reduced conduction loss
D Useful space

Figure 9.2: The environmental functions of atria evolved from open space.

Figure 9.3: Züblin Headquarters Stuttgart 1985 by Gottfried Bohm, in which an atrium links two office blocks vertically and horizontally.

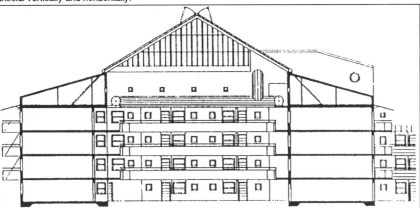

Figure 9.4: Suncourt Project in which apartments are arranged around an atrium which acts as buffer space and source for a heat pump, Stockholm, Sweden, by VBB Sweco 1984-1986.

Figure 9.5: Opreyland Hotel, Nashville, Tennessee, U.S.A. where a 2hectare atrium garden provides walks at any time of year and sites for "outdoor" restaurants.

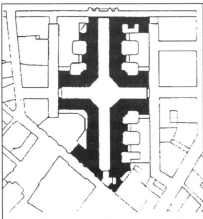

Figure 9.6: Site Plan of the Galleria in Milano, by Mengoni 1867.

the examples studied in the U.S. save energy by comparison with the same building without an atrium; but that optimizing the design can reverse this situation. (see Figure 9.1)

It is helpful to consider the evolution of the atrium from an open space which would provide the parent building with daylight, natural ventilation and useful passive solar gains if it faces south. These natural benefits may be very pleasing to occupants as well as being energy-saving.

If the open space is now glazed over as an atrium or conservatory we must ensure that these natural benefits are not lost. In addition, we must ensure that summer overheating in the atrium or parent building does not occur. If the parent building was mechanically cooled, an overheated atrium would increase cooling costs.

To ensure the original functions of the open space are maintained we should observe the following points:

1. Every effort should be made to optimize the available daylighting levels, using highly reflective surfaces and clear glazing for the roof of the atrium.
2. Consider the fresh air supply to the atrium and the use of atrium to temper (pre-heat) air for the parent building in winter.
3. Provide shading and ventilation for summer use.

Sometimes (1) and (2) conflict. The use of tinted or reflected glass or fixed shading devices to prevent summer overheating, for example, will result in reduced daylight levels in winter and may require the use of additional artificial light in the parent building and even in the atrium itself. Therefore, solar gain control should be variable and ventilation should respond to climatic conditions.

If the principal reasons for constructing an atrium or sunspace are aesthetic or commercial and energy use is ill-considered, the result can easily be an increase in energy consumption of the building.

However, if the original functions of the open space are at least maintained and the atrium does not produce an increase in energy consumption, we can then go on to consider the potential environmental benefits of the atrium or sunspace.

blocking out wind, rain, pollution and noise and modifying the temperature and light levels. Experience has shown that such atria are often attractive and interesting places to be in and to look into from surrounding spaces.

There is, however, a strong belief that the incorporation of an atrium or sunspace in a building design leads automatically to reduced energy consumption. Atria do not always reduce energy consumption and may actually increase it. One of the rare systematic studies of atrium performance [3] suggests that few of

9.2

ADVANTAGES OF ATRIA

The advantages of atria over open courtyards are much more obvious in northern latitudes - hence their prevalence in Scandinavia, Germany and the U.K. and their rarity in Italy, Portugal and Spain. The following advantages are often quoted:

- The provision of semi-outdoor space with protection from cold and wet weather. (Figure 9.3 illustrates the protected courtyard in the Züblin Offices [4].

- The conversion of open courts to daylit and protected space, which can be used for circulation, restaurants, recreation in hospitals, etc.[5]

- The possibility of using the atrium as a useful sink for warm extract air, for example the Suncourt Housing Rehabilitation in Stockholm [6] (Figure 9.4), or as a preheater for ventilation air.

- The reduction of heat loss from building surfaces which would otherwise be exposed to winter weather (Figure 9.5 shows a large hotel, whose wings, ranged around a court, are protected as a result of the glazed roof).

- The reduction of maintenance costs to facades otherwise exposed to the weather.

- The enhanced use of daylighting so that for the majority of the year no electric lighting is required during office hours (see the Mount Airey Library Figure 9.8).

- The provision of vast and entertaining interior gardens. (Figure 9.7 illustrates the gardens in the John Deere Offices).

- The provision of links, both within one building and between streets, (for example The Galleria, Milan by Megoni, Figure 9.6).

Disadvantages of Atria

The problems however, involved in designing atria, are also serious:
- Added fire and smoke risks.
- The provision of ventilation to spaces which would otherwise be open to ambient air.
- The loss of daylight to rooms adjacent to courtyards, caused by roof structures.
- The risk of cross contamination in hospital atria, and the spread of odours in glazed malls.
- The cost of glazing.
- The risk of overheating.

These points are addressed in the following Sections: but designers are urged to read Saxon's book [1], which contains design guidelines, and to examine the design guidelines prepared for the competition WORKING IN THE CITY [7].

9.3

ATRIUM FUNCTIONS

Internal Communication
Vertical and horizontal communication can be greatly improved by linking disparate parts of a building with an atrium. In the Landeszentral Bank, for example Figure 9.9, there are four banking functions which used to be carried out in unconnected offices, whereas in the new building, the four sets of staff constantly meet on their way across the linking atrium to communal areas. Similarly the vertical communication with four glass lifts and the subsequent exposed walkways straddling the atrium at each floor level in the Wiggins Teape HQ, encourage people to mingle, Figure 9.10.

Similarly, hospitals, schools, universities, libraries, museums, shopping centres, hotels and multi-family housing can enjoy the benefits of atria.

Linking Urban Spaces
Atria can be used to link streets, as Mengoni has shown (Figure 9.6). Unfortunately, the modern desire for secure buildings and town planning which does not often give priority to public amenity, preclude such beneficial uses of atria in many cities. Notable exceptions are found in north America, where the severe climate encourages extensive protected arcades joining, and running parallel to, the streets. For example, the public is encouraged to use the Bell Offices, (Parkin, 1983), as a link between City Hall and the Eaton Arcade and to walk through the street level of the Philadelphia Stock Exchange which unites four blocks - Figure 9.11. Similarly, the Hanse Viertal Shopping Centre in Hamburg links some streets, but the opportunities are fewer in such dense urban situations.

Shelter
The sheltered central court of an atrium creates an all-weather semi-public gathering space. The atrium can bring in light but prevents wind, rain and extreme temperatures. The shelter effect is most marked when the atrium is not serviced to reach full-comfort conditions itself, but acts as a buffer space, a transition area between inside and outside.

Figure 9.7: John Deere HQ Moline Illinois, by Roche Dinkeloo 1987. Private indoor garden linking offices and providing restaurant and discussion spaces.

Figure 9.8: Mount Airey Library, N. Carolina, Pease & Associates, Consultant E Maxria. 1982.

Accommodation

The need for accommodation is usually the force which triggers the building process of an atrium. It is usually the space around the atrium which is directly required, and the atrium space is a bonus. However, the two interact and depend on each other. Hotels, shops and offices gain from the people-pulling power of atria and glass-covered arcades.

Cultural Function

Atria appeal to the mind and the senses. They put people at the centre of things in a way lost in recent architecture. They encourage playing as on a stage, people-watching, promenading, moving through space, enjoyment of nature and social life. New hotels have developed this feature to a degree where a visual antidote to the oppressive interiors and the formless external spaces of today is obvious.

All of these atrium functions are interlinked and their benefits, while not applicable in all cases, should be given serious consideration as they may not otherwise be so easily provided.

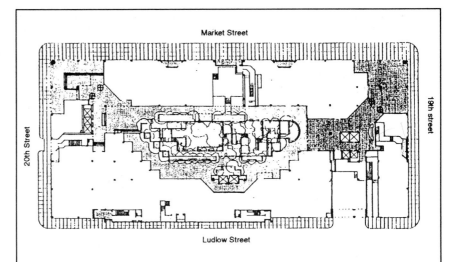

Figure 9.11: Philadelphia Stock Exchange, joining four dtreets with protected walks through the planted atrium. Cope Linder Associates 1981.

Figure 9.9: Landeszentral Bank

Figure 9.10: Wiggins Teape HQ, Basingstoke, Arup Associates, 1983

Figure 9.12: "Slightly more erotic, perhaps, were the atrium and green houses that were favourite settings for romance in Victorian England. The lovers could get lost among the leaves of exotic tropical plants and possibly mistake the hot humid atmosphere for their own concealed passion." (Lisa Heschong, "Thermal Delight in Architecture", 1979).

140

9.4

THERMAL RESPONSE

Atria as Collectors of Useful Solar Heat:

Commercial buildings have small heating loads - they have large volume to surface area ratios and high internal gains. Typically, the heating load will be only 10% to 50% of the total load (Athens to Copenhagen). Figure 9.13 and 9.14 show typical values for domestic and non-domestic buildings.

Added glazing will only save heating energy if the useful solar gain captured exceeds the added thermal loss through the glass, over say a 24 hour period. Excessive solar gain which merely has to be vented cannot be regarded as useful. The "LT method" shown in Appendix 13 illustrates the typical impact of added glazing on space heating energy for a variety of atrium forms and locations in the Community countries.

When a building is in a climate which imposes a heating load, then increased north glazing increases that load, but decreases the lighting load. The detrimental effect on the heating is less for east and west glazing, and for south or horizontal glazing the heat load may be reduced. However, the gains by reducing the small heating load are almost always overtaken by the gains achieved by reducing electric lighting and may be outweighed by the loss due to added cooling induced by excessive glazed areas.

Winter Performance

In winter there will be a marked increase in the temperature of the space, even if unheated, above the ambient temperature. This temperature increase will be dependent on a number of factors:-

1. The ratio of external glazing area of the atrium to the wall area of the parent building which is protected by the atrium (Figure 9.15).
2. The thermal transmittance of the separating wall. Usually this will be dominated by the amount of glazing in this wall.
3. The orientation and inclination of the glazing of the atrium or sunspace (availability of solar gain).

Temperature elevations for an unheated atrium are shown in Figure

9.17. These are predicted for a typical atrium configuration in the U.K. Also shown is the "seasonal shift" in thermal climate. Note that, due to their variability, thermal conditions are described as a "climate" rather than an environmental standard.

The temperatures shown are monthly averages. Daytime temperatures are much higher since the ambient temperature is higher, and all of the solar gains and the conductive gains from the parent building occur in this period. A computer simulation shows the hourly atrium temperatures for a typical February day in Figure 9.18; Figure 9.19 shows the daytime increment above the monthly mean for sunny and cloudy days.

The elevated temperature will have a major influence on the usefulness of the space. For example, in most cases it will be sufficiently high for use as circulation space. But air temperature is not the only environmental parameter influencing the usefulness of the atrium. Air movement will be reduced to virtually zero which will influence the "seasonal shift" further than implied by the temperature elevation curve alone.

Another important factor which will have considerable influence on the enjoyment of the space by the occupants is the penetration of direct sunlight. This provides psychological benefit, and the physiological effect of raising the effective temperature. The degree of sunlight penetration is mainly influenced by the geometry of the atrium. However, imaginative use of mirror surfaces and reflective glazing could refresh places that normal sunbeams cannot reach (Figure 9.20)

Thermal Energy

The temperature increment of the atrium has two main energy benefits. Firstly it reduces the heat losses from the parent building via the wall separating the heated building form the atrium. (This buffer effect allows a larger fraction of glazing to be used in this wall, compensating to some extent for the detrimental effect of the atrium of daylighting).

Secondly, the atrium offers the possibility of providing pre-heated ventilation or "make-up" air to the heated building. This means of

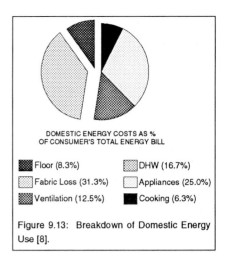

DOMESTIC ENERGY COSTS AS %
OF CONSUMER'S TOTAL ENERGY BILL

Floor (8.3%) DHW (16.7%)
Fabric Loss (31.3%) Appliances (25.0%)
Ventilation (12.5%) Cooking (6.3%)

Figure 9.13: Breakdown of Domestic Energy Use [8].

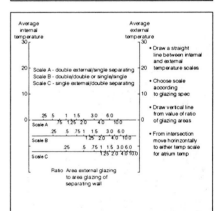

Figure 9.14: Breakdown of Non-Domestic Energy Use in Atrium Buildings [9].

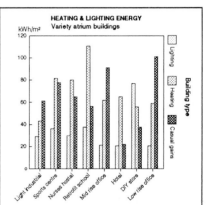

Figure 9.15: A nomogram for estimating the elevation of temperature is given in Figure 9.16

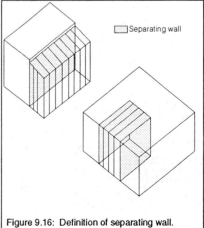

Figure 9.16: Definition of separating wall.

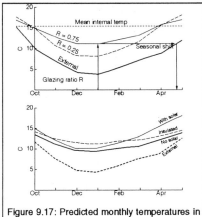

Figure 9.17: Predicted monthly temperatures in a typical atrium.

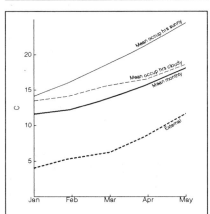

Figure 9.18: Computer simulation of hourly temperatures in an unheated atrium.

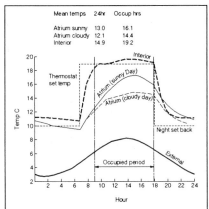

Figure 9.19: Daytime temperature increment over monthly average.

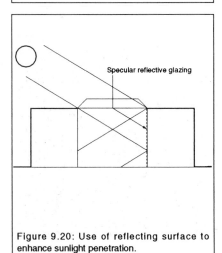

Figure 9.20: Use of reflecting surface to enhance sunlight penetration.

thermal energy saving is particularly appropriate to larger buildings since ventilation heat loss assumes a greater proportion of the total heat loss than in smaller buildings. Furthermore, where in a normal building ventilation air would have to be mechanically handled and pre-heated, to avoid the local chilling effect of cold incoming air, the atrium climate will provide the opportunity of "natural" ventilation to the surrounding rooms by casual window opening.

Figure 9.22 shows the effect on the annual heating load of a hypothetical atrium building, with various ventilation options. (A) represents the building with the atrium removed, (B) with the atrium present but no ventilation coupling with the building. (C) shows the effect of ventilation pre-heating as described above. In (D) ventilation air from the building is dumped into the atrium, elevating its temperature and reducing conductive losses further. (E) is an interesting possibility - recirculating exchange with the atrium such as might be provided by simply opening windows. The large atrium volume acts as a reservoir of fresh air, refreshed by infiltration over the 24 hour period and perhaps assisted by the effects of vegetation.

Whilst the heating energy savings look very encouraging, it must be pointed out that heating energy forms a relatively small proportion of the total energy cost of larger buildings. The largest savings are often to be made in lighting, cooling, and fan and pump energy, usually all requiring expensive on-peak electricity.

Summer Performance

In summer the main concern is to prevent overheating. It is difficult to reduce air temperatures in the atrium below ambient temperature throughout the day; although this may be possible during the morning, following ventilation overnight when the air temperature is below the daily average. However, it is possible to provide shade, since the atrium envelope provides a structure upon which to install movable shading devices. Shading provides a way of rejecting solar gains as soon as they enter the atrium (or even before, with external devices), thereby reducing the heat buildup. A combination of

shading and ventilation can keep the temperature increment to within 3K or 4K above ambient (Figure 9.23).

A second benefit is that the shading of a person from direct sun has the effect of reducing the effective temperature by between 4K and 8K.

The reason for recommending movable shades must be stressed. The problem is that the sky brightness varies over a wide range. On cloudy days in winter, the atrium glazing should provide a minimum reduction to the sky illumination, in order to maintain the daylighting in the rooms and indeed, to provide enough light in the atrium for the planting. Furthermore, the ingress of direct sunbeams would be welcome.

On a sunny day in summer, however, the situation is reversed . With the sky brightness twenty times greater, there is now no shortage of daylight and the presence of direct sun will overheat the atrium in general, and increase surface temperatures of both plant and human occupants. Obviously if there is sufficient shading in the summer, then there must be too much in winter. Movable shading is the answer.

Ventilation is the other means by which summer temperatures can be limited. This may be generated by either wind pressure or thermal buoyancy (stack-effect). In either case, ventilation will be more effective if openings are provided top and bottom. In most cases, breezes cannot be relied upon, and thus the limiting case for design will be to provide sufficient stack effect ventilation.

An atrium can be used to provide stack-effect ventilation for itself or the parent building. However, it must be emphasised that this works only when the average temperature of the air in the stack is greater than the external air temperature. In summer conditions, this may already be too hot. Also, note that the driving pressure is dependent upon the average temperature of the air column, not the maximum. Heating up the top of the stack only will not be very effective.

Figure 9.21 shows the predicted stack effect ventilation for a four storey roof-glazed atrium with and without shading. Note that to keep the average temperature inside the atrium down to about 4°C above

	No atrium: open courtyard	Exhaust airatrium	Supply airatrium	Buffer atrium: no ventilation
Heat loss from building thousands of kWh/year	640	410	396	437
% reduction in heat loss	0	36	38	32

Table 9.1: Energy savings from options in the retrofit overglazing of a hospital courtyard.

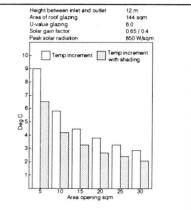

Figure 9.21: Calculated temperatures for atrium cooled by stack effect and shading.

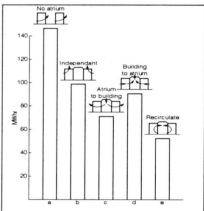

Figure 9.22: Annual heating energy consumption for a building with an atrium.

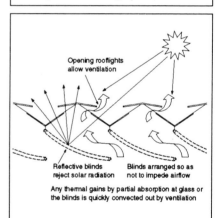

Figure 9.23: Shading and ventilation to prevent overheating.

Figure 9.24: State of Illinois building Chacago.

ambient it is necessary to have an open area top and bottom of about 12% of the roof area with no shades, and about 7% when shades are present. At the air-flow rates that these openable areas allow, there will be almost no stratification.

At night, heat retained by massive elements in the atrium will cause a greater temperature difference between the atrium and the cool night air, and will drive stack ventilation to provide useful night cooling. This must be considered by the designer since providing large openings at ground level when the building is unoccupied may present security problems. Security grilles as an alternative to doors (necessary in the winter) offer a solution.

Atria are often required to have power operated vents for the purpose of smoke venting in event of fire. Usually the openable areas required for this will be adequate for summer ventilation, in conjunction with shading.

Prediction of Overheating

Most of the dynamic simulation models will give a designer a good measure of overheating problems. None yet embraces sufficient convection modelling to provide complete, accurate data about stratification. Examples such as the State of Illinois Building in Chicago (Figure 9.24) which exposes a huge glass cylinder truncated at the top and facing south, to high ambient temperatures and solar radiation and which severely overheats [10], illustrate the problems but offer no solution.

Remedies

External shading, as previously mentioned, is far more effective than internal shading, and movable devices prevent the loss of useful daylight and heat when these might be obscured by fixed devices. Natural ventilation can readily achieve 20 air changes per hour.

9.5

VENTILATION DESIGN

Ventilation through the Atrium

Make-up air for mechanical ventilation systems or for natural ventilation to adjacent offices if drawn through the atrium will collect heat on its way; however such a process cools the atrium in winter and may overheat offices in summer. Thus for large atrium buildings using the atrium as an outlet for ventilation air does ensure slightly higher temperatures without heating in winter as shown in Figure 9.25

In January, for example, the elevation of atrium temperature above ambient is highest for the exhaust air design (8K) and lowest for the supply air atrium (5.6 K).

In both cases the savings are small. Table 9.1 shows data from a hospital case study where the open courtyard shown in Figure 9.26 was to be glazed [11].

Measured Performance

Architects and engineers, often have a healthy scepticism concerning the computer model predictions of academics. Fortunately, however, we can quote here the results of a modest monitoring programme on a real atrium which does generally adopt the principles outlined above. The atrium at Cambridge Consultants Limited on the Science Park at Cambridge is unheated and has natural ventilation of both atrium and surrounding rooms. The atrium finishes are all highly light reflective; this, together with clear roof and window glazing, gives good daylighting to the surrounding rooms. Cream canvas "sails" shade the south sloping segments of the pyramidal rooflights. These are electrically operated by a temperature sensor. The opening of the vents in the roof glazing is also operated automatically.

The results have been encouraging. Figure 9.27 shows the temperatures recorded over two seven-day periods in June 1985 and February 1986. For the period in June, the maximum daily temperature stayed within 4 K of the external temperature. For the winter period, which was exceptionally cold, the minimum atrium temperature was about 7 K, 18 K above the ambient temperature. The average atrium temperature was about 14 K above the average ambient.

Clearly this free-running atrium has a climate rather than an environmental standard. This actually seems to suit the plants extremely well. The plants have been investigated by Ian Woodward of the University Department of Botany who concludes that their excellent health is due to a combination of good light levels and the seasonal stimulation of the atrium climate.

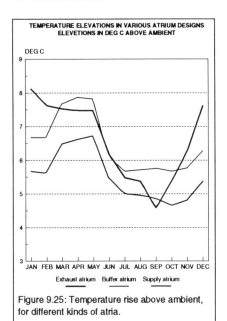

Figure 9.25: Temperature rise above ambient, for different kinds of atria.

9.5

VENTILATION DESIGN

Few large atria in the US are naturally ventilated, and indeed the outdoor climate is frequently more demanding; but in Europe, the tendency is to avoid providing active cooling or heating to atria which are transiently occupied, such as foyers and lobbies, and to shopping malls where people are on the move most of the time.

For example, the linking atrium in the Züblin offices passively receives air, actively supplied to the surrounding offices, and allows buoyancy to remove the air via roof vents. The Wiggins Teape HQ goes further and allows natural ventilation from the atrium to draw air through the offices from outside. Both are examples of exhaust air atria.

Where for example lobbies contain seated receptionists, local heating or cooling can be provided, sufficient to temper conditions to a height 3m above the floor. However the demands on planted atria are frequently more stringent, since plants can be up to 10 or more metres tall and will suffer badly if they exist in two climatic regions or suffer from perpetual draughts. Frequently designers and clients want to put coffee terraces or meeting areas high up and looking into the atrium. If these are open to the void, the implications for climate control can be expensive.

The Hanse Viertl shopping centre in Hamburg [12] allows natural ventilation in the malls but conditioned air in the shops. To prevent cold air falling from the glazing, heaters can be placed at the foot of the glazing or, as in the case of the very tall glazed walls in the Hyatt Hotel at Cambridge Massachusetts and the new National Gallery in Ottawa, heaters can be placed at intervals all the way up the wall. In general there is more apprehension about discomfort caused by falling cold air than there is about dripping condensation. As a rule of thumb, designers should aim to keep the inner glass temperature less than 4K below the prevailing indoor temperature.

Some atrium buildings which have not been purpose designed for a client but speculatively built, use brute force and abandon natural ventilation. The Deutsche Bank in Frankfurt is an example, which may be compared with the Landeszentralbank [13] a block away, where conditioned air from offices is led between two layers of atrium roof glazing to prevent cold down-draughts; but the atrium itself is naturally ventilated.

Calculation of natural air flow rates.

The designer may wish to gain an approximate idea of natural air flow rates. It has been pointed out already that accurate tools are not readily available. The following guidance is offered:

- use a building thermal analysis program to calculate the atrium temperatures in summer and winter.
- use these temperatures to calculate the natural exfiltration rates.
- use the calculated exfiltration rates as inputs to the thermal program to recalculate air temperatures.
- if the temperatures change repeat the procedure.

Thermal programs such as SERI-RES, ESP, DOE 2.1C etc. would be appropriate, and BREEZE [14], VENT or other methods summarised in Reference [15], might be used for the ventilation calculation. However Owens suggests a simple hand calculation method [16], which is repeated here:

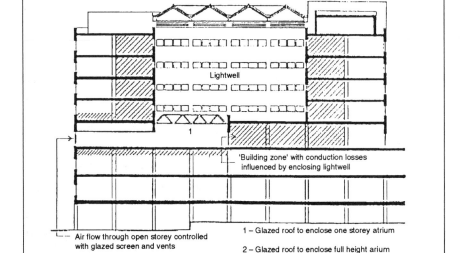

Figure 9.26: Section through John Radcliffe hospital.

1 – Glazed roof to enclose one storey atrium
2 – Glazed roof to enclose full height atrium

Air flow through open storey controlled with glazed screen and vents

Airflow =

$$0.827 \times \frac{A1^*A2}{(A1^2 + A2^2)0.5} \; Dp^{0.5} \; m^3/s$$

where:

A1 and A2 = inlet and outlet free areas in m^2

D_p = difference in pressure

= 3462 h [$1/t_0+273 - 1/t_1+273$]

h = height between openings [m]

t_0, t_1 = inside and outside temperatures in deg. C.

It should be noted that the temperature difference is roughly proportional to the atrium height, and thus the buoyancy forces are similarly related to height; but since the flow through an opening is roughly proportional to the area of the opening, the air change rate for tall atria is less than short atria, although the volumetric flow is greater. Figures 9.28 and 9.29 illustrate the response to height and open area for atria on hot day in a temperate climate.

Substantial flow rates can be achieved, and if the 10% roof opening recommended for smoke removal is also used for natural ventilation, then more than 20 air changes per hour may be available for cooling in the summer. (The air outlets through the glazing at roof level, may fail to work effectively in cold weather, if the surrounding glass temperature is very low, causing the air suddenly to cool down. It is recommended that insulating glass be used.)

In the design of a Law Court at Canterbury, Perera et al [17] show that inlet duct flow rates of 1000 to 2000 m^3/h are readily achieved using natural ventilation, and that even under varying wind and temperature conditions, the desired flow rates are provided for over 90% of the time.

However, natural ventilation is not capable of cooling vast areas of glass in hostile climates. The very highly glazed State of Illinois Building severely overheats. By contrast, the Bateson Building (Figure 9.30) in Sacramento, which also has a very hot summer, uses large numbers of roof vents successfully to keep the building cool. Night flushing in the summer, and destratification using fans blowing air down yellow decorative fabric tubes, provide year-round comfort conditions.

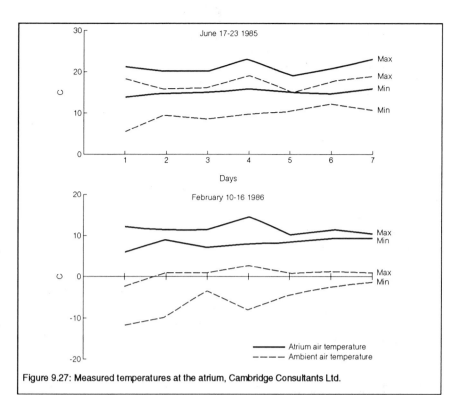

Figure 9.27: Measured temperatures at the atrium, Cambridge Consultants Ltd.

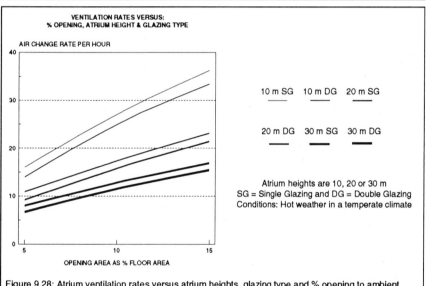

Figure 9.28: Atrium ventilation rates versus atrium heights, glazing type and % opening to ambient.

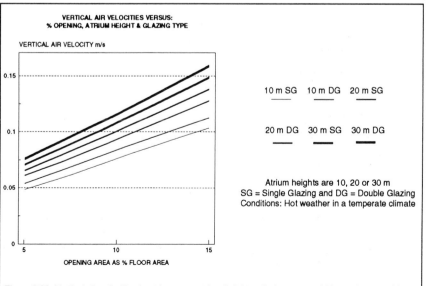

Figure 9.29: Vertical air velocities in atria versus atrium heights, glazing type and % opening to ambient.

145

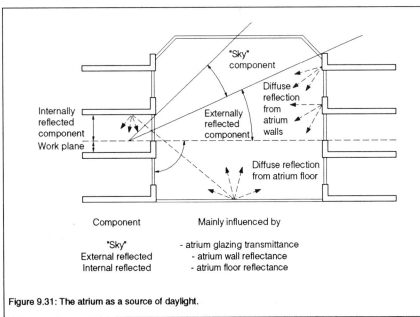

Figure 9.30: Bateson building, Sacramento.

Figure 9.31: The atrium as a source of daylight.

Component	Mainly influenced by
"Sky"	- atrium glazing transmittance
External reflected	- atrium wall reflectance
Internal reflected	- atrium floor reflectance

9.6

DAYLIGHTING IN ATRIA

An atrium can make a very significant contribution to energy savings in the parent building by providing a source of natural light, often deep in the building, which replaces artificial light. Both electricity consumption and cooling loads (even where high-efficiency lighting is used) can be reduced. By comparison with an external wall, the area of glazing between the parent building and the atrium can be increased to admit more daylight while energy losses in cold climates are greatly reduced.

The importance of daylight for psychological reasons is now recognised. For normal heights we can expect a room to be daylit for a maximum depth of about 6 m from the facade, provided the facade is glazed for about 50% of its wall area. External obstructions will reduce this, and the surrounding walls of an atrium are, of course, obstructions. However, these obstructions can reflect light into rooms provided they have light coloured finishes. Indeed there could be cases in dense urban sites where rooms facing a highly reflective atrium could receive better reflected light than from external building surfaces dirtied by the urban atmosphere.

Usually daylit rooms will also receive light direct from the sky. In the case of the atrium, the 'sky' is now the glazing, and thus its brightness is reduced by the absorption of the glass and the fraction of the fixed shading. The reflected light also has to pass through the glazing. Thus, where daylighting of surrounding rooms is required, it is essential to provide as much light from the sky into the atrium and the rooms, by the geometry and the use of high transmittance glass, and as much reflected light as possible by the use of highly reflective finishes to the atrium walls (Figure 9.31). The subject of Daylighting is treated in more detail in Chapter 7.

Design parameters

Daylight entering the building will pass many obstacles whose design will affect the quantity and quality of natural light which eventually reaches the desk or 'workplace':

- geometry of the courtyard/atrium
- roof construction
- roof glazing
- atrium walls and floor
- glazing to the occupied space
- dimensions of the occupied space
- interior reflectances of the occupied space.

Geometry of the atrium

The proportions of the atrium will affect the amount of direct daylight reaching the atrium floor. The wider and more shallow the atrium space, the better generally the contribution of direct daylight from the sky. The shape of the floor plan is also important (Figure 9.32).

If the atrium is shallow and wide, then the shape of the floor plan is less critical. However as the building height increases the distribution of daylight becomes more dependent on internal reflections and a simpler, quadrangular plan shape often performs best. For a given area of glazing and building height, the amount of daylight reaching the floor of the quadrangular atrium shown in Figure 9.33 is 7% higher in the rectangular atrium.

The quadrangular atrium provides four sides with roughly equal illumination whereas the rectangular atrium provides two different levels of facade illumination.

Roofing of the atrium

The amount of daylight entering the space depends on:

- the roof construction
- the transmission of the glazing
- the shading devices and solar controls.

In atria designed for a climate with mainly overcast skies, one should attempt to minimise the area taken up by the primary and secondary structural members to maximise the area of glazing. This will affect the type of roof construction chosen and the dimensions of structural members.

The following diagram (Figure 9.34) shows different types of roof shape and support structure with a maximum span of 20 m. In each case it is assumed that the atrium roof is sitting on top of the surrounding building.

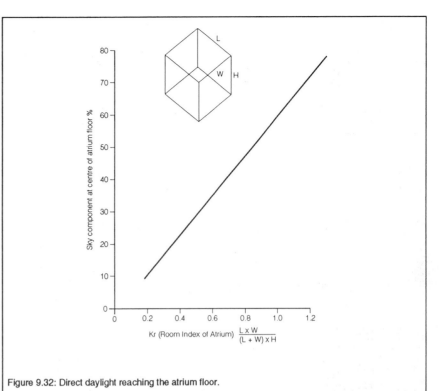

Figure 9.32: Direct daylight reaching the atrium floor.

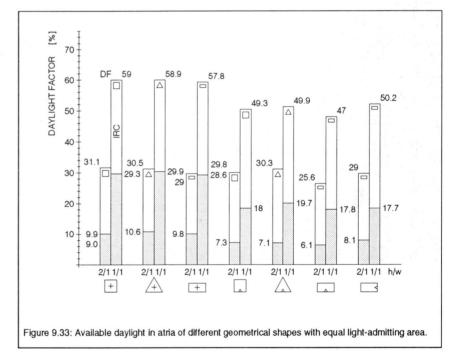

Figure 9.33: Available daylight in atria of different geometrical shapes with equal light-admitting area.

	quadrangular floor area				rectangular floor area			triangular floor area
vertical supporting structure	6.8	7.7	8.8	8.8	7.5	8.0		8.0
vertical + horizontal supporting structure	10.5	11.0	11.7	11.4	11.7	10.5	10.5	11.6

Figure 9.34: Percentage reduction of the light admitting area by different roof constructions.

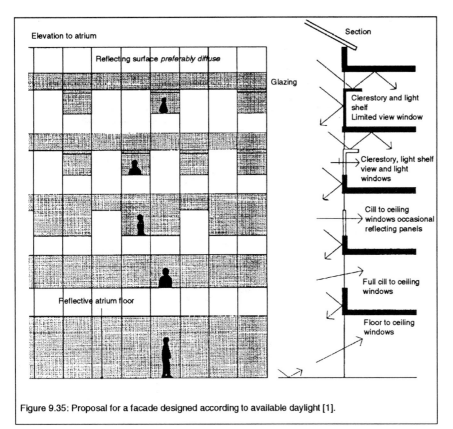

Figure 9.35: Proposal for a facade designed according to available daylight [1].

(Labels within figure: Elevation to atrium; Reflecting surface *preferably diffuse*; Glazing; Reflective atrium floor; Section; Clerestory and light shelf Limited view window; Clerestory, light shelf view and light windows; Cill to ceiling windows occasional reflecting panels; Full cill to ceiling windows; Floor to ceiling windows)

Legend:
- — · — · — highest position of sun
- ▬ ▬ ▬ direct sunlight
- ▬ ▬ ▬ direct sunlight
- ⇨ diffuse daylight
- non-specular, movable sunscreening prismatic panels
- specular surface, fixed sun-screening prismatic panels
- optical control prismatic panels

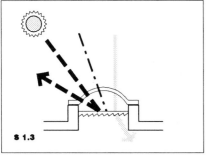

S.2.1

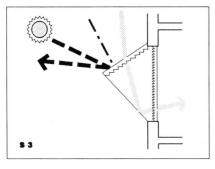

S 3

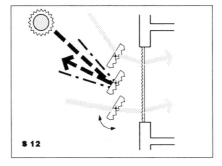

S 12

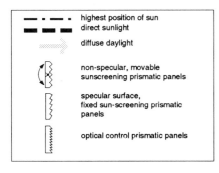

S 1.3

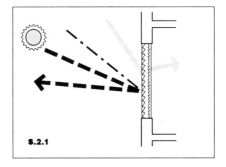

S 2.2

Figure 9.36: Schemes of prismatic structures.

Roof glazing

The optical properties of glazing materials influence daylighting quality and lighting savings. The thermal properties affect heating and cooling loads.

The two major types of glazing are: transparent and translucent, the latter being clear, tinted or reflective.

Transparent, clear glazing transmits the most daylight and provides the most natural view of the sky. However these materials admit the strong direct beams of sunlight to the building. Beam sunlight may be blocked, redirected, scattered or diffused by interior objects (shades, banners, structural members, etc.) to prevent possible discomfort for occupants.

Translucent glazing materials which diffuse sunlight do not allow a direct view of the sky. Major differences in weather outside can however be detected. To provide a uniform light quality under direct sun conditions the glazing material should be highly diffusing. Highly diffusing materials however tend to have a lower light transmittance which reduces the light levels under overcast sky conditions drastically. Diffusing materials with higher transmittance properties are not as good at diffusing the light uniformly.

To reduce downdraughts and condensation and to ensure adequate natural ventilation in case of fire, it is desirable to use double glazing. The future appearance of 'smart glass' which will act as electronically controlled shading and daylight control will make designers' tasks easier. Tensioned fabric roofs are becoming more common (for example [21]) and can provide excellent daylighting quality to the atrium without danger of overheating; but their low transmissivity reduces light in deep atria and adjacent rooms.

None of the readily available daylight models handles specular reflection well and designers may find it easier to build scale models to examine the intensity of the daylighting and the effects of surface reflectivities. For example see the Building 2000 Project [19] and Figure 9.34.

The lower light transmission properties of transparent tinted glazing are commonly used to reduce sunlight and the 'sky view'. However the colour of the tint affects the perceived colour of objects in the space below and therefore the better light transmission properties of reflecting glazing materials are often preferred.

For daylighting purposes transparent glazing materials with high transmission properties are the best - for example single glazing with no tint.

Atrium walls and floor

The design of the atrium walls will affect the distribution of light in the space. The amount of reflected light will depend on the average reflectance of the walls and the type of reflection. Diffuse reflecting materials perform better but tend to increase glare for the occupants especially when sunlight strikes the walls .

Walls covered with a light, highly reflective finish would direct most light into the building, but the need for internal glazing to light adjacent spaces of necessity diminishes the amount of reflected light.

The way light enters the occupied spaces varies at different levels of the atrium: the upper part receives predominantly direct light from the sky and the lower part receives predominantly reflected light from the opposite walls and the floor of the atrium. As a consequence as much light as possible should be reflected to the lower floor; since glazing does not reflect as much light as white walls the amount of glazing in upper storeys should be reduced to the minimum sufficient to provide good daylight for the occupants. Each floor will therefore have different window sizes, being largest at the lowest floor and smallest at the top. Another design option is to alter the ceiling heights for each floor having the highest spaces at the bottom of the atrium.

The atrium floor reflectance will also influence light levels in lower floors. Light colours such as white marble provide the greatest amount of reflected light. This may be reduced by the presence of plants or other obstructions which should therefore ideally be located towards the centre of the floor allowing a band of clear floor around the perimeter to reflect light.

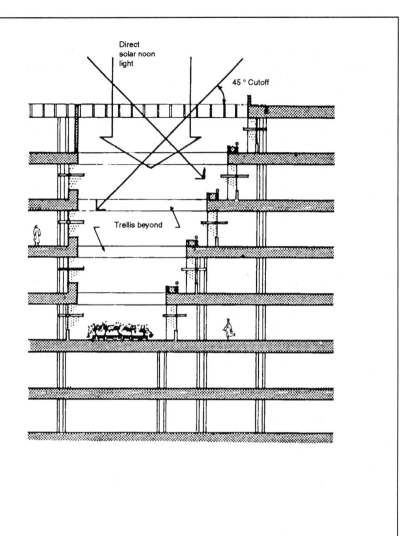

Figure 9.37: Trellised light court section.

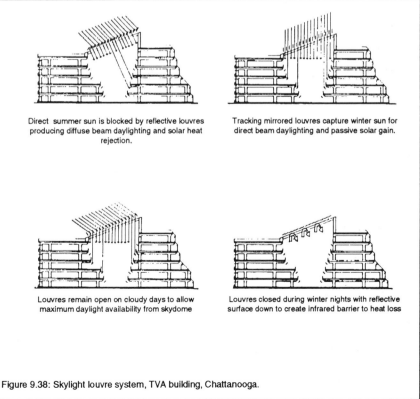

Direct summer sun is blocked by reflective louvres producing diffuse beam daylighting and solar heat rejection.

Tracking mirrored louvres capture winter sun for direct beam daylighting and passive solar gain.

Louvres remain open on cloudy days to allow maximum daylight availability from skydome

Louvres closed during winter nights with reflective surface down to create infrared barrier to heat loss

Figure 9.38: Skylight louvre system, TVA building, Chattanooga.

Glazing to adjacent spaces

For thermal savings, glazing is recommended between the atrium and the adjacent space. The glazing used should, for daylighting purposes have high light transmission properties, single glazing being generally best if building and fire regulations permit.

Dimensions of adjacent spaces

Light levels in rooms with conventional windows fall off quite rapidly. Normally, daylight can be quite adequate for most tasks up to a depth of twice the ceiling height. As the available daylight in the atrium decreases towards lower levels, the depth of adjacent spaces being daylit from the atrium should also decrease otherwise some artificial light will be required.

Adjacent spaces can be doubly sidelit leaving only a small 'dark' zone at the centre which may be used for circulation, etc.

Interior reflectances should be as high as possible. Depending on the direction of the 'exterior' light sources in the atrium the parts of the room receiving the most light from the atrium should have the highest reflectance, i.e. the ceiling on the lowest floor due to the reflectivity of the bottom of the atrium.

The levels of illuminance can be best improved by designing the atrium facades in such a way that light from the zenith (which enters through the atrium roof) is optimized for lower floors. Two strategies may be applied are:
- step back the floors so that each 'sees' some sky.
- use control devices which direct light from the zenith towards the ceilings especially at lower floors. Lightshelves, reflectors and prismic structures are examples of such devices, and have the added benefit of diminishing direct glare.

Sunlighting

Developments in the use of sunlight in the U.S. suggest similar possibilities for Europe, but these may not be directly transferable due to climate differences. In climates with mainly overcast sky conditions sunlighting strategies may work, but as the area of glazing is reduced by comparison with conventional atria, they may not be so effective under all conditions.

Sunlight in atria can cause glare, either directly or by reflection. Shading is often necessary either on the atrium walls or roof. Roof shading devices may be fixed or movable, the latter offering greater flexibility for seasonal and daily variations, but fixed shading is cheaper.

There are two main categories of shading:
- Diffusing shading which intercepts direct sunlight and creates a bright diffuse 'sky'.
- Redirectional shading which consists of reflectors, mirrors, prismatic and hydrographic structures.

Diffusing the sunlight can be achieved by using fabrics; white fabrics have little impact on the colour of the light source. They can be designed as horizontal shades, or they may follow the slope of the roofing or may be used as vertical banners. If the sunlight strikes these fabrics their luminance increases and this may cause discomfort for the occupants. The facades of the atrium must be designed correspondingly, e.g. lightshelves to reduce glare from the sky.

Light directing elements such as mirrors or reflectors are fixed shades and reduce the light-admitting area. Prismatic structures can be used if the desire for a sky view through the glazing is less important than the day/sunlighting issues. These elements are often extremely expensive.

FIRE

Predicting Fire and Smoke Spread.

Atria give rise to added fire and smoke risks [22]. Thermal simulation modelling is now well established and predictions can be made about overheating and the benefits of various heat recovery schemes, with confidence; but the level of confidence in prediction falls fast as we consider daylighting and then natural ventilation and the spread of smoke.

The interaction of daylighting measures on building thermal loads is readily assessed using such models as ESP or SERI-RES and other large simulations; but buoyancy-driven air flows are now well modelled and rigourous modelling with computational fluid dynamic methods (CFD) requires large amounts of computer time. If CFD methods are coupled to thermal models, then, unlike daylighting calculations, they interact with each other, and the solution will either involve iteration or simultaneous handling of both topics. Modellers do not have tools ready for us to use.

Similarly the spread of smoke models, for example PHOENICS [23] are not suitable for everyday use.

Thus models for natural ventilation and fire spread are the least well founded, and least well validated of the simulation methods used by designers. Fire spread shares with weather forecasting the butter-fly effect, in which an insignificant change in one part of the world (or fire zone) may, by amplification over time, completely alter the course of the weather (or fire). Weather and fire spread may be inherently unpredictable.

We therefore have to proceed by experience, simple rules of thumb and simple models (see for example [24]).

Smoke Suppression and Removal

Deaths from fire in buildings occur by inhaled smoke and toxic chemicals rather than by burning. The aim must therefore be to keep escape routes free of smoke. The implication is that atria themselves should not be the escape routes; but secure escape routes should exist outside the atrium space.

Atria can accelerate the spread

of fire and smoke by channelling fire between separate zones, for example on different floors, with the result that the Fire Officers may not view separate zones as isolated one from another, and may refuse to allow routes to be planned with staged evacuation in mind. Naturally, simultaneous evacuation requires larger escape passages with an impact on costs.

Fire calculations generally consider a fire size of 3 m by 3 m as the maximum likely event, producing heat at a rate of 5 MW. In a zone of area 2000 m^2, and in the absence of a means of removing the smoke, people would have less than ten minutes to escape before being enveloped in smoke. There have been fortunately few atrium fires with which to establish rules, but it is also true that new mechanisms for fire spread are coming to light. For example the 1988 London Underground fire was spread by velocity added in the confined space of upwardly sloping escalator shafts. Thus the suggestions made here are only indications of issues which designers should sort out with fire authorities at the early stages of design, since fire precautions in atria can have major impacts on costs. In addition to checks made with fire and building control officers, it is advisable to review the design with insurance companies.

Some States in the US have fire codes which are lax by comparison with Europe, and US atria cannot always be taken as guides to good practice in this regard.

Design for Fire Safety

Two publications are of particular interest to designers in this area: Fire and the Atrium [25] and Fire Safety for the Atrium [26]

The atrium floor must be kept clear of smoke, and thus the production of smoke must be kept to a minimum by avoiding combustible materials. This limits the activities allowed on the atrium floor, and excludes bookstores and food serveries /kitchens; but in the UK it does not preclude the use of the atrium floor as office space in a building which is entirely devoted to offices. Even this restriction may be relaxed as experience accumulates.

Sprinklers are not recommended. If located at roof level, they will cool the smoke and it will not naturally pass out through roof vents, and may actually descend. If located under overhangs into the atrium, the absence of a complete deflecting roof makes their operation ineffective.

Smoke production in adjacent spaces should be kept to a minimum and be restricted by sprinklers or an extract device. A grid of 3m is commonly accepted for the sprinklers, and downstands should divide large areas into compartments each with its own extract of area less than 1000m^2.

There are limits to the volumes which can be kept clear even with mechanical extract, and atria open to surrounding spaces should not be more than 3 storeys high (24m); however various designs can be found which do not conform to this suggestion. The Toronto City Library is a case in point. The Lloyds Building in London on the other hand, has fire stop glass between the upper office floors and the atrium, and the design was influenced by the early attitude towards atria, in London, where the 'sterile tube' approach was recommended (in which no spaces communicate directly to the atrium). It is now accepted that controlled extract from the atrium is a better solution, and open floors at lower levels are permitted. However there is no general agreement on the use of drenchers to spray water at glazing into an atrium. If such sprays cause the glass to crack, they may be worse than useless, and for this reason very expensive fire stop glazing has sometimes been used.

Basements should not form part of the atrium.

Walkways at upper levels across atria are not allowable escape routes. They must not collapse into the atrium during the period allowed for evacuation.

Doors from adjacent spaces should be self-closing, roof decks above atria should be non combustible, not made of aluminium, and the glazing should be non-splintering, and mounted on a structure which will not collapse. Walls and floors separating zones from atrium should have the same fire resistance as other walls and floors to those zones.

Smoke Removal / Natural Ventilation

Smoke can be removed by buoyancy effects through roof openings of adequate size, or by mechanical extraction.

Above a certain height the hot gas will have cooled and will not have sufficient buoyancy to flow through the openings fast enough, and mechanical extract will be necessary. The height depends on the fire size; but the conservative recommendation is that natural ventilation is allowed at heights up to 12m.

Natural ventilation should be adequate if the roof openings equal 10% of the maximum area of the atrium floor. Half of these openings should operate automatically. The openings must direct the smoke away from openings in the remainder of the building, and other openings in the atrium roof should be avoided. In order to optimize extraction, the openings should be distributed over the whole of the atrium roof, and their orientations should be such as to make best use of prevailing wind directions and aid extract.

Openings at the base of the atrium need to be automatic, and to be at least twice as great in area as the roof openings. If sprinklers are not used in the rest of the building, all the openings should be automatic.

Of the two well documented atrium fires, the mechanical smoke venting equipment at the Hyatt O'Hare Chicago was being serviced and did not come into action, and at the International Monetary Fund in Washington, the automatic roof vents failed to open. Hoglund [27] draws attention to the 'belt and braces' approach adopted in the Skarholmen Shopping Centre where two back-up systems open the vents. Vent failure is not uncommon - the Crystal Cathedral near Los Angeles is reputed to have hired an acrobat with a trapeze to go round the enormous glazed space, kicking closed recalcitrant windows! Regular maintenance is indeed essential.

Mechanical Extract

Above heights of 12 m mechanical extract may be recommended, at 6 air changes per hour with the requisite inlet openings at the floor of the atrium.

9.8
ACOUSTICS

The large volume of the atrium and the hard finishes which tend to prevail, will lead to a long reverberation time. This will have two detrimental effects. Firstly it may lead to noise problems both within the atrium, and for the surrounding rooms if naturally vented by opening windows. Secondly, the long reverberation will not give the right 'feel' to the environment, and can be unsuitable for casual conversation.

For a given volume, the way to reduce reverberation is to incorporate sound absorbent surfaces. This may conflict with the aesthetic design, typically requiring surfaces such as brick, stone and, of course, glass. However, some absorbtion can be introduced into the space without conflict - decorative hangings, blinds and shutters, even planting can all help.

If there are opaque sections of roof, the underside of these can be treated with a sound absorbing material. Or absorbing surfaces may be disguised by using, for example, timber strips over an absorbent, porous plaster, or microporous lacquered panels. The latter options give the appearance of a hard stucco finish, but give good sound absorption, especially in the case of the microporous panels.

Lastly, if the atrium does end up slightly reverberant, it becomes all the more important to keep noise sources, such as traffic, or mechanical plant, isolated from it.

9.9
HEATING IN ATRIA

In spite of the apparent success of the unheated atrium, there may often be a legitimate need to heat or partially heat an atrium. For example, the space may be used for purposes other than circulation - a reception area, a cafe, or a temporary meeting area.

Provided certain rules are observed, effective heating can be applied without incurring large energy costs, and without negating the principle of the 'atrium climate'. We list below a number of guidelines which will reduce the need for heating to a minimum.

1. Make use of 'free' heat gains as much as possible, e.g. solar gains stored in massive elements, dumping exhaust air from the parent building.
2. Consider making movable shading devices function also as night insulation.
3. If heat is to be applied, avoid warm air heating. This causes stratification, maximising heat loss and removing the heat from the occupants (Figure 9.39)
4. Apply heat to the occupants (or plants), e.g. use underfloor heating or directional radiant sources. Consider soil heating for plants.
5. Choose plant species which can tolerate low temperatures. Remember the plants occupy the buildings 24 hours a day, seven days a week, four times longer than the human occupants.

9.10
PLANTING

The choice of plants which are suitable for an atrium climate is important. Most will need daylight for longer than normal office hours, typically 12 to 14 hours per day. The minimum illuminance required will vary with plant species, but the following broad guidelines may be useful:

- trees with a height of 1.5 to 10 m require 750-2000 lux.
- bushes with a height of 0.6 to 1.8 m need 250-750 lux.
- small plants can often live with less than 250 lux.

Some plants need direct sunlight for growth whereas others can survive with diffuse light.

9.11
MAINTENANCE

Accessible atrium elements such as lightshelves and atrium windows are relatively easy to clean and maintain, as are interior surfaces which are protected from the weather. However, dirt or dust on the glazing of the atrium room can significantly affect light transmission and such roofs are by their nature often inaccessible and may require special cleaning devices particularly for the internal surfaces of the glazing.

9.12
SUBJECTIVE FACTORS

The perception of comfort is complex and responds to many stimuli and expectations, apart from the purely physical thermal environment. For example, people wearing indoor clothes will happily gather on a pleasant veranda or terrace, on a spring day although the temperature may be only say 13°C. Subject them to this temperature in their office or their lounge and they would complain bitterly.

If the design of an atrium creates the expectation of a plush interior environment - carpeted floors, soft furnishings - occupants will expect 'interior type' thermal standards. The image of palm trees and parrots often reinforces this notion of sweltering warmth. (At the Cambridge Consultants' atrium they wisely have an inflatable parrot).

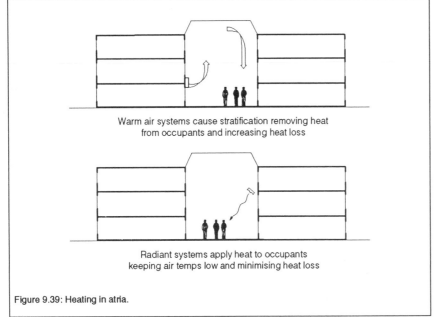

Warm air systems cause stratification removing heat from occupants and increasing heat loss

Radiant systems apply heat to occupants keeping air temps low and minimising heat loss

Figure 9.39: Heating in atria.

Consider instead an atrium formed by walls of normal external finishes, and a paved floor: bright and (sometimes) sunny in winter, the sky and clouds clearly visible, and shady in summer. Temperate species of plants, even permitted to lose their leaves in winter, bursting with new growth when the warmth and increased daylight hours signal the approach of the 'atrium spring'. When and where heating is provided, a radiant source is used, evocative of the warming effect of winter sunshine rather than the warmth of a cosy living room.

The recognition of these principles will not only ensure that atria are both comfortable and energy-saving, but also that they possess distinctive architectural and environmental qualities.

REFERENCES

[1] Atrium Buildings - Developments and Design', R. Saxon. Architectural Press 1986.
[2] Building Study, R. Saxon. Architects Journal, 13 February 1985.
[3] Design Strategies for Energy Efficient Atrium Spaces. Landsberger, R., Misuriello, H.P., and Moreno,S. ASHRAE 2996 (Technical Committee 9.8 Research Project-315)1988.
[4] Züblin Construction Company Brochure, Stuttgart, W. Germany.
[5] Daylighting and Thermal Gains - a case study of a retrofit atrium. Erwine, B., Hancock,C., Little.,J. National Lighting Conference, CIBSE, 1988.
[6] Suncourt Low Energy Multifamily Housing, Swedish Council for Building Research, Svensk Byggtjanst, Box 7853, S-10399 Stockholm, Sweden.
[7] Working in the City Results of CEC architectural ideas competition. Lewis, J.O., O'Toole, S. (Eds).Energy Research Group, University College, Richview, Clonskeagh, IRL - Dublin 14.
[8] The Proposed Building Regulations: Real Progress or a Shabby Compromise?. Hancock, C,. , Oreszczyn Littler, T.and Robinson, P. Atrium Journal, July 1989.
[9] Data taken from:'Passive Solar Design Studies for Non Domestic Buildings', Ove Arup Partners 1988. UK D.Energy ETSU 1157-P2.
[10] S. Selkowitz at the Workshop on Atria, held by the American Solar Energy Society, Boston Annual Meeting, July 1988.
[11] Lightwell to Atrium Conversions, Energy in Buildings,Erwine, C. Hancock and J. Littler, September 1987 p.24-35.
[12] Covered Courtyards, Atriums and Winter Gardens', HL. Technik Munich, 1988.
[13] Architects Journal, 14 September 1988. Article by Jourdan, J., Duffey,F.
[14] Perera,E., Walker,R. Building Research Establishment, Garston, UK.
[15] Interzone Air Movement, Final Report to CEC DG12, Walker ,J and Little,J. August 1989.
[16] Natural Ventilation of Atria, Owens, P. Pilkington Glass, Prescot Road, St.Helens, England.
[17] Natural Ventilation for a New Crown Court, Perera,E., Walker,R., Tull,R. - 9th AIVC Conference Ghent, September 1988, from Air Infiltration & Ventilation Centre, Warwick Science Park, Warwick, UK.
[18] The Passive Solar Energy Book, Mazria,E. Rodale Press, 1979, and a sequence of papers by Anderson, B. et al from Lawrence Berkeley Laboratories, Berkeley, California.
[19] Newbury Library - Building 2000 - an example of a daylight design study, Hancock,C., Littler,J. 'ISES Conference on Daylighting Buildings, 1989, London.
[20] Copenhagen School of Economics and Business Administration (Larsens, H).1989.
[21] Straffordshire House, Architects Ron Herron Ptnrs.1989, discussed in ABC & D .May 1989 and Buildings. April 28 1989.
[22] Invaluable assistance was provided by Wright J. and Corcoran M. of Building Design Partnership, who have been compiling a Report on Fire Safety in Atria for the UK Department of Energy.
[23] PHOENICS from CHAM Ltd. ,Bakery House 40 High Street, Wimbledon, London.
[24] Modelling Building Airflow and Related Phenomena, Hammond, G. BEPAC Meeting June 14 1988, from BEPAC Secretariat BRE Garston WD2 7JR, UK.
[25] Fire and the Atrium, Ferguson,A. Architects Journal, 13 February 1985.
[26] Greater London Council, Department of Architecture and Civic Design, 'Fire Safety for the Atrium', July 1985.
[27] Øverglasning av stora byggnadsvolymer - Skärholmens centrum, Höglund,I., Roman,G. Ottoson Bulletin 150,Department of Building Technology, The Royal Institute of Technology, Stockholm, Sweden, 1987.
[28] Proceedings of Building 2000 Workshop on Daylighting, Atria and Cooling, held in Barcelona, December 1988. The workshop was part of a concerted action of the Commission of the European Communities' programme 'Solar Energy Applications in Buildings', and was coordinated by Professor Cees Den Ouden, EGM Engineering BV, P.O. Box 1042, 3300 BA NL-Dordrecht, who published the proceedings.

Additional Reading

The thermal performance of large glazed spaces, Baker, Nick. Proceedings of the CEC Solar Architecture Conference, Cannes, 1982.

Atria and conservatories 1: introduction and case study, Hawkes, Dean. Architects Journal, Vol.177, No. 20, 1983.

Atria and conservatories 3: principles of design, Baker, Nick. Architects Journal, Vol. 177, No. 21, 1983.

Atrium Buildings: The Third Generation - Climate and Energy Use', Roelandts, Dirk. Proceedings of the Int. Climatic Architecture Congress, Louvain-la-Neuve, Belgium, 1986.

Glazed Courtyards: an Element of the Low-Energy City. Hawkes, Dean and Baker, Nick. in Energy and Urban Built Form, 1987.

Design Aspects of Atria and Conservatories, Project Profile, No. 030, 1988. Renewable Energy Enquiries Bureau, ETSU, Building 156, Harwell Laboratory, GB-Oxfordshire OX11 03A.

Siemens Product Information Sunlighting as Formgiver for Architecture, Lam, W. 1986.

The New Atrium, Bednar, M. 1986.

DESIGN GUIDELINES

Contents

10.1

INTRODUCTION

Guidelines for designers

Buildings in Europe account for 28% to 45% of total energy consumption and approximately two thirds of this is used in housing. A recent CEC study [1] indicates that 13% of primary energy used in buildings is currently provided by solar energy. With positive actions such as outlined in the following guidelines, the overall use of solar energy in the European Community could be increased by more than 50% by the year 2010. In some individual countries the increase could be as much as 87%.

Objectives of these design guidelines

The aim is to provide appropriate information at strategic decision stages to make better use of energy in urban planning and building design. To be fully effective these guidelines should ideally be considered by everyone involved in the shaping of our environment, including politicians, local authorities, developers as well as planners, architects and engineers.

Opportunities and limitations of general guidelines

General guidelines are not meant to be prescriptive, but instead aim to provide a set of targets which will encourage innovative, individual designs solutions (i.e., showing the direction to go without specifying which route to take).

Given the differences in Europe between the climatic regions and the range of building types and methods of construction employed these guidelines can be most useful in the early stages of design.

GUIDELINES FOR URBAN PLANNERS

Considerable energy is invested in providing infrastructure in urban areas (for the construction and maintenance of transport systems, for distribution of energy and in the pattern of use of towns and cities). Therefore decisions taken by city planners have a significant effect upon the use of energy. During the planning process awareness of how and where energy is used can lead to improvements in energy efficiency and environmental quality by the substitution of conventional fuels with less environmentally-damaging alternatives and by a reduction in consumption.

Land use and density:
Objective
To determine land use and development densities in order to maximize energy savings.

Actions
1. Locate developments where microclimatic conditions can be exploited to supply the maximum possible energy needs for heating, lighting and cooling of the buildings.
2. Integrate living, work, recreation and other activities at a local scale to minimize energy consumption through daily travel. Encourage the use of walking and bicycles by the provision of fast and pleasant routes separated from road traffic.
3. Over longer distances make public transport more attractive than travelling by car. Encourage the better use of car transport by the provision of park-and-ride schemes and by favouring car-sharing.
4. Identify those areas providing good solar access during the heating season. South-facing slopes can be more densely developed than level areas. Lower density development can be located on north-facing slopes. In southern Europe, west-facing slopes are the least favourable for energy efficiency. (see Figure 10.1)

Integration of energy supply and infrastructure:
Objective
Planners should consider heat sources and heat flows at each scale, from city to street, as an important factor in their planning decisions.

Actions
1. In cities, minimize distribution losses from large centralised power stations located outside urban areas by creating a network of decentralised combined-heat-and-power plants (CHP). Surplus heat and electricity should be fed into regional distribution systems.
2. Use the water supply system only for potable water. Water used for washing, toilets and gardens etc., can be supplied by collected rainwater or 'grey-water'. Reduce the surface water drainage infrastructure by 'unsealing' hard paved areas.
3. Separate all waste into its recyclable components, including organic material. Arrange for the collection of materials that cannot be processed on site and use unrecyclable waste for localised power and heat generation.

Street layout, site form and size:
Objective
To take into account, when making planning decisions, the daily and seasonal variations in solar radiation and the consequent energy-use implications.

Actions
1. Protect solar access where practicable, so that at no time during the heating season will the surrounding vegetation or structures overshadow any part of the actual solar collecting areas between the following angles east and west of south:

Northern Europe	30°-40°
(10 a.m. to 2 p.m.)	
Middle Europe	40°-50°
(9 a.m. to 3 p.m.)	
Southern Europe	50°-60°
(8 a.m. to 4 p.m.)	

2. Locate buildings on a site without encroaching on the solar access of neighbours. (see Figure 10.2)
3. Arrange building plots and streets to suit the topography of the site without compromising the optimum building positions. (see figures 10.3 a + b)

Vegetation
Objective
To use trees and plants to enhance climatic conditions along streets and in public open spaces.

Actions
1. Reduce the heat-island effect in densely built-up areas by maintaining natural cold air flows or by the provision of green parks including water features to aid evaporative cooling. (see Figure 10.4)
2. Estimate the shading caused by

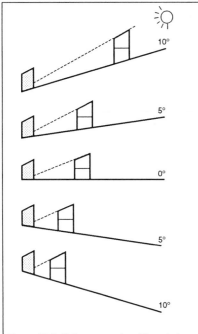

Figure 10.1. Solar access for different slopes and development densities [7].

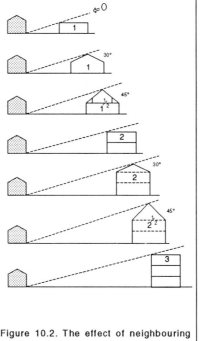

Figure 10.2. The effect of neighbouring buildings on solar access [7].

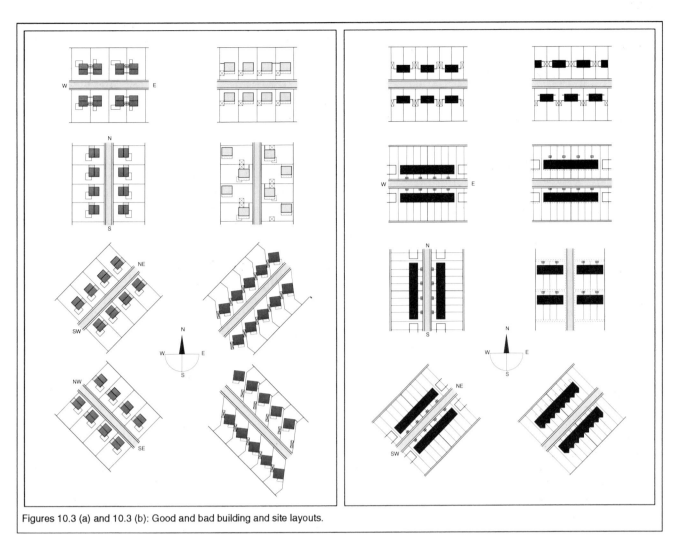

Figures 10.3 (a) and 10.3 (b): Good and bad building and site layouts.

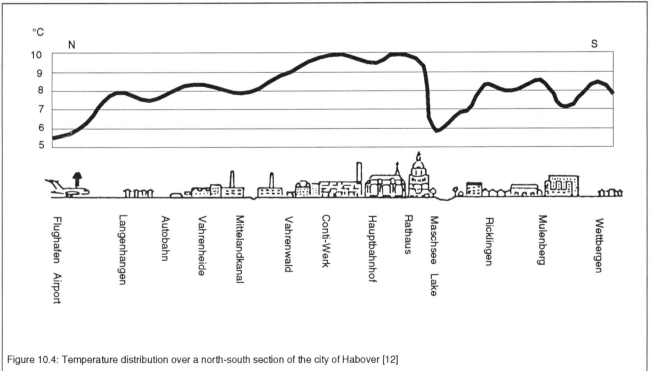

Figure 10.4: Temperature distribution over a north-south section of the city of Habover [12]

trees and plants during the heating and cooling season (as appropriate) and locate accordingly. Select the most appropriate species taking into account mature height, crown shape and seasonal variations in foliage and branch density (winter sun penetration can vary from 20% to 85% between deciduous species and by up to 20% for each species). (see Figure 10.5)

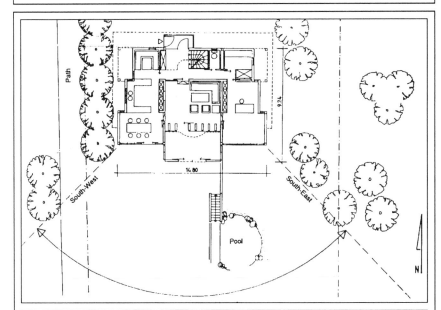

Figure 10.5: Relative light penetration through barren trees [15]

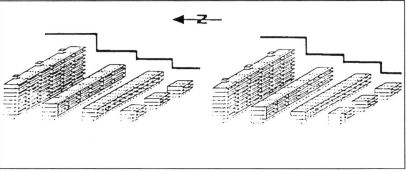

Figure 10.6 (a): Positioning of a passive solar building and surrounding trees to secure solar access during the heating season [4]
Figure 10.6 (b): Development of building height to secure solar access [11]

10.3

GUIDELINES FOR RESIDENTIAL BUILDINGS

In residential buildings the main use of energy is in space heating. In general in northern Europe there is no requirement for cooling. In Southern Europe, natural cooling techniques can provide most of the cooling required in buildings which use heavyweight construction materials. However, there is an increasing trends towards the use there of lightweight construction in buildings.

Site layout:
Objective
To locate buildings upon a site in such a way as to improve the microclimate to provide better external and internal comfort conditions.

Actions
1. Analyse the shading produced by surrounding existing and proposed buildings, plants, trees etc. during the heating season and place the building in the zone with least overshading during the following hours of the heating season:

Northern Europe – 10 a.m. to 2 p.m.
(30° east to 40° west of south)

Middle Europe – 9 a.m. to 3 p.m.
(40° east to 50° west of south)

Southern Europe – 8 a.m. to 4 p.m.
(50° east to 60° west of south)

2. Arrange buildings and vegetation on the site so that access to solar gains during the heating season is possible. Where possible, locate tall buildings to the north of lower ones, or where their shadow does not affect the solar access of surrounding buildings. (see Figures 10.6 a + b)
3. Buildings sited as close as possible to the northern end of the site will maximize the potential control of shading constraints over the rest of the site.
4. Where excessive heat loss occurs due to diurnal, seasonal or prevailing winds, use the topography or vegetation barriers of suitable species, density and height to deflect or reduce air flow without reducing solar

access. This will also improve conditions for external activities at ground level related to the building use.

5. During cooling periods reduce heat gains by providing shade from plants and trees located close to the building without compromising winter solar access.

6. Where summer cooling is needed use the topography or vegetation to channel wind around the building so that natural ventilation can be increased.

Built form:

Objective

To design the shape and form of the building to maximize solar collection and to minimize heat losses through the fabric, where heating is the predominant need.

Actions

1. Maximize the area available for solar collection and minimize the remaining external surface areas. A southern facade which is 1.5 to 2 times the length of the east or west facades can give good results for detached and semi-detached houses. This proportion does not apply to apartments or terraced houses.

2. Reduce the ratio between surface area and volume in order to create compact buildings. Detached houses have up to twice the energy consumption per unit of floor area compared to apartments in a multi-family building. However the reduction in potential solar gains for apartments remains approximately in proportion to their requirements because of lower heat losses. (see Figure 10.7)

Orientation:

Objective

To carefully orient the building with due regard to microclimate and solar insolation thereby increasing the potential for energy savings.

Actions

1. Orient the longest side of building to face south to maximize the potential for solar collection.

2. In multi-family housing, apartments with more than one external wall will have greater

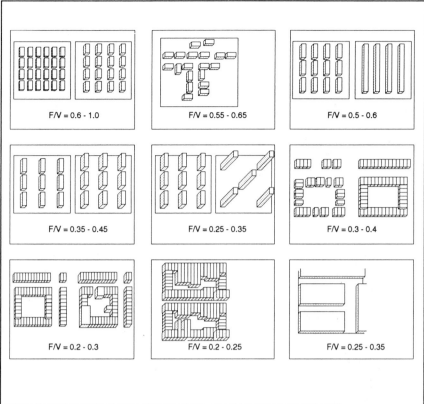

F/V = 0.6 - 1.0   F/V = 0.55 - 0.65   F/V = 0.5 - 0.6

F/V = 0.35 - 0.45   F/V = 0.25 - 0.35   F/V = 0.3 - 0.4

F/V = 0.2 - 0.3   F/V = 0.2 - 0.25   F/V = 0.25 - 0.35

Figure 10.7: Site layouts showing different surface area (F) to volume (V) ratios [11]

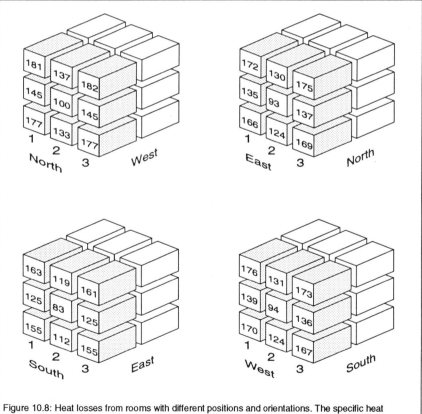

Figure 10.8: Heat losses from rooms with different positions and orientations. The specific heat consumption of the base-case room with a nortyh orientation is 83kW/m2; a = 100%. [14]

heat losses than those with only one external wall. These should be offset by increased insulation or solar gains. The losses from an apartment situated at the northwest corner of the top floor of a conventional block can be up to twice those of an apartment in the middle of the south facade. (see Figure 10.8)

Zoning of rooms:

Objective

To maximize the effective use of energy by locating rooms according to their heating or cooling requirements.

Actions

1. Place spaces with the greatest need for heating closest to insolated facades and spaces such as stairs, utility rooms and less frequently used rooms furthest away. Place unheated spaces on the north side of the building. Reverse this arrangement for buildings requiring cooling.
2. Separate sunspaces from the adjacent heated spaces by tight-fitting doors or windows.
3. Locate all main external doors away from building corners and protect them from prevailing winds. Separate all heated spaces from unheated spaces by draught lobbies. (see Figure 10.9)

Insulation:

Objective

To reduce heat losses through the building skin by using better standards of insulation.

Actions

1. Be aware of the local climatic conditions and insulate buildings accordingly. The transmittance (U-value) of a facade element multiplied by a climate-specific factor provides a rough estimate of the yearly amount of energy (in litres of oil or cubic meters of gas) lost through the element:

Heating degree days (Base 20°C)	Multiplication factor
1,700	4
2,600	6
3,400	8
4,300	10

Example: In a climate with 3,400 heating degree days, a wall of one square meter with U = 0.8 w/m2·K loses heating energy equivalent to 0.8 x 8 = 6.4 litres of oil per year.

2. Avoid heat bridges. These occur where insulation is bridged or interrupted by materials with higher transmittance values, due to poor design or poor workmanship. Air-tight buildings with high insulation standards are particularly sensitive to heat bridges and this can lead to 'pattern staining' and condensation.

Uncontrolled air infiltration

Objective

To provide an appropriate level of controlled ventilation.

Actions

1. Reduce uncontrolled air infiltration by proper design and good workmanship, sealing cracks, pipe and cable holes and selecting continuous finishes rather than jointed ones where possible.
2. Use draught lobbies at all entrance doors. Entrances should be kept away from corners where high wind speed and pressure fluctuations increase the heat loss rate and sometimes cause discomfort.
3. Teach occupants to develop energy saving ventilation habits. For example opening windows fully for a short period provides more efficient ventilation than leaving them slightly open for a long period.
4. Consider the possibility of installing a controlled ventilation system incorporating an air-to-air heat exchanger.

Passive solar heating:

Objective

To use solar radiation to provide the maximum possible proportion of the heating requirements of the building.

Actions

1. Effective passive solar heating can be achieved with a similar area of glazing as in a conventional building provided that 60 to 70% is located on the south facade and about 10% to 15% on each of the others. Further increases in glazing area will only yield small improvements in performance. The total amount of glazing may be around 20% of the floor area in northern Europe.
2. Minimize heat losses through glazed units by using special gas-filled double glazed units with low emissivity coatings in the northern half of Europe and standard double glazed units in the southern half of Europe. Use only clear glass to ensure maximum transmission of solar radiation.
3. Design the openings and fixed shading devices on the south facade to admit the maximum possible solar gains throughout the heating season and to minimize solar input during the summer. Appropriately designed fixed shading devices may be the most effective in practice because they don't require user or mechanical operation and are therefore 'self-regulating'.
4. Dark colours are not essential for internal surfaces in order to absorb solar gains in rooms with windows because only a very small proportion of the solar radiation is lost to the outside through the windows, especially

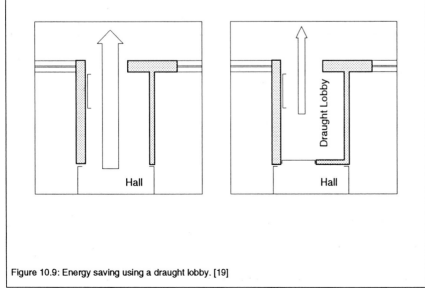

Figure 10.9: Energy saving using a draught lobby. [19]

where gas-filled double glazing with a low-emissivity coatings is used.

5. Prevent overheating by minimizing the window area in the west facade and by avoiding inclined glazed areas on the south and west facades.

6. The collection of solar gains by sunspaces is most effective if you do the following things:

- separate from heated spaces by doors, windows, etc.

- for heat storage purposes the separating wall and floor should consist of massive, dark coloured materials.

- partial enclosure is better than attachment.

- the floor must be well insulated from the ground to avoid excessive heat losses, (typically 50 to 100 mm rigid foam insulation underneath and at the edges of the slab).

- minimise shading of heat collecting surfaces caused by carpets or furnishings, where possible.

- sunspaces should only be heated by solar radiation and therefore will only be usable during certain periods of the year. Heating a sunspace in northern Europe in winter can double the energy consumption of the building.

- the most effective use of glazing is to provide double glazing for the separating doors and windows and single glazing to the outside.

- if the sunspace is to be heavily planted then the external glazing should be doubled to reduce condensation problems.

- only vent the sunspace to the outside if the trapped heat can't be used inside the house, (consider the use of automatic vents).

- in cases of overheating, comfortable conditions can be achieved more easily through natural ventilation by providing openings at the very top and bottom, each equal to about 10% of the glazed area, rather than by adjustable shading, which may block views.

7. Additional types of passive solar heat collectors, such as Trombe walls, solar walls, thermo-siphon walls, water-walls and roof-ponds should only be used after the more effective window and sunspace gains have been exploited. Designing these requires a thorough understanding of their physical performance, detailing and maintenance.

An advantage of these passive solar systems is that, provided the internal surfaces are not obstructed, they offer higher radiant surface temperatures thereby allowing occupants to reduce internal air temperatures and so reduce energy losses, while maintaining comfortable indoor conditions. See Chapter 4 - Thermal Comfort.

8. Use massive interior walls, floors and ceilings surrounding the main living spaces for storage of solar heat to ensure that heat will be radiated to occupied spaces. Some of the solar heat stored in external walls will be lost to the outside. New materials such as transparent insulation can help to overcome this problem.

In the last five years prototype transparent insulation materials have been applied to buildings using 'high-tech', intelligent shading devices for protection from overheating in summer. In these circumstances the costs of complete transparent insulation systems will not yield adequate energy savings during their lifetime.

9. Maximize the retention of heat within occupied spaces at night by the use of draught-sealed external shutters/roller shutters to all external glazed areas. Ensure that the shutters can be easily operated from inside without opening windows and do not obscure any solar collection surface, when open. Internal insulated shutters are easier to use but can create condensation on the internal surface of the glazing.

10. Locate ventilation openings (in addition to doors) between solar heated spaces and other cooler areas of the house to allow for the distribution of heat by natural thermocirculation as required.

Auxiliary heating:

Objective

To use auxiliary heating only where the maximum possible collection and retention of solar gains has been clearly predicted to fall short of the heating required to provide suitable comfort conditions.

Action

1. Auxiliary heating is most effective when it complements passive solar heating. Therefore select systems which have a fast response to temperature variations due to solar gains and can redistribute excess solar gains to cooler rooms in preference to their removal by ventilation.

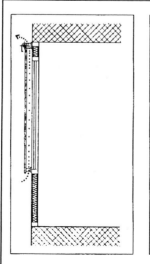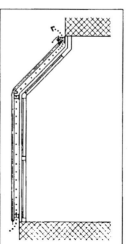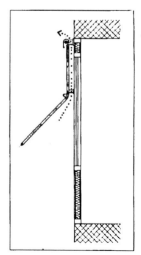

Figure 10.10. Essential ventilation gaps for fixed shading devices.

Passive cooling:

Objective

To use non-energy consuming techniques to prevent the accumulation of heat gains above comfortable conditions appropriate to the needs of the space.

Actions

1. The most effective means of preventing overheating due to solar gains is by external shading from direct sunlight to all insolated apertures, while ensuring internal lighting levels and winter heating gains are not compromised. Fixed horizontal overhangs provide the best shade to south facades and vertical fins to the east and west facades. Appropriate sizes can be easily calculated by using sun angle tables for the site latitude. Ensure that all fixed horizontal shading devices are separated from the wall face by a ventilation gap of at least 100 mm to prevent heat build up below. (see Figure 10.10)

2. Use light coloured, reflective surfaces, wall shading (for example ivy or creepers on trellises) and improved insulation to all external surfaces exposed to excessive solar radiation to control the absorption of heat by the building fabric.

3. Add thermal mass in lightweight buildings to reduce cooling loads and to cut peak indoor temperatures. Greater solar aperture areas require greater thermal mass.

4. Use passive night-time ventilation to remove excess heat from the building interior taking advantage of the cooler air available (the air temperature difference between night and day can be up to 15 K), ensuring that the air drawn through the building is below the internal air temperature.

   Cool air during day-time may be provided by a range of techniques including drawing air over water, drawing air from a cooler area or space (such as a cellar or from under vegetation) or through tubes buried in the ground (which should be accessible for cleaning).

   Ensure that during the day warm air is not inadvertently drawn in through windows or gaps in the building fabric. Provide measures which allow the occupants to control the air flow through the building by proper location of interior walls and adjustable openings. (see figure 10.11)

5. Reduce all internal gains to the minimum, using the most energy-efficient equipment where its use cannot be avoided. The collection of all sources of heat production in one zone or space, provided with its own cooling mechanisms, will greatly reduce the cooling requirements for the remaining areas. For example, equipment may be located 'down-stream' in the internal ventilation path.

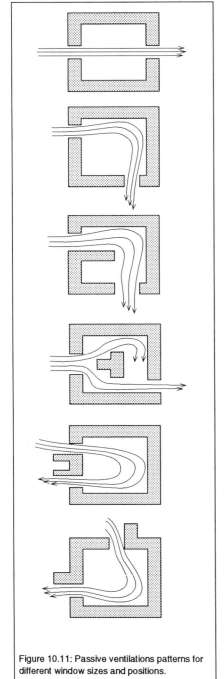

Figure 10.11: Passive ventilations patterns for different window sizes and positions.

Natural ventilation:

Objective

To maximize the use of controllable passive methods to provide ventilation for cooling in summer and to reduce ventilation in winter to supply the minimum fresh air required.

Actions

1. In the summer period make optional use of ventilation to provide passive cooling by the appropriate design of natural air circulation within and around the building and by the use of pre-cooled air.

2. During the heating season, reduce the number of air changes to the minimum fresh air supply required (0.7 to 1 airchange/hour) by ensuring that windows and doors are air-tight. Higher airchanges may be required if any sources of air contamination within the building, such as dense occupancy, smoking, dust or the release of fumes or odours from the building fabric and furnishings exist

3. Minimize ventilation heat losses and draughts by pre-heating incoming fresh air during the heating period. This can be achieved by drawing outside air through a space or over a surface heated by solar energy, such as a sunspace or by using heat recovery techniques.

4. Reduce uncontrolled heat losses due to cold air infiltration by ensuring that all external openings are sealed and the fabric is air-tight. Also, provide draught-lobbies to frequently used external doors and shelter them from prevailing winds. (see Figure 10.9)

5. The provision of small openings separate from windows can give greater control over ventilation while providing security. Ensure that all openings can be easily adjusted from inside, have well insulated covers and are air-tight when closed. In many countries baffles are available which help to reduce noise transmission through ventilation openings.

6. Locate all ventilation inlets and outlets where the use of wind pressure and the stack effect can be maximized (ventilation in houses is usually a combination of both).

Place inlets at low level on the windward side of the building or, where this is not possible due to noise or lack of pre-heating/pre-cooling potential, in a position where positive wind pressure can be generated by external features such as fin walls.

Place outlets at high level, where the stack effect is most effective, or on the lee-ward side of the building to improve cross-ventilation. Minimize the distribution of smells and water vapour around the house by locating outlets from kitchens and bathrooms suitably.

7. Maximize the effectiveness of ventilation by using adjustable louvres or panels, and permanent guides such as walls or fins, to direct air flows through the spaces to be ventilated.

Mechanical ventilation:

Objective

Mechanical ventilation in residential buildings will not normally be required except under a few special circumstances, in which case specify mechanical systems which minimize energy consumption.

Actions

1. The circumstances under which mechanical ventilation systems are likely to be used will be
   - bringing fresh air to internal bathrooms, toilets and kitchens,
   - protection from adverse external conditions such as noise and pollution.

   In these cases use a controlled ventilation system, preferably in combination with an air-to-air heat exchanger.

2. Mechanical ventilation should be controlled to ensure that only those rooms with dense occupation and/or air quality problems are ventilated for the time necessary to provide fresh air.

3. Where mechanical systems are already provided they can be used to enhance summer night cooling by flushing the building with cold outside air.

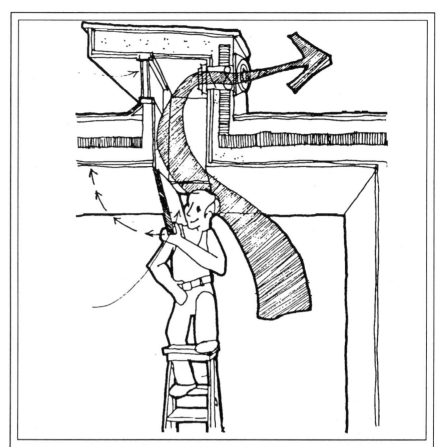

Figure 10.12 (a): High level ventilation [20]

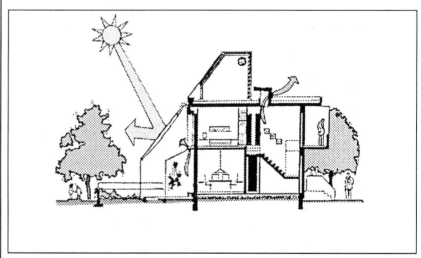

Figure 10.12 (b): North-south building section [20]

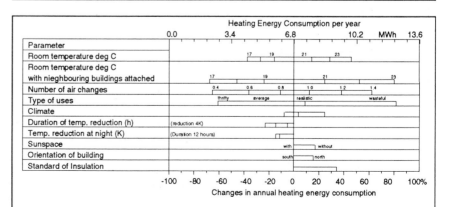

Table 10.2: Simulation of the influence of different parameters on the annual heating energy consumption of a passive solar terraced house in Freiburg 1986. Architect - Rolf Disch [17], [18]

Natural lighting:
Objective
To maximize the use of available solar radiation to provide suitable lighting conditions for activities within and around the building.

Actions
1. Locate all activities requiring the highest lighting levels closest to the best naturally lit areas. Where possible ensure that activities can be relocated at different times of the day and year to adjust to the varying lighting conditions.
2. In residential buildings natural lighting conditions will be sufficient for most daylight-hours of the day. Smaller windows on north walls should be located close to the main activity areas.

    The typical lighting level of 100-200 lux for a living room will be exceeded under natural conditions for an average Daylight-factor of 5% for the following percentage of time between the hours of 9 a.m. and 5 p.m.:

    Northern Europe   90-95%
    Middle Europe   95-100%
    Southern Europe   100%
3. Convert sunlight to diffuse light for better distribution.
4. Use light coloured internal surfaces within occupied spaces to improve the distribution and evenness of internal light levels. This will also help to reduce contrast and, if necessary, the amount of glazing required, which will in turn help to reduce any overheating problems.
5. For all glazed openings, use light-coloured splayed reveals and sills externally, and light-coloured splayed reveals and heads internally to improve the penetration of daylight and to reduce contrast. The use of dark window frames on the inside will exacerbate contrast problems.
6. Where deeper plan rooms are required or higher lighting levels are needed further back from the glazing, increase the penetration of light by using taller windows or clerestorey windows, both taken up to ceiling level. Good lighting levels can be achieved at distances of up to twice the height of the window head from floor level.
7. Avoid the risk of glare by ensuring that there are no direct views of the sun.

Artificial lighting:
Objective
To minimize the energy consumption of any artificial lighting system used.

Actions
1. Only use artificial lighting when natural lighting levels fall below the minimum level appropriate for each activity. Where high levels are required for specific activities, use task lighting. Automatic occupancy sensors may be used to mimimise wasted lighting energy.
2. Use low energy, high efficacy lamps and fittings with reflectors located to avoid glare and to minimize unwanted light spillage. Ensure that only lamps with good colour rendition are used where the quality of light is important.
3. Avoid the risk of glare by ensuring that there are no direct views of the light source.

Hot water supply:
Objective
To minimize the energy used in heating water, which forms 10% to 17% of the energy consumption in European housing.

Actions
1. Integrate active solar systems into new buildings rather than adding them after the design of the building is largely complete, with south-facing sloped roofs or sunspaces being the best places for complete integration. The most efficient angles for active solar collectors are as follows:

    swimming pools
    angle = (Latitude -20°)
    washing/cleaning
    angle = (Latitude -10°)
    space heating
    angle = (Latitude +10°)
2. Use active-solar hot water systems for multi-family houses (eight apartments and more). They are already cost-effective in Middle and Northern Europe when built as central systems. Active solar hot water systems for all types of housing are generally cost-effective in Southern Europe, if properly installed and maintained.
3. Careful sizing and detailing of the collector area, storage volume and controls will be required in order to guarantee high efficiency. The CEC provides software for designing active solar systems [10].

Performance goals:
Achievement of the following standards will ensure substantial energy savings and can be used to compare the effectiveness of different design solutions. They include better insulation as well as higher solar passive and active heating and cooling input. The figures quoted are based on total primary energy consumption for space and hot water heating in KWh/m²/yr for heated areas, and refer to dwellings in Middle Europe.

Detached + Semidetached	
Current	Future
150-200	50-100

Terraced + Multi-family	
Current	Future
100-150	30-50

Generation and distribution of heat:
Objective
To select heating, ventilation and control systems which support the operation of the passive solar strategies employed in the building.

Actions
1. The heating device should ideally be centrally located to minimise installation costs and running costs. A central flue loses some heat to the inside rooms and is therefore better than a flue on one of the exterior walls.
2. Passive solar houses have varying levels of solar gains. Even with high thermal mass they need a fast-responding heating and control system.
3. Allow for separately controlled heating of zones (groups of rooms) or rooms, for example, north-facing or south-facing rooms.
4. In heavily insulated buildings with low-emissivity double-glazing the heat source (radiators, etc.) can be mounted on interior walls. If mounted on exterior walls this part of the wall will need additional insulation.
5. If heating devices are mounted on exterior walls, extra insulation will be required
6. Heaters, radiators etc., should ideally not be placed in front of glazing. If this is unavoidable, then an insulated panel should be placed between the heater and the glazing.

Fuel selection and environmental pollution:

Objective

To select a fuel with minimal environmental impact.

Actions

1. Note that reducing fuel consumption is the most effective means of reducing environmental pollution. Substitution of fossil fuels by solar energy is the next best strategy.

2. Where the use of fossil fuels is still required, selection criteria will depend on a range of local conditions, heating/cooling technology available, and the availability of cost and security supply. The following emissions (in kg per 100 MWh usable energy) of different heating systems in Germany (from [13]):

Behaviour of users:

Objective

The effective use of a low-energy solar building will depend to a great extent upon passive features designed by the architect and the understanding of their purpose and performance by the user.

Keep in mind that the primary goal is to provide comfortable living and working conditions, rather than the construction of an energy-saving machine.

Actions

1. Architects should ensure that all users fully understand how the building and its energy-saving techniques can be used and adapted to achieve the minimum energy consumption.

2. Ensure that all energy efficiency mechanisms are robust, easy to operate and cheap to maintain.

3. Comfort conditions and energy consumption in low-energy buildings are extremely sensitive to misuse by occupants. Therefore maximize the provision of self-regulating passive features to reduce the inefficiencies caused by user errors or absence.

    Low-energy buildings will require a different understanding and pattern of use by their occupants because, unlike conventional buildings, they do not have such a high tolerance of misuse.

4.. Ensure that there is some form of monitoring of energy consumption and building use so that further improvements can be made.

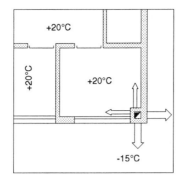
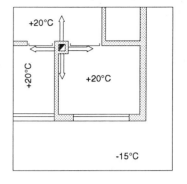

Figure 10.13 (a): Location of chimney [19]

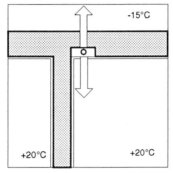

Figure 10.13 (b): Location of heating pipes [19]

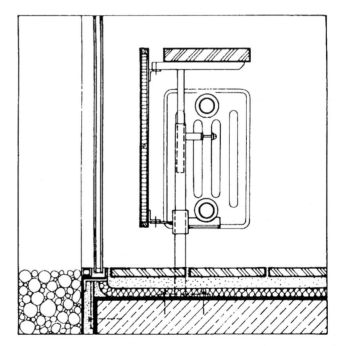

Figure 10.13 (c): Insulated panel between heater and windows [19]

	SO2	NOx	Dust	CO2 (Kg/100 MWh)
Heating system using oil	40	31	1.4	37.000
Heating system using gas	3	16	0.4	27.200
Electric storage heating (80% coal, 20% nuclear)	81	67	8.3	80.400
District heating (coal and oil)	10	10	0.7	11.500

165

GUIDELINES FOR NON - RESIDENTIAL BUILDINGS

In general, non-residential buildings are more densely occupied by equipment and people and have smaller external surface area-to-volume ratios than houses. This results in higher internal gains and lower heat losses which can generate considerable cooling loads throughout the year. This is compounded by the typically, almost permanent use of artificial lighting, especially in deep-plan buildings.

However, non-residential buildings offer a great potential for the use of natural lighting and also for solar heating and passive cooling due to their daytime occupancy.

Siting and orientation:
Objective
The building should be located so that energy consumption for artificial lighting and mechanical cooling are minimized and access to solar energy for natural lighting is maximized.

Action
1. Overshadowing by surrounding buildings will significantly reduce the availability of natural lighting, therefore analyse the effects of present and future planned buildings particularly to the east, south and west-sides of the site.

Built form: and shape for solar heating
Objective
To maximize the available solar radiation.

Actions
1. Maximize the amount of south facade area available for natural lighting, while minimizing excessive solar gains during the cooling period.
2. Where deep-plan buildings are required use the roof as the fifth facade, through which openings can offer potential for natural lighting and cooling. Overheating in summer should be avoided by the proper design of roof openings.

Zoning of rooms:
Objective
To provide comfortable, well-lit working spaces.

Actions
1. Locate spaces requiring natural light during working hours along the perimeter of the building and place areas with lower lighting requirements furthest away.
2. Separate areas with high internal equipment heat gains from areas of low gain and locate on the north facade.

Natural lighting:
Objective
To maximize the use of available solar radiation to provide comfortable lighting conditions, avoiding glare and contrast for day-time activities within the building without increasing cooling loads.

Actions
1. Locate all activities requiring high lighting levels closest to the best naturally lit areas, which are south and north faces, rooflit top floors and perimeters of atria.
2. Convert sunlight to diffuse light for improved light distribution.
3. Use light coloured internal surfaces within internal spaces to improve the distribution and evenness of internal light levels. This will also help to reduce contrast and, if necessary, the amount of glazing required.
4. Provide a minimum daylighting level of 300 lux for general office areas. Under natural conditions this will be exceeded for an average Daylight-factor of 5% by the percentage of time, shown in the table below between the working-hours of 9 a.m. and 5 p.m.:
5. For all glazed openings, use light-coloured splayed reveals and sills externally, and light-coloured splayed reveals and heads internally to improve penetration of daylight and reduce contrast. The use of dark window frames internally will cause contrast problems.

6. Where deeper plan rooms are required or higher lighting levels are needed to the rear of a room, increase the penetration of light by using taller windows or clerestorey windows, both taken up to ceiling level. Good lighting levels can be achieved at distances of up to twice the height of the window head from floor level.

Similar improvements can also be obtained by the use of reflecting devices such as external light shelves and internal reflective horizontal blinds above eye-level. In these cases a sloped ceiling will help to increase the evenness of light distribution.
7. For very deep plan spaces (i.e. over approximately 5 m from a naturally lit perimeter) introduce lighting from the roof either in the form of an atrium or lightwell for multi-storey buildings, or roof lighting in single-storey spaces.
8. All natural lighting methods carry the risks of overheating and glare because the constantly changing position of the sun. Overheating can be avoided by proper sizing, shading and location of windows. Glare can be eliminated by preventing direct views of the sun from working positions.

Artificial lighting:
Objective
To minimize the energy consumption of any artificial lighting system used.

Actions
1. Integrate artificial and natural lighting strategies by taking both into account during the design of the building and by the use of appropriate controls.
2. Only use artificial lighting where natural lighting levels fall below the minimum level required for each activity. Install automatic (continuous, not step) dimmers to control artificial light use on bright days. Such daytime

Distance from window:	< 3 m	3 - 6 m
Northern Europe	70 - 85 %	< 5 %
Middle Europe	80 - 90 %	< 5 %
Southern Europe	85 - 95 %	< 5 %

savings can help to reduce electricity demand peaks during high tariff periods, as well as reducing cooling costs.

3. Maximize the efficient use of electric lighting by ensuring that the levels provided do not exceed those required for each activity. Where high levels are required for specific activities, use task lighting and reduce background levels to the minimum required to avoid excessive contrast.

4. Use low energy, high efficacy lamps and fittings with reflectors located to avoid glare and to minimize unwanted light spillage. Ensure that only lamps with appropriate colour rendition are used where the quality of light is important.

Passive cooling:

Objective

To use non-energy consuming techniques to prevent the accumulation of heat gains above required comfort conditions appropriate to the needs of the space.

Actions

1. Minimize the ingress of direct solar radiation by the provision of adjustable, external shading particularly to windows facing west and, to a lesser extent, to the east.

2. Minimize internal heat gains from artificial lighting by optimising the availability of daylight. Separate those areas requiring high artificial lighting levels either due to special requirements or because they are used largely outside of daylight hours. Daylighting produces less heat energy than artificial lighting for the same level of illumination. The use of low-energy lighting can significantly reduce heat output.

3. Minimize internal heat gains by using low-energy equipment or where this is not possible, locate equipment in a separate area with high ventilation rates, ideally 'down-stream' in the internal ventilation path through the building.

4. Minimize heat gains from occupants by reducing the occupation density.

5. Minimize the amount of excess solar radiation falling on the building fabric, especially the roof and east and west facing walls by providing shading fixed to the surface itself or preferably, by using shading from adjacent buildings or trees. Ensure that the shading provided does not compromise either the daylighting to the building or solar access during the heating season.

6. For northern summer, or southern conditions minimize solar heat gains through the building fabric by using light coloured external finishes to reflect radiation; placing insulation on external surfaces exposed to the sun; or incorporating ventilated voids in roofs with low emissivity surfaces facing the cavity.

7. Minimize the transfer of heat absorbed by the building fabric to the internal spaces during the cooling period by locating elements of high thermal capacity in areas of highest heat gains.

8. Maximize the removal of excess heat by passive ventilation. For cooling to occur, the temperature of the incoming air must be below that of the maximum comfort temperature for the space, see Chapter 6 - Natural Cooling. The greater the temperature difference, the greater the cooling effect. Night-time cooling is an effective technique for thermally massive structures provided the cooled mass does not reduce internal temperatures below comfort conditions during occupancy.

Mechanical cooling:

Objective

To minimize the energy consumption of any mechanical cooling systems used.

Action

Consider the use of mechanical ventilation and heat pumps in preference to the use of conventional air-conditioning systems.

Natural ventilation:

Objective

To maximize the use of controllable non-energy consuming methods to provide the ventilation required, particularly for cooling.

Actions

1. During the cooling season, provide sufficient ventilation to remove excess air and fabric heat without causing excessive cooling during periods of occupation. Ensure that the temperature of the incoming air does not exceed the maximum comfort air temperature for the space.

2. The effectiveness of cooling by ventilation can be improved (or alternatively the ventilation rate can be reduced) by increasing the temperature difference between the internal air and air drawn in. Sources of cooler air include night-time air and air drawn from adjacent areas of vegetation or over water.

3. During the heating season, minimize the required number of air changes by eliminating, or where this is not possible by reducing or isolating any sources of air contamination within the building. These include smoking, dust or the emission of fumes or smells from the building fabric and furnishings.

4. During the heating season, minimize ventilation heat losses and draughts by pre-heating incoming fresh air. This can be achieved by drawing outside air through a space or over a surface heated by solar energy, such as a sunspace.

5. During the heating season, minimize heat losses due to uncontrolled air infiltration by ensuring that all external openings are sealed and that the fabric is air-tight. Also, provide draught-lobbies to all frequently used external doors and shelter them from prevailing winds.

6. The provision of openings separate from windows enables greater control over ventilation and the avoidance of draughts, especially during the heating season. Ensure that all openings can be easily adjusted from inside, have well insulated covers and are air-tight when

closed. The provision of a fine mesh screen will prevent insect ingress and improve security. Lining the reveals and installing baffles will help to reduce noise transmission.

7. Locate all ventilation inlets and outlets where the use of wind pressure and the stack effect can be maximized. Place inlets at low level on the windward side of the building or, where this is not possible due to noise; ingress of dust or pollution or lack of cooling potential, in a position where positive wind pressure can be generated by external features such as fin walls. Place outlets at high level where the stack effect is most effective. In buildings containing an atrium or some other form of roof glazing, the stack effect can be enhanced by the use of internal blinds below the glazing to form a solar chimney.

8. Maximize the effectiveness of ventilation by using adjustable louvres or panels, and permanent guides such as walls or fin walls, to direct air flows through the spaces to be ventilated.

Mechanical ventilation:
Objective
To minimize the energy consumption of any mechanical system used.

Action
1. Use the most energy-efficient, adjustable equipment available with a heat-exchanger and humidity and temperature operated switching.

Hot water supply:
Objective
To minimize the energy used in heating water.

Action
1. Integrate active solar systems into the building fabric rather than adding them. South-facing sloped roofs are the best places for complete integration.

Performance goals:
Achievement of the standards set out in the table below will ensure substantial energy savings and can be used to compare the effectiveness of different design solutions. The figures quoted are expressed in kWh of primary energy:

Selection and sizing of mechanical equipment:
Objective
Address the design and specification of all major components of energy use within the building (lighting, electric demand, cooling, heating, ventilation) together with the objective of minimizing their energy consumption.

Action
1. Give preference to daylighting strategies and equipment which reduces the amount of energy needed for artificial lighting, cooling and ventilation.
2. HVAC systems, including plant, ducts and fans, should be downsized to meet the reduced cooling and ventilation load.
3. Make allowance for future changes and improvements (see life-cycle of mechanical systems versus life-cycle of a building).

Fuel selection and pollution of the environment:
Objective
Select fuels according to their least polluting impact on the environment.

Actions
1. Energy saving has the most favourable effect on the environment. Substitution of fossil fuels by solar energy is the second best strategy.
2. Where use of fossil fuels is still required, selection criterion will depend on a range of local conditions, heating/cooling technology, plant efficiency, and the availability, costs and security of supply of fuel. The following example shows the emissions (in kg per 100 MWh usable energy) of different heating systems in Germany [13]:

Behaviour of users:
Objective
To maximize potential energy savings through the correct use of the building by its users.

Actions
1. Ensure that all users fully understand how and why the building can be used and adapted to achieve the minimum energy consumption. Prepare clear operating manuals which explain the users' role, how the systems should work, potential savings and the effects of poor manual control and use patterns.
2. Ensure that all energy efficiency mechanisms are robust, easy to operate and cheap to maintain
3. Ensure that there is some form of monitoring of energy consumption and building use so that further improvements can be made. Ask maintenance staff to take frequent, detailed records of energy consumption.

Space heating/cooling energy consumption in kWh/m2yr						
	Offices		Schools		Gymnasiums	
	Current	Future	Current	Future	Current	Future
Northern Europe Middle Europe Southern Europe	>250	150-250	>250	100-150	>400	150-200
Performance goals						

	SO2	NOx	Dust	CO2 (Kg/100 MWh)
Heating system using oil	40	31	1.4	37.000
Heating system using gas	3	16	0.4	27.200
Electric storage heating (80% coal, 20% nuclear)	81	67	8.3	80.400
District heating (coal and oil)	10	10	0.7	11.500
Fuel selection and pollution of the environment				

10.5
SELECTION OF MATERIALS

Objective

To minimize the energy used in the materials and construction of buildings.

Action

1. Use materials that require low energy input during extraction, manufacture, transport, construction, use, demolition and disposal. Where a high energy content material has to be used, ensure that it is used in its most efficient form and can be easily recycled. (see Table 10.1)

2. Use materials in a form that can be easily reclaimed and reused with the minimum of reprocessing. Avoid the use of composite materials which make separation for recycling difficult.

3. Use materials with minimum impact upon the environment throughout their life cycles, especially atmospheric pollution. Use materials from sustainable sources in preference to non-renewable ones and recycled materials in preference to new.

4. Maximize the design life of a building, ensuring that elements with shorter design-lives than the structure can be easily replaced and maintained.

Element	typical design-life	number of cycles in 100 years
structure	100 years	1
services	25 years	4
finishes	5 years	20

5. Request detailed information from manufacturers on the energy content of their products as well as their environmental impacts during manufacture and use.

	MJ/t
Concrete, Normal	.540
Precast concrete	1.413
Reinforced concrete	2.001
Chipboard panels	2.001
Bricks, light	2.205
Bricks, heavy	2.610
Precast, reinforced concrete	2.897
Gypsum board	3.213
Mineral wool	18.000
Glass	21.852
Steel for reinforcement	25.884
Polystyrene foam	126.314
Aluminium sheets	260.820

Table 10.1: Primary energy used in the materials and construction of buildings.

REFERENCES

[1] Passive Solar Energy as a Fuel 1990 - 2010. A study of current and future use of passive solar energy in buildings in the European Community. Burton, Simon, et al. Brussels-Luxembourg, 1990.

[2] Energy Conscious Design: A Primer for European Architects. Goulding, John R., Lewis, J. Owen, Steemers, Theo C., (Eds.). Batsford 1992 for CBC

[3] European Passive Solar Handbook. Preliminary edition. Achard, Patrick, Gicquel, Renaud, (Eds). Brussels and Luxembourg, 1986.

[4] Energie und umweltbewußtes Bauen mit der Sonne. Architects Lohr, Alex & Gabi, Ein BINE Informationspaket, Köln, 1991.

[5] Design with climate. Olgyay, Victor: Princeton, 1963.

[6] Protecting Solar Access for Residential Development: A Guidebook for Planning Officials. Jaffe, Martin; Erley, Duncan. Washington, 1979.

[7] Byggeri og Okologi - Idékatalog. Gade, Torben; et al. Kobenhavn, 1988.

[8] Working in the City. - Results of CEC architectural competition Lewis, J. Owen, O'Toole, Eblana / Gandon, Dublin, for CEC 1988.

[9] Bauen mit der Sonne - Vorschläge und Anregungen. Hebgen, Heinrich. Heidelberg, 1982

[10] A suite of four programmes (ESM 1 + 2, EURSOL and EMGP3) for IBM PC, developed within the CEC's OPSYS programme and distributed by the Energy Research Group, School of Architecture, Richview, Clonskeagh, IRL-Dublin 14.

[11] Rationelle Energiewendung in der Bauleitplanung, Roth, Ueli; et al, BMBau, Heft 03.102, 1984.

[12] Lebensraum Stadt, Fellenberg, Günter, Zürich, 1991.

[13] Energiewende in der Neubauplanung - Handbuch für eine kommunale Neubaupolitik, 1990.

[14] Rationelle Energieverwenndung im Hochbau, Epinajeff, Peter, Weidlich, Bodo, Berlin, 1986.

[15] Don't Let the Trees Make a Monkey of You, Holzerlein, Thomas, Proceedings of 4th National Passive Solar Conference, Kansas City 1979. Also mentioned in Passive Principles; Let the Sun Shine In, Kohler,J., Lewis, D., Solar Age Nov 1981

[16] Der Primärenergieinhalt von Baustoffen, Marmé, W., Seeberger, J., Bauphysik, Issues No. 5 & 6, 1982

[17] Parameterstudien und Simulation des thermischen Verhaltens eines Mittelhaus der Reihenhaussiedlung "Am Lindenwäldle", Stanzel, B., Hahne, E., Proceedings of 7th International Sonnenforum, Volume 1, München 1990.

[18] Project Monitor, Issue No. 6, June 1987, Commission of the European Communities DG XII

[19] Normengerechter und Wirtschaftlicher Wärmeschutz, Meinert, S., Köln, 1978

[20] Solardorf Lykovrissi bei Athen, BINE Projekt Info-service, Fachinformationszentrum Energie, Physik, Mathematik, GmbH, Karlsruhe, May 1987. Architect:- A. Tombazis, Athens.

SCHEDULE OF APPENDICES

Contents

PREDICTION OF SOLAR GEOMETRY

A1.1

SOLAR TIME AND LOCAL TIME

The trigonometric prediction of the position of the sun is carried out using solar time and not local time. However, in some design situations, it may be necessary to consider problems in terms of local time rather than solar time.

The world is divided into time zones, based on standard longitudes (meridians): the European Community time zones are based on 0° (Greenwich Mean Time - GMT - Ireland, Portugal and UK), 15° (Belgium, Denmark, France, Germany, Italy, Luxembourg, The Netherlands, Spain - one hour earlier than GMT) and 30°E (2 hours earlier than GMT - Greece). In summer all the European countries move onto daylight saving time, which is an hour earlier than these standard times. This makes sunset one hour later. However the days on which the changes occur vary from year to year and state to state. There are moves to standardize the dates of change within the Community.

The time specified in reference to sun angles is usually solar time, which rarely coincides with local clock time. It is possible to convert local time to solar time by applying three corrections. First, there is the correction needed if daylight saving time is in use. Second, there is a constant correction for difference in longitude between the location and that of the standard meridian of the time zone (based on the four minutes that the sun takes to cross 1° longitude). The earth rotates towards the east, so that if the site location is west of the standard longitude, solar time is later than local time, while conversely to the east of the standard meridian solar time is earlier than local time. The third correction is known as the equation of time. It allows for the small precession of the earth's motion around the mean polar axis. The correction is given in minutes in Figure A1.1. Thus, solar time can be obtained from local time by the following equation:

Solar time = LMT + E/60 + (l_{loc} - l_{st})/15 - D hours.

where E = Equation of time in minutes, from Figure A1.1.

LMT = Local mean time in hours.
l_{st} = the standard meridian for the local time zone in degrees, east positive, west negative.
l_{loc} = the longitude of the location in degrees, east positive, west negative.
and
D = the difference due to summer daylight saving in hours. (Usually 1 hour forward in local mean time)

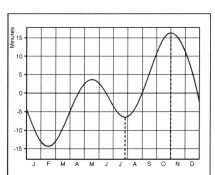

Figure A1.1:
The equation of time, E, in minutes as a function of time of year.

The difference between solar time and local mean time in minutes is given by:- E+4 (l_{loc} - l_{st}) - D * 60.
Thus, using Figure A1.1 to determine the equation of time, at Valentia, longitude 10°15' west, the correction to local mean time on 27 July in minutes is -6.5 + 4(-10.25 - 0) - 60 = -107.5 minutes.
For Brindisi, longitude 17°55' east, the correction to local mean time on 31 October is +16.5 + 4(17.92 - 15) - 0 = +28 minutes.

A1.2

ESTIMATION OF THE SOLAR ALTITUDE AND ASIMUTH

Three inputs are necessary to calculate the solar altitude:-
1. the latitude of the site, f.
2. the solar declination, d.
3. the time expressed as an hour angle from solar noon, w.

The solar declination is the angle the sun's rays make with the equatorial plane. It varies from a maximum value of +23°27' at

Dates in year	Solar declination Degrees
21 Jun	23.44
21 May/23 Jul	20.13
21 Apr/22 Aug	11.83
21 Mar/23 Sept	0.84
21 Feb/21 Oct	-10.63
21 Jan/21 Nov	-19.91
22 Dec	-23.44

Table A1.1 Declinations at different dates used for plotting solar charts in this book.

midsummer to a minimum value of -23°27' at midwinter. At the equinoxes, the value is zero. The solar declination may be calculated from the formula:-
$$d = Sin^{-1} (0.3978 \, Sin(J' - 80.2 + 1.92 \, Sin(J' -2.89)))$$

where $J' = 360 * J/365.25$ degrees and J is the day number in the year, 1 Jan = 1, 1 Feb = 32, etc.
Table A1.1 gives the approximate values of the declinations used to plot the path lines on the solar charts, presented in the main body of the chapter.

Table A1.2 gives the declination values used in the preparation of the CEC European Solar Radiation Atlas for Inclined Surfaces together with the appropriate dates associated.

The values in Table A1.1 were derived using the declination formula above, adjusting the choice of days in the second half of the year until the closest predicted declination match was achieved with the predicted declination values on 21st of the month in the first half of the year. Then the means of the two values were taken, and entered accordingly into the Table. Because of leap year effects there are variations from year to year. The above values are representative means for the four year leap year cycle. The declination is, of course, a continuously varying function even across a day.

The solar hour angle, w, expresses the time of day in terms of the angle of rotation of the earth at any place referenced to solar noon. As the earth rotates 360 degrees about its axis in 24 hours, it rotates 360/24 degrees in one hour i.e. 15 degrees. By convention the hour angle is negative before solar noon and positive after solar noon, i.e. at 0900 solar time, the hour angle is -45 degrees, while at 1500 hours the hour angle is 45 degrees.

The solar altitude, g, is calculated from the formula:-

$$sin \, g = cos \, f * cos \, d * cos \, w + sin \, f * sin \, d$$

The solar azimuth (a_s), is calculated from the formula:-

$$sin \, a_s = cos \, d * sin \, w/cos \, g$$

In the afternoon (here defined in terms of solar time) the solar azimuth will be positive, i.e. to the west.

At solar noon, the solar altitude = 90 - f + d degrees.

A1.3
CALCULATION OF THE ANGLE OF INCIDENCE

The angle of incidence on a plane of tilt, b , and wall azimuth a_w is calculated using the following formula:-

$$cos \, i = cos \, g * cos \, (a_s - a_w) * sin \, b + sin \, g * cos \, b$$

where g is the solar altitude, degrees.

a_s is the solar azimuth in degrees measured from due south, east negative, west positive.

a_w is the azimuth angle of the inclined plane, degrees, measured from due south, east negative, west positive.

b is the tilt of the plane from the horizontal plane, degrees.

The sun will be perpendicular to the surface if i = 0.

If cos i is negative, it indicates the beam is incident on the back side of the inclined plane. If this situation cannot occur, then for negative values of cos i, set cos i to zero.

A south-facing surface tilted to achieve maximum noontime beam irradiance should have a tilt of f - d degrees. At midwinter, at latitude of 57N, this implies a tilt of 80.45 degrees compared with 59.45 degrees at latitude 36N.

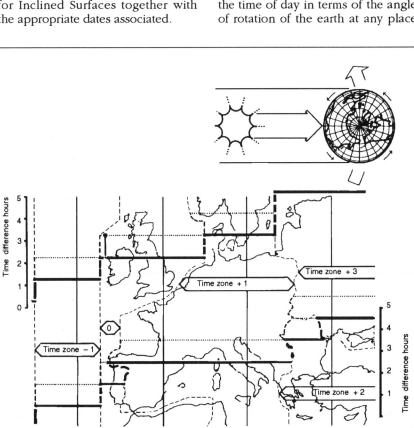

Figure A1.2. Time zone map for Europe, indicating the four reference longitudes used for time systems, and the boundaries for the time zones.

COMPUTATION OF VERTICAL AND HORIZONTAL SHADOW ANGLES

Horizontal shadow angle

The horizontal shadow angle HSA is easily determined. It is the angle between the direction of the facade azimuth and the solar azimuth.

Thus $HSA = a_s - a_w$ degrees

where a_s = solar azimuth in degrees, east negative, west positive.

a_w = wall azimuth in degrees, east negative, west positive.

If HSA>180 then HSA = HSA - 360 degrees
If HSA<-180 then HSA = HSA+180 degrees.

Sign convention: If sun lies in an anticlockwise direction from the normal to the plane projected on the horizontal plane, sign negative. If sun lies in a clockwise direction from the normal to the plane projected on the horizontal plane, sign positive.
In the case of a vertical surface, if HSA<-90 or HSA>90, then the sun lies behind the vertical plane, and is incident from the back.

Vertical shadow angle

The vertical shadow angle, VSA, is calculated from the solar altitude, g, and the horizontal shadow angle, HSA.

$VSA = \tan^{-1}(\tan g/\cos(HSA))$ degrees.

If computed on a computer and VSA<0 and g>0 then VSA = 180 + VSA degrees.

Table A1.2
Recommended values of solar declination (δ) with representative dates and associated day numbers (J) for use in the calculation of monthly mean and mean monthly maximum levels of solar radiation together with values of the correction factor to mean solar distance kd.

| | Computation of monthly means of global radiation | | | | Computation of mean monthly maxima of global radiation | | | | | | | |
| | | | | | Northern hemisphere | | | | Southern hemisphere | | | |
Month	Date	Day No.	δ^* deg.	k_d	Date	Day No.	δ^{**} deg.	k_d	Date	Day No.	δ^{**} deg.	k_d
Jan	17	17	-20.71	1.032	29	29	-18.16	1.030	4	4	22.80	1.033
Feb	15	46	-12.81	1.025	26	57	-9.04	1.020	4	35	16.48	1.028
Mar	16	75	-1.80	1.011	29	89	3.43	1.003	4	64	6.39	1.017
Apr	15	105	9.77	0.994	28	119	14.19	0.986	4	95	-5.74	1.000
May	15	135	18.83	0.978	29	150	21.64	0.973	4	125	-16.00	0.983
Jun	11	162	23.07	0.969	21	173	23.43	0.967	4	156	-22.45	0.971
Jul	17	198	21.16	0.967	4	186	22.87	0.967	29	211	-18.73	0.970
Aug	16	228	13.65	0.975	4	217	17.22	0.971	29	242	-9.73	0.981
Sep	16	259	2.89	0.990	4	248	7.15	0.984	28	272	2.06	0.997
Oct	16	289	-8.72	1.007	4	278	-4.39	1.001	29	303	13.50	1.015
Nov	15	319	-18.37	1.022	4	309	-15.42	1.018	28	333	21.32	1.028
Dec	11	345	-22.99	1.031	4	339	-22.25	1.029	22	357	23.46	1.033

The date and day numbers above refer to a 365 day year.

* The values of montly mean solar declination are the average of the individual daily values. For use in the southern hemisphere the sign should be reversed. The day is the day on which this declination occurs.

** The representative values of solar declination used to estimate the mean monthly maxima of global radiation were obtained using Dogniaux's Algorithm which is based on a 366 day year, adopting the representative dates shown in the Table.

SITE LAYOUT & WIND FLOWS AT GROUND LEVEL

These diagrams are based on those provided by Gandemer (1988), using reference 36 of Chapter 2 - Climate and Design. Additional references by Gandemer [9, 10 & 11] are given at the end of this Appendix 8. The level of discomfort experienced due to increased wind speeds is indicated by , the 'discomfort index'.

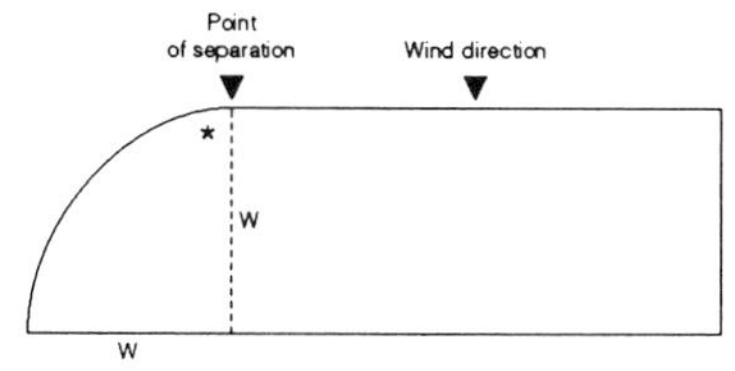

Notes:

1. Disturbed zone is roughly located within an arc of radius equal to the width of the building.

2. * indicates the worst area.

3.
a) For heights of 15 m Ψ = 1.2
b) For heights of 35 m to 45 m Ψ can reach 1.5
c) For heights of 100 m Ψ can reach 2.2 or more

Figure A.2.1 *Corner Effect*

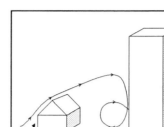

Tall building set among low building

1.
Ψ near base at point 1 can rise to 2.0

2.
Low buildings upwind can make flow worse

Figure A.2.2
Tall Building Set among Low Buildings

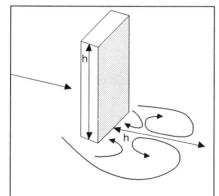

1.
In corner effect for towers Ψ varies from 1.4 to 2.2 according to height (16 to 30 floors)

2.
For relatively low slabs (less than 16 floors with the building width greater than 3 times the height), Ψ varies from 0.5 to 1.4.

3.
Seriously disturbed area approximately extends a distance H.

Figure A.2.3 *Wake Effect*

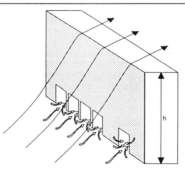

Gap effect

1.
Discomfort depends on height of building.

2.
(a) if less than 5 floors high, usually no discomfort.
(b) at 7 floors discomfort index is 1.2.
(c) at 60 floors discomfort index reaches about 1.5 regardless of length of building.

3.
Emerging jet stream produces discomfort zone of about the same area as the passageway.

Figure A.2.4 *Gap Effect*

Schedule of diagrams of particular layouts and situations

A.2.1 *Corner effect*

A.2.2 *Tall building set among low buildings*

A.2.3 *Wake effect*

A.2.4 *Gap effects*

A.2.5 *Bar effect*

A.2.6 *Pressure connections*

A.2.7 *Venturi effect*

A.2.8 *Pyramid effect*

A.2.9 *Buildings of continuously varying height*

A.2.10 *Mesh effect*

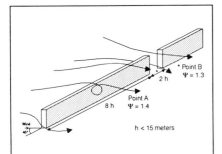

1.
45° wind on corner. Wind flows over top of roof and spins down middle of leeward facade of building, Ψ - 1.4 at point A.

2.
Wind perpendicular to building : wind flows through gap strongly, Ψ = 1.3, critical width 1-2h.

Figure A.2.5
Bar effect - long narrow buildings, w<10 m, length>8 h, where h is building height, between 15-25 m, max 30 m

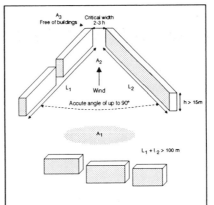

where l₁ and l₂ > 100m
if h = 15-30m then Ψ = 1.3
if h = 50m approx. then Ψ = 1.6
if curved to enhance effect Ψ = 2.0

Notes:

Area A_2 approximately equal to A_1
Similar area A_3 approximately equal to A_1

} free of buildings.

Figure A.2.7 *Venturi Effect*

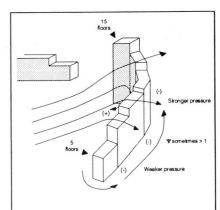

Notes:

The wind at the back (leeward side) of the building, moves from a zone of moderately reduced pressure to a zone of strongly reduced pressure.

Figure A.2.8
Buildings of Continuously Varying Height

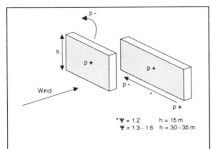

Wind flows from positive zone to negative zone behind leading slab.

Ψ = 1.2 if h = 15m
Ψ = 1.3-1.6 if h = 30-35m

Narrow space 1/4 of width

Ψ up to 1.8

Figure A.2.6 *Pressure Connections*
(excluding passage ways under buildings).
Tall buildings displaced sideways in relation to the windward tall building

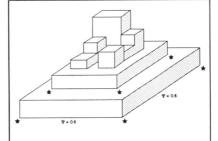

* Critical areas at corners
Notes: Ψ varies - some strong flows, in general about 0.8, but can be higher in places.

In general reasonable protection is provided particularly if the building has less than 9 floors.

Shape encourages an upward deflection of the wind power rather than downward.

Figure A.2.8 *Pyramid Effect*

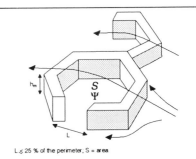

L ≤ 25 % of the perimeter, S = area

Relatively closed - L ≤ 25% of perimeter s Ψ varies within courtyard.
h_m = mean height

1.
Protection starts if h_m is 15m-20m, regardless of wind direction, so long as $s/h_m2 < 10$
Ψ = 0.4 - 0.8

2.
If h>30m, wind direction becomes important
(a)leeward opening with $s/h_m2 < 30$, Ψ < 0.5
(b)opening parallel to wind, Ψ < 1.0
(c)opening facing wind with $s/h_m2 < 2.0$, Ψ =0.7 Æ 1.1

3.
When wind is incident at 45° to the gap all the air in the courtyard is in rotational motion.

Figure A.2.10 *Mesh Effect*

MAPS FOR SHADING ANALYSIS IN MOUNTAINOUS AREAS.

A3.1

INTRODUCTION

The following method can be used for the calculation of sunshine hours in hilly or mountainous areas.

A contour map, to a scale that allows both a suitable degree of detail (e.g. for mountainous area 50-100 m contours) covering a sufficient area of land to enable accurate examination of shading of the site by its surroundings is needed.

First assemble the solar geometry data needed for construction of the graph by obtaining the latitude of the site, then estimating solar azimuths and altitudes for each hour of the day for the monthly dates being considered. Each month to be examined, and each latitude needs a different transparent graph. Because solar geometry is symmetrical, there is no need to plot more than six months if information for both summer and winter is required. For sun angles refer Figure 2.39 in main chapter.

The method involves the combined use of two transparent graphs scaled to the scale of the map used. One graph is above the horizontal axis and the other below.

Construct the transparent graphs as follows:

In the graph below the horizontal axis, solar azimuth hourly values are plotted with a superimposed series of concentric circles representing scaled distances from the centre point. The centre point represents the site in question. In the graph above the horizontal axis, hourly solar altitude angles and scaled heights of obstructions (mountains, peaks, ridges, etc.) are plotted. The height of an obstruction is really a height difference between the top of the obstruction considered and that of the site considered. Examples of constructed graphs for winter solstice and equinoxes are shown in Figure A3.1. The elevation scale at the top of the combined graphical overlay has to be drawn to the scale of the map, in the same units as the distance scale circles.

Step 1
Draw the geographical transparent graphs for the months and latitudes to be examined to the correct scale.

Step 2
Superimpose them on the contour map, making sure that the east-west axis is orientated correctly (Figure A3.2).

Step 3
Locate, in the area below the east-west axis and along the azimuth lines, all the peaks and ridges of mountains which could cast shadows on the site in question.

Step 4
For each peak or ridge, follow the arc on which it lies up to the east-west axis. At this point, plot vertically the height difference between this peak and the site in question, to the same scale as the map.
If this height difference is less than the height difference at that point to the altitude line of the sun for the hour considered, then the site is not shaded. If the height difference drawn crosses the altitude angle of the sun, then the site will be in shade at that hour.

Step 5
Proceed in this way for all the peaks and ridges and for all hours of that day in the month.

To examine other months new graphs will have to be drawn because the angles of azimuth and altitude change.

The outputs of this method provide systematically appropriate information on shading on the site. One such presentation is given in Figure A3.3

Contents

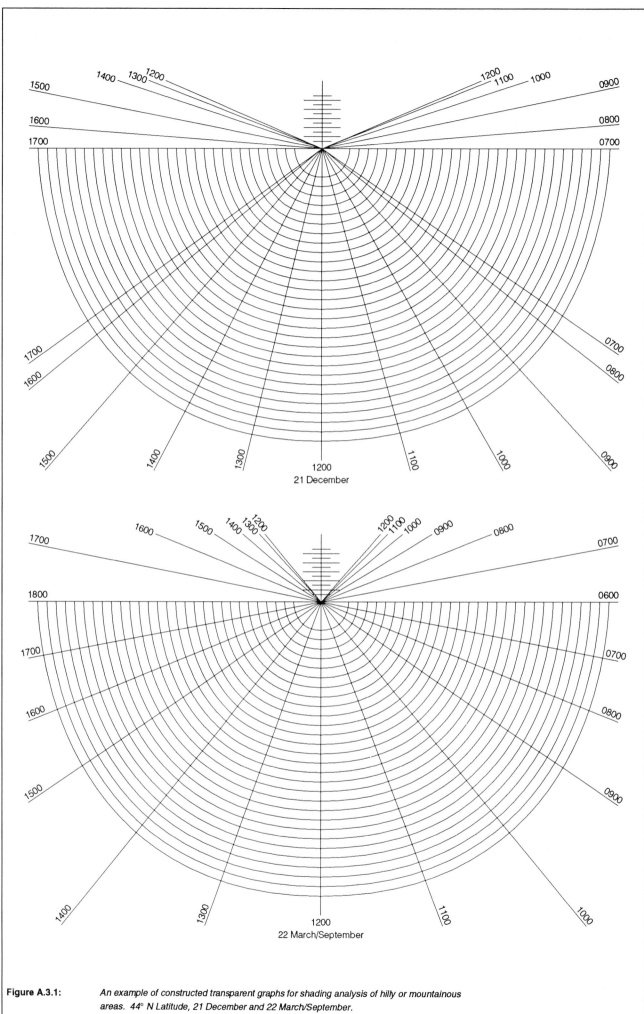

Figure A.3.1: *An example of constructed transparent graphs for shading analysis of hilly or mountainous areas. 44° N Latitude, 21 December and 22 March/September.*

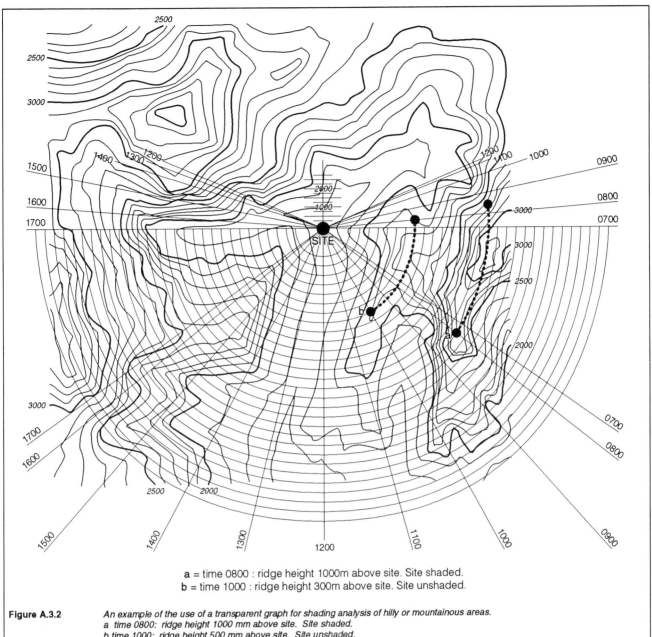

a = time 0800 : ridge height 1000m above site. Site shaded.
b = time 1000 : ridge height 300m above site. Site unshaded.

Figure A.3.2   *An example of the use of a transparent graph for shading analysis of hilly or mountainous areas.*
a time 0800: ridge height 1000 mm above site. Site shaded.
b time 1000: ridge height 500 mm above site. Site unshaded.

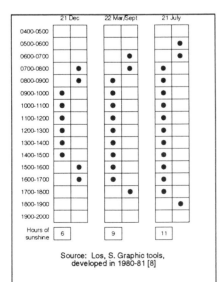

	21 Dec		22 Mar/Sept		21 July	
0400-0500						
0500-0600					●	
0600-0700			●		●	
0700-0800		●	●		●	
0800-0900		●	●		●	
0900-1000	●		●		●	
1000-1100	●		●		●	
1100-1200	●		●		●	
1200-1300	●		●		●	
1300-1400	●		●		●	
1400-1500	●		●		●	
1500-1600		●	●		●	
1600-1700		●	●		●	
1700-1800				●	●	
1800-1900						●
1900-2000						
Hours of sunshine	6		9		11	

Source: Los, S. Graphic tools,
developed in 1980-81 [8]

Figure A3.3.: *Tabulation, showing the hours of sun or shade, for the winter and summer solstices and the equinoxes at the site shown on the map in Figure A3.2.*

SIMPLE SITE ANGULAR SURVEY TECHNIQUES

A4.1
ANGULAR SURVEYS

Tools needed:

*   a map and compass (to find the azimuth angles of prominent features on the skyline),
*   a hand level and protractor (to find the altitude angle of prominent features on the skyline),
*   a copy of the sun chart for the site location,
*   an ordnance survey map to locate true South.

Step 1
Stand at the approximate location on the site where you want to put the building. Record the skyline as follows:

Step 2
Use the ordnance survey map to locate a prominent feature in the distance and use this to establish true south.

Step 3
Aiming the hand level true south, determine the altitude (angle above the horizon) of the skyline in the due south direction. Plot this point on the sun chart above the azimuth angle 0° (true south).

Step 4
Similarly, determine and record the altitude angle of the skyline for each 15° (azimuth angle) along the horizon, both to the east and west of south, to at least 120° in both directions. This is a total of 17 altitude readings. Plot these readings above their respective azimuth angles on the sun chart and connect them with a line.

Step 5
For isolated objects that block the sun during the winter, such as buildings or tall evergreen trees, find both the azimuth and altitude angles for various points on each object.

Step 6
Finally, record the deciduous trees on the skyline. These are of a special nature, because by losing their leaves in the winter, they let some of the sun pass through as long as they are not densely spaced. The outputs can then be plotted as described in the main body of the Chapter. The deciduous trees can be plotted with a dotted outline.

A4.2
SOLAR SITE SELECTOR

The Solar Site Selector provides a base which can be easily levelled and oriented and which holds a magnetic compass, a viewer, and transparent cylindrical sun-charts. This is an extremely useful tool for site selection, giving instant views of the skyline and solar obstructions for any time of year, at any selected point on the site. The selector can be mounted on a tripod and the skyline, etc., marked on the transparent chart.

A4.3
TNO SUNLIGHT METER

The TNO Sunlight meter was originally developed for environmental surveys of low cost housing in The Netherlands. It is however a useful instrument for solar site surveys. The instrument basically consists of a perspex lens mounted over a solar chart. The instrument is mounted in gimbals, so the instrument levels itself. When correctly oriented, the user can view from the defined viewing point both the general scene reflected from the upper surface and simultaneously the appropriately scaled solar chart below. This enables the user to see immediately the available periods of sunshine at any time of year. The image can be photographed with a suitable reflex camera. Details of the use of the TNO sunlight meter for solar site surveys may be found in reference [1].

Contents

A4.4

SIMPLE PHOTOGRAPHIC SURVEYS

An alternative technique makes use of a site photograph taken using a lens of a known focal length. It is possible then to superimpose an overlay which allows the angles subtended at the reference point by all obstructions to be read-off directly and then be plotted on the sun chart.

A camera (24 x 36 mm negative) with a wide-angle lens of 21 mm has an aperture angle of approximately 90°. It is possible, therefore, to encompass a full 180° arc in two exposures. Since the 90° angle will generally not be indicated on the print, it is convenient to know the angle between a few prominent spots on the skyline in order to enable assessment of the enlargement factor.

The camera mounted on a tripod can be accurately levelled and oriented. The photographic record avoids the necessity to make further site visits for this purpose. The overlay chart for a 21 mm lens is shown in Figure A4.1 and should be copied onto a transparent film.

Input:

Take a site photograph with a level camera, facing due South, using 21 mm wide angle lens (or two photographs, facing precisely south-west and south-east).

Operation:

Place the acetate copy of the computer-generated overlay (Figure A4.1) over a print of the whole photograph, 242 x 158 , (or over a print of each photograph, 242 x 158 mm).

Output:

Read-off angles of obstruction at each angle of azimuth. These may then be plotted on to a solar chart as above.

REFERENCES

[1] University of Sheffield Building Science note.

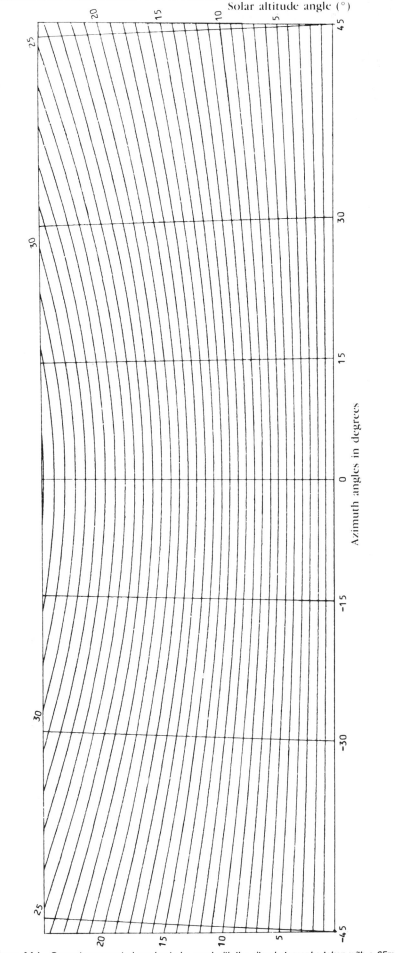

Figure A4.1. *Computer generated overlay to be used with the site photographs taken with a 35mm camera and a standard 21mm wide angle lens. Camera altitude angle 0°. Photo size = 242 x 158mm*

SOLAR VIEW

A5.1

PLOTTING PROJECTED VIEWS OF BUILDINGS AS SEEN BY THE SUN

This appendix is based on a paper by R.H. Byrd [12]. The original paper has been modified to relate better to the information in the main body of the Climate chapter. It describes in detail how to plot an isometric sun's eye view of a building.

Stage 1

Considering the design objectives, determine the critical times of day and year for which the building should be viewed. Extract the relevant solar altitudes and azimuths for these selected times from the solar charts, using interpolation if necessary, to obtain values for intermediate latitudes.

Stage 2

Select the first time plot. Then, using Table A5.1, extract the horizontal and vertical corrections factor for that given solar altitude.

A5.2

METHOD FOR PLOTTING A SPECIFIC VIEWPOINT

Step 1.

Extract data to plot sun's eye isometric for following sun position

Solar altitude 25 degrees
Solar azimuth -50 degrees

Step 2.

.Extract correction factors from Table A5.1
Horizontal correction factor 0.42
Vertical correction factor 0.91

Step 3

Moving to the drawing board, draw plan correctly oriented to the sun's direction, making the sun's azimuth vertical on plan. Figure A5.1 top left drawing.

Step4

Draw below, leaving a convenient space, a horizontal base line. Drop vertical lines from the key edge points to the base line.

Contents

Solar altitude degrees	Horizontal correction factor	Vertical correction factor
0	0.00	1.00
5	0.09	1.00
10	0.17	0.98
15	0.26	0.97
20	0.34	0.94
25	0.42	0.91
30	0.50	0.87
35	0.57	0.82
40	0.64	0.77
45	0.71	0.71
50	0.77	0.64
55	0.82	0.57
60	0.87	0.50
65	0.91	0.42
70	0.94	0.34
75	0.97	0.26
80	0.98	0.17
85	1.00	0.09
90	1.00	0.00

Table A5.1   *Horizontal and vertical correction factors.*

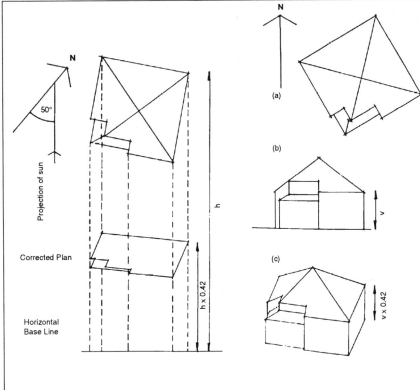

Figure A5.1 *Projected sun's eye view of the building drawn in plan in (a) and in section at (b), shown isometrically in (c), using the geometric construction on left hand side, as explained in the text [12].*

Step 5

For each corner, measure the distance equivalent to distance h. Plot each corner for base plan of the isometric at a distance = horizontal correction factor x h, i.e. in Figure A5.1 at 0.42h from base line. Remeasure the appropriate distance h each time.

Step 6

Turning to vertical section, Figure A5.1(b), extract vertical height, v for each key point. Multiply v by the vertical correction factor, i.e. in this case by 0.91. The vertical dimensions remain vertical and are plotted in this case at 0.91v. Figure A5.1(c). Join up key points to give the completed isometric sun's eye view.

As it is a sun's eye view, no shadows will be seen on the exposed surfaces drawn.

Proceeding in this way, a series of views may be built up, as shown in Figure 2.46 in Chapter 2..

Reference Source [8] [12]

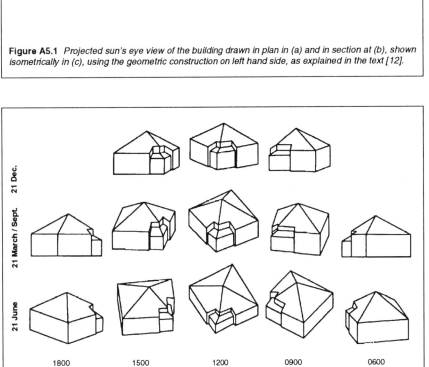

Figure A5.2 *Sun's eye view of the building at different times of the day and year.*

DEFINITION OF THE SOLAR ENVELOPE

A6.1

INTRODUCTION

The solar envelope defines the greatest volume which a building can fill on a site without causing significant overshadowing of adjacent sites. This would obviously be of major importance in the development of a solar townscape, though it is likely to be more easily applicable in rural or suburban settings than on high density urban sites.

A6.2

CONSTRUCTION OF THE SOLAR ENVELOPE

Input:

a.
an ordnance survey map, showing the plan of the site at a reasonable size, with contours across the site and ideally the heights of surrounding buildings.
b.
solar altitude and azimuth data in December.

Step 1

Construct the solar pyramid. This is selected for low winter sunshine, and is based on a minimum useful solar altitude of 15° for southern Europe and 12.5° or 10° for northern Europe. The vertex of the pyramid is the top of a vertical pole at the same scale at the site plan. The northern extremes of the pyramid base are the parts reached by shadows cast by the vertical pole when the sun is at the relevant minimum altitude, in the morning and afternoon of the winter solstice. The solar azimuth at this time can be read from the Sun Chart for the correct latitude (Figure 2.39 in Chapter 2 - Climate and Design). The southern edge of the pyramid is given by the orthogonal projection of the northern edge on to the east-west axis passing through the base of the vertical pole. Figure A6.1 shows the simple pyramid on a flat site.

If the pole is subdivided (into storey heights or metres, for example), the pyramid can also be subdivided into contours representing the shadows cast by poles of different heights (Figure A6.2).

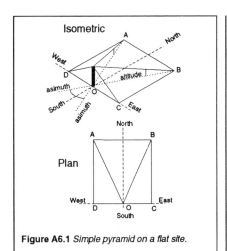

Figure A6.1 *Simple pyramid on a flat site.*

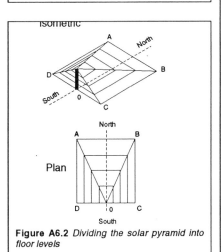

Figure A6.2 *Dividing the solar pyramid into floor levels*

If the site is not level, the pyramid will be distorted and must be constructed as shown below projected on to the plane of the slope.

The direction of the slope (the line maximum slope) is at azimuth angle w to the north-south axis through the base of the vertex pole. The slope along this line is b (Figure A6.3).

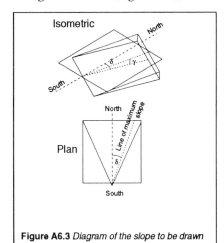

Figure A6.3 *Diagram of the slope to be drawn*

With the direction of the slope horizon project the vertices of the pyramid (0, A, B, C, D) onto a horizontal plane.(0_1, A_1, B_1, C_1, D_1) (FIgure A6.4).

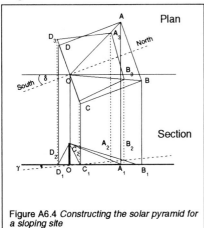

Figure A6.4 *Constructing the solar pyramid for a sloping site*

Draw the elevation of a pyramid, with the same pole height. When the edges of the pyramid meet the plane of the slope, 0_2, A_2, B_2, C_2, D_2 are defined and then the vertical projection of these back on to the marginal edges (or their extensions) gives the vertices of the base of the new pyramid 0_3, A_3, B_3, C_3, D_3 (Figure A6.5).

This enables the solar envelope pyramid to be plotted as in Figure A6.5.

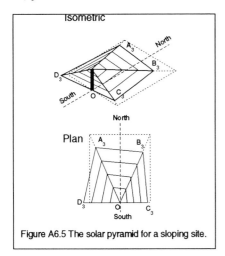

Figure A6.5 The solar pyramid for a sloping site.

If the site is not even, it must be divided into a number of areas of simple sloping planes - referred to as a slope-orientated domains - and a pyramid constructed for each.

Step 2

Either the site plan or the plan of the pyramid must be transparent, so that they can be overlaid and the result traced. Moving the pyramid around the perimeter of the site will define the solar envelope(Figure A.6.6).

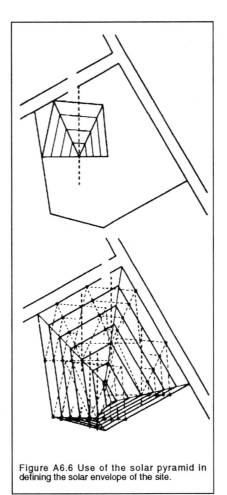

Figure A6.6 Use of the solar pyramid in defining the solar envelope of the site.

Output:

A plan projection of the maximum building volume for a given site, which will not cast significant shadows beyond the site boundary. The method can also be extended to investigate overshadowing of the site by neighbouring buildings.

THE EUFRAT DATA BASE

A7.1
PERSONAL COMPUTER DATABASE

Table A7.1 gives the list of the European locations in the data base. The files are available on diskettes, which can be read into a personal computer (IBM or IBM Compatible). These files contain the Cumulative Frequency Curves and the Utilizability Curves for different collector types: non tracking collectors, both flat plate and CPCs in a range of orientations.

A7.2
EURFAT DATA TABLES

Tables are also available with summarized versions of the data contained in the data files described above.

For information on availability, contact the editors.

Contents

Country	Location	Latitude	KT$_y$
Denmark	Vaerlose	55.77	0.40
Ireland	Valentia	51.58	0.38
	Dublin	53.26	0.37
	Birr	53.05	0.36
Germany	Hamburg	53.63	0.37
United Kingdom	Brackwell	51.47	0.37
Belgium	Uccle	50.80	0.36
France	Trappes	48.77	0.39
	Nancy	48.68	0.39
	Rennes	48.06	0.42
	Macon	46.30	0.41
	La Rochelle	46.15	0.48
	Limoges	45.82	0.43
	Embrun	44.57	0.56
	Agen	44.18	0.45
	Carpentras	44.08	0.53
	Nice	43.65	0.53
	Carcassonne	43.22	0.46
	Ajaccio	41.91	0.52
Switzerland	Zurich	47.45	0.39
	Locarno	46.17	0.50
	Davos	46.80	0.50
Spain	Oviedo	43.37	0.41
	Palma	39.37	0.53
	Tortosa	40.97	0.52
	Madrid	40.68	0.56
	Badajoz	38.88	0.56
	Murcia	37.98	0.55
	Seville	37.42	0.60
	Izana	28.30	0.73
Portugal	Bragança	41.80	0.53
	Porto	41.13	0.53
	Coimbra	40.20	0.52
	Lisboa	38.72	0.55
	Evora	38.57	0.57
	Faro	37.02	0.61
Greece	Athens	37.58	0.53

Table A7.1. European locations included in the EUFRAT data base. KTy is the yearly clearness index of the location.

LONG WAVE RADIATIVE EXCHANGES

A8.1
INTRODUCTION

This appendix deals in detail with the estimation of the long wave radiation exchange on inclined surfaces for clear sky conditions, for monthly mean conditions and for overcast conditions, using observed screen dry bulb air temperatures and cloud cover as the primary data inputs. The appendix has been developed from a section of a report prepared by J.K. Page for the U.K. Department of Energy Climate Handbook, Page and Lebens (Eds) [1]. It takes into consideration input data available in the EUFRAT Climatic Data Handbook, discussed in the main text.

Terminology
The following basic terminology is used:-

H_a the long wave radiation from the atmosphere on a horizontal surface.

$H_a(S)$ the long wave radiation from the atmosphere on a slope of tilt b degrees.

$H_g(S)$ the long wave radiation received from the ground on a slope of tilt b degrees.

Additional subscripts
In order to distinguish between different cloudiness conditions, the following additional second subscripts are used:-

Subscript c referring to cloudless conditions.

Subscript m referring to monthly mean conditions.

Subscript b referring to overcast conditions.

E_g the long wave emittance of the ground

E_s the long wave emittance of the receiving surface.

t_a the screen air temperature, deg. C.

t_e the surface temperature of the external surface under consideration, deg. C.

t_g the ground surface temperature, deg. C.

A8.2
INCOMING LONG WAVE RADIATION

There are two external sources of incoming long wave radiation impacting on buildings:-

1.   the atmosphere of the earth,

2.   the surfaces of the ground and the various objects standing on it, like obstructing buildings.

The appendix only considers unobstructed sites.

A two way radiative exchange process takes place, long wave radiation being simultaneously absorbed on and being emitted from any surface. The difference between the two fluxes is the **net long wave radiation exchange** with that surface. For horizontal and near horizontal surfaces, at or above surface air temperature, this net exchange is nearly always negative. The long wave absorptance/emittance properties of the surfaces involved are very important in determining the precise magnitude of the exchange.

A8.3
INCOMING LONG WAVE RADIATION FROM THE ATMOSPHERE

The band of wavelengths under consideration is from 3 to 30 microns. The radiation from the cloudless sky is not blackbody radiation, and there is an important "window" in the atmospheric radiation spectrum between 8 microns and 15 microns. The window is most pronounced under clear sky conditions and is very much less evident in the spectrum of overcast skies. Figure A.8.1 shows typical long wave spectral characteristics of the long wave radiation emitted from clear skies, and from overcast skies.

Contents

The cloudless atmosphere emits this incoming radiation from a range of layers of differing composition at different temperatures. A significant part of the atmospherically-absorbed solar energy is re-emitted as downward atmospheric radiation.

If clouds are present, the base of the clouds becomes an important source of long wave radiation. The lower the clouds, the higher the typical cloud base temperature and consequently the greater the amount of downward long wave radiation emitted from their base. Natural radiative cooling becomes much less effective, as cloud cover increases.

The long wave radiance of the sky vault is not uniform as Figure A8.1 shows. The path length through the atmosphere at the zenith is the shortest, and is associated with the lowest long wave radiance. Radiatively, for long wave radiation, the zenith has the lowest effective sky temperature, i.e. it is the coldest point radiatively in the sky.

An inclined surface will also "see" the ground, which will reflect some atmospheric radiation towards it. However this reflected ground radiation is normally not large, as the ground long wave absorptance is normally about 0.95. Most of the radiation from the ground is thermal radiation emitted by the ground itself. This flux is discussed later. However it should be noted that normally the ground is radiatively warmer than the sky. The consequence is that the net radiative cooling exchange is greatest for horizontal surfaces, and least for vertical surfaces.

LONG WAVE RADIATION FALLING ON HORIZONTAL SURFACES

Systematic measurements of long wave radiation are difficult to make on a continuous basis, so observed long wave radiation data is rarely available.

A representative value of the downward long wave flux onto a horizontal surface, H_{ac} may be estimated from a number of formulae. Here the formulation of Monteith and Unworth is adopted (3.)

$$H_{ac} = 1.06\, St(t_a + 273.15)^4 - 119 Wm^{-2}$$
(Equation 1)
where St is the Stefan Boltzmann constant,

$5.6697 \times 10^{-8}\ m^{-2}\ K^4$
t_a is the screen air temperature, deg. C.

Clouds will substantially increase this downward flux. If the effects of cloud are to be considered, it is scientifically more convenient to define a quantity, e_a called the apparent admittance of the sky, to estimate H_a using the following relationship:-

$$H_a = e_a St(t_a + 273.15)^4$$
(Equation 2)

For the clear sky, we have:-

$$e_{ac} = H_{ac}/St(t_a + 273.15)^4$$
(Equation 3)

The effects of cloud can then be estimated as follows. The apparent horizontal emittance of a partially clouded sky, with a cloud fraction of N, (N = 1, overcast, N = 0, cloudless) is given by :-

$$e_{an} = (1 - 0.84N)\, e_{ac} + 0.84N$$
(Equation 4)

which, for N = 1 the overcast sky, reduces to

$$e_{ab} = 0.16\, e_{ac} + 0.84$$
(Equation 5)

Therefore, substituting the appropriate apparent emittance into Equation 2, the required horizontal flux may be obtained for different amounts of cloud cover from the screen air temperature.

An alternative formulation for estimating e_{ac} from the dewpoint temperature has been put forward by Berdahl and Fromberg [2]. For the clear sky, at night, they recommend the estimation of e_{ac} from the dewpoint of the air, using their correlation formula:-

$$e_{ac} = 0.741 + 0.0062\, T_{dp}$$
(Equation 6)

where T_{dp} is the surface dewpoint, deg. C.

For the day, their correlation formula became

$$e_{ac} = 0.727 + 0.0060\, T_{dp}$$
(Equation 7)

Unfortunately systematic dewpoint temperature data is rarely readily available. This is why the Monteith and Unsworth formulation was chosen, Equation 1.

LONG WAVE RADIATION ON INCLINED PLANES FROM THE ATMOSPHERE ALONE

The incoming atmospheric radiation from the sky falling on slopes is greater than that on horizontal surfaces. Setting the slope angle as b degrees, the following expressions were developed by Sharples (4) from Monteith and Unsworth's work [5]. The formulae may be used to estimate the long wave radiation flux from the atmosphere falling on any unobstructed slope.

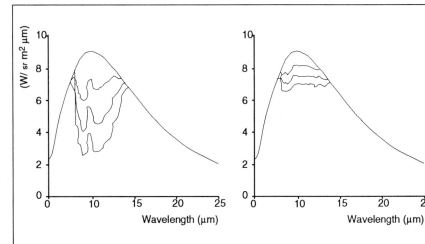

Figure A8.1 Estimated spectral radiance of clear sky (left figure) and cloudy skies (right figure) for mid-summer mid-latitude atmosphere as a function of wavelength. Note the significant departure from blackbody radiation in atmospheric window between 8 and 14 microns. Curve marked 0 represents zenith values, curve marked 90, the horizon values. (after Berdahl and Fromberg [2]).

Cloudless conditions

$$H_{ac}(S) = (e_{ac} \cos^2 (b/2) + 0.03(\sin b)^{1.4}) \times St(t_{ac} + 273.15)^4 \ Wm^{-2}$$
(Equation 8)

Partially clouded sky

$$H_{am}(S) = (e_{am} \cos^2 (b/2) + 0.03(1 - .084N) \times (\sin b)^{1.4}) \times St(t_{am} + 273.15)^4 \ Wm^{-2}$$
(Equation 9)

Overcast conditions

$$H_{ab}(S) = (e_{ab} \cos^2 (b/2) + 0.0048 \times (\sin b)^{1.4}) \times St(t_{ab} + 273.15)^4 \ Wm^{-2}$$
(Equation 10)

where t_{ac} t_{am} and t_{ab} are the screen air temperatures for clear sky, monthly mean and overcast conditions in degrees C.

In order to make the calculations relatively simple, graphs for the estimation of H_{ac} on any slope have been prepared as a function of screen air temperature (Figures 8.2 and 8.3.)

Sometimes the radiative flux is used to derive an "equivalent blackbody radiative temperature" of the sky, T, in degrees Kelvin. This is simply done for horizontal surfaces by equating the flux to the equivalent blackbody radiation, StT^4_{eq}, thus

$$T_{eq} = (H_a(S)/St)^{1/4}$$
(Equation 11)

Its value will vary with slope, and will be lowest for horizontal planes. By subtracting the screen air temperature, one obtains the depression of equivalent radiative temperature below screen air temperature. For cloudless nights, the depression is typically about 30 deg. C. below air temperature for horizontal surfaces when the air is very dry (dewpoint -20 deg. C.) falling to a value of about 10 deg. C. under hot humid conditions (dewpoint about 20 deg.C.) (2). Under overcast conditions, with low cloud, the equivalent radiative temperature of the sky is relatively close to screen air temperature.

A8.6
ESTIMATING THE MONTHLY MEAN CLOUDINESS

It is sometimes possible to obtain cloudiness data at different times of day and night from national meteorological services. Rough estimates of monthly mean cloudiness fraction can be derived from the CEC European Solar Radiation Atlas, Vol. 1, Horizontal Surfaces, using the published values of the relative sunshine duration, S/S_0. (Refer Table 2.1 in Chapter 2 - Climate and Design). As a very rough approximation

$$N = 1 - S/S_0$$
(Equation 12)

However this simple relationship does not relate well to actual reported data. The sunshine recorder is an imperfect instrument. Furthermore the radiation passing through the thinner edges of clouds and holes tend to burn the card. Surface observers have a tendency to over-exaggerate cloud cover, often giving undue weight to the lower portion of the sky, which is the part most easily seen. As a consequence, a better formula has the form:-

$$N = a - S/S_0$$
(Equation 13)

where a typically has values in the region 1.05 to 1.15

The higher values tend to occur in the summer months which experience a lot of cumulus cloud, or when there is a lot of high thin cloud.

The Atlas value for S/S_0 for Hamburg for August is 0.50. Assuming a = 1.15, the approximate estimate of the mean cloudiness fraction is 1.15 - 0.50 = 0.65. The corresponding value of S/S_0 for Seville is 0.85, yielding a cloudiness fraction of 1.15 - 0.85 = 0.30. The potential for effective nightime radiative cooling is obviously much greater at Seville.

A8.7
THE GROUND AS A LONG WAVE RADIANT SOURCE

The long wave radiation reaching inclined surfaces from the ground now has to be considered. The long wave emittance of the ground is typically around 0.93 to 0.96. This value does not vary very much from surface to surface, so the key factor determining the long wave radiation received from the ground is the temperature of the surface. The ground temperature may be above the screen air temperature, for example, in sunny weather, or below the screen air temperature, for example on clear winter nights. However, in general, especially at night, the equivalent radiant temperature of the ground is usually much closer to the dry bulb air temperature than to the equivalent radiative temperature of the sky. The ground radiation is particularly important in the case of steeply tilted planes, because the ground with its obstructions begins to fill around half or more of the overall hemispherical view seen from the surface. One has also to consider the small amount of atmospheric long wave radiation reflected from the ground.

In the absence of obstructions above the horizon, the long wave radiation from the ground on a slope of b degrees may be estimated as follows:-

$$H_g(S = (E_g St(t_g + 273.15)^4 + H_a(1 - E_g)) * \sin^2(b/2) \ Wm^{-2}$$
(Equation 14)

where E_g is the long wave emittance of the ground

St is the Stefan Boltzmann constant

t_g is the ground surface temperature, deg.C.

H_a is the incoming long wave radiation, on the horizontal surface, Wm^{-2}

It will be noted that there are two ground components, a directly emitted component and a ground reflected component of atmospheric radiation.

A8.8

THE EMISSION OF LONG WAVE RADIATION FROM THE EXTERNAL RECEIVING SURFACE.

On the other side of the long wave radiation exchange is the outward rate of thermal radiation emission from that surface. This emitted energy depends on the surface temperature and the long wave emittance of the receiving surface.

Calling this outgoing flux F, using Stefan's law, one obtains

$$F_s = E_s St(T_e + 273.15)^4$$
(Equation 15)

where E_s is the long wave emittance of the receiving surface

St is the Stefan Boltzmann Constant

t_e is the external surface temperature, dec.C.

The values of F for a perfect blackbody (E_s = 1) are given in Table 8.2.1 for a range of surface temperatures. The long wave radiation emitted from any specific surface can be estimated from this Table by multiplying the blackbody value by the surface long wave emittance, E_s.

A8.9

NET LONG WAVE RADIATION EXCHANGES WITH INCLINED PLANES

The net long wave radiative exchange will be the difference between the long wave radiation absorbed and long wave radiation emitted from the surface. Provided the radiative temperatures are the same, the long wave absorbtance will equal the long wave emittance. The net flux on a slope is thus

$$H_{net} = E_s(H_a(S) + H_g(S) - F)$$
(Equation 16)

As F is normally greater than the sum of the other two terms, unless the surface is artificially cooled, the balance will be negative.

A8.10

GRAPHS FOR THE ESTIMATION OF INCOMING LONG WAVE RADIATION ON SLOPES

Figures A8.2 and A8.3 plot the long wave radiation from sky and ground incident on inclined planes as a function of screen air temperature. Figure A8.2. is plotted for cloudless conditions. Figure A8.3 is plotted for overcast conditions. The values for partially clouded conditions can be estimated using Equation 9.
Figure A8.2 Incoming long wave radiation for cloudless conditions for different slope inclinations as a function of shade air temperature. Note the increase of incoming radiation with increase of slope. Ground emittance set at 1 and ground surface temperature set equal to air temperature. The ground contribution has to be separately estimated, using equation 14.

Figure A8.3 Incoming long wave radiation for overcast conditions for different slope inclinations as a function of shade air temperature. Note the smaller increase of incoming radiation with increase of slope, compared with cloudless conditions. Ground emittance set at 1 and ground surface temperature set equal to air temperature. The ground contribution has to be separately estimated using equation 14.

Surface Temp Deg. C.	+0.0°C	+1.0°C	+2.0°C	+3.0°C	+4.0°C	+5.0°C
-20.0	232.8	236.5	240.3	244.1	247.9	251.8
-15.0	251.8	255.7	259.7	263.7	267.8	271.9
-10.0	271.9	276.0	280.2	284.5	288.8	293.1
-5.0	293.1	297.5	302.0	306.5	311.0	315.6
0.0	315.6	320.3	325.0	329.7	334.5	339.4
5.0	339.4	344.3	349.2	354.3	359.3	364.4
10.0	364.4	369.6	374.8	380.1	385.5	390.0
15.0	390.9	396.3	401.8	407.4	413.0	418.7
20.0	418.7	424.5	430.3	436.1	442.0	448.0
25.0	448.0	454.1	460.2	466.3	472.6	478.8
30.0	478.8	485.2	491.6	498.1	504.6	511.2
35.0	511.2	517.9	524.6	531.4	538.3	545.2
40.0	545.2	552.2	559.3	566.4	573.6	580.9
45.0	580.9	588.2	595.6	603.1	610.7	618.3
50.0	618.3	626.0	633.7	641.5	649.5	657.4
55.0	657.4	665.5	673.6	681.8	690.1	698.4
60.0	696.4	706.8	715.3	723.9	732.6	741.3
65.0	741.3	750.1	759.0	768.0	777.0	786.1
70.0	786.1	795.3	804.6	814.0	823.4	833.0

Table A8.1 Blackbody long wave radiation emitted as a function of surface temperature in degrees Celsius.

	June 1925	January 1925
Tar Macadam	32.6	6.8
Sand	25.9	5.4
Earth	25.0	5.4
Gravel	21.1	5.7
Grassy ground	16.0	3.3
Clay soil	11.5	5.0
Screen Air Temperature	14.2	6.6

Table A8.2 Monthly range of temperature, deg. C. 10 mm below surface at Salisbury Plain, U.K. Source Johnson and Davies.

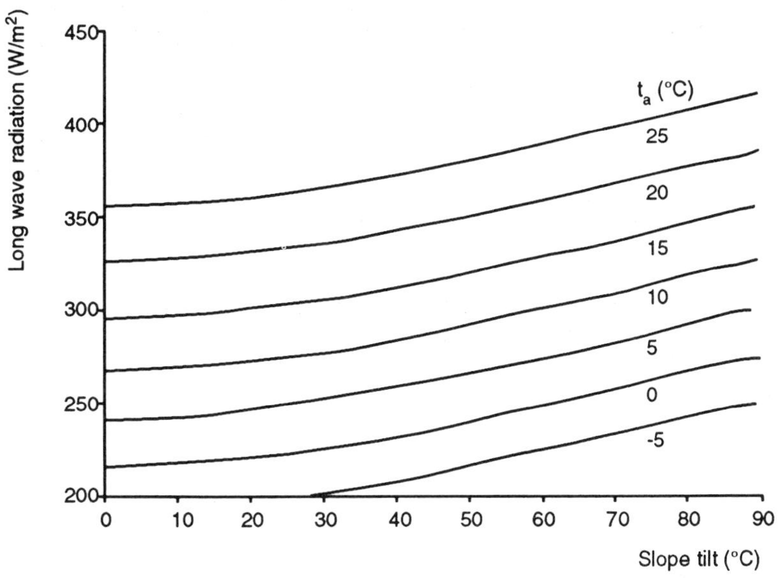

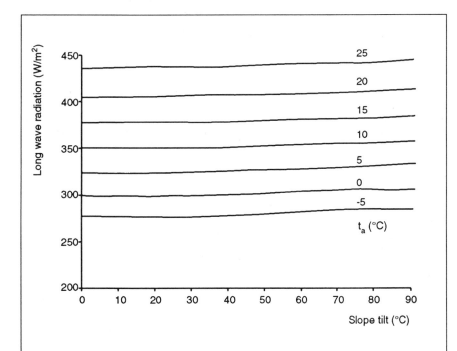

Figure A8.2 *Incoming long wave radiation for cloudless conditions for different slope inclinations as a function of shade air temperature.*

Figure A8.3 *Incoming long wave radiation for overcast conditions for different slope inclinations as a function of shade air temperature*

A8.11
GROUND SURFACE TEMPERATURES

A difficult practical problem is the estimation of ground surface temperatures, which on sunny days may be substantially above air temperatures and on clear nights may be somewhat below air temperatures. At night time, the negative differences may be 2-3 degrees C. Table A8.2 gives some representative summer and winter observations that illustrate the issues implicit in attempting to assess the impacts of ground cover.

Figure A8.4 illustrates measured results for surface temperatures measured in front of the walls of a building 14 metres high in September in Budapest on a site surrounded with buildings. The walls faced on bearings of 75 degrees, designated east, 165 degrees designated south and 255 degrees, designated west. The measurement points were set 5 metres from the walls. The periods of insolation were for the point in front of the east wall, 0900 - 1100 hours, in front of the south wall 0600 - 1700 hours, and in front of the west wall 1100 to 1230 hours. Being approximately the equinox, the sun was above the horizon between 0600 and 1800 hours. It will be noted that the grass on the east side is warmer than the grass on the west side until 1130 hours. During the morning the grass on the west side was about 3 deg. C. below air temperature. As soon as the sun comes onto the west side grass, the temperature starts to climb, reaching 5 deg. C. above air temperature. The point on the south side is insolated over the longest period, and the grass is warmer than the air, until about an hour after sunset. In the hour before sunset, the surface temperatures fall back towards the air temperatures and will continue to cool further throughout a clear night, so by dawn, a grass surface temperature may be several degrees below air temperature. The impact of microclimate on radiative exchanges is clearly complex.

It appears therefore reasonable, for assessing overheating risks, to assume that the temperature on hot sunlit ground surfaces, like asphalt, desert sand, etc. may be 20 deg.C. above shade air temperature in the middle of the day under cloudless conditions. The temperature rises on growing green vegetation supplied with adequate water are much smaller.

195

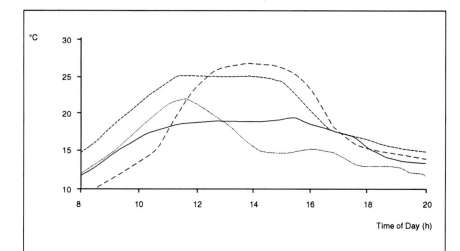

Figure A8.4 *Measured grass surface temperatures measured 5m in front of the walls of a building 14 metres high in September in Budapest.*

While very simplified estimation methods do exist, the implications tend to conceal the real radiative interactions. For example the CIBSE Guide (7) uses the following simplified relationships:-

For horizontal surfaces

$$H(S) = 93 - 79N \ \ W/m2$$
(Equation 17)

For vertical surfaces

$$H(S) = 21 - 17N \ \ W/m2$$
(Equation 18)

where N is the cloudiness fraction.
To estimate the actual exchanges, the expressions of course have to be multiplied by the appropriate emittances.

REFERENCES

[1] Page, J.K. & Lebens, R. (1986), Climate in the United Kingdom, a handbook of solar radiation, temperature, and other data for thirteen principal towns, HMSO, P.O. Box 276, London SW8 5DT.

[2] Berdahl, P. and Fromberg, R. (1982), The thermal radiance of clear skies, Solar Energy, Vol. 29, pp 299-314.

[3] Unsworth, M.H. and Monteith, J.L. (1975), Long wave radiation at the ground (1), Angular distribution of incoming radiation, Q.J. Roy, Met. Soc., Vol 101, pp 13-24.

[4] Sharples, S. Unpublished computer programs for computing sol-air temperatures, Dept. of Building Science, University of Sheffield.

[5] Unsworth, M.H. (1975), Long wave radiation at the ground (II), Geometry of interception by slopes, Q.J. Roy, Met. Soc., Vol 101, pp 25 -34

[6] Gajzago, L. (1972), Outdoor microclimates and human comfort in Proc. Conf. Teaching the teachers on Building Climatology, Statens Institut for Byggnadsforskning, Stockholm, Vol 1, Paper 16.

[7] CIBSE Guide, Vol. A, Design data, CIBSE, Delta House, 222 Balham High Road, London SW12 9BS. 1986

[8] Graphic Tools, developed in 1980-81, Los, S.

[9] Gandemer, J. (1973), Inconfort du vent aux abords des batiments: etude aerodynamique du champ de vitesse dans les ensembles batis: CSTB, Nantes. ADYM 12-73.

[10] Gandemer, J. (1974), Etude de la simulation des structures gonflables, la simultude aerodynamique, CSTB, Nantes. ADYM 110-74.

[11] Gandemer, J. & Barnaud, G., (1974), Simulation des proprietes dynamiques du vent en stabilite neutre dans la sofflerie a couche limite due C.S.T.B., Nantes. ADYM 1 -74.

[12] Drawing a projected view of a building as seen by the sun, Byrd, R.H., Lighting Research and Technology, 22(1), pp. 53-54. 1990

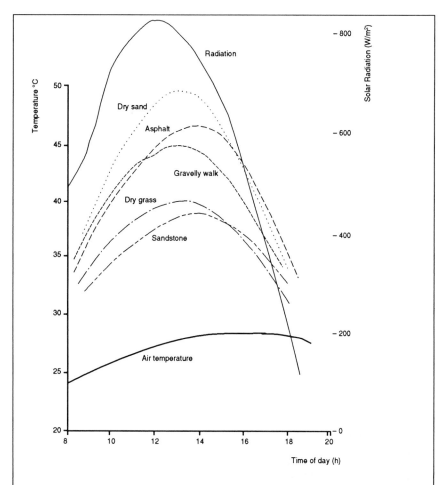

Figure A8.5 Measured surface temperatures on different horizontal surfaces on a sunny day, 17 July, 1969, Budapest, Hungary, together with air temperature and solar radiation.
Source: Gajzago, L. Teaching the teachers on Building Climatology [6]

FUNDAMENTALS OF HEAT TRANSFER

A9.1

INTRODUCTION

When solar radiation is absorbed by the surface of a material, heat is released. Hence the issue with passive solar systems is not the conversion of solar radiation to heat, but the transfer of heat to the appropriate location and, if necessary, its storage for later use. The transfer of heat can take three possible routes: conduction, convection and radiation.

Conduction is the transfer of heat in a solid or a fluid at rest. This transfer of heat is by direct molecular interaction.

Convection is the process of heat transfer by flow and mixing motions in fluids. If the fluid movement is caused by density differences (as a result of temperature differences) it is called free convection. Where this is not the case - for instance where a fan or pump cause the flow - forced convection results.

Radiation is the process of heat transfer by means of electromagnetic waves. All molecules emit radiation, depending on their temperature, and they absorb radiation from their environment. This form of heat transfer takes place without any transport medium. The absorption of radiation by the air can be neglected in the case of average distances between walls in a building.

A9.2

RADIATION

All bodies emit radiation. The energy given off per unit of time and per unit of surface area is termed exitance. As well as emitting radiation, materials will absorb radiation. A body completely absorbing the incoming radiation of all wavelengths is termed a black body. When the temperature of a material rises, the exitance increases in proportion to the fourth power of the absolute temperature (Stefan-Boltzmann law). The exitance of a black body is formulated as follows:

$$M_b = \sigma \times T^4 \quad (W/m^2)$$

where:
T = absolute temperature (K)
σ = Stefan-Boltzmann constant
   = 5.67×10^{-8} W/m^2K^4
M_b = the exitance of a black body (W/m^2)

The spectral exitance of a black body is exclusively dependent on temperature. When the temperature rises, the spectrum moves towards shorter wavelengths (Figure A.9.1). If one were to regard the sun as a black body, its temperature would amount to approximately 6000°C. This is so much higher than temperatures in the atmosphere, at ground level and in buildings, that the wavelengths are at totally different intervals. This difference is important, because nearly all materials have different properties for short wave radiation than for long wave radiation.

Black bodies do not exist in nature. One can tell by the spectrum of the sun, for instance, that it is not a black body. For a specific body, the exitance of thermal radiation may be described as a fraction of the exitance of a black body at the same surface temperature. This fraction is termed the emissivity ε. Thus the exitance of a surface is:

$$M = \varepsilon \times \sigma \times T^4$$

The fraction of incoming radiation absorbed is termed the absorptance. For thermal radiation, the absorptance is at each wavelength equal to the emissivity:

$$\alpha = \varepsilon$$

Furthermore, a fraction of the incoming radiation is reflected (R = reflectance) and a fraction is transmitted (t = transmittance). Therefore:

$$\tau + R + \alpha = 1$$

In the case of short-wave radiation it is important to realize that these factors may be dependent on the angle of incidence (angle between radiation direction and a line normal to the surface). Moreover, reflectance can be partially specular and diffuse, whereas transmittance can be direct and diffusing.

Contents

The properties of a material or a surface regarding solar radiation are partly visible, because the visible radiation forms an important part of the solar spectrum. Thus it is straightforward to determine whether a material has a high transmittance (i.e. transparent material), absorbs much solar radiation (matt black surface, α = 0.9) or whether it reflects to a significant degree (white or specular surface, R = 0.8). For thermal radiation these properties are not so easy to determine. Most building materials (and glass) do not transmit thermal radiation, and have a high absorptance (= emissivity). Thus τ = 0 and emissivity ranges from 0.9 to 0.95. Exceptions are some polymers, which slightly transmit thermal radiation, and polished metal surfaces which have a low emissivity (0.1 to 0.15). However, these properties cannot be determined with the naked eye. A minutely thin layer of varnish or dirt on a polished surface will result in a high emissivity. Absorption of radiation begins to take place when the layer has a thickness of one fourth of the wavelength. For thermal radiation this is around 3 μm. The expression "spectral selective" is used to describe foils or coatings with a low emissivity, and a high absorptance or a high transmittance for solar radiation.

It has already been pointed out that black bodies do not exist in nature. However, a black body is easily simulated by means of an aperture in a cavity. All radiation entering the aperture is reflected many times and is retained for the greater part within the cavity by which it is absorbed (in particular when the aperture is small in relation to the cavity). With regard to radiation entering the window aperture of a room, the apparent absorptance is much greater than the absorptances by the walls. In the case of thermal radiation there is virtually total absorptance (ε = 1). The apparent absorptance for solar radiation has been plotted in Figure A9.1 as a function of the ratio aperture area - total wall surface area, for various absorptances of the interior surface.

All objects continually emit and absorb thermal radiation. When an object is placed in a room of which the walls have the same temperature as the surface of the object, no net heat transfer takes place, because the object emits as much radiation as it receives. Heat transfer by radiation occurs only when surface temperatures differ.

This is expressed in the following formula:

$$\Phi_{12} = A_1 \times \varepsilon_1 \times \varepsilon_2 \times \Psi_{12} \times s \times (T_1^4 - T_2^4)$$
(W)

where:

A_1 = the surface area of plane 1
ε_1 = the emissivity of plane 1
ε_2 = the emissivity of plane 2
T_1 = the absolute temperature of surface 1
T_2 = the absolute temperature of surface 2
Ψ_{12} = the exchange factor
Φ_{12} = heat flow rate from surface 1 to 2

The exchange factor depends very much on the view factor between surface 1 and 2 (Φ_{12}) and the reflection and view factors of other surfaces: radiation of surface 1 may reach surface 2 via reflection by a third surface. If reflections are not taken into account,

$$\Psi_{12} = \Phi_{12}$$

Several conclusions relating to heat transfer by radiation can be drawn from this formula:

• heat transfer does not vary linearly with temperature difference, but this non-linearity is not so noticeable at near room temperatures;
• heat transfer increases with a better "view" between the surfaces;
• a lower emissivity of one of the surfaces considerably reduces the heat transfer.

The heat transfer coefficient for radiation hr W/m^2.Kg is defined in the following formula:

$$\Phi_{12} = A_1 \times h_r \times (T_1 - T_2) \text{ (W)}$$

At room temperatures hr will be approximately:

$$h_r \cong 5.8 \times \varepsilon_1 \times \varepsilon_2 \times \Psi_{12} \quad (W/m^2.K)$$

The calculation of view and exchange factors is to be found in various handbooks.

There are two special cases for which simple expressions exist:
• heat transfer coefficient in a wall cavity:

$$h_r = 5.8/((1/\varepsilon_1 + 1/\varepsilon_2) -1) \ (W/m^2.K)$$

• heat transfer coefficient from a small surface to a larger surrounding surface (e.g. window in a room).

$$h_r = \varepsilon_1 \times 5.8 \qquad (W/m^2.K)$$

There is no emissivity from the larger surface, the room apparently being a black body.

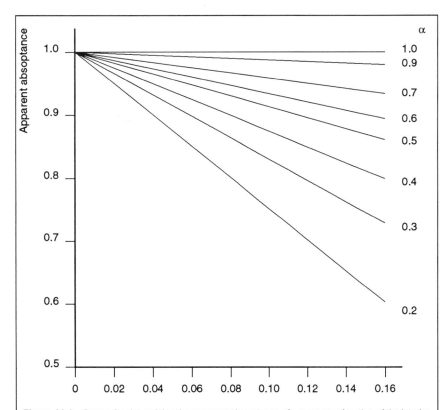

Figure A9.1: Curves for determining the apparent absorptance of a room as a function of the interior surface absorptance a, the aperture to interior surface area ratio, A1/(A2 - A1), and the transmittance t of the window. A1 = area of aperture, A2 = internal surface area, including aperture [1]. The curves are based on the assumption that A1 is considerably smaller than A2. This would be true for most direct gain rooms but not for highly glazed conservatories.

A9.3
CONVECTION

The convective heat transfer of a wall is expressed by means of the heat transfer coefficient h_c. This is defined in the following formula:

$$\Phi = A \times h_c \times (t_1 - t_2) \qquad (W)$$

where:

Φ = heat flow rate from surface area to air
A = surface
h_c = convective heat transfer coefficient
t_1 = surface temperature
t_2 = air temperature

In a cavity wall, heat is transferred by convection from one wall to the other. In the formula F stands for this heat, while t_1 and t_2 stands for surface temperatures.

Generally the heat transfer coefficient of free convection depends upon the temperature difference $(t_1 - t_2)$ and in case of forced convection it depends on the speed along the surface. So far there is little agreement between the experimental values for h_c, relating to horizontal and vertical walls.

The surface heat transfer coefficient is the sum of the heat transfer coefficient for radiation and convection. Its reciprocal is the heat transfer resistance; in case of a wall cavity, it is airspace resistance. Values for surface and airspace resistances are presented in appendix 10.7.

Whenever (dry) air enters a room at temperature t_o and leaves it at temperature t_r, then convective heat transport to the room is:

$$\Phi = \rho \times C_p \times q \times (t_o - t_r) \qquad (W)$$

where:

ρ = density (kg/m³)
C_p = specific heat capacity (J/kg.K)
q = air flow rate (m³/s)

If in the same case the air is moist, then the formula reflects the transport of sensible heat. The total amount of heat transported is the sum of sensible and latent heat (heat released when vapour in the air condenses).

Air movements are caused by pressure differences which in turn are occasioned by fans (forced convection), by wind or by density differences (free convection).

The factors playing a part in free convection can be clarified by considering a thermosiphonic loop (Figure A9.2).

The heat supplied by a thermosiphon is expressed in the formula given above, where tr is the air temperature at the entrance aperture and to the temperature of the outflowing air. If the average temperature in the thermosiphon equals tc $(t_r < t_c < t_o)$, then the pressure difference that caused the airflow is approximated by ΔP:

$$\Delta P = \rho \times g \times H \times (t_c - t_r /T_r \ (P_a)$$

where:

ρ = density (kg/m³)
g = acceleration due to gravity (m/s²)
H = column height (m)
T_r = absolute temperature of the room (K)

This sheds light upon two important factors affecting the air flow:

• the column height;
• the difference between the average temperature in the thermosiphon and the room.

The flow caused by the pressure difference depends on the flow resistances in the system. Hence the third factor affecting the mass flow is the cross-sectional area of the flow channel and of the inlet and outlet vents.

Optimal values exist for the cross-sectional areas of the inlet and outlet vents and the flow channel. If these cross-sectional areas are small, there will be too much flow resistance; conversely, where these areas are very large, air circulating in the flow channel and vents may be the result. Inlet and outlet ventholes should have approximately the same cross-sectional area as the channel. However, the overall performance of a collector is reduced by only 10% when this surface area is halved. The cross-sectional area of the channel should be at least 1/20 and at the most 1/10 of the surface area of the collector.

In very high buildings in winter the pressure difference on the external facade surface caused by the temperature difference may be an important cause of infiltration.

A9.4
CONDUCTION

The stationary heat transport in a wall in the direction perpendicular to the wall equals:

$$\Phi = A \times (t_1 - t_2)/R \quad (W)$$

where:

Φ = heat flow rate (W)
A = surface area of the wall (A)
R = thermal resistance of the wall
$(t_1 - t_2)$ = temperature difference over the wall

The thermal resistance of a homogeneous layer depends on the thermal conductivity (k) and the thickness (d):

$$R = d/k \ (m2k/W)$$

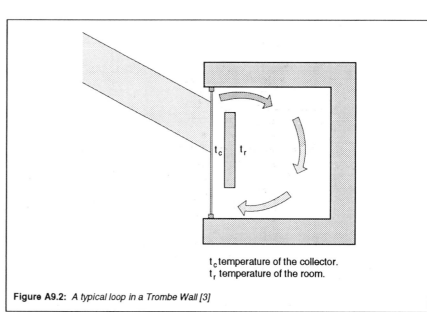

t_c temperature of the collector.
t_r temperature of the room.

Figure A9.2: *A typical loop in a Trombe Wall [3]*

A low conductivity results in a considerable thermal resistance (see Appendix 10.8 - 'Thermal properties of materials'). The thermal transmittance or U-value is the reciprocal of the total resistance i.e. the sum of the resistance of every layer (including surface and airspace resistances).

Calculations for non-stationary heat transport in walls are only possible by means of a computer. For temperature fluctuations occurring sinusoidally in time, analytical calculations are also possible, but the results so rapidly become complicated that no better insight into the matter is gained. There are, however, two cases for which the solutions are relatively simple and which play an important role in passive solar heating. In both cases one starts from the principle of a homogeneous wall which is heated on one side (front) by means of a heat flow which occurs sinusoidally in time. The two cases are:

- the heat flow on the other side (back) is equal to zero: this is in first approximation a well-insulated wall;

- the temperature variation at the back equals the temperature variation at the same distance from the front in an infinitely thick wall of the same material. This approximation improves as dampening (= temperature variation front) becomes greater, and may be illustrative of Trombe and Mass walls.

The storage effect of a well insulated wall for a sinusoidal temperature variation with a given period can be equated to the effect of a hypothetical isothermal wall (infinite conductivity) with the same density and specific heat capacity but with a smaller thickness.

This thickness is called the effective thickness. This effective thickness is plotted in Figure A9.3 for a period of 12 hrs. The value d^* of the asymptote in this figure equals:

$$d^* = \sqrt{k \times T_0 / (\rho \times c \times \pi)}$$

where

ρ = density (kg/m^3)
c = specific heat (W/kg.K)
T_0 = time period of the variation (s)

The effective thickness is roughly equal to the real thickness if the real thickness is smaller than d^*; if the real thickness is greater than d^*, the effective thickness is equal to d^*. In case of a rapid temperature variation, only a thin layer of the surface will apparently assist in the storage. This has considerable implications for the storage in walls as far as direct systems are concerned.

The timelag of a temperature wave when passing from the front to the back of a thick wall amounts to:

$$\text{timelag} = \frac{T_0 \times d}{2\pi \times d^*}$$

where:

d = thickness of the wall
d^* is as defined above
The dampening of the temperature variation equals:

$$D = \exp(-d/d^*)$$

where D is the dampening factor.

In case of a mass wall with a thickness equal to d^* and for a time period of the variation T_0 of 24 hrs by approximation, the timelag equals 3.8 hrs and the dampening factor equals 0.37. From the formulae it appears that a large timelag corresponds automatically with a substantial dampening. The delaying effect of the thick mass wall has little influence since there is only a small temperature variation left. This also appears clearly from the table below [5] showing the figures for a concrete Trombe wall on a clear day:

Thickness (cm)	Temperature fluctuations Internal surface area	Timelag (hrs)
20	22	6.8
30	11	9.3
40	51	1.9
50	31	4.5

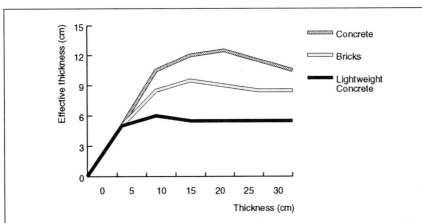

Figure A9.3: *Effective thickness as a function of the real thickness of a homogeneous slab, assuming in this figure that the mass is directly irradiated.*

Assumptions:
It is assumed in this figure that the mass is directly irradiated (i.e. primary thermal mass) and has the following properties:

	Conductivity k (W/mK)	Density p(kg/m^3)	Specific Heat c (J/KgK)
Concrete	1.7	2300	880
Lightweight concrete	0.15	510	1050
Brick	0.58	1500	840

Units, Symbols and Properties of Materials

A10.1

The SI Base Units

Name of Unit	Symbol
metre	m
kilogram	kg
second	s
ampere	A
Kelvin*	K*
mole	mol
candela	cd

The unit Kelin and its symbol K are used to express an interval or a difference of temperature, such that the difference between a temperature of 21° C and a temperature of 18° C, for example, is 3K.

A10.2

Names and Symbols for the SI Prefixes

Factor	Prefix	Symbol
x 10^{18}	exa	E
x 10^{15}	peta	P
x 10^{12}	tera	T
x 10^{9}	giga	G
x 10^{6}	mega	M
x 10^{3}	kilo	k
x 10^{2}	hecto	h
x 10^{1}	deca	da
x 10^{-1}	deci	d
x 10^{-2}	centi	c
x 10^{-3}	milli	m
x 10^{-6}	micro	μ
x 10^{-9}	nano	n
x 10^{-12}	pico	p
x 10^{-15}	femto	f
x 10^{-18}	atto	a

A10.3

Units which are outside the SI system but are recognised for use within it

Name of unit	Symbol
minute	min
hour	h
day	d
degree	°
minute	'
second	"
litre	l
tonne	t
degree Celsius	°C

Contents

A10.4

COMMON UNITS IN PASSIVE SOLAR DESIGN

Quantity	Unit	Symbol
Area	m^2	A
Density	kg/m^3	ρ
Energy	J or kWh*	Q
Heat capacity	J/K	C
Heat flow rate	W	Φ
Insulation value of clothing	clo	
Mass	kg	m
Pressure	Pa or N/m^2**	p
Radiation direct	$W/m2$	
diffuse	"	
global	"	
Specific heat capacity	J/kg.K	c
Temperature	°C	t
Temperature difference	K	Δt
inside air	°C	t_i
mean radiant	"	t_{mr}
outside air	"	t_o
Thermal conductivity	W/m.k	λ
Thermal resistance	m^2.K/W	R
Thermal resistance inside surface	"	R_{si}
Thermal resistance outside surface	"	R_{so}
Thermal transmittance	W/m^2.K	U
Thermodynamic temperature	K	T
Velocity	m/s	v
Ventilation rate	m^3/s	q

* $1kWh = 3.6 \times 10^6 J$

** $1 Pa = 1N/m^2$

A10.5

COMMON UNITS FOR ENERGY ECONOMICS

1 Mtoe = 10^6 toe

1 toe   = 1 ton of oil equivalent

         = 4.53×10^{10} J

A10.6

RADIATIVE PROPERTIES OF MATERIALS

Ground Cover [2]	Solar absorptance	Solar reflectance
Asphalt pavement	0.93	0.07
Desert ground surface	0.75	0.25
Grass	0.67	0.33
Ice with sparse snow cover	0.31	0.69
Oak leaves	0.71	0.29
Sand, dry	0.82	0.18
Sand, wet	0.91	0.09
Sand, white powdered	0.45	0.55
Snow, fine particles, fresh	0.13	0.87
Snow, ice granules	0.33	0.67
Water, solar zenith angle:		
50°	0.90	0.10
60°	0.84	0.16
70°	0.74	0.26
80°	0.53	0.47
Agricultural crops	0.75	0.25
Deciduous forest	0.85	0.15
Coniferous forest	0.95	0.05
Concrete, clean	0.55	0.45
Concrete, averaged	0.70	0.30
Concrete, weathered	0.80	0.20

Building finishes [4]

Surface	Solar absorptance	Thermal absorptance and emissivity
Aluminium, polished	0.15	0.08
Whitewash	0.20	-
Chromium plate	0.28	0.20
White lead paint	0.29	0.89
White marble	0.46	0.95
Light green paint	0.50	0.95
Aluminium paint	0.55	0.55
Limestone	0.57	0.95
Wood, pine	0.60	0.95
Asbestos cement, aged 1 year	0.71	0.95
Red clay brick	0.77	0.94
Gray paint	0.75	0.95
Galvanised iron, aged (oxidized)	0.90	0.28
Black matt	0.97	0.95

Colour [2]	Solar absorptance	Solar reflectance
For white, smooth surfaces, use	0.25 to 0.40	0.60 to 0.75
For grey to dark grey, use	0.40 to 0.50	0.50 to 0.60
For green, red and brown, use	0.50 to 0.70	0.30 to 0.50
For dark brown to blue, use	0.70 to 0.80	0.20 to 0.30
For dark blue to black, use	0.80 to 0.90	0.10 to 0.20

Specular surface solar reflectance [5]

Specular surface	Solar reflectance
Electroplated silver, new	0.96
High-purity aluminium, new clean	0.91
Sputtered aluminium, optical reflector	0.89
Brytal processed aluminium, high purity	0.89
Back-silvered water, white plate glass, new, clean	0.88
Aluminium, silicon oxygen coating, clean	0.87
Aluminium foil, 99.5% pure	0.86
Commercial Alzac process aluminium (plastic w/ aluminium surface film)	0.85
Back-aluminized 3M acrylic, new	0.85
Aluminized Type C Mylar (from Mylar side)	0.76

Solar protection [5]

Shading	Type of sun protection	solar gain factors* for the following types of glazing	
		Single	Double
None	None	0.76	0.64
	Lightly heat absorbing glass	0.51	0.38
	Densely heat absorbing glass	0.39	0.25
	Lacquer coated glass, grey	0.56	-
	Heat reflecting glass, gold (sealed unit when double)	0.261	0.25
Internal	Dark green open weave plastic blind	0.62	0.56
	White venetian blind	0.46	0.46
	White cotton curtain	0.41	0.40
	Cream holland linen blind	0.30	0.33
Mid plane	White venetian blind	-	0.28
External	Dark green open weave plastic blind	0.22	0.17
	Canvas roller blind	0.14	0.11
	White louvred sunbreaker, blades at 45°	0.14	0.11
	Dark green miniature louvred blind	0.13	0.10

*   *The solar gain factor is the fraction of incident radiation to enter the room, by transmission and by absorption and longwave re-radiation from the glazing.*

All glazing clear except where stated otherwise. Factors are typical values only and variations will occur due to density of blind weave, reflectivity and cleanliness of protection. (Values are strictly for UK only, approximately correct world wide).

SURFACE AND AIRSPACE RESISTANCE [3]

Internal resistance

Building element	Surface emissivity	Heat flow	Surface resistance (m².K/W)
Walls	High	Horizontal	0.123
	Low	Horizontal	0.304
Ceilings or roofs (flat or pitched) floors	High	Upward	0.106
	Low	Upward	0.218
Ceilings and floors	High	Downward	0.150
	Low	Downward	0.562

External surface resistance

Building element	Surface emissivity	Sheltered	Normal (standard)	Severe
			Surface resistance (m².K/W)	
Wall	High	0.08	0.055	0.03
	Low	0.11	0.067	0.03
Roof	High	0.07	0.045	0.02
	Low	0.09	0.053	0.02

Standard thermal resistance of ventilated air spaces

(Airspace thickness 20mm minimum)	Thermal resistance* (m².K/W)
Airspace between asbestos-cement or black painted metal cladding with unsealed joints and high emissivity lining	0.16
As above, with low emissivity surface facing airspace	0.30
Loft space between ceiling and unsealed asbestos-cement or black metal cladding pitched roof	0.14
As above with aluminium cladding instead of black metal or with low emissivity upper surface on ceiling	0.25
Loft space between flat ceiling and unsealed tiled pitched roof	0.11
Airspace between tiles and roofing felt or building paper on pitched roof	0.12
Airspace behind tiles on tile-hung wall	0.12

*   *Including internal boundary surface.*

Standard thermal resistance of unventilated air spaces

Type of airspace or thickness	Surface emissivity	Thermal resistance* (m².K/W)	
		Heat flow horizontal or upwards	Heat flow downwards
5 mm	High	0.11	0.11
	Low	0.18	0.18
20 mm or more	High	0.18	0.21
	Low	0.35	1.06
High emissivity gated sheets in contact	0.09	0.11	
Low emissivity multiple foil insulation with airspace on one side		0.62	1.76

*   Including internal boundary surface

A10.8

$\mathbf{T}$*THERMAL PROPERTIES OF MATERIALS [1]*

Structural materials

	Density (kg/m³)	Conductivity (W/m.K)		Specific heat (J/kg.K)	(Wh/Kg.K)
Timber					
Softwood	630	0.13		1,360	(0.38)
Hardwood		0.15		1,250	(0.35)
Plywood	530	0.14		1,214	(0.34)
Chipboard	800	0.15		1,286	(0.28)
Wood wool slabs	600	0.11			
Strawslab compressed	350	0.11			
Stone					
Granite	2,600	2.50		900	(0.25)
Limestone	2,180	1.49		720	(0.20)
Sandstone	2,000	1.30		712	(0.20)
Marble	2,500	2.00		802	(0.22)
		Protected	exposed*		
Concrete lightweight	1,200	0.38	0.42	1,000	(0.28)
Gravel concrete					
no fines	1,800	0.87	0.96	1,000	(0.28)
dense	2,100	1.28	1.40	820	(0.23)
Autoclaved aerated					
concrete	500	0.16	0.18	1,000	(0.28)
Brickwork					
light	1,300	0.36	0.49		
medium	1,700	0.62	0.84	790	(0.22)
dense (engineering)	1,900	0.81	1.09		

*   moisture contents % by volume
protected = 1% for brickwork, 3% for concrete
exposed = 5% for brickwork and concrete

External/cladding materials

	Density (kg/m^3)	Conductivity (W/m.K)	Specific heat (J/kg.K)	(Wh/Kg.K)
Window glass	2,500	1.05	1,800	(0.5)
Asbestos cement sheet				(0.29)
Rigid PVC	1,350	0.16		
Polyester - glass reinforced	1,450	0.23		
Silicone - glass reinforced	1,800	0.34		
Roofing felt	960	0.19		
Felt/bitumen layers	1,700	0.5	1,000	(0.28)
Metals				
aluminium	2,800	160	896	(0.25)
copper	8,900	200	413	(0.12)
carbon steel	7,800	50	512	(0.14)
Hollow glass blocks		0.67		
Concrete roof tiles	negligible due to gaps.			
Clay roof tiles	Use airspace resistances			
Mastic asphalt roofing	2,325	1.15	1,000	(0.28)
Glass reinforced cement				
moulded	2,200	1.3	840	(0.23)
render	1,700	0.5	840	(0.23)
Sand cement render				
1% moisture cement	1,570	1.13	1,000	(0.28)
5% moisture cement		0.79		

Finishes

	Density (kg/m3)	Conductivity (W/m.K)	Specific heat (J/kg.K)	(Wh/kg.K)
Carpeting		0/055	1,360	(0.38)
Cellular rubber underlay	180	0.1	1,360	(0.38)
Wool felt underlay	180	0.04	1,220	(0.34)
Granolithic	2,085	0.87	840	(0.23)
Hardboard	600	0.08	2,000	(0.56)
Timber flooring				
strip or block	650	0.14	1,200	(0.33)
Hardboard -				
standard	900	0.13	2,000	(0.56)
Screed -				
concrete	2,100	1.28	1,000	(0.28)
lightweight	1,200	0.41	840	(0.23)
Linoleum	510	0.52		
Rubber flooring	1,600	0.3	2,000	(0.56)
Tiles				
burnt clay - quarry	1,900	0.85	780	(0.22)
cork	540	0.085	1,000	(0.28)
plastic	1,050	0.5	1,070	(0.34)
pvc asbestos	2,000	0.85	1,000	(0.28)
Plasterboard				
gypsum	950	0.16	820	(0.23)
Plastering				
gypsum	1,120	0.38	1,000	(0.28)
vermiculite	640	0.28	1,000	(0.28)
sand cement	1,570	0.53	1,000	(0.28)

Insulation materials

	Density (kg/m³)	Conductivity (W/m.K)	Specific heat (J/kg.K)	(Wh/kg.K)
Fibreboard	300	0.06	1,000	(0.28)
Mineral fibre (rock				
mat or quilt)	25	0.040	960	(0.27)
semi-rigid felted	130	0.036	960	(0.27)
loose, felted slab or mat	180	0.042	1,000	(0.28)
Foamed plastics				
phenolic foam board	30	0.038	1,400	(0.39)
expanded polystyrene board	15	0.037	1,400	(0.39)
expanded polystyrene board	25	0.034	1,210	(0.34)
extruded polystyrene board smooth skin	36	0.029	1,210	(0.34)
extruded polystyrene board cut skin	28	0.036	1,400	(0.39)
Polyurethane foam board (aged)	30	0.026	1,570	(0.44)
Urea formaldehyde foam	10	0.04	1,400	(0.39)
Sprayed urethane foam	32	0.02		
Brown cellulose fibres	37-51	0.045	1,350	(0.38)
Thatch				
reed	270	0.09	1,000	(0.38)
straw	240	0.07	1,000	(0.38)
Vermiculite granules	100	0.065		
Insulating render (sand cement and polystyrene beads)		0.66		
Cellular glass	175	0.17	1,000	(0.38)

Air/moisture barriers

Building paper Kraft	negligible effect except when
Aluminium foil	forming air space: see Appendix 10.7
Sarking felt	for air space resistances
Polyethelene	

Water

		Density (kg/m³)	Conductivity (W/m.K)	Specific heat (J/kg.K)	(Wh/Kg.K)
Ice	-10°C	920	1.7	1,980	(0.55)
	-1°C	920	1.7	2,052	(0.57)
Water	0°C	1,000	0.56		
	20°C	998	0.60	4,176	(1.16)
	40°C	993	0.63		
	80°C	970	0.67		
Snow:	fresh	190	0.17		
	compacted	400	0.43		

Latent heat of melting = 92.7 Wh/kg
Latent heat of evaporation = 627 Wh/kg

THERMAL PROPERTIES OF BUILDING ELEMENTS

Alternative methods for calculating ground floor losses
Edge loss factors for concrete slabs* [2]

R-value of insulation (m²·K/W)	Depth of perimeter insulation (m)		
	0.3	0.45	0.6
1.76	0.346	0.312	0.294
1.18	0.554	0.485	0.467
0.88	0.744	0.675	0.640
0.70	0.987	0.866	0.831
0.58	1.212	1.073	1.021
0.44	1.731	1.523	1.437

*   *multiply by perimeter in metres to obtain steady state loss (W/K).*

U-values for solid and suspended floors [2]		U-Value (W/m²·K)		
Length (m)	Breadth (m)	Four edges exposed	Two perpendicular edges exposed	Suspended floor
Very long	100	0.06*	0.03**	0.07
Very long	60	0.09*	0.05**	0.11
Very long	40	0.42*	0.07**	0.15
Very long	20	0.22*	0.12**	0.26
Very long	10	0.38*	0.22**	0.43
Very long	6	0.55*	0.33**	0.60
Very long	4	0.74*	0.45**	0.76
Very long	2	1.19*	0.74**	1.04
100	100	0.10	0.05	0.11
100	60	0.12	0.07	0.14
100	40	0.15	0.09	0.18
100	20	0.24	0.14	0.28
60	60	0.15	0.08	0.16
60	40	0.17	0.10	0.20
60	20	0.26	0.15	0.30
60	10	0.41	0.24	0.46
40	40	0.21	0.12	0.22
40	20	0.28	0.16	0.31
40	10	0.43	0.25	0.47
40	6	0.59	0.35	0.63
20	20	0.36	0.21	0.37
20	10	0.48	0.28	0.51
20	6	0.64	0.38	0.65
20	4	0.82	0.49	0.79
10	10	0.62	0.36	0.59
10	6	0.74	0.44	0.71
10	4	0.90	0.54	0.83
10	2	1.31	0.82	1.08
6	6	0.91	0.54	0.79
6	4	1.03	0.62	0.89
6	2	1.40	0.87	1.1
4	4	1.22	0.73	
4	2	1.51	0.95	1.15
2	2	1.96	1.33	1.27

*   *these values can be used for a floor with two parallel edges exposed taking the breadth as the distance between the exposed edges.*

**   *These values can be used for a floor with one exposed edge taking the breadth as the distance between exposed edge and the edge opposite it.*

These values are for uninsulated floors. The value when a layer of insulation of thermal resistance is added, can be calculated form the formula:

$$U_{new} = I/C(I/U_{old} + R)$$

where R = thermal resistance of insulating layer.

Note: *These corrections are based on thermal insulation having a minimum thermal resistance of 0.25 m².K/W.*

Percentage relation in U-value for edge insulations of different depths

Dimensions of floor		Percentage reduction in 'U-value' for edge insulation extending to a depth (m)		
		0.25	0.5	1.0
Very long x	100 m broad	2	6	10
	60 m	2	6	11
	40 m	3	7	11
	20 m	3	8	11
	10 m	4	9	14
	6 m	4	9	15
	4 m	5	12	20
	2 m	6	15	25
100 m x	100 m	3	10	16
60 m x	60 m	4	11	17
40 m x	40 m	4	12	18
20 m x	20 m	5	13	19
10 m x	10 m	6	14	22
6 m x	6 m	6	15	25
4 m x	4 m	7	18	28
2 m x	2 m	10	20	35

U-values for windows for various frame construction and glass configurations [3]

Frame type	Percentage of area used by frame	U-value for stated exposure (W/m2.K)								
		Single glazing			Double glazing			Triple or Low E* glazing		
		Sh	N	Se	Sh	N	Se	Sh	N	Se
Wood or PVC	10	4.7	5.3	6.3	2.8	3.0	3.2	1.9	2.0	2.1
	20	4.5	5.0	5.9	2.7	2.9	3.2	1.9	2.0	2.1
	30	4.2	4.7	5.5	2.7	2.9	3.1	1.8	1.9	2.0
Aluminium	10	5.3	6.0	7.1	3.3	3.6	4.1	2.3	2.4	2.5
	20	5.6	6.4	7.5	3.9	4.3	4.8	2.6	2.8	2.9
	30	5.9	6.7	7.9	4.4	4.9	5.6	3.0	3.1	3.3
Aluminium with thermal break	10	5.1	5.7	6.7	3.1	3.3	3.7	2.1	2.2	2.4
	20	5.2	5.8	6.8	3.4	3.7	4.0	2.4	2.5	2.6
	30	5.2	5.8	6.8	3.7	4.0	4.4	2.6	2.7	2.8

Sh = Sheltered exposure, N = Normal exposure, Se = Severe exposure

*   *Indicates double glazing with a low emissivity coating which reduced the transfer of longwave radiation from the inside space resulting in a lower U-value.*

GLOSSARY

Absorbent - a material which, due to an affinity for certain substances, extracts one or more such substances form a liquid or gaseous medium with which it contacts and which charges physically or chemically, or both, during the process. Calcium chloride is an example of a solid absorbent, while solutions of lithium chloride, lithium bromide, and ethylene glycols are liquid absorbents.

Absorber - the blackened surface in a solar collector, which absorbs solar radiation and converts it to heat.

Absorptance - the ratio of absorbed to incident solar radiation on a surface.

Absorptance of a Glazing Material - the absorption of radiant energy in glazing occurs as a continuous process throughout the thickness of a glazing material. The overall fraction of the incident radiant energy removed by absorption on its passage through the material (assumed to have parallel sides) which involves multiple reflections, is known as the absorptance of the glazing. This depends on wavelength and angle of incidence, the transmittance + absorptance + reflectance = 1.

Absorptance of an Opaque Surface - the ratio of the radiation absorbed at a surface to the radiation incident on that surface. The absorptance varies for different radiation wavelengths and angles of incidence, but the absorptance of normally incident radiation is always numerically equal to the emittance of the surface for the same wavelength. A 'black body' is a theoretical perfect absorber with an absorptance of 1 for all wavelengths: a typical matt black paint has an absorptance of 0.97 for normal solar radiation and 0.95 for long-wave (thermal) radiation. For any wavelength and angle of incidence, for an opaque surface, absorptance + reflectance = 1.

Absorption - a process whereby a material extracts one or more substances present in an atmospheric or mixture of gases or liquids accompanied by the material's physical and/or chemical changes.

Absorption Factor - the fraction of solar radiation transmitted through a glazing system that is absorbed inside the building.

Activated Alumina - a form of aluminium oxide which adsorbs moisture readily and is used as a drying agent.

Adfreezing - the process whereby wet soils freeze to below grade materials such as fountains walls or insulation, forcing movement of the material.

Adiabatic Process - a thermodynamic process during which no heat is extracted from or added to the system.

Adsorbent - a material which has the ability to cause molecules of gases, liquids, or solids to adhere to its internal surfaces without changing the adsorbent physically or chemically. Certain solid materials, such as silica gel and activated alumina, have this property.

Air Barrier - a material carefully installed within a building envelope assembly to minimise the uncontrolled passage of air into and out of a dwelling.

Air Change - introducing new, cleansed, or recirculated air to conditioned space, measured by the number of complete changes per unit time.

Air Change Per Hour (ach) - a unit that denotes the number of times a house exchanges its entire volume of air with outside air in an hour. This is generally used in two ways: (1) under natural conditions and (2) under a 50 Pascal pressure difference. The R-2000 Home Program requires a house with 1.5 ach at 50 Pascals.

Air Changes - a measure of the air exchange in a building. One air change is an exchange of a volume of air equal to the interior volume of a building.

Air Cleanser - a device used to remove airborne impurities.

Air Cooler, Natural Convection - an air cooler depending on natural convection for air circulation.

Air Leakage - the uncontrolled flow of air through a component of the building envelope, or the building envelope itself, when a pressure difference is applied across the component. Infiltration refers to inward flowing air leakage and exfiltration refers to outward flowing air leakage.

Air Permeability - the property of a building component to let air pass when it is subjected to a differential pressure.

Air Pressure - the pressure exerted by the air. This may refer to static (atmospheric) pressure, or dynamic components of pressure arising from air flow, or both acting together.

Air Sealing - the practice of sealing unintentional gaps in the building envelope (from the interior) in order to reduce uncontrolled air leakage.

Air Tightness - the degree to which unintentional openings have been avoided in a building's structure.

Air, Ambient - Generally, the air surrounding an object.

Air, Outside - external air, atmosphere exterior to refrigerated or conditioned space, ambient (surrounding) air.

Air, Recirculated - return air passed through the conditioner before being resupplied to the conditioned space.

Air, Reheating of - in an air conditioning system, the final step in treatment in the event the temperature is too low.

Air, Return - see air recirculated.

Air, Saturated - moist air in which the partial pressure of water vapour equals the vapour pressure of water at the existing temperature. This occurs when dry air and saturated water vapour coexist at the same dry-bulb temperature.

Air, Standard - dry air at a pressure of 101.325 kPa at exactly 20°C (68 F) temperature. At these conditions, the density is 1.2041 kg/m^3

Altitude Angle - the angular distance from the horizon to the sun.

Altitude Angle - the angular height of a point above the horizontal plane, i.e. solar altitude - the angle between the centre of the solar disc and the horizontal plane.

Ambient Temperature - temperature of the air, (1) outside the building or (2) within a room.

Anemometer - an instrument for measuring the velocity of a fluid.

Annual LCR Method - application of the tables of annual solar savings fraction versus load collector ratio (the 'LCR tables') to a calculation of the annual auxiliary heat.

Atomize - reduce to fine spray.

Attached Sunspace - solar collector that doubles as useful building space; also attached greenhouse, solarium. The term 'attached' specifically implies a space that shares one common wall with the associated building. compare with semienclosed sunspace.

Automatic Flue Damper - a damper added to the flue pipe downstream of a furnace or boiler and connected with automatic controls to the burner. Its function is to reduce heat loss up the chimney when the unit is not operating, consequently it provides the greatest savings during relatively mild weather when the furnace is on infrequently.

Auxiliary Heating - the conventional (i.e. non-solar) contribution to the total load (e.g. gas boiler etc.).

Awning - an exterior, movable and usually flexible element. Protects detaining or diffusing solar radiation at certain angles.

Azimuth Angle - in the northern hemisphere the angular distance between true south and a point on the horizontal plane; negative to the east, positive to the west, i.e. solar azimuth - the angle between due south and the point on the horizon vertically below the sun (negative before noon, positive after noon). In the southern hemisphere the azimuth angle is measured from due north.

Azimuth Angle - the angular distance between true south and the point on the horizon directly below the sun (negative before noon, positive after noon).

Backdrafting - (flow reversal) the reverse flow of chimney gases into the building through the barometric damper, draft hood, or burner unit. This can be caused by chimney blockage or it can occur when the pressure differential is too high for the chimney to draw.

Base Temperature - a fixed temperature in the definition of degree-days.

Beam or Direct Radiation - theoretically it is the solar radiation that reaches the surface from only the solid angle of the sun's disc. In practice instruments for measuring direct radiation have to have a wider aperture so the beam radiation as normally measured contains some diffuse energy from the immediate region of the sky around the sun (see diffuse radiation).

Beam Radiation - as 'direct sun radiation'.

Bearing - the angular distance between true north (not magnetic north) and a point on the horizontal plane, 0-360°, with east 90°, south 180°, west 270°, i.e. solar bearing - the angle between due north and the point on the horizon vertically below the sun.

Berm - a man made mound or small hill of earth, either abutting a house wall to help stabilise the internal temperature, or positioned to deflect wind from the house.

Bimetallic Element - an element formed of two metals having different coefficients of thermal expansion, used as a temperature control device.

Black Body - a perfect absorber and emitter of radiation. A cavity is a perfect black body. Lampblack is close to a black body, while aluminium (polished) is a poor absorber and emitter of radiation.

Bright Sunshine - sunshine of sufficient intensity to be registered on a Campbell-Stokes sunshine recorder. The threshold is imprecisely defined and varies according to the type of recorder, but generally refers to sunshine bright enough to cast a shadow. Burning typically starts when the beam intensity is 200W/m^2.

Building Orientation - the siting of a building on a lot, generally used to refer to solar orientation which is the siting of building with respect to solar access.

Calorific Value - the energy content per unit mass (or volume) of a fuel, which will be released in combustion. (kWh/kg, MJ/kg, kWh/m^3, MJ/m^3)
.

Candela (cd) - SI unit of luminous intensity: the candela is the luminous intensity, in a given direction, of a source that emits monochromatic radiation of frequency 540 x 1012 hertz and has a radiant intensity in that direction of 1/683 watt per steradian. (16th General Conference of Weights and Measures, 1979)

$$1\ cd = 1\ lm \cdot sr^{-1}$$

Candela Per Square Metre ($cd \cdot m^{-2}$) - SI unit of luminance. (Note: this unit was sometimes called the nit [nt] (name discouraged). Other units of luminance:

metric, non-SI: Lambert (L) $= \dfrac{10^{-4}}{\pi} cd \cdot m^{-2}$

Casual Gains - see internal gains

Chimney Effect - the tendency of air or gas in a duct or other vertical passage to rise when heated due to its lower density in comparison with that of the surrounding air or gas. In buildings, the tendency towards displacement (caused by the difference in temperature) of internal heated air by unheated outside air, due to the difference in density of outside and inside air.

CIE 1974 Special Colour Rendering Index (Ri) - measure of the degree to which the psychophysical colour of a CIE test colour sample illuminated by the test illuminant conforms to that of the same sample illuminated by the reference illuminant, suitable allowance having been made for the state of chromatic adaption.

CIE Standard Overcast Sky - completely overcast sky for which the ratio of its luminance L_y in the direction at an angle γ above the horizon to its luminance L_z at the zenith is given by the relation

$$L_y = \frac{L_z(1+2\sin \gamma)}{3}$$

CIE Standard Sources - artificial sources specified by the CIE whose radiations approximate CIE standard illuminants A, B and C, (see CIE Publication N°15).

CIE Standard Clear Sky - cloudless sky for which the relative luminance distribution is described in CIE Publication N°22 (1973).

Clear Sky - clear sky in this document has two separate meanings. In the radiation Tables clear sky refers to days producing the mean maximum monthly daily global radiation on a horizontal surface. These days on average contain some cloud but the precise cloud amount is not known. In the temperature Tables, the term clear skies refers to summaries of measured temperature data for days when the mean total amount of cloud during hours of daylight was less then 2/8. Night-time temperature data for such days are related to the same 'clear' days i.e. night-time cloud cover was not taken into account.

Clearance Pocket - see capacity reducer.

Clerestory - a window that is placed vertically (or near vertical) in a wall above one's line of vision to provide natural light in a building.

Clo - a unit measuring the insulating effect of clothing on a human subject. 1 clo = 0.155 $(K \cdot m^2)/W$.

Collector, Flat Plate - an assembly containing a panel of metal or other suitable material, usually a flat black colour on its sun side, that absorbs sunlight and converts it into heat. This panel is usually in an insulated box, covered with glass or plastic on the sun side to take advantage of the greenhouse effect. In the collector, this heat transfers to a circulating fluid, such as air, water, oil or antifreeze.

Collector, Focusing - a collector that has a parabolic or other reflector which focuses sunlight onto a small area for collection. A reflector of this type can obtain considerably higher temperatures but will only work with direct beam sunlight.

Collector, Solar - a device for capturing solar energy, ranging from ordinary windows to complex mechanical devices.

Colour Rendering - effect of an illuminant on the colour appearance of objects by conscious comparison with their colour appearance under a reference illuminant. (Note: in German, the term 'Farbwiedergabe' is also applied to colour reproduction.)

Colour Rendering Index (R) - measure of the degree to which the psychophysical colour of an object illuminated by the test illuminant conforms to that of the same object illuminated by the reference illuminant, suitable allowance having been made for the state of chromatic adaptation.

Combustion Air - the air required to provide adequate oxygen for fuel burning appliances in the building. The term 'combustion air' is often used to refer to the total air requirements of a fuel burning appliance including both air to support the combustion process and air to provide chimney draft (dilution air).

Comfort Chart - a chart showing effective temperatures with dry-bulb temperatures and humidities (and sometimes air motion), by which the effects of various air conditions on human comfort may be compared.

Comfort Line - a line on the comfort chart showing relation between the effective temperature and the percentage of adults feeling comfortable.

Comfort Zone - average - the range of effective temperatures over which the majority (50% or more) of adults feels comfortable; extreme - the range of effective temperatures over which one or more adults feel comfortable.

Condensation - beads or drops of water (and frequently frost in extremely cold weather) that accumulate on the inside of the exterior covering of a building (most often windows) when warm, moisture-laden air from the interior reaches a point where the temperature no longer permits the air to sustain the moisture it holds.

Conditions, Standard - a set of physical, chemical or other parameters of a substance or system which defines an accepted reference state or forms a basis for comparison.

Conditions, Steady - an operating state of a system, including its surroundings, in which the extent of change with the time of all the significant parameters is so small as to have no important effect on the performance being observed or measured.

Conductance - a measure of the ease of heat transfer between two points, namely, the heat flow rate per degree of temperature difference per unit of area.

Conductance, Surface Film - time rate of heat flow per unit area under steady conditions between a surface and a fluid for unit temperature difference between the surface and fluid.

Conductance, Thermal - time rate of heat flow through a body (frequently per unit area) from one of its bounding surfaces to the other for a unit temperature difference between the two surfaces, under steady conditions.

Conduction - the heat moving from a warmer to a colder region in the same substance without mass transfer, this type of heat transfer depends on the thermal conductivity of the material.

Conduction, Thermal - process of heat transfer through a material medium in which kinetic energy is transmitted by the particles if the material from particle to particle without gross displacement of the particles.

Conductivity, Thermal - time rate of heat flow through unit area and unit thickness of a homogeneous material under steady conditions when a unit temperature gradient is maintained in the direction perpendicular to area. Materials are considered homogeneous when the value of the thermal conductivity is not affected by variation in thickness or in size of the sample within the range normally used in construction.

Contrast - (1) in the perceptual sense: assessment of the difference in appearance of two or more parts of a field seen simultaneously or successively (hence: brightness contrast, lightness contrast, colour contrast, simultaneous contrast, successive contrast, etc.) (2) in the physical sense: quantity intended to correlate with the perceived brightness contrast, usually defined by one of a number of formulae which involve the luminances of the stimuli considered, for example: $\Delta L/L$ near the luminance threshold, or L_1/L_1 for much higher luminances.

Controlled Ventilation - ventilation brought about by mechanical means through pressure differentials induced by the operation of a fan.

Convection - heat transfer between a surface and adjacent fluid (usually air or water) and by the flow of fluid from one place to another. See also natural convection and forced convection.

Convection - the mechanism of heat transfer between a surface and a fluid.

Convection, Forced - convection resulting from forced circulation of a fluid, as by a fan, jet or pump.

Convection, Natural - circulation of gas or liquid (usually air or water) due to differences in density resulting from temperature changes.

Convector - an agency of convection. In heat transfer, a surface designed to transfer its heat to a surrounding fluid largely or wholly by convection. The heated fluid may be removed mechanically or by gravity (gravity convector). Such a surface may or may not be enclosed or concealed.

Cooling Effect, Sensible - difference between the total cooling effect and the dehumidifying effect, usually in watts (Btuh).

Cooling Effect, Total - difference between the total enthalpy of the dry air and water vapour mixture entering the cooler per hour and the total enthalpy of the dry air and water vapour mixture leaving the cooler per hour, expressed in watts (Btuh).

Cooling Element - heat transfer surface containing refrigeration fluid where refrigerating effect is desired.

Cooling Medium - any substance whose temperature is such that it is used, with or without a change of state, to lower the temperature of other bodies or substances.

Cooling Range - in a water cooling device, the difference between the average temperatures of water entering or leaving the device.

Cooling Water - water for condensation of refrigerant; condenser water.

Cooling, Direct Method of - a system in which the evaporator is in direct contact with the material or space refrigerated or is located in air-circulation passages communicating with such spaces.

Cooling, Evaporative - involves adiabatic heat exchange between air and water spray or wetted surface. The water assumes the wet-bulb temperature of the air, which remains constant during its traverse of the exchanger.

Cooling, Indirect Method of - a system in which a liquid, such as brine or water, cooled by the refrigerant, is circulated to the material or space refrigerated or is used to cool air so circulated.

Cooling, Regenerative - process of utilizing heat which must be rejected or absorbed in one part of the cycle to function usefully in another part of the cycle by heat transfer.

Cooling, Surface - a method of cooling air or other gas by passing it over cold surfaces.

Corresponding Values - simultaneous values of various properties of a fluid, such as pressure, volume, temperature, etc., for a given condition of fluid.

Counterflow - in heat exchange between two fluids, opposite directions of flow; the coldest portion of one meeting the coldest portion of the other.

Cylindrical Irradiance (at a point, for a direction) ($E_{e,z}$; E_z) - quantity defined by the formula

$$E_{e,z} = \frac{1}{4\pi sr} \int_{\Phi} L_e \sin \varepsilon \cdot d\Omega$$

where $d\Omega$ is the solid angle of each elementary beam passing through the given point, L_e, its radiance at the point and ε the angle between it and the given direction; unless otherwise stated, that direction is vertical. (Unit $W \cdot m^{-2}$)

Dampproofing - the process of coating the interior or exterior of a foundation wall, floor, etc., with bituminous emulsions and plastic cements. The purpose of dampproofing is to prevent or interrupt the capillary draw of moisture into the wall or floor system and to the interior of the foundation.

Daylight - visible part of global solar radiation. (Note: when dealing with actinic effects of optical radiations, this term is commonly used for radiations extending beyond the visible region of the spectrum.)

216

Daylight Factor [D] - ratio of the illuminance at a given point on a given plane due to the light received directly or indirectly form a sky of assumed or known illuminance distribution, to the illuminance on a horizontal plane due to an unobstructed hemisphere of this sky. The contribution of direct sunlight to both illuminances is excluded. (Note: (1) glazing, dirt effects, etc. are included. (2) when calculating the lighting of interiors, the contribution of direct sunlight must be considered separately.)

Declination - the angular position of the sun with respect to the plane of the equator, north being positive. The declination varies between -23.45° at the winter solstice and +23.45° at the summer solstice.

Declination of Sun - the angle of the sun above or below the equatorial plane. It is plus if north of the plane and minus if below and varies day by day throughout the year from 23.47° on June 21 to -23.47° on December 21.

Degree Days - the product of the number of degrees below a given base temperature and the number of days when that difference occurs. The base temperature is usually defined between 15.5° to 21°C.

Degree days for Cooling - the sum, over a stated period of days, e.g. month, of positive values of the arithmetic difference between a stated reference 'base temperature' and the daily mean dry bulb air temperature. (Positive values, i.e. daily mean air temperature below base temperatures are set to zero.) The base temperature for assessing cooling requirements is normally set above the base temperature used for assessing heating requirements to allow for the fact that acceptable comfort temperatures in summer are higher than those adopted for winter heating assessments.

Degree days for Heating - the sum, over a stated period of days, e.g. month, of positive values of the arithmetic difference between a stated reference 'base temperature' and the daily mean dry bulb air temperature. (Negative values, i.e. daily mean air temperature above base temperature, are set to zero). This is the generally accepted definition. However, in this publication the degree day total is assessed using the method adopted for routine Department of Energy

calculations, which includes a (small) contribution towards the total even when the hourly temperature rises above the base temperature for most of the day. Different base temperatures are required to estimate heating demands, frost protection, and for relating insulation standards to internal gains.

Dehumidification - (1) condensation of water vapour from air by cooling below the dew point; (2) removal of water vapour from air by chemical or physical methods.

Dehumidifier, Surface - an air-conditioning unit, designed primarily for cooling and dehumidifying air through the action of passing the air over wet cooling coils.

Dehumidifying Effect, Air Cooler - product of the weight of moisture condensed in the cooler by the constant 1060.

Dehydration - (1) removal of water vapour form air by the use of absorbing or adsorbing materials; (2)removal of water from stored goods.

Density (p) - the mass of a unit of volume of material .

Desiccant - any absorbent or adsorbent, liquid or solid, that removes water or water vapour from a material. In a refrigeration circuit, the desiccant should be insoluble in the refrigerant.

Design Heat Losses - a term expressing the total predicted envelope losses over the heating season for a particular house design in a particular climate.

Dewpoint - the temperature at which a given air/water vapour mixture is saturated with water vapour (i.e. 100% relative humidity). Consequently, if air is in contact with a surface below this temperature, condensation will form on the surface.

Differential Thermostat - a thermostat that operates on the basis of the temperature difference between two points. Blowers and pumps turn on or off depending on the magnitude of t.

Diffuse Lighting - a form of lighting where the same intensity of light comes from different directions.

Diffuse Radiation - the component of solar radiation that has been scattered by atmospheric molecules and particles. The diffuse radiation is assumed to be isotropic, that is, equally intense from all points of the sky. Also solar radiation scattered by transmission through diffusing glazing, See diffuse transmission.

Diffuse Sky Radiation - that part of solar radiation which reaches the Earth as a result of being scattered by the air molecules, aerosol particles, cloud particles or other particles.

Diffuse Transmission - the type of solar transmission through a diffusing or translucent glazing, namely, transmission that is scattered by interaction with the glazing material. The diffuse transmitted radiation is assumed to be isotropic, that is, equally intense in all directions.

Diffuser - a device object or surface used to alter the spacial distribution of light by diffusing it.

Diffusing Glazing - a translucent glazing, or a glazing that produces diffuse transmission. See diffuse transmission.

Diffusion, Scattering - process by which the spatial distribution of a beam of radiation is changed when it is deviated in many directions by a surface of by a medium, without change of frequency of its monochromatic components. (Note: a distinction is made between selective diffusion and non-selective diffusion according to whether or not the diffusion properties vary with the wavelength of the incident radiation.)

Dilution Air - the air required by some combustion heating systems in order to isolate the furnace form outside pressure fluctuations and to maintain an effectively constant chimney draft.

Direct Gain - the transmission of sunlight directly into the space to be heated where it is converted to heat by absorption on the interior surfaces.

Direct Glare - glare caused by self-luminous objects situated in the visual field, especially near the line of sight.

Direct Solar Gains - direct solar radiation passing through glass areas (mainly south facing) which contributes to space heating (kWh).

Direct Solar Radiation - that part of extraterrestrial solar radiation which as a collimated beam reaches the Earth's surface after selective attenuation by the atmosphere.

Direct Solar Radiation - the solar from the solid angle of the sun's disk.

Direct Sunlight - portion of daylighting coming directly from the sun at a specific location which is not diffused on arrival.

Directional Lighting - a form of lighting where light is received from a single direction.

Draught - a current of air, when referring to the pressure difference which causes a current of air or gases to flow through a flue, chimney, heater or space; or when referring to a localised effect caused by one or more factors of high air velocity, low ambient temperature, or direction of air flow, whereby more heat is withdrawn from a person's skin than is normally dissipated.

Driving Rain Index - the product of rainfall amount and wind speed, usually expressed in units m^2/s. The average annual driving rain index is useful for assessing the exposure of a site. In the past, only all-direction indices were available with values greater than $7m^2/s$ indicating 'severe' exposure and values less than $3m^2/s$ indicating sheltered sites (for 10m above ground in open, level terrain).

Ecliptic - the great circle cut in the celestial sphere by an extension of the plane of the earth's orbit. The great circle drawn on a terrestrial globe makes an angle of about 23.27° with the equator.

Effect, Chimney - tendency of air or gas in a duct or other vertical passage to rise when heated due to its lower density compared to that of the surrounding air or gas; in buildings, tendency towards displacement (caused by the difference in temperature) of internal heated air by unheated outside air due to the difference in density of outside and inside air.

Effect, Dehumidifying - heat removed in reducing the moisture content of air, passing through a dehumidifyer, from its entering to its leaving condition.

Effect, Humidifying - latent heat of water vaporization at the average evaporating temperature times the number of pounds of water evaporated per hour in Btuh.

Emissivity - the ratio of the radiant energy emitted from a surface at a given temperature to the energy emitted by a black body at the same temperature.

Emittance e - the ratio of the radiant energy emitted (in the absence of incident radiation) from a given plane surface at a given temperature to the radiant energy that would be emitted by a perfect black body at that same temperature.

Enthalpy - a thermodynamic property of a substance defined as the sum of its internal energy plus the quantity Pv/j; where P=pressure of the substance, v=its volume, and J=the mechanical equivalent of heat; formerly called total heat and heat content.

Enthalpy, Specific - enthalpy per unit mass of a substance.

Entropy - the ratio of the heat added to a substance to the absolute temperature at which it was added.

Entropy, Specific - entropy per unit mass of a substance.

Envelope - the exterior surface of a building including all external additions e.g. chimneys, bay windows, etc.

Equation of Time - part of time adjustment applied to local standard time (clock time) to adjust local standard time to solar time. It is used in conjunction with information about the longitude of the site in relation to the standard time zone reference longitude for that area, to find solar time. (For UK the time reference longitude of 15°E). The equation of time allows for certain irregularities in the daily rotation of the earth about its axis across the course of the year.

Equivalent Leakage Area (ELA) - the total area of all the unintentional openings in a building's envelope, generally expressed in square centimetres.

Evaporation - change of state from liquid to vapour.

Evaporative equilibrium (of a wet-bulb instrument) - the condition attained when the wetted wick has reached a stable and constant temperature. When the instrument is exposed to air at velocities over 4.6m/s, this temperature may be considered to approach the true wet-bulb temperature.

Exfiltration - the uncontrolled leakage of air out of a building.

Externally Reflected Component of Daylight Factor [D_e] - ratio of that part of the illuminance at a point on a given plane in an interior which is received directly from external reflecting surfaces illuminated directly or indirectly by a sky of assumed or known luminance distribution, to the illuminance on a horizontal plane due to an unobstructed hemisphere of this sky. The contribution of direct sunlight to both illuminances is excluded.

Extinction Coefficient - a property of glazing material that characterizes the solar absorption in the material, namely, the fraction of radiation that is absorbed per unit of path length in the material.

Extraterrestrial Radiation - solar radiation impinging on the earth's outer atmosphere.

Fan Depressurization - a large fan is used to exhaust air from a building in order to create a pressure difference across the building envelope; an analysis of the flow rate through the ran at different pressure differences provides a measurement of air-tightness

Forced Convection - heat transfer between a surface and an adjacent fluid (air in the present context) stream that is produced by external means such as wind or a fan. Compare with natural convection.

Frost Heaving - the movement of soils caused by the phenomenon known as ice lensing or ice segregation. Water is drawn form the unfrozen soil to the freezing zone where it attaches to form layers of ice, forcing soil particles apart and causing the soil surface to heave.

Glare - condition of vision in which there is discomfort or a reduction in the ability to see details or objects, caused by an unsuitable distribution or range of luminance, or to extreme contrast.

Glare by Reflection - glare produced by reflections, particularly when the reflected images appear in the same or nearly the same direction as the object viewed. (Note: formerly 'reflected glare'.)

Glare, Disability - glare that impairs the vision of objects without necessarily causing discomfort.

Glare, Discomfort - glare that causes discomfort without necessarily impairing the vision of objects.

Glazing - transparent or translucent material (glass or various plastics) used to cover the solar aperture.

Global Illuminance (E_g) - illuminance produced by daylight on a horizontal surface on the earth.

Global Radiation - the total solar radiation falling on a surface, i.e. the sum of the direct and diffuse radiation.

Greenwich mean Time (GMT) - an international reference time scale, based on mean astronomical time at Longitude 0° (the Greenwich Meridian).

Heat - form of energy that is transferred by virtue of a temperature difference.

Heat Capacity - the amount of heat necessary to raise the temperature of a given mass one degree. Numerically, the mass multiplied by the specific heat.

Heat Exchanger - a device, usually consisting of a coiled arrangement of metal tubing used to transfer heat through the tube walls from one fluid to another.

Heat Load - the heat loss from a building in a designated time period.

Heat Loss - heat flow through building envelope components (walls, windows, roof...).

Heat of Fusion - latent heat involved in changing between the solid and the liquid state.

Heat Pump - a thermodynamic device that transfers heat from one medium to another. The first medium (the source) cools while the second (the heat sink) warms up.

Heat Recovery - the process of extracting heat, (usually from a fluid) that would otherwise be wasted. For example heat recovery in housing generally refers to the extraction of heat from exhaust air.

Heat Transmission - any time rate of heat flow; usually refers to conduction, convection, and radiation combined.

Heat Transmission Coefficient - any one of a number of coefficients used in calculating heat transmission by conduction, convection, and radiation, through various materials and structures.

Heat, Latent - change of anthalpy during a change of state, usually expressed in J/kg (Btu per lb). With pure substances, latent heat is absorbed or rejected at constant temperature at any pressure.

Heat, Latent, of Condensation or Evaporation (specific) - thermodynamically, difference in the specific enthalpies of a pure condensible fluid between its dry saturated vapour state and its saturated (not subcooled) liquid state at the same pressure.

Heat, Sensible - heat which is associated with a change in temperature; specific heat exchange of temperature; in contrast to a heat interchange in which a change of state (latent heat) occurs.

Heat, Specific - ratio of the amount of heat required to raise the temperature of a given mass of any substance one degree to the quantity required to raise the temperature of an equal mass of a standary substance, usually water at 15°C one degree.

Heat, Vital - heat generated by fruits and vegetable; in storage, due to ripening.

Heating Degree-days (DD) - the summed differences between a fixed base temperature and the daily mean outdoor temperature. Only positive differences are counted, that is, when the outdoor mean is less than the base temperature. The daily mean is computed as the mean of the daily minimum and maximum temperatures.

Heating Season - the period of the year during which heating the building is required to maintain comfort conditions.

Humidifier - a device to add moisture to air.

Humidify - to add water vapour to the atmosphere; to add water vapour or moisture to any material.

Humidistat - a regulatory device, actuated by changes in humidity, used for automatic control of relative humidity.

Humidity - water vapour within a given space.

Humidity Ratio (or alternatively, The Mixing Ratio) - the ratio of the mass of water vapour to the mass of dry air contained in the sample.

Humidity, Absolute - the weight of water vapour per unit volume.

Humidity, Percentage - the ratio of the specific humidity of humid air to that of saturated air at the same temperature and pressure, usually expressed as a percentage (degree of saturation; saturation ratio).

Humidity, Relative - the ratio of the mole fraction of water vapour present in the air to the mole fraction of water vapour present in saturated air at the same temperature and barometric pressure. Approximately, it equals the ratio of the partial pressure or density of the water vapour in the air to the saturation pressure or density, respectively, of water vapour at the same temperature.

Hybrid Solar Heating System - solar heating system that combines active and passive techniques.

Hygrometer - an instrument responsive to humidity conditions (usually relative humidity) of an atmosphere.

Hygroscopic - absorptive of moisture, readily absorbs and retains moisture.

Incident Angle - the angle between the sun's rays and a line perpendicular (normal) to the irradiated surface.

Indirect Gain - the indirect transfer of solar heat into the space to be heated from a collector that is coupled to the space by an uninsulated, conductive or convective medium (such as thermal storage wall or roof pond).

Infiltration - the uncontrolled movement of outdoor air into the interior of a building through cracks around windows and doors or in walls, roofs and floors. This may work by cold air leaking in during the winter, or the reverse in the summer.

Infrared Radiation - electromagnetic radiation having wave length above the wave length range of visible light. This is the preponderant form of radiation emitted by bodies with moderate temperatures such as the elements of a passive building.

Internal Gains - the energy dissipated inside the heated space by people (body heat) and appliances (lighting, cooker, etc.). A proportion of this energy contributes to the space heating requirements (kWh).

Isolated Gain - the transfer of heat into the space to be heated from a collector that is thermally isolated from the space to be heated by physical separation or insulation (such as a convective loop collector or an attached sunspace with an insulated common wall).

Illuminance (at a point of a surface [E_v; E]) - quotient of the luminous flux $d\Phi_v$ incident on an element of the surface containing the point, by the area dA of that element. Equivalent definition. Integral, taken over the hemisphere visible at a given point, of the expression $L_v \cdot \cos \Theta \cdot d\Omega$ where $L_v \cdot$ is the luminance at the given point in the various directions of the incident elementary beams of solid angle $d\Omega$, and Θ is the angle between any of these beams and the normal to the surface at the given point.

$$E_v = \frac{d\Phi_v}{dA} \qquad = L_v \cdot \cos \Theta \cdot d\Omega \atop 2\pi sr$$

Impermeable - not permitting water vapour or other fluid to pass through.

Incidence Angle - the angle of incidence here refers to the angle between the centre of the solar disc and a line normal to any irradiated surface.

Index of Refraction - a property of glazing materials that determines the reflection/refraction characteristics of the glazing.

Indirect Gain - the indirect transfer of solar heat into the space to be heated from a collector that is coupled to the space by an uninsulated, conductive, or convective medium (such as a thermal storage wall or roof pond).

Indirect Lighting - lighting achieved by reflection, usually from wall and/or ceiling surface.

Induced Draft Flue System - a term referring to a type of gas heating system equipped with a fan downstream of the furnace. The fan pulls gases from the furnace and propels them to the outside, thereby eliminating the requirement for dilution air.

Infiltration - the movement of air from the outdoors into the heated space of a building.

Infiltration - the uncontrolled movement of outdoor air into the interior of a building through cracks around windows and doors or in walls, roofs and floors. This may work by cold air leaking in during the winter, or the reverse in the summer.

Infrared Emittance - a measure of the ability of a surface to emit infrared radiation, namely, the ratio of the infrared radiation emitted at a given temperature to the radiation emitted at that temperature by a perfect emitter.

Infrared Radiation - electromagnetic radiation having wave length above the wave length range of visible light. This is the preponderant form of radiation emitted by bodies with moderate temperatures such as the elements of a passive building.

Insulation, Sound - acoustical treatment of fan housings, supply ducts, space enclosures, and other parts of system and equipment to isolate vibration or to reduce noise transmission.

Insulation, Thermal - a material having a relatively high resistance to heat flow and used principally to retard heat flow.

Internal Gains - the energy dissipated inside the heated space by people (body heat) and appliances (lighting, cooker, etc.). A proportion of this energy contributes to the space heating requirements (kWh).

Internal Heat - heat generated inside the building by sources (such as appliances, lights and people) other than the solar system or the space heating equipment.

Internal Sources - the sources of internal heat such as appliances, lights and people. See internal heat.

Internally Reflected Component of Daylight Factor [D_i] - ratio of that part of the illuminance at a point on a given plane in an interior which is received directly from internal reflecting surfaces illuminated directly or indirectly by a sky of assumed or known luminance distribution, to the illuminance on a horizontal plane due to an unobstructed hemisphere of this sky. The contribution of direct sunlight of both illuminances is excluded.

Intrinsic Heat - heat from human bodies, electric light bulbs, cooking stoves, and other objects not intended specifically for space heating.

Irradiance (at a point of a surface) (E_e; E) - quotient of the radiant flux $d\Phi$, incident on an element of the surface containing the point, by the area dA of that element. Equivalent definition: Integral, taken over the hemisphere visible from the given point, of the expression $L_e \cos \Theta \cdot d\Omega$ where L_e is the radiance at the given point in the various directions of the incident elementary beams of solid angle $d\Omega$ where Θ is the angle between any of these beams and the normal to the surface at the given point.

$$E_e = \frac{d\Phi_e}{dA} \qquad = L_e \cdot \cos \Theta \cdot d\Omega \atop 2\pi sr$$

Irradiation - see radiation.

Isentropic - an adjective describing a reversible adiabatic process; a change taking place at constant entropy.

Iso - (1) prefix meaning constant: at isothermal, constant temperature; isentropic, constant entropy; isobaric, constant pressure; etc.; and (2) in chemicals, one having different characteristics but with the same number and kind of atoms.

Isobaric - an adjective used to indicate a change taking place at constant temperature.

Isolated Gain - the transfer of heat into the space to be heated from a collector that is thermally isolated from the space to be heated by physical separation or insulation (such as a convective loop collector or an attached sunspace with an insulated common wall).

Isothermal - an adjective used to indicate a change taking place at constant temperature.

Jalousie - an exterior fixed element made up of a perforated frame which covers the whole window. It allows natural ventilation and protects against direct solar radiation and the view from the exterior.

Joint, Mechanical - a gas-tight joint obtained by joining of metal parts through a positive holding mechanical construction (such as flanged joint, screwed joint, flared joint).

Lag - delay inaction of the sensing element of a control device due to the time required for the sensing element to reach equilibrium with the property being controlled.

Latent Heat - the energy required to change the state of a unit mass of material from solid to liquid or liquid to gas, without a change in temperature. This energy is then released, again without change of temperature (provided super cooling does not occur), when the material reverts from gas to liquid or from liquid to solid. (Wh/kg, kJ/kg)

Latent Heat of Fusion - the amount of heat required to change the state of a substance from solid to liquid at constant temperature (e.g. ice to water at $0°$ - 335 kJ/kg). The same amount of heat is liberated with a change of state in the reverse direction.

Latitude - the angular distance north (+) or south (-) of the equator, measured in degrees of arc.

Light, Perceived - Universal and essential attribute of all perceptions and sensations that are peculiar to the visual system. (Notes: (1) light is normally, but not always, perceived as a result of the action of a light simulation on the visual system.)

Lighting, Diffused - lighting in which the light on a working plane or on an object is not incident predominantly from a particular direction.

Lighting, Indirect - lighting by means of luminaires having a distribution of luminous intensity such that the fraction of the emitted luminous intensity such that the fraction of the emitted luminous flux directly reaching the working plane, assumed to be unbounded is 0 to 10 %.

Lightwell Absorption Fraction - the fraction of solar radiation that directly heats the air after it is transmitted through the glazing and after each reflection from interior surfaces. It is intended to simulate the presence of lightweight objects that absorb solar radiation and rapidly convect heat to the air.

Load - see heat load.

Load Collector Ratio (LCR) - the ratio of the building load coefficient to the projected area.

Load Factor - the ratio of actual mean load to a maximum load of maximum production capacity in a given period.

Long-Wave Radiation - radiation emitted between roughly 5 and 30 μm wavelength, as in thermal radiation from the surfaces of a room, or from the outside surface of the roof.

Longitude - the arc of the equator between the meridian of a place and the Greenwich meridian measured in degrees east or west.

Louvre - an assembly of sloping vanes intended to permit air to pass through and to inhibit transfer of water droplets.

Lumen (lm) - SI unit of luminous flux: luminous flux emitted in unit solid angle (steradian) by a uniform point source having a luminous intensity of 1 candela (9th General Conference of Weights and Measures, 1948). Equivalent definition: luminous flux of a beam monochromatic radiation whose frequency is 540 x 1012 hertz and whose radiant flux is 1/683 watt.

Luminance - The physical measure of the brightness of a surface such as a lamp, a reflecting material or the sky in a given direction. Luminance is the luminous intensity emitted by an area of a surface. Units : candela per square meter (cd/m^2)

Luminance Efficacy of Radiation (K) - quotient of the luminous flux Φ, by the corresponding radiant flux Φ,

$$K = \frac{\Phi_v}{\Phi_e}$$

Unit: $lm \cdot W^{-1}$

Note: when applied to monochromatic radiations, the maximum value of $K(\lambda)$ is denoted by the symbol K_m. K_m = 683 $lm \cdot W^{-1}$ for v_m - 540 x 10^{12} Hx ($\lambda_m \approx 555nm$) for photopic vision. K_m' - 1700 $lm \cdot W^{-1}$ for λ_m - 507 nm for scotopic vision.

For other wavelengths:

$K(\lambda)$ - $K_m V(\lambda)$ and $K'(\lambda)$ - $K_m V'(\lambda)$.

Luminous Environment - lighting considered in relation to its physiological and phychological effects.

Luminous Flux (Φ_v; Φ) - quantity derived from radiant flux Φ_e by evaluating the radiation according to its action upon the CIE standard photometric observer. For photopic vision

$$\Phi_v = K_m \int_0^\infty \frac{d\Phi.(\lambda)}{d\lambda} \cdot V(\lambda) \, d\lambda$$

where $\dfrac{d\Phi.(\lambda)}{d\lambda}$

is the spectral distribution of the radiant flux $V(\lambda)$ is the spectral luminous efficiency. (Unit: 1m)

Luminous Intensity (of a source, in a given direction) (I_v; I) - quotient of the luminous flux dF_v leaving the source and propagated in the element of solid angle $d\hat{I}$ containing the given direction, by the element of solid angle.

Lux (lx) - SI unit of illuminance: illuminance produced on a surface of area 1 square metre by a luminous flux of 1 lumen uniformly distributed over that surface 1 lx = 1 $lm \cdot m^{-2}$.

Masonry - concrete, concrete block, brick, adobe, stone, and other similar materials.

Mass - the quantity of matter in a body as measured by the ratio of the force required to produce a unit acceleration to the acceleration.

Mass-Area-to-Glazing-Area Ratio - the ration of the total surface area of massive elements in a direct-gain building to the direct-gain glazing area. Massive elements including in this definition are all floors, walls, ceilings, or other interior objects

221

with densities comparable to high-density concrete provided their surfaces are exposed and located in a room that is at least partially illuminated by direct solar gains.

Mean Radiant Temperature - the area weighted mean temperature of all surrounding surfaces, i.e. ‰ (s x t) / ‰ (s), where 't' is the temperature of each surface of area 's'. It is an approximate indication of the effect that the surface temperatures of surrounding objects have on human comfort.

Mechanical Systems - a term widely used in commercial and industrial construction, referring to all the mechanical components of the building: i.e. plumbing, heating, ventilation, air conditioning and heat recovery.

Mixed Systems - a solar heating system that combines two or more passive solar heating types.

Monthly SLR Method - application of the monthly solar load ratio (SLR) correlation to a calculation of the monthly auxiliary heat.

Natural Convection - heat transfer between a surface and adjacent fluid (usually air or water) and by the flow of fluid from one place to another, both induced by temperature differences only rather than by mechanical means, also called free convection. Compare with forced convection.

Natural Lighting System - component or a series of components joined in a specific building for natural daylighting.

Negative Pressure - a pressure below atmospheric. In residential construction negative pressure refers to pressure inside the house envelope that is less than the outside pressure: negative pressure will encourage infiltration.

Net Load Coefficient (NLC) - net reference load per degree of indoor-minus-outdoor temperature difference per day

Net Reference Load - steady-state heat loss from a building, excluding the solar wall, assuming constant indoor temperature. Compare with net load coefficient (Btu).

Night Insulation - movable insulation that covers a glazing at night and is removed during the day.

Normalized Leakage Area (NLA) - the NLA is calculated by dividing the ELA from the fan test by the area of the exterior envelope of the house.

Obstruction - anything outside a building which prevents the direct view of part of the sky.

Orientation - the orientation of a surface is in degrees of variation away from solar south, towards either the east or west. Solar or true south should not be confused with magnetic south which can vary due to magnetic declination.

Overcast Sky - the measured temperature data and solar radiation tables for overcast skies within this Handbook refer to days when the cloud cover was 8/8 during all the daylight hours. Night-time temperature data are related to the same 'overcast' days, i.e. night-time cloud cover was not taken into account.

Ozone - triatomic oxygen (O_3) sometimes used in air conditioning or cold storage as an odour eliminator; can be toxic in certain concentration.

Pascal - a unit measurement of pressure. House air tightness tests are typically conducted with a pressure difference of 50 Pascals between the inside and outside. 50 Pascals is equal to 5.008 mm (.2 in) of water at 12.9°C (55°F).

Percent of Possible Sunshine - the actual number of hours of sunshine expressed as a percentage for the month.

Percent Radiation Received - often referred to as the KT value. It is the monthly mean Global Radiation on a horizontal plane, expressed as a percentage of the monthly mean radiation on a horizontal plane outside the atmosphere.

Permeance - water vapour permeance is the rate of water vapour diffusion through a sheet of any thickness of material (or assembly between parallel surfaces). It is the water vapour flow to the differences of the vapour pressures on the opposite surfaces. Permeance is measured in perms (m^2°C/W).

Permeability - water vapour permeability is a property of a substance which permits passage of water vapour. When permeability

varies with psychrometric conditions, the 'spot' or 'specific permeability' defines the property at a specific condition.

Phase Change Material (P.C.M.) - a heat storage medium which relies on a plane surface. When the solar radiation coming from the sun's disk is obscured from the instrument, a pyranometer can be used to determine diffuse solar radiation.

Phase-Change Materials - materials with a phase change point close to the desired room temperature which can be used either to reduce temperature fluctuations within a room or to store energy in the phase change process (usually melting) and re-releasing it to the system when the system temperature falls below the transition point. Phase change materials may be surface mounted as in phase change tiles or storage based as in latent heat stores for active solar systems (see latent heat).

Photometry - measurement of quantities referring to radiation as evaluated according to a given spectral luminous efficiency function, e.g. $V(\lambda)$ or $V'(\lambda)$.

Pond, Spray - arrangement for lowering the temperature of water in contact with outside air by evaporative cooling of the water. The water to be cooled is sprayed by nozzles into the space above a body of previously cooled water and allowed to fall by gravity into it.

Positive Pressure - a pressure above atmospheric. In residential construction this refers to pressure inside the house envelope that is greater than the outside pressure: a positive pressure difference will encourage exfiltration.

Preheating - in air conditioning, to heat the air ahead of other processes.

Pressure - the normal force exerted by a homogeneous liquid or gas, per unit of area, on the wall of the container.

Pressure Difference - the difference in pressure of the volume of air enclosed by the building envelope and the air surrounding the envelope.

Pressure, Absolute - pressure referred to a perfect vacuum. It is the sum of gage pressure and atmospheric pressure.

Pressure, Critical - vapour pressure corresponding to the substance's critical state at which the liquid and vapour have identical properties.

Pressure, Saturation - the saturation pressure for a pure substance for any given temperature is that pressure at which vapour and liquid, or vapour and solid, can coexist in stable equilibrium.

Pressure, Total - the sum of the static pressure and the velocity pressure at the point of measurement.

Pressure, Vapour - the pressure exerted by the molecules at a given vapour.

Profile Angle (Vertical Shadow Angle) - the angle between the direction of the sun resolved in a vertical plane perpendicular to the face of the solar aperture, and the normal to the solar aperture. This angle is used to predict the effect of shading devices on sectional drawings.

Projected Area - the principal net glazing area projected on a vertical plane.

Properties, Thermodynamic - basic qualities used in defining the condition of a substance, such as temperature, pressure, volume, enthalpy, entropy.

Psychrometry - instrument for measuring relative humidities by means of wet- and dry-bulb temperatures.

Pyrometer - an instrument for measuring high temperature.

'R' Value - the thermal resistance of a unit area of a material of known thickness to heat flow caused by a temperature difference across the material: it is often used to give a comparative value of the effect of different insulating materials of differing thickness. (m^2°C/W)

Radiance of the Sky L_s - the radiant flux emitted in a small solid angle from a specific area of the sky in the direction of the observer expressed as the flux received normal to the beam per unit solid angle. (W/m^2sr)

Radiant Flux, Radiant Power (F_e; F; P) - power emitted, transmitted or received in the form of radiation. (Unit: W)

Radiant Heat Transfer - the transfer of heat by beat radiation. Heat radiation is a form of electromagnetic radiation. Radiant heating is due to infrared radiation is very prevalent in passive systems.

Radiant Heat Transfer - the transfer of heat energy from a location of higher temperature to a location of lower temperature by means of electromagnetic radiation.

Radiation - see thermal radiation or solar radiation.

Radiation, Electromagnetic - (1) emission or transfer of energy in the form of electromagnetic waves with associated photons. (2) these electromagnetic waves or these photons. (Note: the French term 'radiation' applies preferably to a single element of any radiation, characterized by one wavelength or one frequency.)

Radiation, Optical - Electromagnetic radiation at wavelengths between the region of transition to X-rays (l = 1nm) and the region of transition to radio waves (l = 1mm).

Radiation, Visible - any optical radiation capable of causing a visual sensation directly. (Note: there are not precise limits for the spectral range of visible radiation since they depend upon the amount of radiant power reaching the retina and the responsivity of the observer. The lower limit is generally taken between 360 mm and 400 mm and the upper limit between 760 mm and 830 mm.)

Reference City - a city for which solar radiation and weather data are used to generate sensitivity data.

Reference Design - a detailed specification of the passive solar features of a hypothetical passive solar building used as the subject of performance analysis.

Reference Illuminant - an illuminant with other illuminants are compared. (Note: a more particular meaning may be needed in the case of illuminants for colour reproduction.)

Reference Nonsolar Building - a building similar to the solar building but with an energy-neutral wall in place of the solar wall and with a constant indoor reference temperature.

Reference Temperature - a fixed indoor temperature in the definition of the net reference load. also the fixed indoor temperature in the reference nonsolar building used in the definition of solar savings.

Reflectance (for incident radiation of given spectral composition, polarization and geometrical distribution) (p) - ratio of the reflected radiant or luminance flux to the incident flux in the given conditions.

Reflectance - the ratio or percentage of the amount of light reflected by a surface to the amount incidence. Good light reflectors are not necessarily good heat reflectors.

Reflectance of a Glazing Material - radiant energy is reflected from both inside and outside surfaces of any sheet of parallel sided glazing. Multiple inter-reflections occur within the glazing. The ratio of the overall reflected radiant energy to the incident radiant energy is known as the reflectance. Its value depends on the angle of incidence. Energy inter-reflected from the inside face is dependent on the absorptivity of the material and so is also dependent on wavelength. For any wavelength and angle of incidence, absorptance + reflectance + transmittance = 1.

Reflectance of Opaque Materials - the ratio of radiation reflected by a surface. the reflectance varies for different wavelengths and angles of incidence: white paint has an average reflectance of 0.71 for normal incidence solar radiation and 0.11 for long-wave (thermal) radiation. For any wavelength and angle of incidence, absorptance + reflectance = 1.

Reflected (Global) Solar Radiation - radiation that results from reflection of the global solar radiation by the surface of the Earth and by any surface intercepting that radiation.

Reflection (light) - process by which radiation is returned by a surface or a medium, without change of frequency of its monochromatic components. (Note: (1) part of the radiation falling on a medium is reflected at the surface of the medium [surface reflection]; another part may be scattered back from the interior of the medium [volume reflection]. (2) the frequency is unchanged only if there is no Doppiler effect due to the motion of the materials from which the radiation is returned.)

Relative Humidity - the ratio of the amount of water vapour in the atmosphere to the maximum amount of water vapour that could be held at a given temperature.

Relative Sunshine Duration - ratio of sunshine duration to possible sunshine duration within the same period.

Resistance Value (RSI) - thermal resistance value. Measurement of the ability of a material to resist heat transfer.

Resistance, Thermal - the reciprocal of thremal conductance.

Resistivity l/k - the thermal resistance of unit area of a material of unit thickness to heat flow caused by a temperature difference across the material. (m° C/W)

Respiration - production of carbon dioxide and heat by ripening of perishables in storage; also the breathing process of animals.

Reverse Thermocirculation - thermocirculation in the reverse direction, that is, from the heated space to the solar collector (sunspace or Trombe wall). This can occur at night when the heated space is warmer than the collector. It is assumed in the reverence designs to be prevented by dampers.

Rock Bed - a container filled with rocks, pebbles, or crushed stone, to store energy by raising the temperature of the rocks, etc.

Saturation, Degree of (Saturation Ratio) - the ratio of the specific humidity of humid air to that of saturated air at the same temperature and pressure, usually expressed as a percentage.

Sealants - flexible materials used on the inside of a building to seal gaps in the building envelope thereby preventing uncontrolled air infiltration and exfiltration.

Selective Coating - finishes applied to materials to improve their performance in relation to radiation of different wavelengths: those applied to solar absorbers have a high absorptance of solar radiation by a low emittance of long-wave (thermal) radiation, while those for glazing have a high transmittance to solar radiation and high reflectance of long wavelengths.

Selective Surface (Absorber) - a surface absorbing essentially all incident solar radiation (short wave - high temperature source), while emitting a small fraction of thermal radiation (long wave - low temperature source).

Semienclosed Sunspace - a sunspace that shares three common walls with the associated building.

Sensitivity Data - date that express the dependence of the heating performance of a passive solar heating system on individual parameters of the system design.

SI - (Systeme International d'Unites) the International System of Units being adopted throughout the world.

Sky Component of Daylight Facteur [D_s] - ratio of that part of the illuminance at a point on a given plane which is received directly (or through clear glass) for a sky of assumed or known luminance distribution, to the illuminance on a horizontal plane due to an unobstructed hemisphere of this sky. The contribution of direct sunlight to both illuminances is excluded.

Skylight - visible part of direct solar radiation. (Note: when dealing with actinic effects of optical radiations, this term is commonly used for radiations extending beyond the visible region of the spectrum.)

Sol-air Temperature - the hypothetical temperature which would give the same temperature distribution and heat flow in a building element that would result from the impact of the actual combination of the absorbed short wave radiation, long wave exchange with sky and ground, forced convection due to wind at the given external air temperature. It is a temperature to better access the transfer of heat through a specific building element of a given surface colour. Different sol-air temperatures are encountered on differently orientated surfaces for materials of the same surface properties.

Solar Absorptance - the fraction of incident solar radiation that is absorbed upon striking a surface.

Solar Aperture - that portion of the solar wall covered by glazing.

Solar Collection Area - see projected area.

Solar Constant - the irradiance of solar radiation beyond the earth's atmosphere at the average earth-sun distance on a surface perpendicular to the sun's rays. The value for the solar constant is 1.353 kWh/m^2.

Solar Energy, useful - the amount of solar energy contributing to the total heat load. It is expressed in absolute figures (kWh) or per unit collector area (kWh/m^2).

Solar Fraction (or percentage solar) - the percentage of the total heat load supplied by the solar heating system, including useful losses from the storage.

Solar Heat Gain - in passive solar heating a term referring to the amount of heat gained through windows over the heating season. Net solar gain refers to the solar heat gain less the heat losses through the windows.

Solar Load Ratio (SLR) - ratio of solar gain to building load used in SLR correlations.

Solar Load Ratio (SLR) Correlations - correlations between monthly solar savings fractions and monthly solar load ratios.

Solar Radiation - radiation emitted by the sun, including infrared radiation, ultraviolet radiation, and visible light.

Solar Savings Fraction (SSF) - the ratio of the solar savings to the energy requirement of the reference nonsolar building.

Solar Spectrum - the radiation emitted by the sun outside the atmosphere approximates to that emitted by a black body at 6000K. At the surface of the earth, the majority of the radiation falls within a range of wavelengths from 0.3 to 2.5 µm and typically about half the energy comes in the form of visible light (0.38 to 0.70 µm wavelength). Ultra violet radiation has wavelengths shorter than 0.38 µm and only forms a very small proportion of the spectrum. Due to water vapour and carbon dioxide and other absorbing gases in the atmosphere, several bands of energy are substantially removed in the infra red region. The proportion of infra red increases as the sun gets lower.

Solar Time - a time scale based on the movement of the sun, such that noon is when the sun is at the highest point of its arc and is true south, as opposed to local standard time.

Specific Heat Capacity - a measure of the amount of energy required to raise a unit mass or volume of a material through a unit temperature change. (kWh.kg.K, J/kg.K. kWh/m^3.K, J/m^3.K).

Spectrum (of a Radiation) - display or specification of the monchromatic components of the radiation considered. (Notes: (1) there are line spectra, continuous spectra, and spectra exhibiting both these characteristics. (2) this term is also used for spectral efficiencies [excitation spectrum, action spectrum].

Specular Reflectance - the percentage of solar energy input to the heat storage, subsequently used in the heat distribution system (i.e. excludes unwanted heat losses from the storage device). (%)

Specular Surface - a surface with reflective properties where the angle of visible incidence is equal to the angle of reflection. For example finishes such as polished aluminium, stainless steel or tin.

Spherical Irradiance, Radiant Fluence Rate (at a point) $E_{e,o}$; E_o) - quantity defined by the formula

$$E_{e,o} = \frac{L_e d\Omega}{2\pi sr}$$

where $d\Omega$ is the solid angle of each elementary beam passing through the given point and L_e its radiance at that point. (Unit: W·m^{-2})

State, Gaseous - one of three states of matter characterized by the greatest freedom of molecules and lack of any inherent fixed shape or volume.

State, Liquid - one of three states of matter characterized by limited freedom of molecules and by substantial incompressibility.

State, Solid - one of three states of matter characterized by stability of dimensions, relative incompressibility, and molecular motion held to limited oscillation.

Storage Volume Ratio - ratio of the volume of thermal storage material to the projected area.

Sun Effect - solar energy transmitted into interior spaces through windows and building materials.

Sunlight - visible part of direct solar radiation. (Note: when dealing with actinic effect of optical radiations, this term is commonly used for radiations extending beyond the visible region of the spectrum.)

Sunshine (possible percentage) - the actual number of hours of sunshine expressed as a percentage for the month.

Sunshine Duration [S] - sum of time intervals within a given time period (hour, day, month, year) during which the irradiance from direct solar radiation on an plane normal to the sun direction is equal to or greater that 200 watts per square metre.

Sunspace - see attached sunspace or semienclosed sunspace.

Suntempered Building - a minimal solar building derived from a convectional building by orienting its long axis east-west and placing a substantial fraction of its window area on the south side.

Surface Resistance - the surface resistance is the resistance to heat flow at the surface of a material. It has two components, the surface resistance for convection/conduction R_c whose value depends on the rate of air flow over the surface and radiative surface resistance for long wave radiation heat transfer R_r whose value depends on the surface emittance and the surface temperature. These two components are assumed to act in parallel so that the overall surface resistance R_o is give by

$$1/R_o = 1/R_c + 1/R_r$$

Temperature - the thermal state of matter with reference to its tendency to communicate heat to matter in contact with it.

Temperature Difference, Diffusion - temperature difference between the air temperature at supply opening and design outdoor temperature.

Temperature Difference, Effective - difference between the room air temperature and the supply air temperature at the outlet to the room.

Temperature Difference, Mean - mean of difference between temperatures of a fluid receiving and a fluid yielding heat.

Temperature Swing - the range of indoor temperature in the building between day and night.

Temperature, Absolute Zero - the zero point on the Kelvin temperature scale, -273.16°C

Temperature, Critical - the saturation temperature corresponding to the critical state of the substance at which the properties of the liquid and vapour are identical.

Temperature, Dry-bulb - the temperature of a gas or mixture or gases indicated by an accurate thermometer after correction for radiation.

Temperature, Effective - the dry-bulb temperature of a black enclosure at 50% relative humidity (sea level), in which a solid body or occupant would exchange the same heat by radiation, convection, and evaporation as in the existing nonuniform environment.

Temperature, Mean Radiant (MRT) - the temperature of a uniform black enclosure in which a solid body or occupant would exchange the same amount or radiant heat as in the existing nonuniform environment.

Temperature, Room - the temperature of any room, e.g.: (1) a room in which a refrigerator is being operated or tested; (2) a room being conditioned for the occupants' comfort. Room temperature used colloquially to mean the ordinary temperature one is accustomed to find in dwellings.

Temperature, Saturation - of a fluid, the boiling point corresponding to a given pressure; evaporation temperature, condensation temperature.

Temperature, Wet-bulb - thermodynamic wet-bulb temperature is the temperature at which liquid or solid water, by evaporating into air, can bring the air to saturation adiabatically at the same temperature. Wet-bulb temperature (without qualification) is the temperature indicated by a wet-bulb psychrometer constructed and used according to specification.

Thermal Break - a material of low conductivity used in an assembly to prevent flow of heat by conduction from one side of the assembly to the other. Often used to refer to materials used for this purpose in the frame of metal windows.

Thermal Bridge - a low thermal resistance path of connecting two surfaces; for example, framing members in insulated frame walls or metal ties in cavity wall and panel construction.

Thermal Conductance - the thermal transmittance through 1 square meter of material of a given thickness for each K temperature difference between its surfaces. $(W/m^2.K)$

Thermal Conductivity - the thermal transmission through a material 1 meter thick for each K temperature difference. $(W/m.K)$

Thermal Mass - the mass of the building within the insulation, expressed per volume of heated space $(kg.m3)$. 'Primary thermal mass' receives direct sunlight; 'secondary thermal mass' is in sight of the primary thermal mass and so receives radiative and convective energy from the primary thermal mass; 'remote thermal mass' os hidden from view of both the primary and secondary thermal mass and so receives energy by convection only.

Thermal Radiation - energy transfer in the form of electromagnetic waves from a body by virtue of its temperature, including infrared radiation, ultraviolet radiation, and visible light.

Thermal Resistance or R-value - the reciprocal of thermal conductance - see above. $(m^2.k/W)$

Thermal Resistivity - the reciprocal of thermal conductivity. $(m.K/W)$

Thermal Storage Mass - building elements, usually masonry or water in containers, designed to absorb solar heat during daytime hours for release later when heat is needed.

Thermal Storage Volume Ratio - see storage volume ratio.

Thermal Storage Wall - a wall of massive material (masonry or water in containers) placed between the solar aperture and the heated space. Heat is transferred into the space by conduction through the masonry or

conduction and convection through the water, and, if openings are provided, by natural convection. See also Trombe wall and water wall.

Thermal Transmittance - the thermal transmission through 1 square metre area of a given structure (e.g. a wall consisting of bricks, thermal insulation, cavities, etc.) divided by the difference between the environmental temperature on either side of the structure. Usually called 'U-value'. $(W/m^2.K)$

Thermocirculation - free convection from a warm zone (sunspace or Trombe-wall air space) to a cool zone through openings in a common wall.

Thermocirculation Vents - openings in a common wall between cool and warm zones through which thermocirculation occurs. The vents are arranged in pairs, one of each pair near the floor and one near the ceiling.

Thermosiphon - the convective circulation of a fluid which occurs in a closed system where warm fluid rises and is replaced by a cooler fluid in the same system.

Tilt - the angle of a plane relative to a horizontal plane.

Toplighting - light which enters through the top part of interior space such as clerestories, light wells or skylights.

Total Cloud Amount - ratio of the sum of the solid angles subtended by clouds to the solide angle $2_{,,}$ steradians of the whole sky. (Note: the total cloud amount is frequently called fractional cloud cover in the USA.)

Total Load Coefficient - total reference load (net reference load plus solar wall load) per degree of indoor-minus-outdoor temperature difference per day $(Btu/°F·day)$.

Total Turbidity Factor (according to Linke) [T] - ratio of the vertical optical thickness of a turbid atmosphere to the vertical optical thickness of the pure and dry atmosphere (Rayleigh atmosphere), related to the whole solar spectrum

$$T = \frac{\delta_R + \delta_A + \delta_Z + \delta_W}{\delta_R}$$

where δ_R is the optical thickness with respect to Rayleigh scattering at the

air molecules, δ_A, δ_Z, δ_W are the optical thickness with respect to Mie scattering and absorption at the aerosol particles, to ozone absorption, and to water vapour absorption respectively.

Transmission (light) - passage of radiation through a medium without change of frequency of its monochromatic components.

Transmission - in thermodynamics, a general term for heat travel; properly, heat transferred per unit of time.

Transmittance τ - the ratio of the radiant energy transmitted by parallel sided plane sheet of a given material to the radiant energy incident on the outside surface of that material: the value depends on the angle of incidence and wavelength. For any wavelength and angle of incidence, absorptance + reflectance + transmittance = 1.

Transmittance, Thermal (U Factor) - the time rate of heat flow per unit area under steady conditions form the fluid on the warm side of a barrier tot he fluid on the cold side, per unit temperature difference between the two fluids.

Trombe wall - a thermal storage wall of masonry placed between the solar aperture and the heated space. Heat is transferred into the space by conduction through the masonry and, if vents are provided, by natural convection.

U-value - see transmittance 'thermal'.

U-Value - the thermal conductance of a composite building element: it is the reciprocal of the total thermal resistance of the element, which is the sum of all the relevant material, cavity and surface air film resistance, across the section of the element, and is a measure of the energy flow through a unit area of the element per °C temperature difference across it. $(W/m^2°C)$

Ultraviolet Radiation - electromagnetic radiation having wavelengths shorter than visible light. This invisible form of radiation is found in solar radiation and plays a part in the deterioration of plastic glazings, paint, and furnishing fabrics.

Vapour - a gas, particularly one near to equilibrium with its liquid phase of the substance and which does not flow the gas laws. It is usually used instead of gas for a refrigerant and, in general, for any gas below the critical temperature.

Vapour Barrier - a component of a construction which is impervious to the flow of moisture and air and is used to prevent condensation in walls and other locations of insulation.

Vapour Diffusion - the movement of water vapour between two areas caused by a difference in vapour pressure, independent of air movement. The rate of diffusion is determined by (1) the difference in vapour pressure, (2) the distance the vapour must travel, and (3) the permeability of the material to water vapour. Hence the selection of materials of low permeability for use as vapour retarders in building.

Vapour Pressure - the pressure exerted by a vapour either by itself or in a mixture of gases. For example, when referring to water vapour pressure is determined by the concentration of water vapour in the air.

Vapour Retarder - any material of low water vapour permeability used to restrict the movement of water vapour due to vapour diffusion.

Vapour, Saturated - vapour in equilibrium with its liquid: i.e., when the numbers per unit time of molecules passing in two directions through the surface dividing the two phases are equal.

Vapour, Superheated - vapour at a temperature which is higher than the saturation temperature (i.e., boiling point) at the existing pressure.

Vapour, Water - used commonly in air conditioning parlance to refer to steam in the atmosphere.

Vapour, Wet, Quality of - the fraction by weight of vapour in a mixture of liquid and vapour.

Ventilation - the process of supplying or removing air by natural or mechanical means to or from any space. Such air may or may not have been conditioned.

Ventilation Losses - the heat losses associated with the continuous replacement of warm, stale air by fresh cold air.

Vertical Shadow Angle - see profile angle.

Viscosity - that property of semifluids, fluids and gases by virtue of which they resist an instantaneous change of shape or arrangement of parts. It is the cause of fluid friction whenever adjacent layers of fluid move with relation to each other.

Viscosity, Absolute - the force per unit area required to produce unit relative velocity between two parallel areas of fluid distance apart; also called coefficient of viscosity.

Visible Spectrum - that part of the solar spectrum which is visible to the human eye: radiation with wavelengths roughly between 0.38 and 0.70 µm.

Visual Performance - performance of the visual system as measured for instance by the speed and accuracy with which a visual task is performed.

Volume, Specific - the volume of a substance per unit mass; the reciprocal of density.

Volumetric Heat Capacity (pc) - a measure of the ability of a unit of volume of material to store heat, namely, the heat stored in a unit of volume of material per degree of temperature rise.

Volumetric Heat Loss Coefficient (G-value) - the total heat loss of a dwelling (through the fabric and ventilation), divided by the heated volume and the temperature at which the loss occurs. $(W/m^3.K)$

Water Cooling - water used for condensation of refrigerant; condenser water.

Water Wall - a thermal storage wall of water in containers placed between the solar aperture and the heated space. Heat is transferred into the space by conduction and convection through the water.

Wind Speed - the speed of the air measured in accordance with the recommendations of the World Meteorological Organisation, normally measured ten metres above ground level.

Work Plane, Working Plane - reference surface defined as the plane at which work is usually done. (Note: in interior lighting and unless otherwise indicated, this plane is assumed to be a horizontal plane 0.85 m above the floor and limited by the walls of the room. In the USA the work plane is usually assumed to be 0.76 m above the floor, in the USSR 0.8 m above the floor.)

Zenith Angle - the angular distance from the sun to the zenith, the point directly above the observer (at noon - latitude - solar declination).

RESOURCE GUIDE

12.1

PUBLICATIONS

DANISH

Beregning af energiforbrug i smahuse, Calculation of energy consumption in houses, Statens Byggeforskningsinstitut, 1984, *SBI Report No.148. 78pp,* ISBN: 87-563-0538-9

Energi i Arkitekturen,(Parallel title) Energy in Architecture, Moltke, Ivar., Energiteknologi, Dansk Teknologisk Institut.,Postboks 141, 2630 Taastrup, Danmark, 1990, *An illustrated introduction to energy efficient building design in Scandinavia. Chapters include '10 commandments', concepts, building geometry, climate zones, insulation,windows, sunspaces, shading, ventilation, passive heat storage, active & auxiliary heat sources.* ISBN 87-7756-073-06

Den Energibeviste Tradition, Energy Research Group, School of Architecture, University College Dublin, Richview, Clonskeagh, IRL-Dublin 14, 1989, *An A4 B&W reproduction of a poster exhibition of traditional responses to energy issues in Europe prepared under the CEC's ARCHISOL programme by the Energy Research Group. The,Original poster exhibition, in colour, was circulated to European schools of architecure.*

Glasdaekkede uderum,Glass covered spaces, Byggeriets udviklingsrad, 1985,*116 pp.* ISBN: 87-503-5735-2

Passiv Solvarme - Glasbyginger,Passive Solar Heating - Glass Buildings, Teknologisk Institut, Varmeteknik, 1988, *Samarbejdsgruppen for Passiv Solvarme, 24pp,* ISBN: 87-7511-813-0

Passiv Solvarme - hvorfor? hvordan? hvornar? Passive solar heating - why? how? and when?, Teknologisk Institut, Varmeteknik, 1984, *Samarbejdsgruppen for Passiv Solvarnme, 16pp.* ISBN: 87-7511-456-9
Passiv Solvarme - Projektering-svejledning Passive Solar Heating - Handbook, Teknologisk Institut, Varmeteknik, Danmarks Tekniske Hcjskole Lab. for Varmeisolering, 1985, *Energiministeriets solvarmeprogram, rapport 30, 312 pp.* ISBN: 87-7511565-4

Sol i Boligen,Solar heated dwellings, catalogue of projects., Teknologisk Institut, Varmeteknik, 1984, *Energiministeriets Solvarmeprogram, Report No.25, 88 pp.* ISBN: 87-7511-372-4

Solvarme og Bebyggelses - planlaegning Passive solar heating and development planning, Teknologisk Institut, Varmeteknik, 1985, *Miljcministeriet, Planstyrelsen. 60 pages.* ISBN: 87-503-5632-1

DUTCH

Architectuur en Klimaat, Realisaties, Gilot, C., De Herde A., Minne A., Opergelt D., Programmatiediensten van het Wetenschapsbeleid, 8 Rue de la Science, 1040 Brussel, Juni 1986,160 F.B. *Deze boek telt 128 bladzijden en aan de hand van een vijftiental nieuwe of gerenoveerde huizen-zowel in de stad als op het platteland - wordt een beeld geschetst van de Belgische klimatische architectuur.*

Congrès International d'Architecture Climatique, Cellule Architecture et Climat, CIACO Editeur, B-1348 Louvain-la-Neuve, June 1986, 600 F.B. *Proceedings of International Congress on Climatic Architecture, 2-3 July 1986. 598 pp.*

De Energiebewuste Traditie,Energy Research Group, School of Architecture, University College Dublin, Richview, Clonskeagh, IRL-Dublin 14, 1989. *An A4 B&W reproduction of a poster exhibition of traditional responses to energy issues in Europe, prepared under the CEC's ARCHISOL programme by the Energy Research Group. The Original poster exhibition, in colour, was circulated to European schools of architecure.*

De Factor Zon 1, Woon / Energie, TNO-IMG, Novem, Postbus 8242, 3503 RE Utrecht, Nederland, 1985, HFL 5, *Mogelijkheden tot energiebesparing door het gebruik van zonne-energie in eengezinswoningen.*

De Factor Zon 2, Woon / Energie, TNO-IMG, Novem, Postbus 8242, 3503 RE Utrecht, Nederland, 1988, HFL 5, *Mogelijkheden tot energiebesparing door het gebruik van zonne-energie in gestapelde woningen.*

Contents

De techniek van zonnewarmtesystemen, Schriftelijk cursuspakket, Lysen, Lentz, de Boon, Wijsma, Gramsbergen, Dettmers en Brouwer, Holland Solar, Korte Elisabethstraat 6, 3511 JG Utrecht, 1988, fl. 60

Documentatie Praktijkexperimenten, F.E. Bakker, M.J.C. Bergmans, J.T.J.M. van Tuijn, M.P.G. Plantinga, R.L.C.J. Trines, J.P.G. Severijns, J.L.A. van Hest. Technische Universiteit Eindhoven, Postbus 513, 5600 MB Eindhoven, Nederland, 1990, HFL 20, - *per deel in elk deel wordt één praktijkexperiment behandeld, Documentatie van Praktijkexperimenten, uitgevoerd in het kader van,het national onderzoekprogramma REGO-IO. Inmiddels zijn meer dan 15 Praktijkexperimenten gedocumenteerd in opdracht van PEO en NOVEM.*

Energiebewust Ontwerpen van Woningen, Een Handreiking, Energy-conscious design of dwellings, an assistance, Bouwcentrum/Universities of Delft and Eindhoven, *Gives an insight into physical aspects of comfort, ventilation, solar radiation, etc. Much attention paid to cost/benefit ratio.*

Gids bij het Bioklimatisch Ontwerpen, Le Paige, M. Gratia, De Herde, A. Programmatiediensten van het Wetenschapsbeleid, 8 rue de la Science, 1040 Brussel, Juni 1986, 120 F.B. *Dit werk telt 136 bladzijden en bestaat uit 3 delen. Het eerste deel geeft aan de hand van een aantal tekeningen en grafieken, de nodige informatie aan architecten. Het 2de gedeelte toont voorbeelden en het 3de herneemt waarden en resultaten.*

Klimaatgegevens, Opfergelt D., De Herde A., Programmatie diensten van het Wetenschapsbeleid, Rue de la Science 8, 1040 Brussel, Juni 1986, 100 F.B. *Deze publikatie telt 64 bladzijden en geeft aan de hand van 3 - dimensionele grafieken de belgische klimaatgegevens weer voor de 15e van elke maand.*

Licht in de Architectuur, Een Bescouwing over Dag-en Kunstlicht, van Santen, Christa., Hansen, A.J., J.H. de Bussy B.V. Amsterdam,1985, 90 60547225/cip

Ontwerpen met passieve en actieve zonne-energie, Verslag van lezingen en discussies symposium 21 April 1988, Holland Solar, Korte Elisabethstraat 6, 3511 JG Utrecht, 1988, fl.25

Ontwerpen van energie-efficiënte kantoorgebouwen; Intergratie van gebouw en installatie, G.A.M. van Schaik, C.L.M. Leenaerts, SBR/ISSO, Postbus 20740, 3001 JA Rotterdam, 1990, ISBN 90-5044-019-3,

Richtlijnen en vuistregels voor architecten en aannemers, Sonnemans, E., Dettmers, W., Holland Solar,Korte Elisabethstraat 6, 3511 JG Utrecht,1988,fl. 25, *zonneboilers,ventilatielucht.*

Thermisch en Hygrisch Gedrag van Bouwconstructies, EBES Editor, Antwerpen, Free to schools & architects. *Deals with insulation and moisture problems in buildings, with reference to energy conservation.*

Vuistregels voor het Bouwwezen, Rules of Thumb in Building, Gasunie, *Provides a great number of rules of thumb on: -solar radiation, insulation index/volume/area ratio and ventilation. Published by Gasunie, The Dutch National Gas supplier.*

ENGLISH

1987 European Conference on Architecture - Proceedings, Palz, W. (Ed.), H.S. Stephens & Associates, Agriculture House, 55 Goldington Road, Bedford MK40 3LS, UK., 1987, £85 (£75 in EC), *Papers presented at Conference 6 - 10 April 1987, Munich, F.R.G. Edited by W. Palz of CEC DG XII. Contains 165 papers (832 pages, 1058 figures, 164 photos, 178 tables). Mostly English (7 German, 23 French, mostly with English abstracts),* ISBN 0 - 9.519271-2-3

A Design Guide for Naturally Ventilated Courtrooms, Penz, F., Property Services Agency, UK, Directorate of Civil Projects 1 / Courts Group, C Block, Whitgift Centre, Wellesbey Road, UK - Croydon CR9 3LY, 1990, *Prepared by Cambridge Architectural Research Ltd. for the PSA, this colour-illustrated brochure offers clear, concise guidelines for the design of natural and hybrid ventilation systems in legal courtrooms for UK conditions. 14 pp. Also available in Hypercard form for use on the Macintosh computer.*

A new Bioclimatic Chart for Passive Solar Design, Arens, E., Proceedings of 5th Nat. Passive Solar Conference, Amherst, USA, 1980

A Simplified Daylighting Design Methodology for Clear Skies, Bryan, H.V., Proceedings of 5th Nat. Passive Solar Conference,1980

Aerogels, proceedings of the 1st international symposium. Fricke, J. (Ed.), Springer-Verlag / University of Wurzburg, Germany, 1985, *Part of the series: Springer Proceedings in Physics. Vol. 6.* ISBN 3-540-19256-9

An analytic model for describing the influence of lighting, parameters upon visual performance. Commission Internationale de l'Eclairage (C.I.E.), 1981, *Volume 1 No. 19/2.1: Technical Foundations (Price 180 DM), Volume 2 No. 19/2.2: Summary and Application Guidelines, (Price 120 DM), French and German,* ISBN 1) 2)9290340193

An Evaluation of Energy Conservation Programs for New Residential Buildings. Lawrence Berkeley Laboratory, Berkeley, California. USA. *Series of report evalating the implementation of programmes promoting energy efficiency in new residential construction. Reports seek to combine current information from both publ/ unpubl. Sources for State regulatory commissions, utilities, etc., to identify design and manage demand-side programs.*

An investigation of Passive Solar Heating and Cooling Work in Europe, Lebens, R.M., Birch, H.J., Commission of the European Communities, 1979

Architec Publishing Database Architec Publishing, 1989, £950 U.K. per year, *A database of British standards, statutory instruments, BRE, guidance & similar semi-official publications following, Barbour Microfiche & including basic product information from about 5000 manufacturers. Updated bi-monthly. Used with a compact disk drive connected to a microcomputer. A.J. 14 Dec. 1988*

Architectural Interior Systems. Flynn, J.E., Segil, A.W., Steffy, G.R. Van Nostrand 1988 (2nd ed.) *Covers lighting design in the context of the acoustic and thermal environment. Uses imperial units - not SI.*

Architecture, Ambiance Energie. M. Retbi, MELTH, DAU, AFME, Thermique et Architecture, Regirex, 54, bis rue Dombasle, 75015 Paris, 1989, *Catalogue of the competition. Synthesis and description of 20 projects. 85 p.*

Atrium Buildings, development and design. (2nd ed.), Saxon, Richard. Architectural Press London, 1986. Prev ed. 1983. £45.00, *Topics: building typology and atrium features for design studies*. ISBN 0 8513 9051 X

Availability of Daylight. Hunt, D.R.G., Department of the Environment (Watford, England), Building Research Establishment, Garston, Watford, UK. 1979. *Useful technical summary of data on the outdoor illuminances recorded at different times of the working day for different times of the year. Particular to UK latitudes but a useful model for corresponding studies elsewhere.*

Blocks of flats with controlled natural ventilation,and recovery of heat. Lennart Eriksson et al., Swedish Council for Building Research, 1986, Ref: D 19.

BS 78 - The Geometry of the Shading Buildings by Trees. Sattler, M. A. Dept. of Building Science, Faculty of Architectural Studies, University of Sheffield, S10 2TN, UK, Tel. +0742 768 555, 1985, *A paper describing a computer-based design tool; (SHTREE) used to assess the shading effects of trees in relation to building. The programs output is both numerical and graphical for any time or location and covers a range of tree shapes. Written in Fortran 77.*

BS 5918, Code of Practice for Solar Heating Systems for Domestic Hot Water. British Standards Institute, 1980, £6.40 stg. *Text available in French as 'Code de bonne pratique pour les systemes de chauffage solaire pour l'eau chaude domestique' and in German as 'Richtlinie für sonnenwarmesystme für haushalts - heisswasseranlagen'.* ISBN 0 5801 1179 2 .

BS 8211: Energy efficiency in Housing Part I 1988, Code of Practice for Energy Efficient Refurbishment of Housing, British Standards Institute, 1988, *Guidance & procedures for energy efficient design in the renovation of housing. Should be read in conjunction with BS 5250, 5720 and 5925. Includes a calculation method based on BREDEM.* ISBN 0 5801 6611 2.

Building 2000 - Vol. 1: Schools, Laboratories, University, Sports & Educational Centres, Vol. 2: Office buildings, Public Buildings, Hotels & Holiday Complexes. den Ouden, C., Steemers, T.C. (Eds.), Kluwer Academic Publishers Group, P.O. Box 322, 3300 AH Dordrecht, The Netherlands, 1991. *Two volumes which reflect the results of the exchange of information between design practitioners and the R+D community on passive solar, low energy buildings throughout the European Community. Covers Heating, Cooling and Daylighting applications.* ISBN 07923-1501-4 & 07923-1502-2

Building Research Station (BRS) Daylight Protractor. Explanatory Booklet. Her Majesty's Stationary Office, UK. (London), 1968. *Manual for a version of the original sky-component protractors with which the protractor method was applied for overcast-sky conditions for different types of sky and types of glazing.*

Buildings. Climate and Energy, Markus, Thomas Andrew., Morris, E.N., (et al.), Pitman, UK, 1980, £8.95 stg (Pbk), *Hardback ISBN - 0 2730 0266 X.* ISBN 0 2730 0268 6 (PBK).

California Energy Commission Passive Solar Handbook, Niles, P.W.B., Howard, K.L., 1980

Catalogue of Films and Television Programmes on Architecture, Town Planning and the environment. MacFarlane, Jane, Oxford Polytechnic, School of Planning, Headington, Oxford, OX3 OBP, UK., 1988, £9.00, *Series: Oxford Papers on Planning, Education and Research. 1200 entries covering British television from 1949- and films worldwide from1902-. Addresses of distributors and libraries. Information on availability of films.*

CIBSE Design Guide. Chartered Institute of Building Services Engineers. *This guide outlines calculation methods and presents data for a number of standard building energy calculations including steady-state heat loss, degree-day annual heat demand, summertime temperatures, air-conditioning cooling loads. Section on comfort theory with data.*

CIBSE Energy Code. CIBSE published this Energy Code for non-mechanically ventilated, non-air conditioned buildings. The calculation method takes solar gains into account.

Climate, Development Planning, Energy Requirement, Regional and local variations. The example of Ängelholm, Municipality. Swedish Council for Building Research, Stockholm, 1989. *Climate dependent energy losses and gains in private dwellings have been calculated and charted for Ägelholm Municipality based on hourly observations over a 25 year period to 1979 of 3 building types - those built to Swedish code, SBN 80, those better and those worse.* ISBN: 91 540 4994 6

Climatic Design. Energy-efficient building principles and practices. Watson, D., Labs, K., McGraw-Hill Book Company, New York, 1221 Avenue of the Americas, New York, NY 10020, US.,1983, $29.50, ISBN 0070684782.

Comfort and Energy Conservation in Buildings, Proceedings of a symposium held for Architectural and Urban Studies - July 1981. Martin Centre for Architectural and Urban Studies, Cambridge, UK, Published in Energy & Buildings. Vol.5. No.2. 1982. Dec.

Comfort: The Desirable Conditions Manual of Tropical Housing & Building. Part 1 Climatic Design, Koeningsberger et al. Longman

Concepts and Practice of Architectural Daylighting. Moore, Fuller, Anderson, Greg (Illustrator). Van Nostrand, Reinhold, New York, 1985, $39.95. *Useful account of daylighting fundamentals for architectural design. Coverage of daylight, sunlight, models and measuring instruments, shelf reflectors, baffles, wells etc. Content relates to North America. Compares cost of daylighting with that of electrical energy replaced. Imperial units - not SI.* ISBN 0 442-264-399.

Concepts in Architectural Lighting. Egan, M. David., McGraw-Hill for College of Architecture, Clemson, New York and London, 1983, $38.00 US, *Chapters on daylighting and design and on lighting models, 265 pp.* ISBN: 0-07-019054-2.

Congrès International d'Architecture Climatique. Cellule Architecture et Climat. CIACO Editeur, B. 1348 Louvain-la-Neuve, Belgique, June 1986, 600 F.B. *Proceedings of International Congress on Climatic Architecture, 2-3 July 1986. 598 pp.*

Conservation and Solar Guidelines. Balcomb, J.D., Los Alamos National Laboratory, New Mexico, 87545 USA, 1986, *Guidelines for selecting R-values, infiltration levels and for determining the size of solar collection area for passive solar buildings based on incremental cost/benefit of conservation and passive solar strategies. Source:- Passive Solar Journal 3(3) pp 221-248.*

Controls to reduce Electrical Peak Demand in Commercial Buildings, Piette, M.A., CASU / CADDET, Netherlands, 1991, US$ 25, *Outlines the need for principles and technologies involved in electrical load management. Emphasis on load management, duty cycling demand limiting, compressor and dynamic building control. Includes 15 case studies. pp 90*, ISBN 90-72647-17-3.

Daylight in Architecture, Evans, Benjamin H. AIA., Architectural Record Books., McGraw Hill Publishing Company, 1221 Avenue of the Americas, New York NY 10020 U.S., 1981, $29.95 U.S., *Presents daylighting design as an architectural matter rather than simply treating the subject technically. A very useful book covering fundamentals, evaluation of design alternatives (with models), case studies, cost effective daylight design, approx 200p.* ISBN:0-07-019768-7.

Daylighting - Design and Analysis. Robbins, Claude L., Van Nostrand Reinhold, New York & Wokingham (USA), 1986, $79.95 US. *'Covers all aspects of daylighting in commercial, industrial institutional and residential buildings from a wide range of geographic areas.' Organised to correspond to the building design process. Many case studies. 877pp, some colour.* ISBN: 0-442-27949-3.

Daylighting as a passive solar energy option - An assessment of its potential in non-domestic buildings, Crisp, V.H.C., McKenna, G.T., Littlefair, P.J., Cooper, I. *Building Research Establishment, Dept. of the environment. Garston, Watford, U.K. 1988, BRE Report No. 129 . 55p.* ISBN 0 8512 5287 7.

Daylighting Design Guidelines for Roof Glazing in Atrium Spaces, American Architectural Manufacturers Association, 2700 River Road, Des Plaines, IL 60018, U.S.A. $10.00, *Focuses on large roof areas and atria (30 - 90% of the roof area) for net annual energy and peak demand as the key measures of performance. Outlines a procedure for early design decisions for large areas of horizontal roof glazing in a prototype atrium configuration.* (DDGA-1)

Daylighting in the Mt.Airy Public Library. Adehran, M., Place, W., Anderson, B.,Building Systems Analysis Group, Lawrence Berkley Laboratory, Berkley, California, 94720 U.S. 1986, *Results of case study evaluating the visual environment of a daylit building - the Mt. Airey public library which uses south facing clerestories & more traditional elements to save energy. Published in Passive Solar Journal 3(4), 349-386*

Daylighting Performance and Design. Ander, Gregg, D., Southern California Edison, Rosemead, California, 1986, Free Publication, *Addresses the more pragmatic issues of daylighting how daylighting strategies impact a building's electronic lighting system as well as total connected load.*

Daylighting. Hopkinson, R.G., Petherbridge, P., Longmore, J., Heinemann, London, 1966. *Probably the classic account of the subject, with background information, references, and the derivations of formulae not easy to find in books mainly dealing with applications in buildings. Although emphasis is on NW European conditions it has a chapter on tropical regions. Uses Imperial not SI units.*

Design Aids. Nationale Woningraad/Eindhoven Univ. of Technology, *Rules of thumb related to a manual calculation method, based on a (supposed) model of decision-making in architectural design. Published by The National Housing Council of The Netherlands*

Design for Energy Conservation with Skylights. American Architectural Manufacturers Association, 2700 River Rd., Des Plaines, IL 60018, U.S.A. $6. *A 20-page guide for optimizing efficient design with manufactured dome skylight units. Identifies major design parameters for the energy efficient application of skylights. Energy balance techniques described.* AAMA TIR-A6-1981

Design Guide for Low Energy Housing, British Standards Institution. *This guide has sections on:- climate/passive solar design/current house design/ventilation*

Design Integration for Minimal Energy and Cost. Halldane, John F., Elsevier Applied Science Publishers Ltd., Crown House, Linton Road, Barking, Essex, 1G11 8JU, U.K.,1989, £45, *Reprinted from Applied Energy Vol 33, Nos. 1-3. Covers requirements for creating alternative energy conserving designs including power parameters, energy and life-cycle costs, analyses of power, energy, cost and control systems.* ISBN: 1 85166 292 8,

Design Primer for Hot Climates. Konys, Allan., Architectural Press, 1984, £8.95 (PBK). *Text includes drawings by Charles Swanepool and a selection of illustrations, charts and maps. Prev. ed. 1980.* ISBN 0 8513 9141 9.

Design with Climate. Olgyay, V., Princeton University Press, New York, 1973

Design with Energy - the conservation and use of energy in buildings. Littler, John G.F., Thomas, Randall. Cambridge University Press, 1984, £30.00 stg. *Part of the series: Cambridge Urban and Architectural Studies - Volume 8.* ISBN 0 5212 4562 1.

Desirable Temperature in Dwellings. Humphreys, M.A.,British Research Establishment, Watford, UK, 1976, *BRE current paper 75/76. Reprinted from 'Building Services Engineer' (Journal) vol. 44, no. 8 Nov. 1976, p.176-180.*

Directory of UK. Renewable Energy Suppliers and Services solar, wind, wave, biogas, biomass, small scale hydro geothermal, tidal, including energy recovery. Solar Energy Information unit, University College, Cardiff.,1984 (2nd ed.) *Directory including details of Trade Organisations & manufacturers of components relating to renewable energy applications under headings - Architectural Energy Saving Fitments, Accessories.*

Discomfort glare in the interior working environment. Commission Internationale de l'Eclairage (C.I.E.), 1984, 120 DM, Ref:TC - 3.4, C.I.E. Publication No: 55. ISBN 92 9034 055 X

DOE 2, User News. The simulation Research Group, Energy + Environment Division, Lawrence Berkley Laboratory, One Cyclotron Road, Berkeley, California 94720,

USA. *A quarterly newsletter for users of DOE 2: A computer program for Building Energy Simulation. Includes new algorithms, design tool listings, hints, articles and general news.*

Domestic Energy Fact File. BRE, Buildings Research Establishment (BRE), Watford, UK., 1989, £10.00, *Charts and tables in this fact file detail some of the most, important data on domestic energy along with the measures taken during 1970-1986 to use it more efficiently. It covers consumption, insulation, population, households, sources and uses of fuels, standards of heating and projection methods.*

Draft Code of Practice for Energy Conservation in Health Buildings. Ford, B., Hawkes, D., Martin Centre, Cambridge, 1983, *Report to U.K. Dept. of Health and Social Security*

Effects of surrounding buildings on wind pressure distributions & ventilation losses for single-family houses. Wiren, Bengt G., National Swedish Institute for Building Research, Gavle, Sweden. 1985, *Part 1: 2-storey detached houses. Part 2: 2-storey terraced houses. Report and research no. TN 2. Ref: M85:19.* ISBN 9 1540 9265 5

Electrical Energy Savings in Office Buildings, Holtz, M., AB Svensk Byggtjänst Litteraturjänst, S-171 88 Solna, Sweden, 1990,. *This study examines the use of energy-saving technologies to reduce energy consumption and peak demand in office buildings. The results show large potential savings for some lighting, HVAC and glazing technologies. 259 pp.* ISBN 91-540-5213-0.

Energi i Arkitekturen (Parallel title) Energy in Architecture, Moltke, Ivar, Energiteknologi, Dansk Teknologisk, Institut, Postboks 141, 2630 Tåstrup, Danmark, 1990, *An illustrated introduction to energy efficient building, design in Scandinavia. Chapters include '10 commandments', concepts, building geometry, climate zones, insulation, windows, sunspaces, shading, ventilation, passive heat storage, active & auxiliary heat sources.* ISBN 87-7756-073-06

Energy and Buildings for Temperate Climates A Mediterranean Regional Approach. Fernandes, Eduardo de Oliveria., Yannas, Simos. (Eds.), Pergamon Press, 1988, $140.00, *Proceedings of the 6th International*

PLEA 88 Conference, Porto, Portugal. Passive and low energy Architectural conference. 950p. ISBN 0 08 036617 1

Energy Catalogue '86 HMSO Books, 49 High Holborn, WC1V 6HB, London, U.K. 1986, *Contains a selection of recent publications on, energy available from H.M. Stationery Office (U.K.), in 7 sections: Statutes, Policy and Perspectives, Energy sources, Research and Alternative Technology, Conservation Statistics and References.*

Energy Conscious Design - A Primer for Architects. Goulding J. R., Lewis J. O., Steemers T. C. (Eds.), BT Batsford Ltd.,4 Fitzhardinge St., UK-London W1H 0AH, 1992, *A product of the CEC DGXII SOLINFO R+D Technology transfer programme. This is a highly graphical introduction to the concepts and fundamentals of passive solar energy-efficient building design for European architects and students. A companion to 'Energy in Architecture - The European Passive Solar Handbook'.*

Energy Conscious Design for Health Care Buildings Continuing Education Unit. Institute of Advanced Architectural Studies, York, U.K. *The National Health Service has commissioned this report which adopts an integrated approach sub-divided into categories: comfort, site factors, daylighting & window design, artificial lighting, passive solar design, infiltration ventilation, insulation, heating, cooling & air conditioning*

Energy Conscious Tradition Catalogue, Energy Research Group, School of Architecture, University College Dublin, Richview, Clonskeagh, IRL-Dublin 14, 1989, *An A4 B&W reproduction of a poster exhibition of traditional responses to energy issues in Europe prepared under the CEC's ARCHISOL programme by the Energy Research Group. The Original poster exhibition, in colour, was circulated to European schools of architecure.*

Energy conscious design for health care buildings: principles & case studies. Willoughby, John. Inst. of Advanced Arch. Studies: Continuing Education Unit, York, England, 1985, *Includes case studies*

Energy Design Guide, Design Guide to BS8207: 1985, British Standard Code of Practice for Energy Efficiency in Buildings, Anthony Williams and Partners, British Standards Institution

Publications Sales Office, Garston, Watford, WD2 7JR, U.K. Tel: + 44 (0923) 674040, 1985. *Sets out the general principles and criteria for energy design as a framework to which other codes concerned with energy in building can relate. Text compiled and designed by Anthony Williams and Partners with Graham Frecknell, and sponsored by the Energy Efficiency Office/Dept. of Energy.*

Energy Efficient Design, Approach to the design of non-air-conditioned office blds., being promoted by the U.K. Electricity Council., Stevans, Ted., RIBA Journal, 1987, December, *The emphasis of this article from RIBA Journal is on the use of electrical space heating; article also contains description of two low-energy business/industrial units which are to be constructed at the Milton Keynes Energy Park*

Energy Efficient Housing, A demonstration of the integrated approach to energy efficient housing at Lawrie Pk. Rd. London SE 26. South London Consortium Group. Dept. of Energy in assoc. with SLC Energy Group, 1985, *This report presents an integrated approach to design based on the application of 4 rules: Insulation, Draught reduction, Solar Gain, Comfort*

Energy Efficient Lighting in Buildings. BRE for the Commission of The European Communities DGXVII, Garston, Watford WD2 7JR, UK, 1991. *Produced by BRECSU, an organisation for the promotion of Energy Technologies within the CEC's THERMIE R+D Programme. This 'Maxibrochure' in the Rational use of Energy series provides an overview of the Energy implications of lighting in buildings in a European context. Thirteen brief case studies are included.*

Energy Efficient Renovation of Houses - a design guide. NBA Tectonics, Department of Environment, Her Majesty's Stationery Office, London, 1986, £4.95 stg. *In this Design Guide there is no reference to passive solar design but the strategy for energy efficiency promotes a co-ordinated approach to insulation, ventilation & heating 4 case studies present suitable approaches to the improvement of typical British dwellings.* ISBN 0 1175 1837 9

Energy in Architecture - The European Passive Solar Handbook, Goulding, J., Lewis, J.O., Steemers, T.C. (Eds.), BT Batsford Limited, 4 Fitzhardinge Street, UK-London W1H 0AH, 1991, £ 45. *A handbook for practitioners in Building Design, which draws on European R+D. Chapters on Climate, Urban Design, Passive Solar Heating, Passive Cooling, Daylighting, Comfort, Control Systems, and Atrium Design. Includes design tools and an extensive bibliography. A companion to 'Energy Conscious Design - A Primer for Architects'.*

Energy Options for Housing Design, RIBA Publications Ltd., 66 Portland Place, London WC1, 1982, *Study carried out under the auspices of the RIBA Energy Group*

Energy Policies and the Greenhouse Effect - Volume 1. Policy Appraisal, Grubb, Michael. Academic Publishing Group, The Marketing Dept., The Academic Publishing Group, Gower House, Croft Road, Aldershot, Hants., GU11 3HR, England. 1990, £29.50, *Contents include Energy trends and country studies, taxes, supply mix, futures in the greenhouse.* ISBN 1 85521 1750

Energy Policies and the Greenhouse Effect - Volume 2. Country studies and technical options, Grubb, Michael. (et al.), The Academic Publishing Group, The Marketing Dept., The Academic Publishing Group, Gower House, Croft Road, Aldershot, Hants., GU11 3HR, England. 1991, c. £35.00, *Contents: Energy resources and systems, efficient energy use supply and conversion technologies, greenhouse effect in UK, US and Japan. Soviet CO2 emissions, China/dominanace of coal.* ISBN 1 85521 198 X.

Energy Resource File, A Guide to Published Material on Energy Efficiency in Health Building Design, Harris, S., Sutcuffe, S. University of York, 1983, *Produced by the Continuing Education Unit for NHS Architectural Staff. List of Advanced Architectural Studies*

Energy Trends and Policy Impacts Statistics from seven countries within the CECP project, Klingberg, Tage. National Swedish Institute for Building Research, 1984, *Ref: M84:1*

Energy Use and Energy Efficiency in United Kingdom Commercial and Public Buildings to the Year 2000. Her Majesty's Stationary Office,London. UK., 1988, *This report is an analysis of the pattern of energy use, and the scope for improvements in energy efficiency, in the United Kingdom's service sector. It provides information on possible future patterns of energy use and the potential for energy efficiency measures in both commercial and public buildings.* ISBN 0-11-412903-7

Energy World. Fender, J. (Ed.), Institute of Energy (UK), 18 Devonshire Street, London W1N 2AU, UK, Tel. 071-580 0008, Bi-monthly, UK£ 46 for 10 issues, *Journal published by the UK Institute of Energy covering a range of energy issues including energy use in buildings.* ISSN 0307-7942,

Environmental Design Manual - Summer Conditions in Naturally Ventilated Offices, Petherbridge, P., Millbank, N.O., Harrington-Lynn, J. Building Research Establishment, U.K. Department of the Environment, Watford. 1988, £15. *This manual presents a method of addressing the effect of the window type and size, the kind of construction and the rate of ventilation on summertime comfort conditions and daylighting in offices in the British Isles. Intended for early design stage. Includes remedies for summer overheating.* ISBN 0 8512 5209 5/ BR 86.15.

Environmental Science Handbook for Architects and Builders, Szokolay, Steven. Vajk., The Construction Press, Lancaster, 1980, £25.00, ISBN 0 8609 5813 2.

Environmental Systems Performance, Clarke, J.A., McLean, D., ABACUS Manual No. 31, Strathclyde University, Glasgow,1986.

Estimating Daylight in Buildings. Building Research Establishment, Garston. UK., 1986, *(See BRS Digests 309 & 310). Compact, technical explanation of the use of the BRS protractors and of the Waldram Diagram for Daylight factor calculation. Includes explanation of Externally-reflected component, and of Internally-reflected component*

Estimation of energy balances for houses. National Swedish Institute for Building Research, 1984, *Ref:M84 : 18*

Euroforum New Energies 1988, International Congress - Commission of the European Communities. DG. for Science Research and Development. Palz, W. (Ed.), H.S. Stephens & Associates, Agriculture House, 55 Goldington Road, Bedford, MK40 3LS, U.K., 1988, £95, *Proc. Euroforum New Energies Congress 1988, Saarbrücken, FRG, 24/28 October 1988. 1264p with 316 papers & reports on solar architecture, photovoltic, mini-hydraulic, agriculture wind, & environmental impact. In 3 vols: 1) Invited papers, 2) Commercialization of renewable energies, 3) Workshop papers.* ISBN: 0951027166. (Eur. 11884).

Daylighting in Architecture - A European Reference Book, Baker, N. V., Fanchiotti, A., Steemers, K. A. (Eds.), *A product of the Commission of the European Communities R+D programme on Daylighting, this major work draws on European research and expertise providing a comprehensive overview of daylighting techniques for European building designers. To be published in 1993.*

European Directory of Renewable Energy Suppliers & Services, biogas, CHP, heat pumps, PV solar, water, wind. Cross, B.M. (Ed.), Energy Equipment Testing Service, Univ. of Wales, University of Wales, Cardiff, P.O. Box 917, Cardiff CF2 1XH,U.K. (Tel: + 44 222 874797. Fax: + 44 222 371921, Telex: 498635), 1988, Free Publication, *Circulation 10,000 worldwide. Charge for Directory entries is £50.* ISBN 0 9073 8355 6.

European Passive Solar Components Catalogue (DRAFT). ECD Partnership, London, Energy Research Group, School of Architecture, University College Dublin, Richview, Clonskeagh, IRL-Dublin 14, 1990, *Categories include: Glazing, Sunspaces, Blinds and Shutters, Ventilation, Controls, Solar Wall Panels, Passive Solar Water Heating, Thermal Storage.* EUR 13055 EN,

European Passive Solar Handbook (Preliminary Edition), Basic Principles and concepts for passive solar architecture, Achard, Patrick., Giquel, Renaud. (Eds.), Commission of the European Communities DGXII, CEC Directorate General for Science Research & Development, Rue de la Loi, B-1049 Brussels, 1986, *Handbook for passive solar building in Europe. Chapters on*

Fundamentals, Thermal analysis, Passive Solar building, P.S. design strategies and tools. Case studies. Includes worked examples and climatic data for a variety of European locations. EUR 10 683.

European Solar Radiation Atlas, Vol. 1, Global Radiation on Horizontal Surfaces, W. Palz, (Ed.) C.E.C., Kasten, F. et al. (compilers), Verlag Tuv Rheinland for CEC (ref EUR 9344), Köln, 1984, *Second Improved and Extended Edition, Compiled by Fasten, K. and Golchert, H.J., Deutscher Wetterdienst Meteorologisches Observatorium Hamburg, and Dogniaux, R and Lemmoine, M., Institut Royal Meteorologique de Belgique*, ISBN:3-88585-194-6

European Solar Radiation Atlas, Vol. II, Global and Diffuse Radiation on Vertical and Inclined Surfaces, W. Palz, (Ed.) C.E.C., Verlag Tuv Rheinland for CEC (ref EUR 9345), 1984, *Compiled by Page, J.K. and Flynn, R.J. - Univ. of Sheffield, UK, Dogniaux, R and Preuveneers, G - Institut Royal Meteorologione de Belgique*. ISBN:3-88585-196-4

European Wind Atlas, Palz, W. (Ed.), C.E.C., Orders to:- Dept. of Meteorology & Wind Energy, Risø National Laboratory, P.O. Box 49, DK-4000, Roskilde, Denmark, 1988, DKK 875.00. *An outcome of European Communities' efforts to promote the market for electricity production from wind and to develop associated technologies and systems. The Atlas provides an overview of the wind resource for EC countries and completes previously published national information.*

Evaluating the Performance of Passive Solar Heated Buildings, Balcomb, J.D., Los Almos National Laboratory, 1983, April, *Report No: LA-UR-83-0003*

Experimental Performance of the Barra Costantini Passive Solar System, Barra, O., et al., Colloque Solaire International, 1980, 11 - 12 December

Fire Precautions in the Design Construction and Use of Buildings. Draft BS. 5588., Part 7 - Code of practice for atrium buildings. British Standards, Milton Keynes. UK., £15.00, *This standard aims to reduce fire risks in atrium buildings. Sets out principles for planning for fire and smoke spread,* mixed-use of developments, means of escape. Text deals with systems testing and sets out a form for a fire safety manual

Green Design, Fox, A., Murrell, R., Architecture Design + Tech. Press, London, 1988. *A guide to the environmental impact of building materials.*

Greenhouses and Conservatories, Aspects of Thermal Behaviour and Energy Efficiency, Norton, B. (Ed.), International Solar Energy Society, UK Section (UK - ISES), 1985, ISBN 0 9049 6340 3.

Guide on Interior Lighting (2nd Edition), Commission Internationale de l'Elairage (C.I.E.), 1986, 158 DM, *Ref: TC-4.1. C.I.E. Publication No: 29.2.* ISBN 3 900 734 01 1.

Guidelines for the Economic Analysis of Renewable Energy Technology Applications, O.E.C.D., Chateau Montebello, Quebec, Canada, 1991, FF 180, *Provides guidelines for calculating the costs of eight renewable energy technologies: active and passive systems, solar thermal and photovoltaic electricity generation, bio-energy, small-scale hydro power, geothermal and wind energy.* ISBN 92-64-13481-6.

Handbook of Fundamentals, ASHRAE, American Society of Heating, Refrig. & Air Conditioning, Engineers, Inc., 345 East 47th Street, 10017 New York, U.S., *Comprehensive source of reference data on air-conditioning, heating, ventilation and refrigeration. Both SI and Imperial unit versions are available*

Handbook of Meteorological Data, Her Majesty's Stationery Office, 1986, *Commissioned by The Energy Technology Support Unit on behalf of the UK Depart. of Energy*

Harnessing Solar Energy. Anderson, B., (Ed.), MIT Press, 1990, UK£ 35.95. *Based mainly on North American experience, this book provides information on passive and active solar systems, including Thermal storage. Well illustrated.*

Healthy Construction, The Rosehaugh Guide to The Design, Construction, Use and Management of Buildings. Curwell, S., March, C., Venables, R., RIBA Publications, 1990, UK£ 120. *Assembled information from experts* which promotes environmentally sensitive design and specification of building materials to provide healthier living and working conditions.

Heat Distribution by Natural Convection, Balcomb, J.D., Los Almos National Laboratory, Report No.LA-UR-83-1872, 1983

Heat Storage and Distribution Inside Passive Solar Buildings. Balcomb, J.D., Los Almos National Laboratory. Report No: LA-9684-MS., 1983, April. *Evaluation of performance of Passive Solar Heated Buildings*

Horizontal Study on Passive Cooling, Santamouris, M. (Ed.),Fleury, B., Lopez d'Asiain, J., Maldonado, E., Tombazis, A., Yannas, S., Commission of the European Communities DG XII, Wetstraat 200, B-1049 Brussels, Belgium, 1990. *Produced within the CEC Building 2000 R+D Programme. This report outlines current development in Passive and Hybrid Cooling for building in Europe. It investigates the potential for developments in cooling and identifies gaps in current R+D. Contains a very useful overview of cooling principles, an extensive bibliography and case studies.*

Housing Climate and Comfort, Evans, M., Architectural Press, London, 1980, £20.00 stg, ISBN 0 8513 9102 8

Improved Thermal Insulation,l Problems and Perspectives. Brandeth, D.A. (Ed.), TECHNOMIC Publishing AG, Missionsstrasse 44, CH-4055 Basel, Switzerland, 1991, Sfr. 108. *Technical and economic analysis of current and future developments in thermal insulation. It examines the problems with materials in use today and looks toward alternatives that are both cost effective and environmentally safe. 395 pp.*

Indoor Climate, McIntyre, D.A., Applied Science Publishers, 1980, £26.00 stg. ISBN 0 8533 4868 5

Inside Out, Design Procedures for Passive Environmental Technologies, Brown, G.Z., Reynolds, John, S., Ubbelohde, M, Susan., John Wiley & Sons, New York, U.S.A., 1982, $18.95. *Introduces passive solar concepts & addresses architectural design implications. Attempts to bridge the*

gap between analysis and design synthesis. Based on US conditions. Approx. 370 pp. Hardbound. ISBN: 0-471-89874-0

Insulation of External Walls in Housing, A Report prepared by An Foras Forbartha, Dublin. McCarthy, J. 1986. *This report has highlighted problems related to the use of insulation in the construction of external walls in two storey houses. The study is being expanded to provide appropriate working details for use in the Irish context.*

International Competition for Solutions on New Technologies for Social Housing Caceres-Brighton, Secretariat General UIA, Spanish National Section,Consejo Superior de los Colegios de Arguitectos de Esña, Spanish Section of UIA, Paseo de la Castellana 12,28006 Madrid, 1988, *Deals with international competitions for solutions on new technologies for social housing organised by Consejo Superior de los Colegios de Arquitectos de España, Min. de Obras Publicas (MOPU), United Nations Organisation and International Year of Shelter for the Homeless*

International Journal of Solar Energy, Palz, W., Editor in chief, Harwood Academic Publishers, Distributed by STBS Ltd., 1 Bedford Street, London WC2E 9PP, UK,$132 per volume (6 issues), *Experimental, theoretical & applied results in the science & engineering of solar energy (photovoltaic, power, radiation data & systems employing optical concentration), covering information from Europe, North America and Japan. Mainly papers in English but some in French & German.* ISSN:0142 5919

Land-use information in Sweden: applications of new technology in urban & regional planning and in the management of natural resources, Bengt Rystedt (Ed.), Swedish Council for Building Research, 1987, *Ref:D 2*

Legislation and Regulations on Solar Energy in the EC Member State, Didier, J.M., and Associates, Commission of the European Communities, 1987. *Final report of CEC contract No. 85-B-7032-11-001*

Light as a true visual quantity: principles of measurement. Commission Internationale de l'Eclairage (C.I.E.), 1978,1 50 DM, *Ref: TC 1.4, C.I.E. Publication No: 41*

Low Energy High Tech, Pawley, Martin. 1987, *Published in Architects'Journal Vol. 186, No 35, 2 Sept '87, pp. 24-29. Examines Richard Horden's competition winning design for a major office block in Stag Place, Victoria*

Low-E Glazing Design Guide, Johnson, T.E., Butterworth, 1990. *Design methods, hints, examples for specifying this new family of products.*

Main Findings of The Commission Review of Member States Energy Policies, Commission of the European Community, 1988, CEC COM(88) 174 Final. *The 1995 Community Energy Objectives*

Man Climate and Architecture, Givoni, B. Applied Science Publishers, London, 2nd Ed. 1982 (1st Ed.1969), £15.00 stg. ISBN 0 8533 4108 7

Manual of Tropical Housing and Building Design, Part 1. Climatic design. Koenigsberger, O., Szokolav, S., Ingersoll, T., Mayhew, A., Longman, London, 1974, £4.95 / £2.95 (Pbk). *Shadow angle protractor in pocket of book.* ISBN 0582445450/pbk-0582445469

Methods for measurement of airflow rates in ventilation systems. National Swedish Institute for Building Research, 1983, Ref: M83:11

Microclimate and The Environmental Performance of Buildings Building Research Establishment - UK, Building Research Station, Garston, Watford WD2 7JR, UK, 1988, *Proceedings of Conference*

Model for Dissemination of Information within the Area of Energy Consumption in Buildings. The Danish Ministry of Energy. *Systematic dissemnation of information on the results of research and development projects concerning the reduction of energy consumption for heating and ventilation of buildings. A practical tool for preparing information plans and explaining research results and their importance.* ISBN 87-7511-664-2

Models of infiltration and natural ventilation. Lyrberg, Mats D., National Swedish Institute for Building Research, 1983, *Ref: M83 : 23*

Monitoring Solar Heating Systems a practical handbook, Ferraro, R., Godoy, R., Turrent, D. (Eds.), Pergamon Press, London, 1983, *Edited by Energy Conscious Design, London for the CE Commission of the European Communities.* ISBN:0-08-029992-X

Movable Insulation a guide to reducing heating and cooling losses through windows in your home. Langdon, William K., Rodale Press, Emmaus, PA. USA. 1980, $12.95 / $8.95 (Pbk). ISBN 0878572988/pbk-0878573100

Natural Energy and Vernacular Architecture principles and examples with reference to hot arid climates. Fathy, Hassan., Abd-el-Rahman, Ahmed Sultan., Shearer, Walter. (Eds.), University of Chicago Press for the United Nations Univ., Chicago / London, 1986, $25.00 U.S. *Covers tools and concepts. Demonstrates the use of traditional materials for building in hot, dry climates. 86 photos and drawings illustrate 'Principles and Examples'. Includes a useful glossary.* ISBN 0226239179/Pbk-0226239187

Natural Lighting in Architecture Windows: factors in the quality of life, G.E.P.V.P. Groupement Europeen des Producteurs de Verre Plat, 89 Avenue Louise, B-1050 Brussels, 1987. *A free 20-page brochure produced by The European Flat Glass Manufacturers Association outlining the advantages of daylighting, availability of daylight and methods for predicting illuminance, giving practical advice on visual & thermal comfort, 19 photos, 16 diagrams.*

North Sun Proceedings, Broman, Lars., Rönnelid, Mats. (Eds.), Swedish Council for Building Research, Stockholm, 1988, *Proceedings of 3rd North Sun Conference in Borlänge, Sweden, August 29 - 31, 1988. Includes all 89 lectures & reports presented covering passive & active systems meteorology, materials, large scale systems, with an emphasis on solar energy at high latitudes.* ISBN 91-540-4973-3

Passive and Low Energy Building Design for Tropical Island Climates. Baker, N.V., Commonwealth Science Council, Commonwealth Secretarial Publications, Marlborough House, Pall Mall, London SW1Y 5HX, 1987, £8.00 U.K.

Passive Cooling of Buildings - An Overview. Givoni, B. 1980, *Passive cooling - buildings, Paper prepared for US Dept. of Energy, International Expert Group on Passive Cooling of Buildings*

Passive Solar Architecture in Europe 2 the results of the 2nd European Passive Solar competition, 1982. Lebens, Ralph M. (Ed.), C.E.C. The Architectural Press, London, 1982, ISBN 0 8513 9957 6

Passive Solar Architecture in Europe, Lebens, R.M. (Ed.), Architectural Press, 1981, *Results of the First European Passive Solar Competition held in 1980.* EUR 7291

Passive Solar Buildings in Use, 30 European Case Studies, 1991. *Based on The Project MONITOR series of brochures which reported on a range of buildings in EC Member States whose performance has been monitored. This representative selection of case studies will also be published in other EC languages.*

Passive Solar Construction Handbook, Steven Winter Associates Inc., 1981, *Prepared for the Southern Solar Energy Center, Atlanta, Georgia and U.S. Dept. of Energy, Washington DC. Large vol. emphasising methods of adapting standard building practices to passive solar applications*

Passive Solar Design Handbook (Vols. I, II, and III), Balcomb, J.D. National Technical Information Service, 5285 Port Royal Road, Springfield, VA, 22161, U.S.A.,1980 and 1982

Passive Solar Energy as a Fuel A study of the current & future use of passive solar energy,in buildings in the European Community. Burton S., et al., ECD Partnership, 11-15 Emerald St, UK-London WC1N 3QL for CEC DG XII, 1990, *Study of current use of Solar energy in EC countries now and estimates for the years 2000 & 2010. Estimates of consequent reduction in pollution, particularly CO_2.* EUR 13094

Passive Solar Energy in Buildings Report No. 17, O'Sullivan, P. (Ed.), Elsevier Science Publishers Ltd., Crown House, Linton Road, Barking, Essex 1G11 8JU, U.K., 1988, £34.00, *Produced by the Welsh School of Architecture, UWIST, Cardiff for the Watt Committee on energy, this report is aimed at architects, builders and*

town planners involved in the design of commercial, institutional and domestic passive solar buildings. 70 p.p. 27 illustrations. ISBN 1 85166 280 4

Passive Solar Heating Analysis A Design Manual. Balcomb, J. Douglas., Jones, R.W., McFarland, R.D., Wray, W.O., ASHRAE/Los Alamos National Laboratory, 1791 Tullie Circle, NE, Atlanta, Georgia, GA 30329, 1984, *Large volume dealing with North American conditions. Covers fundamentals, guidelines, methods (e.g. LCR and SLR) examples and systems. Data for North American locations.* 0-910110 38 7

Passive Solar Heating Design, Lebens, R. Applied Science Publishers, for the Commission of The European Community, London, UK., 1982, $16.00, *Results of CEC Second European Passive Solar Competition 1982.* ISBN 0 8533 4870 7

Passive Solar Heating, Williams, J.R. Ann Arbour Science (Butterworth Group), Michigan, 1983, ISBN 0 2504 0601 2

Passive Solar Housing in Cool Climates, Barret, J., 1982, *International CIB W67 Commission Symposium, Lisbon, Portugal*

Passive Solar Housing in the UK A report to The Energy Technology Support Unit, Harwell, (ETSU). Turrent, David., Doggert, John., Ferraro, Richard., Energy Conscious Design, 11-15 Emerald Street, London WC1N 3QL, 1981, ISBN 0 9507 4090 x

Passive Solar Schools in the UK: Features employed currently and their operation. Norton, B, Hobday, R.A., Int. Journal of Ambient Energy, Vol. II No. 2 (April 1990). *Paper reviewing passive solar schools in the UK (of which there are approx. 50), with an emphasis on the main functional requirements of p.s. systems to reduce energy loads and to improve comfort and amenity.*

Passive Solar Techniques for Energy Conservation in Buildings. *Proceedings of International CIB W67 Commission Symposium held in Lisbon, Portugal.*

Passive solar heating in a Scandinavian climate: Three housing projects with low energy consumption. Eek, Hans, Swedish Council for Building Research, 1987, *Ref: D3 : 1987*

Passive Space Heating Systems Reduced Reporting Format for Solar Heating Systems for Projects with Low Level monitoring. Performance Monitoring Group, 1981, *Produced by the Performance Monitoring Group for the Commission of the European Community.*

Passive Space Heating Systems, Reporting format for Solar Heating Systems, CEC, 1981, *Produced by the Performance Monitoring Group for the Commission of the European Communities.*

Principles of Passive Solar Building Design with microcomputer programs. Carter, Cyril., De Villiers, Johan., Pergamon Press, Oxford / Frankfurt / New York, 1987, *Sections on site & climate, passive solar building types, insulation & sealing, heating & cooling windows & reflectors. Thermal mass components. Overview of calculation methods. Examples of calculation methods & simulation procedures.Microcomputer program listings specimen building plans, climate data & charts.* ISBN: 0-08-033637-X

Proceedings of a Workshop on Design Support for Architects, den Ouden, C (Ed.), Commission of the European Communities, 1990. *Workshop held in Edinburgh, UK, June 1990.* EUR 13130 EN

Property Services Agency Six publications on energy saving through landscaping, UK Property Services Agency, PSA Library Sales Office, Room COO5, Whitgift Centre, Croydon, CR9 3LY, U.K., 1988, £75. *Covers microclimate, shelter, tower blocks, winds, macroclimate, thermal performance of rain wetted walls, shelter planting, existing housing, studies of the urban edge and existing housing.*

Renewable Energy in the Walloon Region. Inter-Environment Wallonie, 1988, *This 48 page brochure adds to information published in last year's edition with more emphasis on passive solar architecture.*

Renewable Energy Resources. Twidell, John., Weir, Tony.,H.S. Stephens & Associates, Agriculture House, 55 Goldington Road, Bedford, MK40 3LS, U.K. Also available from The Promotions Dept. E & F Spon, 11 New Fetter Lane, London EC4P 4EE.,1988, £39 / £16.95

(Pbk), *Gives quantitative answers to questions of resource, availability and limitations. The resources covered are solar radiation, hydro-power, wind, wave, tidal, biomass, photosynthesis, geothermal, energy storage and distribution. 460 pp, 210 figures, 23 tables, 200 references.* ISBN: 0 419 12000 9 / 12010 6

Review of Products and Materials. O'Farrell, F., 1981, *Study presented at the Conference: Design of Buildings for Solar Energy, 1981 organised by SESI/NBST (Solar Energy Society of Ireland)/(National Board of Science & Technology), based on the premise that passive solar buildings will use normally available building materials.*

Scale Models and Artificial Skies in Daylighting Studies BEPAC Technical Note 90/3, Littlefair, P.J., Lindsay, C.R.T., UK Building Research Establishment, BEPAC Administration, Environmental Systems Division, BRE, Garston, UK-Watford WD2 7JR, Nov. 1990, £12. *Outlines the principles of daylight measurement for building models under different types of artificial skies. Includes a checklist and a summary of artificial skies in UK institutions. 30 pp.* ISBN 0-187-212-604-9.

Sensible Building Materials, Construction Materials and The Environment Report, No. 2039. Lorch, R., The Economist Intelligence Unit, 1990, UK£ 150. *Sensible use of building materials can help to protect the environment. Report includes notes on EC legislation, case studies on timber paints and the use of hardwoods and CFCs.*

Sick Building Syndrome, an intelligence report. Vince, Dr Ivan., I.B.C. Technical Services Ltd., I.B.C. House, Canada Road, Byfleet Road, Surrey, KT14 7JR, U.K., £95, *Comprehensive report on the causes and prevalence of SBS and its long term implications for the design, management and maintenance of office buildings. Covers symptoms effects and solutions.* ISBN : 1 8527 1026 8

Simulating Daylight with Architectural Models. Marc Schiller, (Ed.), DNNA (Daylighting Network of North America), Marc Schiller, Associate Professor, School of Architecture, University of Southern California, Los Angeles, CA 90089-0291, U.S.A., 1988, $30, *Fundamentals and principles of scale model techniques for daylighting studies, covering aspects of construction and, testing of scale models and evaluation of data obtained.*

Six Green Solutions, Vale, B., Vale, R., RIBA Publications, London, 1991, UK£ 14.95, *Contains six case studies covering the use of transparent insulation materials, superinsulation, airtight construction and controlled ventilation.*

Skylight Handbook Design Guidelines. AAMA (Ed.), American Architectural Manufacturers Association, 2700 River Rd., Des Plaines, IL 60018, U.S.A., 1987, $50.00, *Based on research conducted by Lawrence Berkeley Laboratories, under AAMA sponsorship, this 120-page manual offers the latest data on maximising skylight energy and daylighting benefits in commercial buildings. Includes examples, worksheets, concepts, design & energy issues.* Ref: SHDG-1

Social habits and energy consumer behaviour in single-family homes, Palmborg, Christer., Swedish Council for Building Research, 1986, Ref: D24

Solar Architecture in Europe, ECD Partnership, UK-London, Prism Press, 2 South Street, Bridport, Dorset DT6 3NQ, UK, 1991, £14.95, *A series of case studies of buidings whose performance has been monitored under the CEC DGXII Project MONITOR. Covers a range of building types wihtin EC countries.*

Solar Architecture, The direct gain approach. Johnson, Timothy E., McGraw-Hill Inc., New York & London, 1981, $21.50, *Reviewed in Passive Solar Journal, 3(2) 207-210,1986 - ' It is a blend of architectural implications and technical facts. The designer of direct gain systems will appreciate having this book on the reference shelf'.* 0-07-032598-7

Solar Collectors - Test Methods and Design Guidelines. Gillett, W.B., Moon, J.E., D. Reidel Publishing Company, Dordrecht & Lancaster, 1985, $120.00, *Provides a comprehensive review of the techniques for collector testing based on 8 years of work under EC Solar Energy R & D Sub Programme, Series A, Solar Energy Applications to Dwellings. Includes guidelines for collector design. Ref: EUR 9778.* ISBN : 9 0277 2052 5

Solar Design Components, Systems, Economics. Kreider, Jan F., Hoogendoorn, Charles J., Kreith, Frank., Hemisphere Publishing Corporation, New York, London.,

1989, £56.00 stg., *Includes contributions by C. Robbins, C den Ouden, and J. Owen Lewis.* ISBN : 0 89116406 5

Solar Energy Applications in Houses, Performance and Economics in Europe. Jaeger, F., Pergamon Press,Oxford., 1981, £14.50 stg, *Published for the Commission of European Communities.* ISBN 0 0802 7573 7

Solar Energy for Ireland, Preliminary analysis of the basic data with tentative proposals for a programme of research and development. Lawlor, E., Government Publications (IRL), Dublin Stationary Office., 1975, *Report to the National Social Council*

Solar Energy Thermal Technology. Norton, B., Springer-Verlag, 1991. *Graduate text covering active and passive solar applications and analysis thereof.* ISBN 3-540-19583-1

Solar Energy Today. Jesch, L.F., UK Section of ISES, London, 1981, £10.00 stg, ISBN : 0 9049 6327 6

Solar Energy, Fundamentals in Building Design. Anderson, Bruce. McGraw-Hill, New York & London, 1977, £16.15 stg, ISBN: 0 0700 1751 4

Solar Engineering of Thermal Processes. Duffine, J.A., Beckman, W.A., Wilay, 1980, $13.90, ISBN:0 4710 5066 0

Solar Heating Performance and Cost Improvements by Design. ECD London, Ferraro, R., Godoy, R.L. (Eds.), CEC,1983, *Performance monitoring of solar heating systems in dwellings. A report prepared by Energy Conscious Design, London for the Commission of the European Communitie.*, Ref: EUR 8948 (CEC).

Solar Heating Systems for the U.K., the design installation and economic aspects. Wozniak, S.J., Her Majesty's Stationery Office U.K., London, UK., 1979, £6.00 stg, *Part of the series of UK 'Building Research Establishment Reports'.* ISBN:0 1167 0762 3

Solar Home Planning, a bibliography and a guide. Atkinson, Steven D., Bailey Bros & Swinfen & Metuchen & Scarecrow Folkstone & New York & London, 1988, £22.15, ISBN: 0 8108 2098 6

Solar Houses in Europe - How they have worked. Houghton, D. & Evans W., Turrent, D & Whittaker, C., Palz, W & Steemers T.C. (Eds.), Pergamon Press, 1981 for the CEC DG XII., Oxford, UK., 1981, $17.00, *A review of the performance of a range of solar houses in Europe*. ISBN: 0 0802 6743 2, EUR Ref: B 8947

Solar Passive Building Science & Design, Sodha, M.S., Kumar, A., Bansal, N.K., Malik, M.A.S., Bansal, P.K.,Pergamon Press, Oxford, New York, Frankfurt, 1986, *International series on building environmental engineering., Vol.2.Chapters on:- thermal comfort, climate, radiation, orientation, shading, building clusters, solar exposure passive concepts & components, heat transmission, mathematical models, performance rules of thumb.*

Solar Space Heating, Godoy, R., Turrent, D., Ferraro, R. (Eds.), Commission of the European Communities, 1986. *An Analysis of design and performance data from 33 systems.* EUR 8005

Solar Thermal Energy for Europe, Turrent, D., Baker, N., Steemers, T.C., Palz, W.,D. Reidel Publishing Company, 1983, *An assessment study carried out for the Commission of the European Communities* , DG XII, ISBN 90-277-1592.0

Solar Thermal Technical Information, Superintendent of Documents, US. Governemnt Printing Office, Washington, D.C. 20402, U.S., 1985 - May, $8.75 US. *Produced by Solar Technical Information Program, Solar Energy Research Institute, US Dept. of Energy. Guide to Solar Thermal Energy development with overviews of each topic. Listings of major technical information sources.* Ref: SERI/SP-271-2511

Standardization of luminance distribution on clear skies. Commission Internationale de l'Eclairage (C.I.E.), 1973, 68 DM, *Ref: TC-4.2, Publication No. 22*

Study for a Eurocode on The Rational use of Energy in Building, Uyttenbroeck, J. et al, Commission of the European Communities, 1987. Report 2973/111/86

Sun Power, 2nd Edition, McVeigh, J.C., Pergamon

Sun Wind and light,Architectural Design Strategies, Brown, G.Z., Cartwright, V., John Wiley and Sons, *Architectural design strategies*

Sunlight and Daylight. UK Dept. of Environment, Her Majesty's Stationary Office, United Kingdom, 1971, *Guidance for UK local authorities on protection of access to daylight and sunlight. Used in conjunction with transparent overlays, also available.*

Sunlighting as Formgiver for Architecture, Lam, William M.C.,Van Nostrand Reinhold, New York U.S., 1986, $74.95, *Comprehensive textbook; large number of case studies for different building types. Introduces theoretical & practical aspects of sunlighting as they apply to building design. Examines ways of reducing energy use & cost. 460 pp. approx.*. ISBN: 0-442-25941-7

Sunspace Primer - A Guide for Passive Solar Heating, Jones, R.W., McFarland, R.D.,Van Nostrand Reinhold, 1984, $32.50, *How to design and estimate the performance of the ,sunspace. Includes basic passive solar design information. 8 chapters, 301pp.* ISBN:0-442-24575-0

Survey of European Passive Solar Buildings. Robert, Jean-Francois., Camous, Roger., Schneider, Franz., Technical Information Branch, Solar Energy Research Inst., 1617 Cole Blvd., Golden, Colorado, 80401, U.S.A., 1982, *Prepared for the U.S. Department of Energy & also available from Superintendent of Documents, U.S. Government Printing Office, Washington D.C., 20402. Covers buildings throughout Europe.* Ref: SERI/SP-281-1692

Tables of Temperature, Relative Humidity and Precipitation for the World, H.M.S.O. (Government Publications), 49 High Holborn, WC1V 6HB, London, U.K., 1975, £3.75, *Part 1 - North America & Greenland, Part 3 - Europe & The Atlantic Ocean, North of 35° N, Part 4 - Africa & The Atlantic Ocean south of 35° N and the Indian Ocean, Part 5 - Asia.*

The Climate near the Ground, Geiger, R.,Harvard University Press, 1980

The Cold Bridge Burberry, Peter., Architects' Journal, 1988, *A useful discussion of the problem of the cold bridge. Both the materials and the geometry of buildings can create cold bridges. Peter Burberry sets out the risks, with methods for quantifying them. AJ 27th Jan 1988*

The Design of Energy Responsive Commercial Buildings, Ternoey, Steven., Bickle, Larry., Robbins, Claude., Busch, Robert., McCord, Kitt., John Wiley & Sons, 1985, $44.95, *Produced by the Solar Research Institute, operated for the U.S. Department of Energy by Midwest Research Institute, Golden, Colorado, U.S.*

The development of a Passive Solar air-heating solar-energy collector for the recladding of buildings. Norton, B., Los, S.N.G., Hobday, R.A., Deal, C.R., Adam Hilger, London, Bristol and New York, 1989. *Paper presented at Applied Energy Research Conference UK-Swansea (Institute of Energy), discussing Thermosiphonic and panels as cladding units for use in the refurbishment of schools.*

The economic margin for alternatives in new energy technology. Enno Abel et al., Swedish Council for Building Research, 1987, *Ref: D4*

The Environmental Impacts of Renewable Energy, O.E.C.D., Her Majesty's Stationery Office (U.K.), P.O. Box 276, London SW8 5DT, 1989, *This report considers possible environmental impacts associated with technological development of renewable energy systems: material use, land use, pollution, noise, visual impact, eco-systems, public and occupational health and safety. Each technology is considered in turn.*

The Expanded Appropriate Technology Microfiche Library, A.T. Microfiche Library, Volunteers in Asia, P.O. Box 4543, Stanford, California 94305, U.S., 1988, $695, *Includes every page from 1000 books and documents reviewed in The New Appropriate Technology Sourcebook - total of 138,650 pages.*

The Experience of Energy Conservation Programs with New Commercial Buildings. Lawrence Berkeley Laboratory, Berkeley, California. USA. *Review and assessment of energy conservation programs with new commercial buildings in U.S. and other countries. Focus on non-mandatory strategies which complement, reinforce or substitute for energy efficiency needs in building codes. IE large-scale demos.*

The Fladie groundwater heatpump plant. Anderson, Oiof., Swedish Council for Building Research, £0.57.

The Greenhouse Consumer Guide, Elkington, John., Hailes, Julia., Victor Gollancz. London, 1981/1989, £4.99, *A guide to the environmental impact of consumers in affluent countries which aims to improve awareness and influence manufacturers. Provides information on 'safer' products and packaging with checklists.*

The Lighting of Buildings. Hopkinson, R.G., Kay, J.D.,Faber & Faber, London,1972.

The Natural House Book. Pearson, David., Conran Octopus, London, 1989, £14.95, *Guidance on living in harmony with the environment. Text in 3 sections: The Natural House & The Dangerous House Constituent Elements of the House of the House. Details on super insulation and ecologically sound retro-fit improvements.*

The New Atrium ,Bednar, Michael, J.,McGraw-Hill Book Co., 1986, $42,00, *Design studies, 213 pp,* ISBN: 0-07-004275-6

The Passive Solar Design of Buildings, Renewable Energy News, 1985, *Published in Renewable Energy News No. 12 July 1985*

The Passive Solar Energy Book, Mazria, E.,Rodale Press, Erasmus, PA,1979

The Potential for Passive Solar Technology in UK Housing. Millbank, N.O., 1986, *Published in Passive Solar Technologies for Energy, Conservation in Buildings. Proceedings of CIB Conference, W67, Lisbon 1986.*

The Skylight Handbook-Design Guidelines with PC Disk, American Architectural Manufacturers Association, 2700 River Road, Des Plaines, IL 60018, U.S.A., $100.00 per set, *A spreadsheet template to be used with Lotus 1-2-3 for working through calculations.* (SHDG-2).

The Solar Home Book Heating, Cooling and Designing with the Sun. Andersen, B., Brick House Publishing Company, 1976

The Sun and Ventilation. Fitzgerald, D., Houghton-Evans, W., George, Stephen & Ptnrs., Univ.

Leeds, U.K. Leeds, LS2, 9JT, 1987, *Paper given at International Solar Energy Society, Solar World Congress, Hamburg, September 1987.*

The Urban Climate, International Geophysics series, Vol. 28. Landsberg, H.E., Academic Press, 1981.

Thermal Environments and Comfort in Office Buildings. Schiller, Gail., Arens, Edward., Benton, Charles., Bauman, Fred., Fountain, Marc., Doherty, Tammy. Center for Environmental Design Research, University of California, Berkley, USA., 1988, *Questions ASHRAE standards for thermal comfort in the light of new work on thermal transients. A study of office workers in San Francisco using new procedures to assess thermal environments and comfort suggests that optimum thermal acceptability is lower than current ASHRAE standards.* Ref CEDR-02-89.

Thermal Shutters and Shades. Shurcliff, W.A., Brick House Publishing Company Inc., Andover, Massachusetts, USA, 1980,

Towards Energy Independence. Solar Energy Society of Ireland, 1978.

Towards Green Architecture, Vale, B., Vale, R., RIBA Publications, 1991. *Practical examples based on five British and Dutch case studies.*

Transparent Insulation in Solar Energy Conversion for Buildings and Other Applications. Proceedings of International Workshop, Freiburg 24/25 March, 1988, L.F. Jesch (Ed.).

Utilization of Building Design for Climatic Control in Glazed Spaces: A parametric study. Wall, Maria, Swedish Council For Building Research (Byggforskningsradet), AB Svensk Byggtjanst, Litteraturjanst, S - 171 88, Solna, Sweden. 1991, *(Document no: D2:1991). Report contains calculations for steady-state conditions and calculations with computer prog. DEROB-LTH of the climate in different models of glazed courtyards. Parameters including number of panes, area of glazing, thermal inertia, air change rate are examined in different configurations.* ISBN 91-540-5265-3.

Window Design Allpications Manual. Chartered Institute of Building Services Engineers, Delta House, 222 Balham High Road,

London SW12 9BS, 1987, £17.00. *A very useful publication for Architects and Engineers covering site considerations, roof & sidelight design, daylight, lighting controls, heat gain/loss, noise, shading, prediction methods for daylight & sunlight, glare, guidelines for shape & position of window, 66 pp.* ISBN: 0-900953-33-0.

Windows and Environment. Turner, D.P. (Ed.), McCorquodale (on behalf of Pilkington Brothers Ltd). 1969. *Produced for a leading glass manufacturer, this is a text with diagrams, colour photographs and transparent overlays . Covers the treatment of noise, thermal environment and characteristics of different glass types. Overlays allow daylight & solar calculations for a range of latitudes.*

Working in the City (Results of) European Architectural Ideas Competition. O'Toole, S., Lewis, J. O. (Eds.), Eblana Editions / Energy Research Group, School of Architecture, University College Dublin, Richview, Clonskeagh, IRL-Dublin 14, 1990, IR£14. *A record of the CEC's Third Architectural Ideas Competition. Carried out under the ARCHISOL programme, it includes much of the extensive package of guidance notes supplied to competitors, together with assessor's reports and reproductions of award-winning entries.*

World Survey of Climatology, Vol.6, Climates of Central and South Europe, Wallen, Elsevier Publishing Co. 1977.

FRENCH

Aménager l'Espace et Maîtriser l'Energie. Re - Source, STU / AFME, 64 rue de la Fédération, 75015 Paris, 27 rue Louis Vicat 75015 Paris, 1989, 70 FF. *A guide for town planning with Energy Conscious Saving. 47 p.*

Analyse et Réduction Modales d'un Module de Comportement Thermique du Bâtiment, Lefebure, G., Ecole des Mines de Paris, 60 B. St Michel, 75272 Paris, CEDEX 06, 1987. *A presentation of the modal method applied to heat transfer in buildings for heat load calculations.*

Archi-bio. Izard, J.L., Guyot, A., Ed. Parentheses,1979.

Architecture - Energie: Inspirations et Contraintes. Parant C., Mazaud J. R., S'PLOIT, 188 rue de la Roquette, 75019 Paris,1988, *Study between Architectural Design and Energy Building requirement, numerous examples.*

Architecture et Climat Réalisations, Gilot, C., De Herde, A., Minne, A., Opfergelt D., Services de Programmation de la Politique Scientifique, Rue de la Science, 8,1040 Bruxelles, Belgique, Juin 1986,160 F.B. *Cet ouvrage de 128 pages présente l'Architecture climatique belge contemporaine au travers d'une quinzaine de maisons neuves ou rénovées, en zone rurale ou urbaine.*

Architecture Solaires en Europe Conceptions, Performance, Usages. ECD Partnership, UK-London, EDISUD, La Calade RN 7, 13090. Aix-en-Provence, France, 1991. *Cet ouvrage présente une série d'étude sous la tutelle du CEC DGXII Project MONITOR, comportant une gamme de construction type à l'intérieur de la CEE.*

Architecture Solaires en Europe: Conceptions, Performances, Usages. ECD Partnership,Commission of the European Communities, EDISUD, La Cala, 13090 Aix-en-Provence, 1991, 240 FF. *30 issues of monitor project with a general introduction on Passive Solar Design. 272 p.* EUR 12738 FR.

Architecture, Ambiance Energie. M. Retbi, MELTH, DAU, AFME, Thermique et Architecture, Regirex, 54, bis rue Dombasle, 75015 Paris, 1989. *Catalogue of the competition. Synthesis and description of 20 projects. 85 p.*

Architecture, Climats, Energie: Outils et démarches pédagogiques. COFEDES, 3 Rue H. Heine, 75116 Paris, 1986. *Software developed in the programme Habitats Climatiques is presented with case studies. Part 1 deals with teaching of Energy and architectural projects. Part 2 deals with thermal concepts in cold climates. 235 pp.*

Architecture, Urbanisme et Energie, bilan et perspectives, Actes de colloque. Plan-Construction H2 E85, Ministère de l'Equipement et du Logement, Paris, 1985, 110 FF. *Energy in Building and Town Planning Design Tools and Practice.*

Atlas Climatique de la Construction, Climatic Atlas for Building, Chemery, L., Ducheme-Marullaz, Ph., C.S.T.B., 4 Avenue du Recteur Poincaré, 75016 Paris. *Hygrothermic comfort, stability, energy saving, air pureness, visual comfort, water tightness, comfort and safety in external sites. More than 100 maps on climatic data.*

Atlas Energétique de la Wallonie. *Produced by the Institut Wallon de Développement Economique et Social d'Aménagement du Territoire, this 23 page report shows the energy consumption in different sectors, the fuel used and the potential of renewable energy.*

Atlas Européen du Rayonnement Solaire. Page, J., Palz, W., Flynn, PYC, 254 rue de Vaugirard, 75015 Paris, 1984. *Meteorological data for passive solar calculations.*

Bioclimatic Architecture - L'Architecture Bioclimatique. ENEA, De Luca, Edizioni d'Arte Spa., Via di S. Anna 16, 00186 Roma Italia, 1989, *Catalogue Exposition Architecture Bioclimatique. 88p,*

Bioclimatisme en Zone Tropicale. Ministère de la Coopération - Programme REXCOOP. *Construire avec le CLIMAT. Dossier: Technologies et développement. Ministère de la Coopération - Programme REXCOOP.*

Cahier Pédagogiques Thermique et Architecture T1: Qualité Thermique des Ambiances T2: Analyse Climatique du Site. J. P. Traisnel, A. Duport, P. Depecker, J. P. Izard, AFME, 27 rue Louis Vicat (Service Formation), 75015 Paris, 1989, 100 FF each. *67+111+30 p.*

Catalogue Critique des Toitures Isolantes. CSTB, 4 Avenue du Recteur-Poincaré, 75782 Paris Cedex 16, 195 FF. *Insulated roof construction, products and materials.*

Chauffage de l'Habitat par l'Energie Solaire Experimentations sur les Maisons CNRS d'Odeillo. Cabanat, M., Sesolis, B., Université Paris VII, Place Juffieu, Paris 5, 1976. *The first experimental evidence showing the thermal behaviour of the Trombe Wall by Prof. Trombe's co-workers.*

Conception Thermique de L'Habitat Guide pour la région Provences - Alpes, SOL A.I.R., Conseil régional PACA, EDISUD, La Calade, 13090 Aix-en-Provence,

1988, 170 FF. *Guide pour la région Provence-Alpes Côte d'Azur.*

Congrès International d'Architecture Climatique. Cellule Architecture et Climat, CIACO Editor, Avenue Einstein, B.1348 Louvain-la-Neuve, Belgique, Juin 1986, 600,-F.B. *Cet ouvrage présente les actes du premier Congrès International d'Architecture Climatique qui a eu lieu les, 2 et 3 Juillet 1986 - 598 pages.*

Constructions Solaires Passives - 50 Realisations Francaises. Augier, H., Editions Apogée, France, 1982.

Des Contraintes Energétiques à des Règles de l'Art Favorisant la Conception. Adlophe, L., Ecole d'Architecture UP6, 1985. *Deals with the integration of energy constraints into the work of the architect.*

Données Climatiques. Opfergelt. D., De Herde, A., Services de Programmation de la Politique Scientifique, Rue de la Science, 8, 1040 Bruselles, Belgique, Juin 1986, 100 F.B. *Cet ouvrage de 64 pages présente graphiquement en 3 dimensions les données climatiques Belges pour le 15 de chaque mois.*

Effets de Serres: Construction des Serres Bioclimatiques. Murpy I., Nicolas F., EDISUD, La Calade, 13090 Aix-en-Provence, 206 p.

Etude de la Rentabilité de la Mise en Place de l'Intermittence dans les Etablissements d'Enseignement. Bertolo, M. et al., CSTB, 4 bis Av du Recteur Poincaré, 75016 Paris, 1985. *The gains due to the installation of various levels of intermittency in school buildings.*

Guide d'Aide à la Conception Bioclimatique. Le Paige M., Gratia E., De Herde, A.,Service de Programmation de la Politique Scientifique, Rue de la Science, 8,1040 Bruxelles, Belgique, Juin 1986, 120, - FB. *Cet ouvrage de 136 pages comporte 3 parties complémentaires. La Ière donne au travers de dessins et de graphes l'ensemble de l'information nécessaire aux architectes. La 2éme montre des exemples et la 3éme reprend un ensemble de valeurs et de résultats.*

Guide de l'Eclairage Naturel et de l'Eclairage Artificiel dans les Etablissements Scolaires. P. Chauvel, Ministère de l'Education National, Direction des Personnels d'Inspection et de Direction, Paris, 1989. *A guide for natural and electric lighting in schools. 79 p.*

Influence de la Nature des Materiaux sur le Chauffage. Brau, J., Rousseau, S., INSA Lyon, Av. A. Einstein, 69 Villeurbaune, 1982. *The influence of concrete thickness on the heating load of buildings and comfort conditions.*

L'Eclairage Naturel et le Parti Architectural en Relation avec l'Economie d'Energie, Barra, O., Dogniaux, R., Duchateau, W., 1981, May, *Proceedings of 4th European Light Congress.*

L'Habitat Bioclimatique de la Conception à la construction. Watson D., Camous R., L'ETINCELLE, 3449 rue St. Denis, Montreal Quebec, CANADA H2X 3L1, 1983, 140 FF. *Basic book on Passive Solar Design. 188 p.*

L'Information numerique et technique en énergies renouvelables, energies fossiles, utilisation rationnelle de l'énergie dans l'habitat et l'industrie. IFE (Institut Francais de l'Energie), 3 Rue H. Heine, 75116, Paris. *Survey of energy related information requirements and services such as periodicals, exhibitions, conferences, workshops and databases.*

La Protection contre le Vent Aerodynamique des Brise-vents et Conseils Pratiques. Gandemar, J., Guyot, A., Centre Scientifique et Technique du Bâtiment, 1981.

La Tradicion Energeticamente Consciente. Energy Research Group, School of Architecture, University College Dublin, Richview, Clonskeagh, IRL-Dublin 14 ,1989. *An A4 B&W reproduction of a poster exhibition of traditional responses to energy issues in Europe, prepared under the CEC's ARCHISOL programme by the Energy Research Group. The Original poster exhibition, in colour, was circulated to European schools of architecure.*

Le Guide de diagnostic de l'A.F.M.E.,Thermal analysis handbook for existing dwellings. A.F.M.E., 1986. *Government produced promotional literature. 8 chapters covering fundamentals, energy*

sources, DHW, ventilation, insulation, heating, cooling, applications. 220pp approx.

Le Soleil,Chaleur et Lumière dans le Bâtiment. EDFL, S.I.A., Siciété Suisse des Ingénieurs et des Architectes, Case Postale, CH-8039 Zürich SUISSE, 1990. *Passive Solar Design Guide.*

Les Logiciels d'Energétique des Bâtiments Développement, Evaluation Technique, Illustrations. Rejon, P.B., Ecole des Mines de Paris, 60 Bd. St Michel, 75272, Paris CEDEX 06, 1988. *A review of the software available on the french market for the thermal appraisal of buildings and methods for evaluation.*

Maisons Solaires. Menard, J.P., Editions du Moniteur, 1980.

Maitrise de l'Energie et Economie de Matières Premières Répertoire des actions de formation continue (annuel). A.F.M.E., *Annual list of on-going programmes covering Building Transport, Manufacturing, Agriculture. Lists courses on: Energy management - Thermal analysis (117 courses in 1986), Thermal retrofitting (63 courses in 1986), New & Renewable energies (55 in 1986), Thermal equipment installation (183).*

Methodes de Calcul pour l'Application des Règlements, Thermiques: TH-K, TH-G, TH-C. Centre Scientifique et Technique du Bâtiment, 4 Avenue de Recteur-Poincaré, 75782 Paris Cedex 16. *French thermal regulation applicable since the 1st January, 1989.*

Principes et Formes de l'Habitat Bioclimatique. Queffelec, C., CSTB in Recherche et Architecture, n. 46, 1981.

Projection Solaires, AFME, CATED, Domaine de Saint Paul, 78470 Saint Remy-les-Chevreuses, 1989, 350 FF. *General information on Solar Projections Devices. Catalogue of products. 404 pp.*

Rendements Energétiques des Systèmes de Chauffage, Brochure techniques - exercises d'intègration. Hannay, J., Dols, J.M., SPPS, Brussels, Belgium, 1982.

Solaire, Eolien, Hydraulique, Biomass. CLER (le Comité de liaison des Energies Renouvelables), 299 rue Granier, 73230 - Saint-Alban-Leysse, France. *Published by CLER/ASDER.*

Suivi des Performances d'Installations de Chauffage solaire dans l'Habitat, Document de Synthèse sur les Opérations & Recommandations. Ferraro, R., Turrent, D., Godoy, R., PVC Edition, 254 rue de Vaugirard, 75740 Paris, cedex 15. *Energies Renouvelables Recherches sur l'Energie Auprés des Communautés Européenes.*

GERMAN

Bau und Energie Bauliche Massnahmen zur verstärkten Sonnenenergienutzung im Wohnungsbau. Nikolic, Vladimir., Rouvel, Lothar.,Verlag TÜV Rheinland GmbH, D-5000 Köln 90, 1983, *Research project for the Ministry of Research and Technology.* ISBN:3-88585-098-2.

Bauen mit der Sonne,Vorschläge und Anregungen. Hebgen, Heinrich ,Energie-Verlag GmbH, Postfach 10 21 40, D-6900 Heidelberg 1 / F.R.G., 1982, ISBN:3-87 200-636-3

Bauen und Energiesparen, Ein Handbuch zur rationellen Energieverwendung im Hochbau für Bauherren, Architekten und Ingenieure. Bundesminister für Forschung, u. Technologie, 5300 Bonn, (Ed.), Verlag TÜV Rheinland GmbH, D-5000 Köln 90, 1979, ISBN:3-921-059-99-2

Der Wintergarten, Wohnräume unter Glas. Timm, Ulrich., Verlag Georg D. W. Callwey, D-8000 München, 1986, ISBN: 3-7667-0817-J.

Der Wintergarten :,Wohnkultur unter Glas ; Einrichtung und Beispiele für Konstruction, Einrichtung und Bepflanzung. Reiners, H., Timm, U.,Verlag Georg, D. W. Callwey, München, 1990, ISBN 3-7667-0991-7

Energie und Komfort in Gebäuden mit Wintergärten Eine p a r a m e t r i s c h e Sensitivitätsanalyse. Filleux, Ch., Riniker, W., Bundesamt für Energiewirtschaft (BEW), Postfach, CH-3003 Bern (Switzerland), 1986. *Results from the Swiss National Program of the IEA Solar Heating & Cooling Task 8. Passive and Hybrid Solar Buildings.*

Energie - und umwelt - bewußtes Bauen mit der sonne. Lohr, A., Behnsen, J., Molitor, K., Willbold-Lohr, G., Verlag TÜV Rheinland GmbH, Köln, 1991, 22 DM. *In dem Informationspaket werden*

242

verschledene Systeme zur passiven Nutzung der Sonnenenergie in Gebäuden anhand realisierter Projekte nicht nur auf die optimale Ausnutzung der Sonnenernergie und die weit gehende Verringerung von Wärmeverlusten durch Wärmedämmung gelegt, sondern vor allem auch auf die Umweltverträglichkeit dieser Gebäude. Neue Materialen. ISBN 3-88585-799-5. Komponenten und Systeme werden berücksichtgt.

Energiesparbuch für das Eigenheim, Eine Anleitung zum energiesparenden Bauen und Heizen, (Alt- und Neubau). Hausladen, Dr. Ing. G., Bundesminister für Raumordnung, Bauwesen & Städtebau. Deichmanns Aue, D-5300 Bonn-Bad Godesberg 2, 1988.

Energiebewusste Bautraditionen. Energy Research Group, School of Architecture, University College Dublin, Richview, Clonskeagh, IRL-Dublin 14, 1989. *An A4 B&W reproduction of a poster exhibition of traditional responses to energy issues in Europe, prepared under the CEC's ARCHISOL programme by the Energy Research Group. The Original poster exhibition, in colour, was circulated to European schools of architecure.*

Glahäuser zum Wohnen. Guenoun, Gabriel., Kalmanovitch, Jean-Claude., Bauverlag GmbH, D-6200 Wiesbaden, 1985. *Translation of French book: Des Serres pour Habiter.* ISBN: 3-7625-2325-8

Glas in der passiven Solararchitektur Glastypen - Eigenschaften - Problemloesungen. Gnan, Karl-Heinz, Bauverlag GmbH, D-6200 Wiesbaden, 1986, ISBN: 3-7625-2327-4.

Glasdocu spezial Planungs- und Bauanleitung mit Bezugsquellenteil für Bauherren, Architekten, Unternehmer. Schweizerisches Institut für Glas am Bau, Zürich, Schweizerisches Institut für Glas am Bau, Badenerstrasse 21, CH-8004 Zürich, 1986, SFr. 25.

Glashäuser im Geschoss-wohnungsbau, Benutzerfreundliche Planung bei Neu- und Altbauten. Niche, Wolfgang, Dr Ing., Bauverlag GmbH, D- 6200 Wiesbaden, 1988. *Ph.D. Thesis published without further editing.* ISBN: 3-7625-2554-4.

Handbuch der passiven Sonnen-energienutzung. Zimmermann, M., SIA, Schweizerischer Ingen. u.Archit.-Verein, SIA, Schweizerischer Ingenieur- u. Architekten-Verein, Postfach CH 8039 Zürich, 1986, ca. 50 Sfr.

Handbuch Passive Nutzung der Sonnenenergie, Schriftenreihe des Bundesministers für Raumordnung, Koblin, Wolfram., Krüger, Eckehard., Schuh, Ulrich., Bauwesen und Städtebau BMBau, 1984. *Handbook on Passive Use of Solar Energy.* 04.097.

IRB (3, 5, 6), Energieeinsparung in Wohnhäusern. *No. 3 - Wärmedämmung (Insulation), No. 5 - Solarenergienutzung (use of solar energy), No. 6 - Energie und Architektur (Energy and Architecture).*

(96, 176, 207, 299, 301), Other Collections of Summaries: *No. 96 - Wintergarten (Wintergardens), No. 176 - Isolierverglasung (double glazing), No. 207 - Solarhauser (solar houses), No.299 - Erdarchitektur (underground architecture), No. 301 - Latentwarmespeicher (latent heat storage).*

(332, 415, 421, 423). *No. 332 - Glaskonstruktionen (glass constructions), No. 415 - Wärmespeicherung bei Gebäuden und Bauteilen (heat storage in buildings), No. 421 - Solarheizungen (solar heating systems), No. 423 - Sonnenkraftwerke (solar power plants).*

(454, 455, 457). *No. 454 - Klimafaktoren in Der Bauplanung (climatic factors in planning buldings), No. 455 - Baukonstruktion und Klima (Building construction and climate),No. 457 - Klima und Energiebedarf (climate & energy demand).*

(478, 490, 543, 661). *No. 478 - Alternative Architektur (alternative architecture), No. 490 - Niedertemperatur Solarheizungsanlagen (low temp. solar heating systems), No. 543 - Stadtökologie (ecology of cities), No. 661 - Kühlung mit Solar-energie (cooling & solar energy).*

(665, 679, 749, 774). *No. 665 - Massiv-und Wandabsorber (massive wall absorber), No. 679 - Ökologisches Bauen und Planen (ecological plan and construction), No. 749 - Energiesparhäuser (energy saving houses), No. 774 - Gläsdacher (glass roofs).*

(813, 814, 816, 825). *No. 813 - Energieokonomie im Bauprozess (ecology of energy in the building process), No. 814 - Solardachdeckungen (solar roofs), No. 816 - Solarwande (solar walls), No. 825 - Sonnenschutzglas (glass for sun protection).*

(931, 932). *No. 931 - Wintergärten - Beispiele aus der BRD (Wintergardens - examples from Germany), No. 932 - Wintergärten - Beispiele aus dem Ausland (Winter gardens - examples from foreign countries).*

(933, 942, 943). *No. 933 - Wintergärten - Planung, Gestaltung, Konstrukti, Bauphysik (wintergardens - planning, design, construction, building physics), No. 942 - Besonnung der Wohnung (insulation of dwellings), No. 943 - Wohnen im Gewächhaus (living in wintergardens).*

(950, 1005, 1006, 1011, 1012). *No. 950 - Temporärer Wärmeschutz (moveable insulati, No. 1005 - Thermik in Gebäuden (Thermic in Buildings), No. 1006 - Treibhauseffect (Greenhouse effect), No. 1011 - Entwerfen mit der Sone (designing with the sun), No. 1012 - Gebäudeorientierung (Orientation of buildings).*

Klimagerechte und energiesparende Architektur. Hillman, Gustav., Nagel, Schreck, Hasso, Müller-Verlag, D-7500 Karlsruhe, FRG, 1982. *Introductory /promotional literature on climate responsive and energy saving architecture.* ISBN:3-7880-7197-4

Kosten und Nutzen energiesparender Baukonstruktionen Tabellen, Diagramme und Ausführungsbeispiele für Neu-und Altbauten. Sören Christensen, Bauverlag GmbH, D-6200 Wiesbaden, 1981. *Energy conservation in buildings, economics & calculations.* ISBN 3-7625-1302-3.

Leben mit der Sonne natuerliche Energie - praktisch genutzt. Jürgen Schneider, Eichborn GmbH & Co. Verlag KG, D-6000 Frankfurt am Main, 1983, ISBN:3-8218-1712-7.

Licht Architektur - The Architecture of Light. Auer, G. (Ed.), Pieper, J., Böhringer, H., Conrads, U.,Feuerstein, G., Middelton, R., Bertelsmann Fachzeitschriften, Carl-Bertelsmann Straße 270, D-4830 Gütersloh, 15.3.1988, 42,50 DM. *Articles* *presented in this issue of Daidalos mainly cover artificial lighting & design aspects, outlining the state of the art in design which combines architecture & natural light.* ISBN 0721-4235.

Licht, Luft, Schall. Eberspächer, J. (Ed.), Freymuth, H., Lenz, H., Lutz, P., Schupp, G., Forum Verlag, J. Eberspächer, Postfach 289, D-7300 Esslingen, *Contains three chapters: Daylighting in the design process, (illumination & thermal considerations for openings), HVAC in industrial buildings, Fundamentals of acoustics.* ISBN 3-8091-1032-9.

Mit der Sonne bauen, Anwendung passiver Solarenergie. Wachberger, Michael u. Hedy, Paulhans Peters (Ed.), Verlag Georg D.W. Callwey, D-8000 München, 1983, ISBN:3-7667-0671-3.

Rationelle Energieverwendung im Wohnungsbau (Rational Energy Use in the Housing Sector). Informationszentrum Raum & Bau der Fraunhofer Gesellschaft, Nobelstrasse 12, D-7000, Stuttgart 80, *A complete collection of short summaries on relevant,literature updated from time to time.*

SES - Report Nr. 13: Energiebewusstes Bauen mit dem Klima und der Sonne Eine Einführung in das energiegerechte konzipieren und Planen von Bauten. Armin Binz, SES, Schweizerische Energie-Stiftung, Sihlquai 67, CH-8005 Zürich, 1983, revised 1987, SFr 30.

Solararchitektur und Energiebewusstes Bauen (Solar architecture and energy conscious buildings). Hemmers, Rosa (Ed.), Burgerinformation Neue Energietechniken (BINE), D-5300 BONN 2, 1987, 15 DM. *2nd Edition 1984.* ISBN 3-88585-387-6.

Sonnenenergie zur Warmwasserbereltung und Raumhelizung, Solar energy for domestic hot water and heating. BINE, D-5300 BONN 2, 1988, 5 DM. *2nd Edition 1985.* ISBN 3-88585-448-1.

Städische Sonnenräume (Urban sun spaces). Glassel, Joakim., Verlag C. F. Müller, D-7500 Karlsruhe, 1985, ISBN:3-7880-7226-1. 48-DM.

Tageslichttechnik. Fischer, U. Dr-Ing, Verlagsegellschaft Rudolf Müller, 1982, 78.00 DM. *Technically orientated book. 161pp.*

GREEK

Application on Solar Design Evaluation and Estimation of Solar Systems. Bazos, E. Prof., PHOIVOS, B. Selloyntos Ltd., Stoyrnary 15, 10683 Athens, Greece, Fax: 1-3613244 Tel: 1-3636432, 1990. *Analysis of solar design with evaluations of solar systems. Tables of climatic data for cities in Greece.*

Bioclimatic Architecture - Passive Solar Systems. Andreadaki-Chronaki, E., University Studio Press, Thessaloniki, Greece, 1985. *General reference on bioclimatic design. Description of passive solar systems and their performance in buildings. Examples of well known applications of passive solar buildings around the world. Basic text book used at the University of Thessalonik.*

Building Thermal Insulation Guide. Papasotiriou, A., 1985. *Useful guide describing methods and materials available in the Greek context for building thermal insulation. Significant reference to Greek thermal insulation regulations.*

Energy Conscious Tradition Catalogue. Energy Research Group, School of Architecture, University College Dublin, Richview, Clonskeagh, IRL-Dublin 14, 1989. *An A4 B&W reproduction of a poster exhibition of traditional responses to energy issues in Europe, prepared under the CEC's ARCHISOL programme by the Energy Research Group. The Original poster exhibition, in colour, was circulated to European schools of architecure.*

Greek Energy Directory,. Heliotechnic (Hellas) Ltd., 3 Alex Soutsou Str., 10671 Athens, Greece, 1988. *First published in 1984, both in Greek & English. New edition in 1988. Contains among other energy information, descriptions of passive solar projects in Greece and a directory of Greek energy & passive solar design consultants.*

Low Energy Building Design. Kontoroupis, G., 1984. *Covering fundamental strategies on low energy building design, it compares conventional & alternative systems describing methods of energy calculations. Basic text book for the course at the National Technical University of Athens.*

Low Energy Design and Passive Solar Systems. Papadopoulos, M., Axarli, C., 1984. *General description of the function of passive solar buildings, development of the different systems, basic strategies and materials for passive energy efficient design. Advantages and disadvantages of passive systems and their possible applications within the Greek context.*

Proceedings of the 1st Seminar on Assembly, Manufacture and Testing of Solar Collectors. E. Metaxa Press, Patras, Greece, 1985. *Used as a textbook on a two semester course at the Technological Education Institution of Patras on detailed design and construction of flat plate collectors for residential and commercial applications.*

Solar and Energy coefficient tables. ELKEPA, (Greek Productivity Centre), ELKEPA, Frantzh 9, 11743 Athens, Greece, Fax: 1-9239430 Tel: 1-9215129. *A series of tables which give quick answers to: - the solar radiation incident on a plane of any tilt or orientation - shading methods used in Greece - solar radiation and heat gain in typical Greek buildings.*

Thermal Protection, Sound Protection, Wind protection of Buildings., Fragoudakis, A., University Studio Press, Thessaloniki, 1985. *Used as a textbook on a two semester course at the Aristotelion University of Thessaloniki (Architectural Technology III), covering the basics of heat transfer in buildings and solar energy.*

ITALIAN

116 Edifici Solari Passivi in Italia (116 Passive Solar Buildings in Italy). Funaro, G., Fanchiotti, A and D'Errico, E., ENEA, Viale Regina Margherita, 125 - 00198 Roma, 1985, Lit 30,000. *The majority of existing passive solar buildings in Italy are catalogued. Information is given on buildings (plans, elevations, sections, etc.), on their energy features (insulation, passive/active solar systems) and on their energetic behaviour (energy consumption).*

Abitare con il Sole: Abc della Climatizzazione Naturale (Living with Sun: Fundamentals of Natural Conditioning). Wright, D.,Franco Muzzio & C Editore, Via Bonporti 36, 35100 Padova, 1981, Lit 15,800. *A basic handbook of passive solar design.* 88-7021-159-2.

Adeguamento alle norme di sicurezza e risparmi energeitci negli edifici publici (Security regulations and energy savings in public buildings). Raggi, F.,Publ. CDA (journal) 1988, PEG Editrice, Via Flli Bressan 2, 20126 Milano, 1988, Lit 5,500 (per issue). *Covers retrofitting to upgrade buildings to the requirements of new Italian legislation. Optimum energy savings and fire prevention are also discussed. CDA journal Feb '88.* ISSN/Ref 0373 - 7772.

Annuario Nazionale dell'Energia 1987 (Italian Annual Book on Energy). Salvi, L.A.,Editrice Inter - Ed, Via Cassia 1134/A, 00189 Roma, Annual - last publicised 1987, Lit 142,000 (year 1987). *Details of organisations, research centres & professionals involved in energy matters. Chapter dealing with P.F.E - the National Energy Project. Another chapter deals with bioclimatic architecture. Edited by Editrice Inter-ed. 992pp.* ISSN/Ref: 0392 - 8403.

Applicazione Pratiche dell'Energia Solare con colletori Piani (Case Studies of Flat Solar Collectors Applications). Coniglio, M.,Pirola Editore, Via Comelico 24, 20135 Milano, 1986, Lit 40,000. *Presents the most common types of slope solar collectors (both air & water) with design & evaluation advice. The main part of the book deals with solar collectors, storage and passive solar heating. Section on heat pumps. 262pp.*

Architettura e Ambiente (Architecture and Environment). Pavoni, G., Publ. Modulo (journal) 1986, BE-MA, Via Teocrito 50, 20128 Milano, 1986, Lit 8,000 (per issue). *Experience of bioclimatic building in Piemonte region is described. Modulo issue 126 /Sept. 1986.*

Architettura e Calore, i Principi Dimenticati (Architecture and Heating: Some Forgotten Problems). Palmizi, A., Publ. Modulo (journal) 1987, BE-MA,Via Teocrito 50, 20128 Milano, 1987, Lit 8,000 (per issue). *Shows how heating requirements have been resolved in terms of architectural considerations in a building designed by Le Corbusier. Modulo issue 137/Dec '87.*

Architettura ed Energia,Sette edifici per l'ENEA. (Architecture and Energy: Seven buildings for ENEA). Carozzi, C., Rozzi, R., Gregotti Ass., Cervellati, P.,Corlaita, A., Monti, C., Scudo, G., Seassaro, A.,

Cannetta, A., Gallo, C., De Luca editore, Via di Sant'Anna 16, 00186 Roma, 1987. *Seven buildings having significant features from an energy point of view are presented in a very elegant way., All the seven buildings have been built on behalf of ENEA for use as offices, research centres etc.* 88-7813-116-4.

Architettura Solare Tecnologie Passive ed Analisi Costi/Benefici. (Solar Architecture: Passive Solar and Cost/Benefit Analysis). Bottero, M., Silvestrini, G., Scudo,G., Rossi, G., Edizioni CLUP, Piazza Leonardo da Vinci 2, 20100 Milano, 1984, Lt 25,000. *A complete description of design methods, components, heating/ cooling systems & simplified simulation codes is given with references to passive solar architecture.* ISBN: 88-7005-598-1.

Benessere Acustico e Visivo (Accoustic and Visual Comfort). Francese, D., Publ. Recuperare (journal), 1987, PEG Editrice, Via flli Bressan 2, 20126 Milano, 1987, Lit 7,000 (per issue). *Covers daylighting problems in retrofitting among other topics. Recuperare issue 30/ Jul-Aug 1987.* ISBN:Ref 0392 - 4599.

Calcola dei Valori Medi Mensili della Radiazione Diffusa. (Calculation of Mean Monthly Values of Diffused Solar Radiation) Verifica Sperimentale dell'Affidabilità dei Principali Metodi Proposti in Letteratura. Butera, F.M., Farruggia, R., Festa, C., Cotto, C.,Publ. HTE (journal) 1987, PEG Editrice, Via Flli Bressan 2, 20126 Milano, 1987, Lit 5,500 (per issue). *Calculation of mean monthly values of diffused solar radiation. Test of main existing methods against experimental values. Energie Alternative - HTE issue 50, Nov-Dec 1987.* ISSN? Ref 0391 - 5360,.

Case Solari in Europa,Come hanno funzionato. Compilato e scritto da Houghton-Evans, W., Turrent D., Whittaker, C., C.S.A.R.E. per conto della Commissione delle Comunita eur.

Componenti Innovativi per il Risparmio Energetico La Sperimentazione di Croce del Biacco: (New components for,Energy Savings: The experience of Croce del Biacco). Del Bufalo, S., Romani, R., Publ. L'Industria delle Costruzioni (journal) 1986, Edilstampa, Via Guattani 24, 00161 Roma, 1986,Lit 6,500 (per issue). *New strategies and*

components for collecting/generating distributing and controlling heat are explained. L'Industria delle Costruzioni issue 180/Oct. '86.

Confronto tra i Vari Metodi di Calcola dei Fabbisogni Termici degli Edifici (Comparison of Different Methode for Calculation of Thermal Needs of Buildings). Cucomo, M., Marinelli, V., Publ. CDA (journal) 1988, PEG Editrice, Via Flli Bressan 2, 20126 Milano, 1988, Lit 5,500 (per issue). *Different calculation methods are analyzed and tested against experimental data. CDA issue Jan '88.* ISBN:Ref 0373 7772.

Consigli Progettuali per l'Edificio e l'Impianto Suggestions for Design of Buildings and Plants. CNR - PFE, Via Nizza 128, 00186 Roma, 1982 Free. *As a result of design strategies according to the Guide for an Energy Conscious Design of Buildings and the other three appendices, detailed suggestions are given with reference to, 15 Italian sites that are representative of all the Italian Climatic conditions. It is the 4th appendix to the Guide.*

Dati Climatici per la Progettazione Edile ed Impiantistica Italian Climatic Data for Design of Buildings and Plants. CNR - PFE, Via Nizza 128, 00186 Roma, 1982, free publication. *Complete set of climatic data (temperatures, humidity rates, solar radiations, velocity and direction of wind, etc.) are reported for many Italian sites. It is the first appendix to the Guide for an Energy Conscious Design of Buildings.*

Edifici Autoenergetici in Italia - Possibili Realta o Mere Utopie?, Low Energy Consumption Building in Italy - What are the Real Possibilities? Dall'O', G., Palmizi, A., Publ. HTE (journal) 1988, PEG Editrice, Via Flli Bressan 2, 20126 Milano, 1988, Lit 5,500 (per issue). *Critical analysis of bioclimatic building in Italy, HTE issue 51/ Jan.-Feb. '88.* ISSN/Ref. 0391 - 5360.

Energia e Forma Un Approccio Ecologico allo Sviluppo Urbano: (Energy and Form: An Ecological Approach to Urban Development). Knowles, R.L., Franco Muzzio & C Editore,Via Bonporti 36, 35100 Padova, 1981, Lit 18,000. *Implications of energy savings/ recovery on form of building both at individual and at urban level are dealt with.* 88-7021-148-7.

La Tradizione Energeticamente Attenta Catalogue. Energy Research Group, School of Architecture, University College Dublin, Richview, Clonskeagh, IRL-Dublin 14, 1989. *An A4 B&W reproduction of a poster exhibition of traditional responses to energy issues in Europe, prepared under the CEC's ARCHISOL programme by the Energy Research Group. The Original poster exhibition, in colour, was circulated to European schools of architecure.*

Estate e Inverno con il Sole (Summer and Winter with the Sun). Dell'O, G., Palmizi, A., Publ. Modulo (journal) 1986, BE-MA, Via Teocrito 50, 20128 Milano, 1986, Lit 8,000 (per issue). *A procedure for introducing passive solar featurs to buildings in retrofitting interventions. Modulo issue 125/Aug. 1986.*

Guida al Controllo Energetico della Progettazione (Guide for an Energy Conscious Design of Buildings). CNR - PFE, Via Nizza 128, 00186 Roma, 1985, free. *A most important Italian handbook. Together with its four appendices it covers a complete energy conscious design of buildings and plants. It is the result of research performed in the first National Energy Project.*

Guida all'applicazione pratica della Legge 30 Aprile 1975 No. 373 (Handbook for Building Design according to law 373/1976). ISOVER, Balzaretti Modigliani, Via E. Romagnoli 5, 0146 Milano, 1987, Free publication. *The brochure is edited by an Italian firm producing insulating materials. It is a guide for design of buildings according to Italian energy standards (law 373/1976) and decrees of application).*

Guida alla lettura della Tabella UNI FA 101, Conduttività apparente dei materiali (Guide to Table,UNI FA 101: Effective Thermal Conductivity of Materials). ISOVER, Balzaretti Modigliani, Via E. Romagnoli 6, 20126 Milano, 1985, Free publication. *The brochure is edited by an Italian firm producing insulating materials. It is a guide to the use of standard tables (UNI) of conductivity of materials.*

Guida Pratica agli Impianti ad Energia Solare (Practical Handbook for Solar Plants Design). Lessieur, P.D., PEG Editrice, Via Flli Bressan 2, 20126 Milano, 1987, Lit 20,000. *8th book of the series Quaderni dell 'Energia (Energy Handbooks).*

Il Clima come Elemento di Progetto nell'Edilizia (Climate as a Design Element for Buildings). Gruppo di Energia Solare dell'Università di Napoli, Liguori Editore, Via Mezzocannone 19, 80134 Napoli, 1977, Lit 9,500. *The book deals with design of buildings and components taking into account the climate as a dominant factor.*

Il Residenziale a basso consumo energetico - un traguado per l'edilizia sovvenzionata (Residential low energy consumption building - a topic for low cost public building). D'Errica, E., Funaro, G.,Publ. L'Industria delle Costruzioni (journal) 1987, Edilstampa, Via Tuattani 24, 00161 Roma, 1987, Lit 6,500 the issue. *Investigates the possibilities of development for bioclimatic building in the low cost public building sector in Italy. L'Industria delle Costruzioni issue 192/October, 1987.*

Illuminazione Naturale, l'Altra Faccia della Luce (Daylighting, the Other Aspect of Illuminance). Rubini, F., Publ. Modulo (journal) 1987, BE-MA, Via Teocrito 50, 20128 Milano, 1987, Lit 8,000 (per issue). *The importance of daylighting is stresed. A calculation procedure is reported. Modulo issue 130/Mar. 1987.*

Influenza dell'Immediata Geometria sul Fattore Finestra (Influence of Immediate Geometry on the Window Factor). Pugno, G.A., Caglieris, G.,Publ. L'Industria delle,Costruzioni (journal) 1987, Edilstampa, Via Guattani 24, 00161 Roma, 1987, Lit 6,500 per issue. *Effects of different geometries of windows on daylighting factors are investigated. L'Industria delle Costruzioni issue 1987/May 1987.*

Influenza della Tipologia Volumetrica sulle Prestazioni, Illuminotecniche degli Edifici (Influence of Volume Typology on Lighting Performance of B / Vio, M. Vio, M., Publ. Recuperare (journal), 1987/1988, PEG Editrice, Via Flli Bressan 2, 20126 Milano, 1987/8, Lit 7,000 + Lit 7,000 - 2 iss. *Investigates daylighting comfort levels in different building types. Recuperare issue 32/Nov-Dec '87 and issue 34/Mar-Apr '88. ISBN:Ref. 0392 - 4599.*

Informazioni Tecniche ISOVER (ISOVER Technical Information Bulletin). ISOVER, Balzaretti Modigliani, Via E. Romagnoli 6, 20146 Milano, free. *A Bulletin of ISOVER. Each issue deals with an aspect of energetic design of buildings.*

Isolamento Termico Dinamica ed Innovativa (Innovations in thermal insulation). Aghemo, C., Dutto, M.G., Publ. Modulo (journal) 1988, BE-MA, Via Teorcrito 50, 20128 Milano, 1988, Lit 8,000 (per issue). *Presents typologies in thermal insulation, in particular dynamic insulation. Modulo issue 138/Jan. 1988.*

Isolamento Termico - Guida pratica alla legge 373 (Thermal Insulation) Guida Pratica all'Applicazione della Legge 373. (Thermal Insulation: Handbook for Design according to Law 373/1976). Arcangeli, S., EDILSTAMPA, Via Guattani 24, Roma, 1987, Lit 15,000. *A step-by-step procedure is supplied for building design according to law 373/1976 which is aimed at containment of energy consumption.*

L'Abbinamento di più Sorgenti di Calore (Integration of different energy sources in different,thermal plants). Gini, M., PEG Editrice, Via Flli Bressan 2, 20126 Milano, 1987, Lit 22,000. *Vol.1 - Fundamentals. Vol.2 - Case Studies. Deals with the problems in integrating traditional energy sources with solar energy in heating, cooling and air conditioning.*

L'Architettura del Regionalismo (Architecture of Regionalism), Guida alla Progettazione Bioclimatica nel Trentino (Guide for Bioclimatic Design of Buildings in Trento Region). Los, Sergio., Pulitzer, Natasha., Provincia Autonoma di Trento, Piazza Dante 15, 38100 Trento, 1985 - April, Lit 25,00. *The book and the companion folder contain charts covering fundamentals, examples of passive solar strategies and buildings, design tools and guidelines.*

L'Edilizia Bioclimactica Residenziale in Italia Alcune Considerazioni Tecnico-economiche. (Bioclimatic Residential Building in Italy: Technical-economical remarks). D'Errico, E., Funaro, G., Publ. L'Industria delle Costruzioni (journal) 1987, Edilstampa, Via Guattani 24, 00161 Roma, 1987, Lit 6,500 (per issue). *The results of a technical/economical analysis and monitoring of realized passive solar buildings are reported. L'Industria delle Costruzioni issue 188/Jun 1987.*

L'Edilizia Bioclimatica in Italia. Situazione attuale e Prospettive Future (Bioclimatic Building in Italy. Current Situation and Perspective of Developments). C.N.R. - P.F.E., Via Nizza128, 00198 Roma, 1987, Free. *Edited by CNR-PFE (National Research Council) on the basis of the results of researches performed in the frame of the, 2nd National Energy Project (PFE 2). Exhautive analysis of bioclimatic bldgs. in Italy, showing that greater efforts are needed to make bioclimatic building more cost effective.*

L'Energia Solare e la Produzione del freddo (Solar Energy and Cooling). Lazzarin, R., PEG Editrice, Via Flli Bressan 2, 20126 Milano, 1987, Lit 25,000. *Many types of cooling systems coupled with solar collectors are dealt with. Also covers plant regulation.*

L'Evoluzione dei Materiali Transparenti (New Developments in Transparent Materials). Mancini, E., Trimarchi, M., Publ. Modulo (journal) 1987, BE-MA, Via Teocrito 50, 20128 Milano, 1987, Lit 8,000 (per issue). *The behaviour of new kinds of transparent materials - (i.e. plastic materials) is investigated. Modulo issue 137 Dec '87.*

L'Influenza della massa nell'isolamento termico degli edifici (Influence of mass in thermal insulation of buildings). Di Cesare, G., Publ. Costruire in Laterizio, (Journal) 1988, PEG Editrice, Via Flli Bressan 2, 20126 Milano, 1988, Lit 6,000 (per issue). *The importance of the heat storage masses for a correct energetic behaviour of buildings is stressed.* ISBN:Ref. 0394 - 1590.

La Disponibilità Annua di Luce Naturale,(Annual Values of Daylighting and Sunlighting). Pollini, F., Publ. Luce (journal) 1987 ,A.I.D.I., Viale Monza 259, 20126 Milano, 1987, Lit 4,500 (per issue). *The use of Dresler's diagram is explained in Luce issue Jul.-Aug. 1987.* ISBN:Ref. 0024 - 7189.

La Progettazione dell'Architettura Bioclimatica (Design of Bioclimatic Architecture). Various Authors, Franco Muzzio & C. Editore, Via Bonporti 36, 35100 Padova, 1980, Lit 18,000. *Papers presented at an international Conference held in Bair are reported. 88 - 7021 - 123 - 3,.*

La Progettazione Termica degli Edifici con il Personal, Computer, (CAD Thermal Design of Buildings). Giaccone, A., Rizzo, G.,Franco Angeli Editore, Viale Monza 106, 20127 Milano, 1987, Lit 24,000. *Simplified calculation routines for thermal insulation design (residential & industrial), boilers & heating plants. Guide to evaluation of energy use. Stresses heat bridges optimisation of insulation, shading & solar radiation.*

Manuale di Climatologia I Modellie le Tecniche per l'Analisi del Terziario nella,progettazione energetica (Climatology Handbook). Guzzi, R., Franco Muzzio & C Editore, Via Bonporti 36, 35100 Padova, 1981, Lit 12,000. *By analysing various environmental elements (climate, atmosphere, solar radiation) in different conditions the author gives physical - mathematical models for a building design aimed at saving and recovery of energy.*

Metodologia di Misura dei Flussi Energetici in Edifici Solari Methods for Measurements of Energy Fluxes in Solar Buildings. Carderi, A., Funaro, G., Pagani, R., Caudana, B., Publ. L'Industrial delle,Costruzioni (journal) 1987, Edilstampa, Via Guattani 24, 00161 Roma, 1987, Lit 6,500 per issue. *Strategies for monitoring passive solar buildings. L'Industria delle Costruzioni issue 186 /Apr. 1987.*

Ottimizzazione Energetica di Edifici di Nuova Costruzione per Civile Abitazione (Energetic Optimization of New Residential Buildings). Arneodo, P., Publ. HTE (journal) 1987, PEG Editrice, Via Flli Bressan 2, 20126 Milano, 1987, Lit 5,500 (per issue). *Optimization of orientation, area of transparent surfaces, thermal insulation and thermal inertia of structures is discussed in relation to design of buildings best using radiation. HTE issue 49/Sept.-Oct. 1987.* ISSN/Ref 0391 - 5360.

Per una Valutazione Sintetica delle Prestazioni Illumino-techniche ed Energetiche di Alcune Tipologie Edilizie (A synthetic evaluation of energetic & lighting performances of some building typologies). Vio, M., Publ. HTE (journal) 1988, PEG Editrice, Via Flli Bressan 2, 20126 Milano, 1988, Lit 5,500 (per issue). *Thermal & lighting behaviour of different building types. HTE issue 51 Jan-Feb '88.* Ref 0391 - 5360.

Progettare con il Clima (Design according to Climate) Un Approccio Bioclimatico al Regionalismo Architettonico (A Bioclimatic Approach to Architectural Regionalism). Olgyay, V., Franco Muzzio & C. Editore,Via Bonporti 36, 35100 Padova, 1981, Lit 18,000. *The book takes into account four climates (cold, warm, dry hot, wet hot, and discusses their influence on orientation shadings, form, etc. of buildings in view of a good energetic design. Effects on materials are dealt with too.* 88-7021-166-5.

Progetto ed Energia (Design and Energy). Benedetti, C., Bacigalupi, V., Edizioni Kappa, Piazza Borghese 6, 00100 Roma, 1980, Lit 25,000. *Design of buildings with reference to different energy sources, with a section on passive solar. A useful handbook.*

Quando la Serra è Passiva,(When Sunsapce is a Passive Component). Funaro, G., Gambardella, O., Publ. Modulo (journal) 1986, BE-MA, Via Teocrito 50, 20128 Milano, 1986, Lit 8,000 (per issue). *A proposal for construction of sunspaces in retrofitting of existing buildings. Modulo issue 125/ Aug. 1986.*

Raggi Solari Sotto Controllo (Control of Sunlight). Grespan, O., Caglieris, G., Publ. Modulo (journal) 1986, BE-MA, Via Teocrito 50, 20128 Milano, 1986, Lit 8,000 per issue. *Types of glazing and shading elements for controlling sunlighting and achieving improved comfort with energy saving. Modulo issue 127/ Oct. 1987.*

Regolazione Automatica degli Impianti ad Energia Solare (Automatic Regulation of Solar Energy Plants). Sanguineti, R., PEG Editrice, Via Flli Bressan 2, 20126 Milano, 1987, Lit 12,000. *Both strategies of regulation and suitable appliances are dealt with. Information is given with reference to simple schemes. Large plants are presented showing different storage systems.*

Repertorio delle Caratteristiche Termofisiche dei Componenti Edilizi Opachi e Trasparenti - Collection of Thermo-physical properties of Opaque and, Transparent Components. CNR - PFE, Via Nizza 128, 00186 Roma, 1982. free publication. *Complete set of data for thermo-physical properties of transparent and opaque components are reported. It is the* second appendix to the *Guide for an Energy Conscious Design of Buildings.*

Risparmio di Energia nel Riscaldamento degli Edifici Atti dei Seminari Informativi del PFE (Energy Saving in Heating of Buildings: Proceedings of the Seminars of PFE). CNR - PFE, PEG Editrice, Via Flli Bressan 2, 20126 Milano, 1978 - 1988, Lit 6,000 - Lit 25,000. *Results of research performed in the frame of the National Energy Project and suggestions to designers. Six books available. SI 1 - March 1978, SI 2 - March 1979, SI 3 - March 1980, SI 4 & 4b - March 1981, SI 5 - February 1982, SI 6 - February 1988.*

Risparmio Energetico nel Patrimonio Edilizio Esistente (Energy Saving in Existing Buildings). Publ. R & D services of IACP Public Board for low cost building of Milan, I.A.C.P. - Milano, Viale Romagna 26, 20133 Milano, 1986, Lit 20,000. *By various authors for R & D services of IACP. Shows some case studies. New technologies for energy saving are explained and methods for economic analysis are supplied. 173pp.*

Simulazione Oraria del Comportamento Termico-Energetico degli Edifici - Unsteady State Simulation Code of Thermal Behaviour of Buildings. CNR - PFE, Via Nizza 128, 00186 Roma, 1982, free publication. *The MORE simulation code of thermal behaviour of buildings, is explained. It is an unsteady state simulation code. The book is the third appendix to the Guide for an Energy Conscious Design of Buildings.*

Sistemi Solari Passivi - Passive Solar Systems. Mazria, E., Franco Muzzio & Co. (Editore), Via Bonporti 36, 35100 Padova.

Solare e Scuola: un Binomio alla Verifica (Solar Energy in Schools - Survey of Thermal Behaviour). Arduini, M., Romanazzo, M., Publ. Modulo (journal) 1987, BE-MA, Via Teocrito 50, 20128 Milano, 1987, Lit 8,000 (per issue). *Presents result of a survey carried out by E.N.E.A. on a school at Montefiascone in the early '80s. The school uses active and passive systems. The survey evaluates the performances of the systems and provides recommendations for retrofitting. Modulo issue 136/Dec, '87.*

Solidi Energetici. Proposte di Design e Tecnologia Solare Soffice (Proposals for Architectural and Solar Design). Coniglio, M., Pirola Editore, Via Comelico 24, 20135 Milano, 1985, Lit 30,000. *Passive/active solar buildings. This handbook provides suggestions on the cost effective design of passive and active solar design.*

Tecnica e Tecnologia Solare Strat. & Tech. Equipments for Solar Energy Collection & Use. Granata, G., Paravia & C. Editori, Corso Racconigi 16, 10139 Torino, 1981, Lit 8,800. *Collection, storage and use of solar energy.*

Un Algoritmo per il Calcolo dell'Illuminamento Diretto e Riflesso (An Algorithm for Evaluation of Direct and Reflected Illuminance). Pellitteri, G.,Publ. L'Industria delle Costruzzioni (journal) 1987, Edilstampa, Via Guattani 24, 00161 Roma, 1987, Lit 6,500 (per issue). *An algorithm for evaluation of direct and reflected illuminance is presented. L'Industria delle Costruzioni issue 186/Apr. 1987.*

PORTUGUESE

A Tradição Energeticamente Consciente Catalogue. Energy Research Group, School of Architecture, University College Dublin, Richview, Clonskeagh, IRL-Dublin 14, 1989. *An A4 B&W reproduction of a poster exhibition of traditional responses to energy issues in Europe, prepared under the CEC's ARCHISOL programme by the Energy Research Group. The Original poster exhibition, in colour, was circulated to European schools of architecure.*

Apreciação Técnica das Propostas do Concurso de Aviero - Santiago (Criteria for Thermal Quality Evaluation in Low-income housing Projects). LNEC (National Laboratory for Civil Engineering), 1985. *Report on LNEC's work rules for construction of specific building types. Evaluation of a low-income housing project.*

Coeficientes de Transmissào Térmica de Elementos da Envolvente dos Edifícios - Heat Transfer Coefficients (U-Values) for Building Envelope Components. C. A. Pina Dos Santos, J. A. Vasconcelos Paiva, LNEC, Av. Brasil, Lisboa, 1990, 1000$00. *A comprehensive listing of all typical U-Values for common building elements in Portugese Buildings (walls,*

ceilings, parements and glazings) including a technical description of the methodology for calculations. ITE 28.

Energia Solar Passiva (Passive Solar Energy). Moita, Francisco, Direção Geral de Energia, 1987, 900$00 (Each) 2 volumes. *Produced by F. Moita for the Portugese Energy Department.*

O Vidro na Conservacão da Energia em Edifícios. E de Oliveira Fernandes,Covina (Companhia Vidrera Nacional),1 981, Published by COVINA - National Glass Company - Introduces passive and active concepts related to the use of glass in buildings.

Orientacões - Guia para o Projecto de Sistemas Solares Passivos em Portugal (Guidelines for the Design of Passive Solar Systems in Portugal). Gabinete de Fluidos e Calor, Departamento de Engenharia Mecanica, Faculdade de Engenharia, Universidada do Porto. *Short handbook adapted from The Los Alamos SLR method with the collaboration of Sara and J.D. Balcomb with specific reference to Portugese climatic conditions.*

Projectos do Sector de Construcão do MEREC. Comissão de Coordenacão da Regiào Centro, Coimbra, 1986. *A collection of brochures containing a summary description of building energy conservation related projects, developed within MEREC (Managing Energy and Resource Efficient Cities) in the central region of Portugal.*

Regras de Qualidade Relativas ao Conforto Termico dos Edifícios Escolares (Quality Rules Regarding Thermal Comfort in Schools). LNEC (National Laboratory for Civil Engineering), 1979.

Regras de Qualidade Termica de Edificios (Rules for Building Thermal Quality). CEGENE (Commission for Study of Energy Management in Bldgs), 1984, 1981-84. *Study by CEGENE (a section within the Portugese Ministry of Public Works) which has resulted in a preliminary standard for Thermal Quality in Buildings.*

Regulamento das Caracteristicas de Comportamento Térmico dos Edificios - Manual de Apoio, Applications Manual for the Regulations on Thermal Behaviour of Buildings. E. De Oliveira Fernandez, E. Maldonado, Dir. Geral de Energia, Av. De Outubro, 87, 1000 Lisboa, 1991, (In Press). *A support manual for the correct application of the Portugese Thermal Regulations for Buildings, including a guide describing the principles of energy-conscious design and specific data for common building materials and construction techniques.*

Technical Recommendations for Low-income Housing Construction. LNEC (Laboratorio Nacional de Engenharia Civil). *Second set of rules for low-income housing construction.*

Temperaturas Exteriores de Projecto e Número de Graus-Dia Design Temperatures and Degree-Days for Portugal. J. Casimiro Mendes, J. A. Vasconcelos Paiva, M. Rita Guerreiro, C. A Pina Dos Santos, LNEC, Av. Brasil, Lisboa, 1989, 1 800 PTE. *A detailed compilation of climate data necessary for correct thermal design of buildinga in Portugal. Incorporatea the most up-to-date information in great detail.*

SPANISH

Acondicionamiento Y Energia Solar en Arquitectura Conditioning and solar energy in architecture. Bedoya, C., Neila, J., Colegio Oficial de Arquitectos de Madrid (COAM), Servicio de Publicaciones del COAM, c/Barquillo 12, 28040 Madrid, 1986. *Introduction to Solar Energy and Passive Systems for architects. Covers solar radiation, thermal regulations active and passive systems, heating and cooling.* ISBN:84-85572-96-3.

Aislamiento Termico. (Aplicaciones en la Edificacion Y La Industria Economia de Energia) Thermal Insulation (Applications of Building and Energy Saving Industry). Margarida, M., Editores Técnicos Asociados S.A., C/Maignón 26, 08024 Barcelona, 1984, 3600 pta. *This book is a scientific study of building with special emphasis on thermal insulation as well as Spanish Building Thermal Performance Basic Regulation (NBE-CT-79).* ISBN:84-7146-243-5.

Arquitectura Solar (Aspectos Pasivos: bioclimatismo e Illuminación natural), Solar Architecture (Passive Aspects, Bioclimatism and Daylighting). Yáñez, Guillermo, Secretaria General Tecnica, MOPU, Ministerio de Obras Publicas Y Urbanismo (MOPU), Centro de Publicaciones del MOPU, Paseo de la Castellana 67, 28020 Madrid, 1988, 1500 pts. *A brief reflection on different architectonic concepts: insulation, daylighting, thermal inertia of buildings, passive systems, bioclimatism, cooling, etc.* ISBN 84 - 7433-542-6.

Atlas Climatico de España, Climatic Atlas of Spain. Secrtaria General Técnica, Inst. Nacional de Meteorologia (M.Transp, Turismo y Comun.) Cuidad Universitaria, 28040 Madrid, 1983. *43 illustrations including maps showing altitude, pressure, wind, precipitation, relative humidity etc. for regions in Spain. Sold together with Spanish Solar Radiation Atlas.* ISBN:84-500-9495-X.

Atlas De La Radiacion Solar En España - Spanish Solar Radiation Atlas. Font Tullot, I., Secretaria General Técnica, Inst. Nac. de Meteorologia, Instit. Nacional de Meteorologia (M. Transp.Turismo y Comuic), Instituto Nacional de Meteorologia (INM), Ciudad Universitária, 28040 Madrid, 1984. *Comprises thirty maps with yearly, monthly and daily average values of global solar radiation, insolation, time, and minimum/maximum solar radiation. It is a complement to the Spanish Climatic Atlas.* ISBN:84-505-0501-1.

Cursillo sobre la Aplicación práctica de la NORMA NBE-CT-79 sobre Condiciones Termicas en los edificios, Course and Building Regulation NBE-CT-79 - practical,application of building thermal conditions. Serra, J., Viti, A., Colegio Oficial de Arquitectos de Madrid (COAM), Servicio de Publicaciones del COAM, C/ Barquillo 12, 28001 Madrid, 1980. *Brochure on regulation NBE-CT-79 with practical information about building thermal behaviour.* ISBN: 84-85572-13-0.

El Instalador. Monografia No.13: La Energia Solar II (Sistemas y equipos). El Instalador, C/Navaleno 9, 28033 Madrid, 1982, 1500 Pta. *Journal article about thermal installation, passive systems, solar heating installations. etc.* ISBN: 12.465-1982.

Energia Solar, Edificacion Y Clima (Elementos para una arquitectura solar) Vol. I y II, Solar Energy Buildings and Climate (Elements for solar architecture). Yañez, G., Ministerio de Obras Publicas Y Urbanismo (MOPU), Servicio de Publicaciones del MOPU, Plaza San Juan de la Cruz s/n, 28003 Madrid, 1982, 3000 Pta. *Two volumes introducing solar/bioclimatic architecture covering solar radiation, active and passive systems, heating, cooling, traditional solutions.* ISBN:84-7433-220-6.

Energia Solar Pasiva en Edificacion: Metodos Para Comparar Diseños. Vega, S., Universidad de Valladolid, Secretariado de Publicaciones, Facultad de Medicina, Valladolid, 1987. *Compares different design methods.* ISBN: 84-86192-85-4.

Estavli D'Energia en el Disseny D'Edificis (Aplicació de sistemes d'aprofitament solar passiu) Energy Savings in Design Building (Passive solar systems applications). Mitjà, A., Esteve, J., Escobar, J., Generalitat de Catalunya, Department d'Indústria i Ene, Direcció General d'Energie, Avda. Diagonal 449, 08028, Barcelona, 1986. *Examples of passive solar system & building design with calculation methods, and economic studies for a range of passive systems.* ISBN:84-393-0670-9.

Evaluacion del Aporte Energetico a los Edificios Mediante Sistemas Naturales de Control Climatico en España - Energetic Contribution Evaluation of Buildings with Natural Systems of Climate Control in Spain. Serra, R. et al., Centro de Estudios de la Energia, Ministry of Industry and Energy (MINER), Paseo de la Castellana 160, 28040 Madrid, 1982. *Preliminary study on energy in buildings using natural, systems of climate control.*

La Energia en la Edificacion- Politica Energetica Y Ahorro de Energia - Energy in Buildings, Energy Policy and Energy Savings. Barcelo Rico-Avello, G., Editorial INDEX, Ed., 1978.

La Tradicion Energeticamente Consciente Catalogue. Energy Research Group, School of Architecture, University College Dublin, Richview, Clonskeagh, IRL-Dublin 14, 1989. *An A4 B&W reproduction of a poster exhibition of traditional responses to energy issues in Europe, prepared under the CEC's*

ARCHISOL programme by the Energy Research Group. The Original poster exhibition, in colour, was circulated to European schools of architecure.

Las Energias Alternativas en la Arquitectura,The Alternative Energies in Architecture. Bedoya, C., Carril, A., Macias, M., Neila, J., Colegio Oficial de Arquitectos de MADRID (COAM), Servicio de Publicaciones del COAM, C/Barquillo 12, 28040 Madrid, 1982. *Introduction to Alternative Energies for architects covering passive systems, heat transfer, photovoltaics, wind energy biomass, etc.* ISBN: 84-85572-35-1.

Norma Basica de Edificacion de Condiciones Termicas de los Edificios (NBE-CT-79) Basic Construction Regulations for Building Thermal Performance. Boletin Oficial del Estado, Madrid, Ministerio de Obras Publicas Y Urbanismo (MOPU), 1979, *Spanish Basic Compulsory Building Regulation.*

Propuesta Para La Extension De La Norma Basica De Edificacion De Condiciones Termicas De Los Edificios Proposal for the extension of the building thermal performance basic regulation (NBE-CT-79). Secretaria General Tecnica, CIEMAT, Ministry of Industry and Energy Solar División-IER, Avda. Complutense 22, 28040 Madrid, 1988. *In this proposal the solar gains and passive solar concepts have been considered following the E.C. Guidelines from DG III and XVII (EUROCODE).* ISBN:84-7834-009-2.

Revista del MOPU: Guia de la Arquitectura Popular en España MOPU magazine: Spanish vernacular architecture guide. Ministry Public Works and Urbanisme (MOPU), Servicio de Publicaciones del MOPU, Plaza San Juan de la Cruz s/n, 28003 Madrid. 1986, 450 Pta. *Bi-monthly journal covering vernacular architecture. Produced by The Spanish Ministry of Public Works and Urbanism (MOPU).* ISBN: 0212-7148.

Soleamiento Y Energia Solar (Aplicaciones a la edificación :'Sunny & Solar Energy (Building Applications)'. Herrero, M.A., Universidad Politécnica de Valencia, (Servicio de Publicaciones), 1985. *Theory and calculation methods for flat solar collectors and passive systems with illustrations and case studies.* ISBN: 84-600-3975-7

CLIMA, LUGAR Y ARQUITECTURA: Manual de Diseño Bioclimatico Climate Location and Architecture: Bioclimatic Design Manual. Serra, R., Secretaria General Técnica del CIEMAT, Avda. Complutense, 22, 28040 Madrid, 1989. *Bioclimatic design Manual to help in the design to the architects. The disquette is very useful and easier from predesign process in the passive solar buildings design.*

Comportamiento Energitico De Edificios Solares Pasivos: Plan De Monotorizaciòn Del Instituto De Energias Renovables, Energetic Performance Of Passive Solar Buildings, Monitorization Planning Of Renewable Energies Institute. Heras, M. R., Marco, J., CIEMAT, Avda. Complutense, 22, 28040 Madrid, 1990. *This book recopilated the experience from four buildings evaluated by four spanish teams in collaboration with IER-CIEMAT, and included the global evaluated planing to know the energetic performance of passive solar buildings.*

DESIGN TOOLS

DANISH

EXPERTSYSTEM - Pasive Solar Heating. Teknogisk Institut, Varmeteknik. *Combined calculation and design program under development.*

Heat Losses and Gains in Houses. Indoor Climate, TSBI, Statens Byggeforskingsinstitut, 1985. *Heat losses and gains in houses. Indoor climate. Energy Consumption and its effects. Version modified for passive solar heated buildings under development.*

NS 3031. Norges Byggstandardisenngsrad, Copenhagen 10, N-0566 Oslo 5. *Tel: +47 (2) 355020.*

SBI-148/CSBI. Danish Buildg. Res. Inst., P.O. Box 119, DK-2970 HOERSHOLM, Dänemark. + 45 (46) 286 3533. *Contact: Mr Johnson.*

DUTCH

DYWON. TNO TAP H.A.L. van Dijk, P.O. Box 155, N-2600, AD Delft. +31 (15) 787096. *Contact: Mr van Dann.*

ELAN. Physical Aspects of Built Environment Section, Faculty of Architecture, Building & Planning, Eindhoven University of Technology, Eindhoven NL. *A small dynamic multi-zone model for the calculation of building heating and cooling demands which is intended for use at an early design stage (i.e., requiring only global data). Extensive validation with an advanced thermal model has shown reliable results.*

ENGLISH

BFEP/ESY-BFEP. *Contact: Mr Peter Basnett, ECRC (Env. Cncl. Res. Cntr.), CAPENHURST, Chester, CH1. Tel: +44 (51) 3472421.*

BLAST. Deut.Vertr.: Ing.-Büro Guido von Thun, Kesslerplatz 5, 8500 Nürnbuerg 20. +49 (0911) 550959/533352. *CERL Constr. Eng. Res. Lab., US Dep. of the Army, Champaign, Ill.*

BREADMIT. R. Alphey, or Mr Bloomfield, BRE. Dept. of the Environment, Bucknalls Lane, Garston, Watford WD2 7JR. +44 (923)664477. *Admittance Method.*

BREDEM Worksheet Program. Building Research Estab., BRE Technical Consultancy, Garston, Watford, WD2 7JR U.K. Tel +(0923) 664 800 or +(0908) 510 596. 1989. *Computer emulation of worksheets produced to simplify the use of the BRE Domestic Energy Model, BREDEM (1985). Can be used on an IBM PC or Psion Organiser. Distributed by Energy Advisory Services Ltd., Old Manor House, The Green, Hanslope MK19 7LS. Marketed as 'Energy Calculator'.*

BREDEM-8. G. Henderson, or Mr. Bloomfield, BRE, Dept. of the Environment, Bucknalls Lane, Garston, Watford, WD2 7JR, UK. Tel: + 44 (923) 664517. *Method: time avg. Source: IEA 4, 12/88.*

BREEAM, Environmental Assessment for New Office Design. ECD Partnership, Building Research Establishment, BRE, Environmental Assessment Scheme, Garston, Watford WD2 7JR, UK, 1990. *Method of assessing the environmental analysis of buildings under a scheme operated by BRE. Takes account of environmental issues on three levels: global, neighbourhood and indoor.*

BREEZE. Building Reearch Station, Garston, Watford, WD2 7JR, UK. *An interactive, user-friendly computer programme for calculating the ventilation & inter-cell air flows in multi-celled buildings. Requires a Hercules graphics card reviewed in CIBSE Building Society Journal July 1985.*

BRIS. Teddy Rosenthal, DALAP - Tel: +46 (8) 271160/271161.

CALPAS 3 (4). Berkeley Solar Group, P.O.Box 3289, Berkeley, CA.94703 USA. +1 (415)843-7600, *Output: Hour, Source: IEA 4, 12/88, Bldg, Types: res. small comm. Calculations:heating and cooling Cons. costs.*

CELOTEX EQ HOMEBUILDER. Celotex, Warwick House, St Mary's Road, London W5, UK. 1988, £95. U.K. *New U.K. regulations have increased insulation standards but allow trade-off in U-values between bldg. elements provided overall insulation achieves standard. This program allows rapid 'what-if' calculations to be made.*

COMPUTER PROGRAM FOR CALCULATING THE DAYLIGHT LEVEL IN A ROOM. Jordans, A.A., Institute of Applied Physics TNO-TH, P.O. Box 155, 2600 AD Delft, The Netherlands. *Reviewed in Energy & Buildings No 6, 1984, pp. 207-212. This program calculates the total quantity of daylight in a room from direct, externally & internally reflected light taking account of internal & external geometry. CIE formulae for overcast & clear skies. Computes shadows from buildings.*

DAYLIGHT. Anglia College Enterprises Ltd., External Services, Dept. of the Built Environment, Anglia College, Enterprises Ltd., Victoria Road South, Chelmsford, Essex CM1 1LL, UK. 1990. UK£ 40. *Menu-driven, user friendly, IBM PC-based software which allows individual rooms to be described and evaluated for daylight distributin. Graphical output shows daylight factors or illuminance levels in plan using a grid or isolux plot.*

DEROB. Univ. of Texas., 2604 Parkview Drive, Austin, TX 78757 USA. +1 (512) 471-3148, *Contact: Dr F Aruny, Dept. of Architecture.*

DOE 2.1 C. Aerosoft International Inc., 3120 S. Wadsworth Blvd., Denver CO, 80227. +1 (303) 969 0170, *Source: SERI, Contact: March A Ros, LBL/Los Alamos Scient. Lab.*

DRAN Programme Building - related design tool. ELKEPA, (Greek Productivity Centre), ELKEPA, Franzth 9, 11743 Athens, Greece, Fax: 1-9239430 Tel: 1-9215129. *A computer programme for evaluating the shading caused by neighbouring buildings or other elements incorporated into the building envelope itself. The programme is also used to evaluate the effect of different types of glazing on solar energy penetration and natural lighting.*

DYNAPAS I, II. *Contact: Geoffrey Moore Langdon, Solar Design Specialists, 2106 Mass. Ave., Suite 2C, Troy, N.Y. 12180. Tel: +1 (513) 2745544 or 2745600.*

EEDO (CIRA). LBL (for CIRA), 400 Morgan Center, Butler, Penns. 16001, USA. +1 (412) 285-4761. *Contact: Burt Hill Kosar Rittelmann Associates.*

EMPG3 Simulation of Thermal Systems - A modular program with an interactive preprocessor (EMPG3). Dutré, W., Energy Research Group, School of Architecture, University College Dublin, Richview, Clonskeagh IRL-Dublin 14, 1991. *Prepared within the CEC DGXII OPSYS Programme, this is a modular simulation package for the detailed design of user-defined solar and non-solar systems. A fully interactive preprocessor allows details of the systems to be stored for multiple simulations under different conditions.* Documentation: EUR 13354 EN / ISBN 0-7923-1235-X.

EMPS 2.0. Dick Merian, Arthur D. Little Inc., 3412 Hillview Ave., PALO ALTO, CA 94304. Tel: +1 (415) 855-2168.

Energy Design Advisory service, A Division of the Royal Incorporation of Architects of Scotland (RIAS) Practice Services. West of Scotland Energy Working Group, Free publication, *Advice on condensation, glass selection, plant performance, solar gain & shading, atria, overheating prediction, energy savings, life-cycle cost calculation, energy study analysis, new product appraisal, grants. Information on new research, energy guidelines from CIBSE, building regs & weather data.*

ESM European Simplified Methods for Active Solar System Design. Bourges, B. et al. Energy Research Group, School of Architecture, University College Dublin, Richview, Clonskeagh, IRL-Dublin 14, 1991. *Prepared within the CEC DGXII OPSYS Programme, this is a set of two user-friendly programmes, covered by one manual, based on correlation methods which allows fast calculation of the system performance and sensitivity to the main design parameters.* Documentation: EUR 13355 EN / ISBN 0-7923-1230-9.

ESP. Prof. J.A. Clarke, ESRU, 131 Rottenrow, Glasgow G4 ONG. +44 (41) 552 4400 ext. 3986. *ESRU/ABACUS. Originally for mainframe computers but now available for some microcomputers.*

EURSOL Simulation of Water Based Thermal Solar Systems. Dutré, W., Energy Research Group, School of Architecture, University College Dublin, Richview,

Clonskeagh, IRL-Dublin 14, 1991. *Prepared within the CEC DGXII OPSYS programme, this user-friendly interactive programme allows the designer to simulate active thermal solar systems for space heating and hot water production, based on a pre-defined range of components. The system parameters can be changed with complete freedom.* Documentation: EUR 13353 EN / ISBN 0-7923-1236-8.

EXCALIBUR (PASOLE). Stocker Road, EXETER EX4 4QL, Tel: +44 (392) 264144. *Mr Perman, ESG Phys. Dep. Univ. of Exeter.*

F-Chart. Solar Energy Research, 1536 Cole Blvd., Golden, CO 80401 USA.

F-Load. Beckman-Duffie and Assoc., 4406 Fox Bluff Road, MIDDLETON, WI 53567 USA. +1 (608)263-1590. *Output: M,A cooling for residential buildings.*

FRES. SINTEF, N-7043 Trondheim. +47 (7) 59 38 71. *Contact: Attn. Terjue Jacobsen.*

GFP Programme Building - Related design tools. ELKEPA, (Greek Productivity Centre), Greek Productivity Centre, Franzth 9, 11743 Athens, Greece. Fax: 1-9239430 Tel: 1-9215129. *A computer programme for evaluating the energy performance of a building with or without auxiliary heating or cooling.*

Higbie's Formula. Higbie, H.H., Illuminating Engineering Society, Briaicliff Manor, New York, 1924. *Prediction of Daylight from Vertical Windows: a basic formula for range of simple and complex daylight analysis methods.*

HOT 2000 (HOTCAN 3). Nat. Res. Council of Canada, Div. of Buildg. Res., P.O. Box 7081 Post St. J, OTTAWA, ON/CND K2A 326. *Contact: Energy Analysis Software.*

HTB 2. University of Wales, Institute of Science and Technology, Cardiff. Tel: + 44 222 874000 Ext 5979. *Contact : P. Jones.*

MEPA. Royal Inst. of Technology Archit. - Dept. Building Design, 10044 Stockholm. Tel:+46 (35) 100160. *Contact: Per Mikkel Henrikson, or Prof. Bengt Hidemark, RIT.*

Method 5000. Raoust, M, Claux, P., Franca, J.P., Gilles, R., Pesso, A., and Pouget, A., Energy Research Group, UCD - Richview, Clonskeagh IRL - Dublin 14. Tel: + 353 1 269 2750, Fax: + 353 1 283 8908, 1992. *M5000 is a correlation method for the evaluation of monthly heatings loads in domestic building using European Climatic data. It runs on an IBM PC or compatible.*

MICROPAS. Enercomp, 757 Russell Bd. Suite A3, Davis, CA 95616 USA. *Tel: +1 (916) 753-3400.*

Milton Keynes Energy Cost Index (MKECI). Milton Keynes Development Corporation. *Index of annual cost per sq. metre floor area using BREDEM-5 computer model developed by The British Research Establishment. A house built to current UK codes would have an MKECI of 170. Those built for The Energy World exhibition at Milton Keynes are 120 or less.*

PASCOOL. Dr. Mat.Santamouris, Greek Centre of Productivity (ELKEPA). *PASCOOL is a programme for the evaluation of passive and hybrid cooling components. Radiative, direct and indirect evaporative and earth cooling are considered.*

Passive Solar Resource Guide. Goulding, J.R., Lewis J.O., Steemers, T. C. (Eds.), Energy Research Group, School of Architecture, University College Dublin, Richview, Clonskeagh, IRL-Dublin 14, Tel: + 353 1 269 2750, Fax: + 353 1 283 8908, 1990. IR£ 10. *A guide to publications, design tools and other aids to passive solar design for European architects. On 3.5" computer disk, for use on Apple Macintosh with HyperCard.*

PASSPORT. Santamouris, M., Sigalas, S., Protechna Ltd. for CEC PASSYS Programme, Themistokleous 87, 10683 Athens, GREECE. Fax: 1-3612285 Tel: 1-3612751. *A computer programme evaluating the heating and cooling load of a building taking into account environmental parameters.*

REM/Design. 2540 Frontier Av. Su.201, Boulder, CO 80301 USA. *Tel: +1 (303)444-4149 (Fax 4304).*

SCRIBE SYSTEM. Echotech, 45 Harefield Road, Sheffield, S11 8NU, U.K., 1988. *3-D computer modelling system for bldg. design with links to RoboCAD (drawing system),*

HEATCALC (*beat loss calculations*), SOLPRO (*shadow projection*), SERILUX (*daylight levels*), SPECALC (*schedule of works and costs*), HEATCALC (*heat loss*), SPIEL (*multizone thermal modeller*). *Tel: + 44 (742) 680982.*

SCRIBE USER GROUP. *Contact: Ron Garwood, Garwood Associates, 32 Painswick Road, Cheltenham GL 502HA. Tel: + (0242) 238363.*

SERILINK. Ecotech Design Ltd., 45 Harefield Road, Sheffield S11 8NU (UK). *Tel: +44 (742) 680982.*

Solar 5. Murray Milne., Denwun Lin., Howley, Rosemary E., UCLA Graduate School of Architecture (Urban Planning), Los Angeles, CA 90024, U.S.A. 1986, Free (share ware). *Self instructional tool (2 disks & manual) which runs on IBM PC. Produces 3D graphs predicting energy use. Contains 20 pre-defined building types. Rapid assessments are possible with as few as 4 parameters: climate, building type, floor area, no. of floors.*

Solar Site Selector. Lewis and Associates, 105 Rockwood Drive, Grass Valley, CA 95945, USA.

SPIEL. Ecotech, 26 Botanical Road, Sheffield S11 8RP (UK).+ 44 (742) 660734.

Study on Architectural Design Criteria Through Natural Systems for Climate Control for Spain. CIEMAT, 1988. *Financed by CIEMAT. Carried out by E.T.S. de Arquitectura de Barcelona. This study includes a program which enables the designer to evaluate the energy requirements of a building at the preliminary stage.*

SUNCODE/SERIRES. Ecotope, Inc., BERLIN, IBUS. *Deut. Vertr.*

SUNDAY. 2540 Frontier Av., Su.201, Boulder, CO 80301, USA. Tel: +1 (303) 444-4149 (Fax 4304). *Contact: Architectural Energy Corp.*

Sunmic Solar Angle Finder. Sundance Solar, 24 Dickens Circle, Salinas, California CA 93901 U.S.A. Tel: (408)422-2000, 1985. *Used with a model of the building (any scale) and an artificial light source to show shadowing. Available for even-numbered latitudes from 26 deg - 48 deg.*

SUNPAS. Solarsoft, Inc. 1406 Burlingame Av. Su., Burlingame, CA 94010 USA. *Tel: +1 (415) 342-3338.*

TAS. AMAZON, Lyoner Strasse 44-48, D-6000 Frankfurt am Main 71. Tel: + 49(69) 666 4086. *Simulation programme written for Apollo mini-computer. Additional details from Amazon Energy Ltd., Sunrise Parkway, Milton Keynes MK14 6LQ, U.K. Tel: + 44 908 664123.*

TASS (formerly Build). *Originally for mainframe computers, now available for some microcomputers.*

THERM. *Originally for mainframe computers now available for some micro computers.*

TOWNSCOPE Computer Aided Methodology for Passive Renewal. Lema / Ektenepol, LEMA - Université de Liège, Belgium. *An urban design and renewal computer software programme which takes account of solar gains, shading, wind effects and energy consumption.*

TRNSYS. Univ. of WI., Solar En. Lab., 1500 Johnson Drive, Madison W1 53706. *Tel: +1 (608) 263-1589.*

TWO-TONE. LBL, Bldg. 90, Berkeley, CA 94720, USA. *Tel: +1 (415) 843-2740, ext. 5711.*

FRENCH

B-SOL. Groupe ABC-EAM, Groupe ABC, Ecole d'architecture de Marseille, 70 route Léon - Lachamp, 13288 Marseille Cedex 9, 1985. *Calcul des besoins de chauffage de logements équipés de systémes solaires passifs.*

CASAMO Clim. Centre d'énergetique de l'école des Mines de Paris, DIALOGIC, 1986-88, 5000 FF H.T. *Simulation du comportement thermique du bâtiment destinée: - Au contrôle des conditions de confort thermique d'été sans climatisation - Au calcul des éventuellles charges de climatisation.*

Casamo GB et A. Centre d'énergetique de l'école des mines de Paris, ARMINES, 60 Bd St Michel, 75272 Paris, CEDEX 06, 1982-85, 3000 FF H.T. *Estimation des besoins de chauffage mensuels d'un habitat.*

CODYBA. INSA VALOR - INSA, 20 Avenue Einstein, 69621 Villeurbanne Cedex, 1984-89, 8000 FF H.T. Etude dynamique du comportement thermique d'un bâtiment monozone.

COMFIE. B. Peuportier, I. Blanc Sommeureux, ARMINES Centre d'énergetique, 60 Bd St. Michel, 75272 Paris Cedex 06, 420 FF H.T.

Simulation solaire et thermique em régime variable. Bâtiment multizone sans stratifictaion météologique européenne.

Didacticiels: Vitrage - Ensoleillement Thermique du Mur - Air Humide. Logedic, AFME (Service Formation), 27 rue Louis Vicat, 75015 Paris, 1500 FF pour chaque module. *Educational software for Architectural or Engineering Schools.*

LESOSAI. EPFL, Bâtiment LESO-Ecublens, Ecole Polytechnique Fed., CH-1015 - Lausanne. +41 (21) 6934545.

LPB4/MBDS. Lab.Thermodyn, Fac. Sc.Appl. R. Ernest Solvay-21-Bâtiment 638 - 4000 Liège. *Contact: Marc Sroegnard. Tel: + 32 (41) 520180 Ext. 183.*

MASQUES. E. A. Marseille, Groupe ABC, EAM,70 Avenue M. Lachant, 13288 Marseille Cedex 09, 1987-89, 500 FF H.T. *Tracé des ombres et calcul du facteur d'ombre des masques intégrés.*

METHODE 5000 Version européenne. Claux, Pesso, Raoust, Franca, Gilles, Pouget, DIALOGIC, 70 Bd de Magenta, 75010 Paris, 1984-86, 4000 FF H.T. *Calcul des besoins mensuels de chauffage pour bâtiments d'habitation, analyse du confort, création d'un fichier de données climatiques européennes.*

MICROPAS. ENERCOMP Inc. USA, ADRET, 2 rue Cloris Hugues, 05200 Embrun, 1983-88, 35000 FF H.T. *Simulation énergetique de bâtiments et de leurs équipements.*

MODPAS. Doninik Chuard, Sorane S.A., Lausanne. Tel: + 41 (21) 371175.

OASIS. DIALOGIC - TETA, DIALOGIC, 70 Bd. Magenta, F-75010 Paris, Tel: + 33 1 40345320, 1985-89, 9500 FF. *Software for the estimation of comfort conditions and cooling zones practitioners. Additional Contact: Joachim Roebner, Arndtstr. 3, 6200 Wiesbaden, Tel: + 49 (6121) 374682.*

PASSIM. EPFL Gr. de Rech. en EN. Sol., CH-1015 Lausanne. *Tel: + 41 (21) 693 45 45.*

SIMULA. CERMA - EA de Nantes, IPTIC, 3 rue Léon Bonnat, 75016 Paris, 1989, 10000 FF H.T. *Simulation solaire et thermique multizone d'un bâtiment à régime variable.*

TELEMATIQUE / Thermique Assistée par Minitel : Un Service pour obtenir rapidement un calcul ou une recherche documentaire. *24 hour service provided by Minitel to calculate the energy rating of buildings.*

TRIVEGETAL, Architecture et Végétation - Ambiances Micro-Climatiques. A. Ouvot, EA de Marseille, Groupe ABC, Ecole d'Architecture de Marseille (Groupe ABC), 70 route Léon Lachamp, 13288 Marseille Cedex 9, 1986-89. *Base de données sur les aspects micro-climatiques et végétaux.*

GERMAN

DYNBIL. Ebök - Büro für Energ.berat. & ökol. Konszeote, Dorfackerstr. 12, D-7400 Tübingen. +49 (7071) 82529.

EBIWAN II. Fa. JUPROMA, Herren Poiss, Protaska, HOFERN nR. 14, A-2081. +43 (2949) 2311. *Contact: Dr Gottfried Schaffa.*

ENERGIE-KNOW-HOW. In der Schlade 13, D-5203 Much FRG. +49 (2245) 5024 u.4433. *Contact: Peter Weber.*

GEBA. Lugeck 1-2, A-1010 Wien. +43 (1) 526204. Contact: Dr W. Heindl.

GOSOL-2. Zinsholzstrasse 11, D-7000 Stuttgart 75. + 49 (711) 473994 (privat). *Contact: Dipl. Ing. Peter Goretzki.*

HAUS. ITW Uni Stuttgart. Tel: + 49 (711) 685-3203. *Contact: Herr Stanzel, ITW Univ. STUTTGART.*

HELIOS 1. Herr Zweifel - 4530, Herr Gyggli - 4718, CH-8600 Dübendorf. +41 (1)823 55 11. *Contact: EMPA Sektion Baupysik.*

IGLOU. Fa. Gähler & Partner, Badstraße 16, CH-5400 ENNET-BADEN. Tel: + 41 (56) 209511.

ISO DP 9164. Contact: Dr. H. Werner, Frauhofer IBP Postfach 1180, D-8150, Holzkirchen. *Tel: + 49 (8024) 643 12.*

JULOTTA. IWU, Wolfgang Feist ,Annastrasse 15, D-6100 Darmstadt. *Tel: + 49 (6151) 2904-0. Contact: Kurt Kjelblad, Dept. Bldg. Science.*

TICAD 2 & 3. *Contact: L. Jesch, The Franklin Company Consultants, 192 Franklin Road, Birmingham B30 2HE, U.K. Tel: +44 (32) 459 4853/4826 (Fax: 459 8206).*

254

GREEK

Greek Productivity Centre (EL. KE. PA.) Programme. Athanasakoy, E. *An inter-active programme for evaluation of passive solar designs based on the unutilisability method, developed by the Univ. of Wisconsin, covering heating load, utilisation of solar energy, heat loss coefficient & thermal capacity shading, passive solar system characteristics & climate data.*

THERMAL INSULATION. Civildata.

URBACAD. Santamouris, M., Ektenepol S.A., 1989. *Produced by Protechna Ltd for Ektenepol S.A., programme has been developed to optimize solar and daylighting for urban design using 2-D and 3-D visualisation. Calculations are: Instantaneous & mean monthly shading, heating & cooling load performance of hybrid passive cooling components, & gains.*

ITALIAN

In Laterizio (journal) Series of energy-related computer programmes. Carrier Italiano. *A series of computer programmes have been published by the magazine In Laterzio. Topics:-Unsteady state, thermal behaviour of a wall in steady state condition behaviour of a wall (Glaser method).*

Series of Energy Related Computer Programmes. Publ. in Laterizio (journal), 1986/1987, BE-MA, Via Teocrito, 50 - 20126 Milano, 1986/1987, free lists. *A series of computer programs. Topics: Unsteady state thermal behaviour of a wall steady state conditions behaviour of a wall (Glaser method).*

PORTUGUESE

AUDITORIAS. Faculdade de Engenharia do Porto, Departmento de Mecanica, Gabinete de Fuidos e Calor, 1986. *A building load and systems performance analysis programme based on the ASHRAE method but adapted to Portugese construction and climatic conditions. A detailed manual is available. Has been used in survey of schools in Northern Portugal.*

PRESOP A Passive Solar Building Design Computer Program. Canha da Piedade, A., Moret Rodrigues, A., Santos e Castro, L., Instituto Superior Technico, 1st Technical University of Lisbon. *Passive solar building design*

code, which uses a thermal network solution to simulate building behaviour.

RCCTE, A PC based method to quantify building heating and cooling needs (yearly) for compliance with the Portugese Thermal Regulations. Maldonado, E., Quinta, P., Inegi, Fac. Engenharia, Rua dos Bragas, 4099 Porto Codex, 1991, 25.000 PTE. *A user-friendly spreadsheet containing a detailed data-base and producing the forms necessary to demonstrate compliance with the Portugese Building Thermal Regulations.*

SPANISH

S4PAS: Simulación Simplificada de Sistemas Solares Pasivos S4PAS: Simplified Simulation tool for Passive Solar Systems. Catedra de Termotecnia, E.S.I. Industriales, Universidad de Sevilla, CIEMAT, Division Solar, IER-CIEMAT, Avda. Complutense, 22, 28040 Madrid, 1991. *Financed by CIEMAT*

S3PAS: Metodo de Simulación del Comportamiento Energetico de Edificios Solares Pasivos en Climas Españoles, S3PAS: Simulation of Passive Solar Systems (Simulation Method for the Thermal Behaviour of Buildings in Spanish Climate). Catedra de Termotécnia de la E.T.S.I. Industriales, Universidad de Sevilla, Division Solar, IER-CIEMAT, Avda. Complutense, 22, 28040 Madrid, 1989. *Financed by CIEMAT*

CLA - MANUAL DE DISEÑO BIOCLIMATICO PROGRAMA DE ORDENATOR (incluido en el libro "Clima, Lugar Y Arquitectura") CLA - COMPUTER-PROGRAM FOR THE BIOCLIMATIC DESIGN (included in Climate, Location and Architecture Book). Serra, R., Secretaria General Tecnica del CIEMAT, CIEMAT, Avda. Complutense, 22, 28040 Madrid, 1989. *Passive solar building design tool from pre-design process to simulate the energetic performance of the buildings.*

A12.3

AUDIO VISUAL MATERIAL

ENGLISH

Catalogue of Films and Television Programmes on Architecture, Town Planning and the environment. MacFarlane, Jane, Oxford Polytechnic, School of Planning, Headington, Oxford, OX3 OBP, UK. 1988, £9.00. *Series: Oxford Papers on Planning, Education and Research. 1200 entries covering British television from 1949- and films worldwide from 1902-. Addresses of distributors and libraries. Information on availability of films.*

Energy Effectiveness Part 2 of The Better Building Series. CPD in Construction Group, 26 Store Street, London WCIE 7BT, U.K. *Introduction to the basis of energy conscious design with 3 case studies includes a short VHS video (PAL format).*

Felmore low energy housing. Willoughby, John, Commonwealth Association of Architects, 1981. *Environmental design aids.*

Solar Architecture in Europe An Introductory Video. David Clarke Associates, Energy Research Group, School of Architecture, University College Dublin, Richview, Clonskeagh, Dublin 14, Ireland. Tel +(353 1) 2693244 / 2692750, 1988, IR £25.00. *Produced under the ARCHISOL programme for the Commission of the European Communities (DG XII), this 20 minute video explains how solar architecture can lead to improved standards of thermal and visual comfort. It is intended for those with little previous experience of passive solar design.*

Solar Houses in the UK. Comonwealth Association of Architects, 1981. *Environmental Design Aids. (Tape/Slide).*

FRENCH

Bioclimatisme. Ministère de l'Equipement et du Logement, 2 Avenue du Parc de Passy, 75275 Paris, Cedex 16, 1981. *Video presentation of bioclimatic buildings.*

Construire avec le Climat. Ministère de l'Equipement et du Logement, 2 Avenue du Parc de Passy, 75275 Paris, Cedex 16, 1980. *Film on building and climate.*

Créer le Climat. Vasselin, Harold, CNRS Audiovisuel, 27 rue Paul Bert, 94204 Ivry sur Seine Cedex, 1988. *25 minute video - from the design to the realisation of a building, the management of climatic & energetic parameters to modify the climate.*

Juste une Question A propos d'Energie. Vaye, Marc, AFME - EDRA - ESA, AFME - Cellule Audiovisuelle, 27 rue Louis Vicât, 75737 Paris Cedex, 1987. *Video on the question of architecture and energy, from the analysis of Villa Savoye (Le Corbusier) and Maison Ronde (Mario Botta).*

SPANISH

Arquitectura Bioclimatica "El Sol Tambien Vive en Casa" Bioclimatic Architecture "The Sun Also Lives at Home". Secretaria General Técnica del CIEMAT, Mass Media Department of CIEMAT, Avda. Complutense 22, 28040 Madrid (Spain), 1988. *Fifteen-minute film with an introduction to vernacular Spanish architecture & passive solar bldgs showing 2 solar buildings thermally evaluated by the IER. C.E.M.A.Los Molinos Crevillente, Alicante, (non-residential and) Pedrajas de Esteban, Valladolid (residential).*

COURSES AND EVENTS

ENGLISH

Energy Engineering. International centre for heat and mass transfer, P.O. Box 522, Belgrade, Yugoslavia. Tel + 38 11 455 663, Fax + 38 11 444 0195 or 458 676. *A postgraduate course using video and interactive software (IBM PC) packages sponsored by UNESCO designed for use in academic and industrial institutions. The course structure comprises 3 modules: Basic sciences (140 Hrs), Sciences (120 Hrs) and Energy System Engineering (210 Hrs).*

FRENCH

Gradient Thermique du Bâtiment. Agence Française pour la Maitrise de l'Energie (AFME), 27 rue Louis Vicat 75015 Paris, 1988. *A series of courseware on all aspects of passive solar energy - radiation, glazing, walls, psychometry.*

GERMAN

Principles of Renewable Energy Use. Naumann, E. Dr., Department of Physics, University at Oldenburg, P.O. Box 2503, D-2900 Oldenburg FRG. Annual. *11 month course (held in English), beginning on October 1st each year. Designed for scientists & engineers with at least 4 years of academic training and aims to prepare for a professional occupation in the energy sector of the Third World. Source:- ISES News, Issue 60, Winter 1987.*

GREEK

Low Energy Building Design. Papadopoulos, M., Dr., Axarli, C., Dr. The Aristotelion University of Thessaloniki. *One semester course at the School of Civil Engineering, covering fundamentals of climate, comfort, insulation, shading, physics, passive solar systems and computation methods. Textbook - 'Low Energy and Passive Solar Systems', (1982) by Papadopoulos & Axarli.*

Low-energy design & application of solar strategies to buildings. Kontoroupis, G. Dr., National Technical University - School of Architecture, 42 Patision Str, 106 82 Athens, Greece. *A two semester course covering fundamentals, gain, losses, cooling, configuration, shading, comfort, comparison of conventional & alternative systems, active, passive & hybrid systems, SLR, Method 5000, F-Chart, Yazaki simulation programme. Text used: 'Low Energy Building Design'.*

Special Topics on Building Construction. Papadopoulos, M., Dr., Axarli, C., Dr., Aristotelion University of Thessaloniki. *One semester course in the School of Civil Engineering, covering thermal construction, performance, insulation, energy conservation, moisture control and vapour barriers.*

ITALIAN

Congressi Nazionali (A.T.I. National Conferences) Associazione Termotecnica Italiana (Italian Association of Thermotechnics). A.T.I., Istituto di Fisica Tecnica - Facoltà di Ingegneria - Viale Risorgimento 2, 40100 Bologna. *Conferences on thermotechnics in general Sections are reserved for energy and buildings. They are held yearly in September. The venue varies.*

Convegni (A.I.C.A.R.R.'s Conferences) Associazione Italiana Condizionamento dell'Aria, Riscaldamento, Refrigerazione. A.I.C.A.R.R., Via Sardegna 32, 20146 Milano, Different Ones. *A.I.C.A.R.R. organizes every year three conferences on plant for buildings. Topics are different from one conference to another, but all of them have energy saving as common denominator.*

Corsi del Politecnico di Milano (Courses held at Polytechnic of Milan). Ufficio Instruzione Permanente (Organisation), Facoltà di Ingegneria - Politecnico di Milano, Piazza Leonardo da Vinci 32, 20133 Milano, different ones. *A range of courses, some of which cover low energy architecture for engineers and the course programme varies from year to year.*

Corsi della Scuola die Energetica CISPEL (Courses of the School of Energy, CISPEL). Publitecnica, Via Creta 56B, 25125 Brescia. *The School of Energy CISPEL organizes a range of courses. Some refer directly to energy and architecture. They take 4 - 5 days and they are held at Barbarano di Salò the frequency is not fixed.*

Corso di Risparmio Energetico in Edilizia (Course on Energy Saving in building). E.N.E.A. (Organisation), Centro Ricerche E.N.E.A. Casaccia - C.P. 2400, 00100 Roma. *Every year many editions of the course are given in different Italian cities. Topics dealt with are the strategies for energy savings in buildings .*

Mostra Convegno EXPOCOMFORT (Exhibition/Convention EXPOCOMFORT). A.M.I.C. (Organisation), Via Flli Bressan 2, 20126 Milano. *The most important annual Italian exhibition about energy and buildings. Many conferences are held in connection with the exhibition. It is held in Milan, at the Quartiere Fiere, during the first week in February.*

RIABITAT, Ristrutturazione, Recupero, Manutenziuone (Retrofitting, Recovery, Maintenance). Fiera Internazionale di Genova (Organisation), Ple Kennedy 1, 16129 Genova, yearly. *Materials and components for retrofitting/maintenance of existing buildings. A section is reserved to energy saving and bioclimatic building. It is held in Genova at Fiera Internazionale at end of May - not strictly annually.*

SAIEDUE,(SAIEDUE) Mostra Edilizia di Primavera (Spring building Exhibition). Federlegno/Arredo, Edilegno, UNCSAAL (Organisation), Via Mascheroni 19, 20145 Milano, yearly. *One of the widest European exhibitions of components for building. Mostly dedicated to indoor components. Sections are reserved for energy but components useful for passive solar can be found. Held at end of March each year at the Quartire Fieristico.*

Salie (Saie) Salone Internazionale dell'Industrializzazione Edilizia (International Exhibition on Building Industrialization). Ente Fiere - SAIE, Piazza Costituzione 6, 40128 Bologna, Yearly. *One of the most important Italian exhibitions on building. Sections are usually dedicated to energy problems. Conferences & Seminars are given in the frame of the exhibition. It is held every year in Bologna at the Quartiere Fieristico at the end of October.*

Salone Internazionale delle Nuove Tecnologie e dell'Innovazione (International Exhibition of New Technology and Innovation). Torino Esposizioni,Corse Massimo d'Aseglio 15, 10126 Torino, Yearly. *Exhibition of new technologies and innovation in general. A section is usually reserved to energy including the use of energy in buildings. Held yearly in Turin at Torino Esposizioni beginning of Nov.*

The LT METHOD version 1.2

An Energy Design Tool for Non Domestic Buildings

INTRODUCTION

The energy interactions between heating, cooling and lighting of a building are quite complex and a mathematical model of such a system requires many input parameters. Most of these are unavailable early in the design process, or are often of peripheral interest to the architect. The **LT Method** uses energy performance curves drawn from such a mathematical model, where most of these parameters have been given assumed values. Only a few key design variables, mainly relating to building form and facade design, are left for the user to manipulate.

It is not possible for LT to be regarded as a precision model producing an "accurate" estimate of the performance of an actual building. Rather, the way that LT should be used is to evaluate the energy performance of a number of options and to make comparisons. Furthermore, the energy breakdowns of heating, cooling and lighting, which are evident from carrying out the LT Method will give a picture of the relative importance of various energy components.

The LT Method is a manual method, requiring the use of pencil and calculator; entering values taken from the drawings and the LT curves, on to the LT Worksheet .

The LT Method was developed for the European Architectural Ideas Competition "Working in the City" organised under the SOLINFO project co-ordinated by the Energy Research Group, University College Dublin and funded by the Commission of the European Communities. This competition focussed attention on energy conservation and passive design in non-domestic buildings and in particular on the design of daylighting.

CONTENTS

The Lighting and Thermal value of glazing.
How to use the LT Method
The LT Curves, tables and worksheet.
Worked examples
Design checklist

The LT Method was devised and written by Nick Baker, with important contributions from David Hoch and Koen Steemers, at the Martin Centre for Architectural and Urban Studies, University of Cambridge, and Cambridge Architectural Research Ltd. Graphics and design is by Michael Baker.

Commission of the European Communities
Directorate - General XII for Science, Research and Development
Directorate - General XIII for Telecommunications, Information Technology and Innovation

1.0 THE LIGHTING AND THERMAL VALUE OF GLAZING (LT METHOD)

Amongst the considerations early in the development of a building design, the designer is concerned with two issues: the form of the building - its plan depth, section, orientation etc, and the design of the facades; in particular the area and distribution of glazing. It is useful to know the implications of the designer's early decisions on energy consumption. Any calculation method employed must be quick and easy to use, in order to allow the designer to explore a number of options. Secondly, it must be able to respond to the main **design parameters** under consideration. Energy consumption will also depend upon other parameters such as artificial lighting levels and plant efficiencies. However these can be regarded as **engineering parameters** and to some extent can be considered independently.

The LT Method is an **Energy Design Tool** which has been developed expressly for this purpose. A mathematical model has been used to predict **annual primary energy consumption per square metre** of floor area as a function of:-

1)   local climatic conditions
2)   orientation of facade
3)   area and type of glazing
4)   obstructions due to adjacent buildings
5)   the inclusion of an atrium (optional)

1.1 Climatic zones

The basis of the Design Tool is the sets of graphs, one set of eight graphs for each of four European climatic zones, giving annual primary energy consumption per square metre for North, East/West and South orientations of facade, plus one for horizontal glazing (rooflights). Curves are presented for lighting, heating, cooling and total energy.

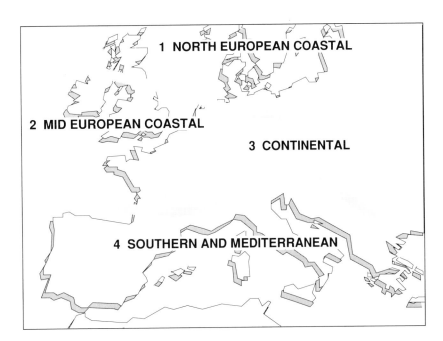

figure 1: European Climate Zones for LT Method

Figure 1 indicates the locations of each of the four European climatic zones:-

Zone 1   North European Coastal,
Zone 2   Mid European Coastal,
Zone 3   Continental,
Zone 4   Southern and Mediterranean

The defining of these zones is not precise and this is indicated by the fuzzy boundaries. LT is influenced by temperature, solar radiation, and sky luminance. It also integrates over the day and the year and is therefore sensitive to day length and heating/cooling season length. This results in certain combinations of these parameters being significant. in defining these zones the most relevant characteristics are as follows:-

Zone 1   Cold winters with low solar radiation and
          short days, mild summers

Zone 2   Cool winters with low solar radiation, mild
          summers

Zone 3   Cold winters with high radiation and longer
          days, hot summers

Zone 4   Mild winters with high radiation and long
          days, hot summers

These characteristics are really more important than the exact geographical location of the site in relation to the zones, and the user may wish to take account of particular local characteristics when choosing the appropriate zones.

zone	mean monthly temperatures°C		mean monthly radiation kWh/m²		December day-length hours
	coldest	hottest	lowest	highest	
1	<3	<18	<1	<5.5	<7.5
2	>3	<18	<1	<5.5	>7.5
3	<3	>18	>1	>5.5	>7.5 <8.5
4	>3	>20	>1.5	>5.5	>8.5

In using this table to identify the appropriate zone for certain locations, it is possible that no zone will comply with all of the conditions. In such cases, bear in mind that the lighting energy is mainly affected by the day length and winter radiation, heating by winter temperature and winter radiation, and cooling by summer temperature and summer radiation.

1.2   The mathematical model

lighting power

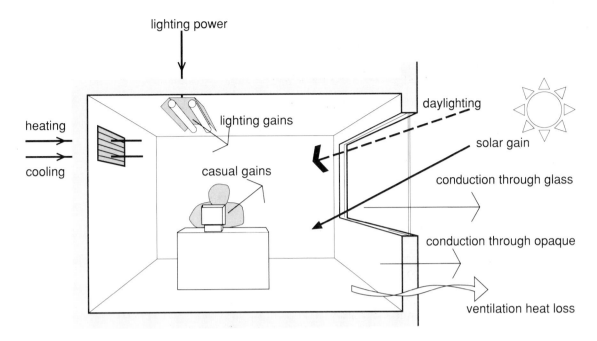

figure 2: Energy flows modelled by LT

The LT Curves are derived by a mathematical model. The energy flows considered are indicated in figure 2. First the model evaluates the heat conduction through the external envelope, and ventilation heat loss (or gain). Using monthly mean temperatures and a *thermal reference point* with a correction factor to allow for intermittent heating, a monthly gross heating load is calculated. The model then evaluates the solar gain and applies a *utilization factor* to this.

At the same time the monthly hours of available daylight are calculated from the average hourly sky illuminance on the facade, the *daylight factor*, and an internal lighting datum value. This gives a monthly electrical consumption for artificial lighting, and a monthly heat gain. A fractional hour is counted when the internal daylight level is above 66% of the datum, to take account of the actual variation about the monthly mean.

The lighting heat gains and useful solar gains, together with a fixed casual gain from occupants and equipment, are then subtracted from the *gross heating load* to establish the *net heating load*. When the gains are greater than the gross load, there is a *cooling load*. This load includes an allowance for fan and pump power within an efficiency factor.

As soon as a cooling load exists, the model eliminates the solar gain due to the direct sunbeams, leaving only diffuse daylight. This corresponds in reality to movable shading devices which are deployed to eliminate all direct radiation as soon as cooling is likely. Note also that the model assumes 'sensible light-switching', i.e. that lights are only on when the daylighting value drops below the datum value. In practice this will almost certainly require automatic light-sensing switching. Thus to a certain extent the model already assumes a considerable degree of good design.

The monthly energy consumption is calculated for a "cell" and then reduced to the value of energy consumption per square metre. These are then totalled for the year and plotted as a function of glazing ratio. A "cell" corresponds to a room surrounded by other rooms. This implies a zero conductive heat loss through all surfaces except the external (window) wall or in the case of rooflighting, the ceiling. Appropriate efficiency factors are applied to reduce all energy to *primary* energy.

The reason that *primary* energy is used is that it allows the different 'fuel' inputs for lighting, heating and cooling to be reduced to one common unit. Primary energy is the energy value of the fuel at source. In the case of fuels such as gas or oil when it is to be used for heating, there is an energy overhead required for extracting, refining and distribution, and then the loss of heat due to combustion losses, at the point of use. In the case of electricity, used for lighting and mechanical power for cooling and ventilation, an energy overhead occurs at the power station due to the thermodynamic efficiency of the conversion of heat to mechanical power. This is a large factor - 1 unit of delivered electricity is equivalent to 3.7 units of primary energy.

For primary energy sources which are fossil fuels, this equivalence is satisfactory. Furthermore it relates well to CO_2 and other pollutant production, and to cost. Problems arise however, where a substantial proportion of the electricity is generated by renewable (mechanical) sources such as hydro or wind. It also presents problems where electricity is generated by nuclear sources since the concept of primary (nuclear) energy is not really equivalent to primary energy of conventional sources.

Further details of the LT Model can be found in reference 1.

(1)   A design tool which combines the energy value of daylight and the thermal value of solar gain. Nick Baker and David Hoch, Energy and Buildings for Temperate Climates, *PLEA 88. Proc. Int. Conf.,* Pergamon 1988.

1.3   Parameter values

The model contains approximately 30 parameters, most of which are fixed in order to produce the LT curves. The assumed values of these parameters have been chosen to correspond to typical modern office or institutional buildings occupied during the day. The parameters include the lighting datum (300 lux), casual gains density (10 W/m^2), occupancy pattern (09.00-19.00 hours). A list of the main parameters and the assumed values is given below.

building envelope	U-value	reflectance
exterior wall	0.6	0.4 (inside)
interior walls	0.0	0.4
ceiling	0.0	0.6
ceiling (roof)	0.6	0.6
floor	0.0	0.2
glazing (1x)	5.7	
(2x)	3.3	
room height	3.0m	

external ground reflectance	0.2 (constant throughout year)
ventilation rate	1 air change / hour
occupancy	10 hr / day (09.00 - 19.00)
non-lighting gains	10 W/m^2
datum illumination level	300 lux
installed lighting density	15 W/m^2

plant efficiency (useful / delivered energy)

heating	0.70
cooling	2.25

energy cost (primary / delivered)

electricity	3.70
heating fuel	1.05

2.0   HOW TO USE THE LT METHOD

2.1   The LT Curves

First study the sets of curves in Section 3 for the different climatic zones including the set most appropriate for your project. This will give you a feel for the influence of local climate. Note how the proportion of heating, cooling and lighting energy changes. The vertical axis represents the Annual Primary Energy Consumption in MWh/m^2, and the horizontal axis is the unobstructed glass area as a percentage of total facade area. '2x' represents double glazing while '1x' is single glazing.

Note that in the colder climates the heating load is much greater than the cooling load, and that the heating load increases with glazing ratio in almost all cases. Note also how steeply the lighting curve drops as daylight becomes available. The cooling energy is greatest and climbs most steeply in the Southern and Mediterranean Zone 4.

Two totals are shown, *heating + cooling + lighting*, and *heating + lighting*. The second total can be used to indicate the energy use of a non air-conditioned naturally ventilated building. In such cases the *potential* cooling load now indicates the probability of overheating. The summation of the curves shows a fairly distinct optimum for glazing ratio. Note that this varies slightly for the different climatic zones, and is much smaller for the horizontal glazing due to the higher illuminance of horizontal surfaces compared to vertical surfaces.

2.2   The Passive Zone

The LT Method relies upon the concept of *passive* and *non-passive* zones. Passive zones can be daylit and naturally ventilated and make use of solar gains for heating. Non-passive zones have to be artificially lit and ventilated.

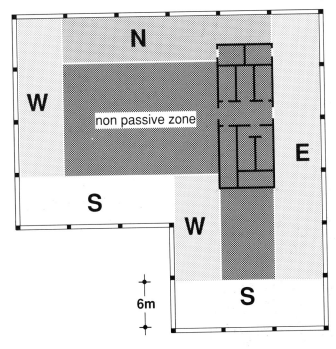

figure 3: Definition of Passive Zones

On a plan of your building, divide it into passive zones and non-passive zones and define their orientation, as in figure 3. Limit the depth of the passive zone to 6m for normal ceiling of 3m or twice the ceiling height for other dimensions. All of the top floor can be a passive zone if rooflit. Work out the zone areas and enter them in the LT Worksheet. When defining the orientation of a passive zone in a corner always choose the best performer - for example, south in preference to east and west, unless this is precluded by a blank wall.

If top floor zones are daylit by roof lights, conductive heat losses through the opaque roof envelope are accounted for correctly. If the top floor is side lit, the LT curve for sidelighting assume no losses through the ceiling and a small error results. This can be accounted for by reading off a heating load from the horizontal glazing curve at zero glazing, and adding it to the heating loads from the appropriately orientated sidelit curve, and the small error is almost eliminated. If a top floor is both side lit and top lit, the floor can be divided into roof lit and sidelit zones. If the top floor is lit with monitor or other rooflight configurations with glazing tilted to more than 45° to horizontal, use the vertical glazing curve of appropriate orientation, and applied to the whole floor area.

If some zones are adjacent to buffer spaces, e.g. atria or sunspaces, enter these separately in the 'buffer adjacent' section of the worksheet. This is explained in more detail in para 2.5

2.3   Envelope glazing and orientation

By looking at the curves and considering other design constraints, decide the ratio of glazed area to *total* external wall (or roof) area, as shown in figure 4. Note that it is not the ratio of the glazing to opaque wall area. Note also that the glass area may be up to 35% less than the aperture area due to the obstruction by framing, glazing bars etc.

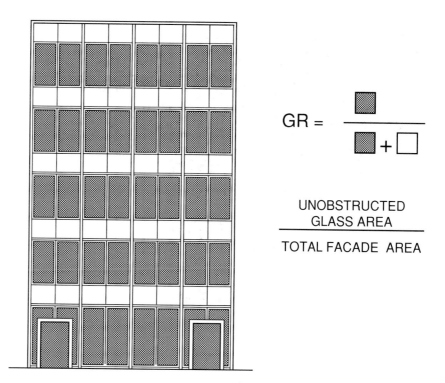

figure 4: Definition of Glazing Ratio

Now enter in the worksheet the glazing type for each zone - that is single glazing (1x) or double glazing (2x). From the appropriate curve, read off the *annual primary energy consumption per square metre* for each zone orientation, and enter the values on the worksheet.

The quick method is to take total energy (the top curve) and enter only these values on the worksheet. Or if you know that there is to be no cooling use the heat+light curve. If, however, you require a breakdown of energy use, (or delivered energy type) then you can read off and enter lighting, heating and cooling separately. This will be of interest if you have low-cost sources of particular kinds of energy, e.g. heat or electricity. Also you will need the separate energy values for taking account of obstructions, (see para 2.4), or for evaluating the heating load reduction gained from an unheated buffer space (see para 2.5).

The energy for non-passive zones can be calculated by reading from the curve at zero glazing area. In this method, non-passive zones must always have a cooling energy allowance. This is to ensure that the fan power needed for ventilation is accounted for.

2.4   Obstructions - the Urban Horizon Angle

In most cases the view of the horizon will not be free of obstructions in all directions; this is particularly true on urban sites where other buildings often block out large areas of the sky. This will affect energy use in three ways. It will reduce the availability of daylight, the useful heating in winter due to solar gain, and the cooling load due to solar gain in summer. The degree to which it affects the annual total will depend upon the height of the sun (latitude), the orientation of the facade under consideration, and of course the angle of elevation of the obstruction relative to the facade.

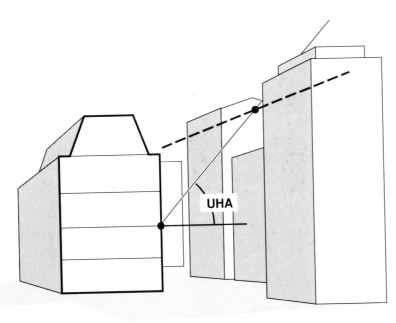

UHA

figure 5: Definition of Urban Horizon Angle

A correction factor has been derived using the mathematical model, to modify the specific energy consumption figure as read from the curves. A table is provided in Section 3 for each climatic zone. Each table gives a correction factor for lighting, heating, and cooling, for three orientations and for two *urban horizon angles* (UHA). The *Urban Horizon Factor* (UHF) is also dependent upon the glazing ratio, quite strongly dependent at angles >45º. It is better to interpolate values here, when intermediate values of glazing ratio are used.

To determine the UHA from a design proposal requires a degree of estimation. It is the average angle from the centre of the facade to the top of the obstruction as shown in fig 5. There are only two angle ranges provided for in the tables. For UHA less than 15º no correction factor is necessary, between 15º and 45º corresponds to moderate obstruction, and above 45º heavy obstruction. Obstructions that are *outside* a line at 60º to a normal (a line at right angles) to the facade *in plan*, can be neglected.

In most cases the UHA will vary significantly from ground to top storey. The quickest method is to us a single average UHA but it is better to divide the building into zones vertically and apply the appropriate UHA to each. This may lead a designer to providing larger glazing ratios on the lower floors than on the higher less obstructed floors. (This will necessitate filling in an LT Worksheet for each floor and adding the sub-totals together for the whole building).

2.5   Buffer spaces

For zones protected by unheated atria or sunspaces (buffer spaces), the LT Method takes account of the fact that a buffer space will reduce heating load, but in most cases will also reduce the daylighting of the adjacent rooms.

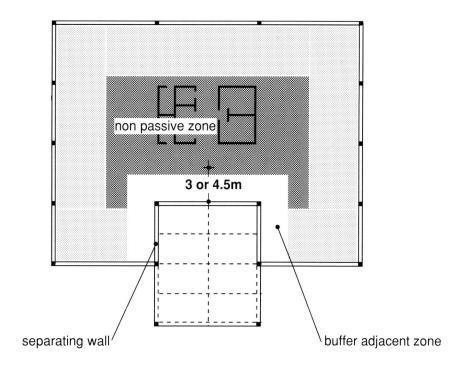

figure 6: Reduced Passive Zone Depth for buffer adjacent space

The effect on lighting

It is not possible to account with precision for the reduction in availability of daylight brought about by the installation of a glazed buffer space. There are too many unknowns - geometry, obstruction, reflections, etc. For the purpose of the LT Method, we simply reduce the depth of the daylight passive zone (from the full 6m), according to the estimated light transmittance characteristics of the buffer-space, as in figure 6. By reducing the depth of the daylight zone, the non-passive zone area is increased and the extra lighting energy accounted for.

Consult Table A to determine passive zone depth. Calculate total area of this zone and enter on worksheet.

characteristic	score zero	score 1
shading glazing internal finishes sky angle*	fixed tinted or reflect dark coloured greater than 45^0	movable clear light coloured less than 45^0
total score	**buffer adjacent passive zone depth m**	
0 or 1 2 or 3 4	0.0 3.0 4.5	

table A

*Sky angle is defined in much the same way as UHA. In the case of an enclosed atrium the effect of the obstructing side walls may have to be estimated to give an *effective sky angle*. Very deep atria or lightwells, where the depth is more than 3 - 4 times the average width should not be considered for the provision of light to adjacent spaces, and thus the passive zone depth becomes zero, irrespective of the other characteristics.

If an atrium is open to the parent building, that is it has no seperating wall, this procedure does not apply. In this case the atrium must be considered as normal internal space with energy consumption predicted from the horizontal or sidelit LT curve as appropriate.

The effect on heating

The heating and cooling energy for the buffer adjacent zone is initially calculated from the LT Curves in the same way as the other passive zones. However, because the penetration of direct radiation via the buffer-space is likely to be slight, the East/West curve is used irrespective of the orientation of the zone.

An unheated atrium or buffer space effects the heating load of the adjacent building in two ways - by reducing the conductive losses through the *separating wall*, and reducing the ventilation heat losses. This is more complex than it may seem since the reduction of conductive loss is strongly influenced by the geometry of the atrium, and the effect on ventilation heat loss is dependent on the mode of ventilation coupling between the atrium and the building.

The LT Method calculates the effect by applying a *Buffer-space Thermal Saving* (BTS) which is dependent upon the length (in plan) of separating wall between the buffer space and the building. A table of these savings has been prepared for a range of buffer-space configurations and ventilation modes and for each climatic zone, by running another computer based model called ATRIUM (2)

To calculate the reduction in heating load, first determine the *separating wall plan length* as in figure 6, and enter on the worksheet. Then consult the BTS table for the appropriate climatic zone, included with each set of LT Curves. Choose the closest configuration type or interpolate.

The different ventilation modes are (A) independent (B) buffer space to building, (C) building to buffer space and (D) re-circulated. These values are further subdivided into single glazing (1x) and double glazing (2x) referring to the external glazing of the buffer-space. The types of buffer-space configuration and orientation are illustrated in the left hand column and are numbered 1 to 7.

Considering the ventilation type, note that types B and C will only occur all the time with mechanical ventilation. Natural ventilation may be able to create these flow paths but the effect of varying wind direction and strength, and temperature difference, will effect the flow.

Type D is most likely to occur passively, but only in a shallow zone around the atrium, and it is important to realise anyhow, that a small conservatory or atrium could not provide fresh air for a large building for more than a small fraction of the day, unless the atrium itself was being refreshed at very high rates of infiltration.

Note that the in deriving the savings value, it was assumed that the fraction of ventilation involved was proportional to the separating wall area (or length assuming constant floor to ceiling height) in relation to the external wall of the building.

The tables give the Buffer-space Thermal Savings in annual primary energy per metre separating wall length. Enter the value on the worksheet in the "Buffer-space Thermal Saving" box.

(2)   The thermal performance of large glazed spaces. N Baker. *Architecture Solaire. Proc.Int.Conf., Cannes.* Lavoisier Tec. and Doc., 1983.

2.6 Buffer-space Average Temperature Nomogram

The average temperature of an unheated buffer-space will depend upon the heat flow from the parent building, and solar gains. The former will depend upon the relative areas of glazing in the external wall and the separating wall of the buffer-space, defined as the 'protectivity ratio'. In absence of solar radiation the temperature will lie somewhere between the outdoor temperature and the temperature of the heated building.

Table B below, offers a way of estimating the average temperature due to gains from the parent building by conduction, excluding the effects of solar gains and ventilation exchange. It can be used as a fairly good indicator of the minimum temperature that would exist in the buffer-space on cloudy days, by using the internal set temperature and the average daytime temperature. It is meant for use in winter conditions.

It is not part of the LT Calculation.

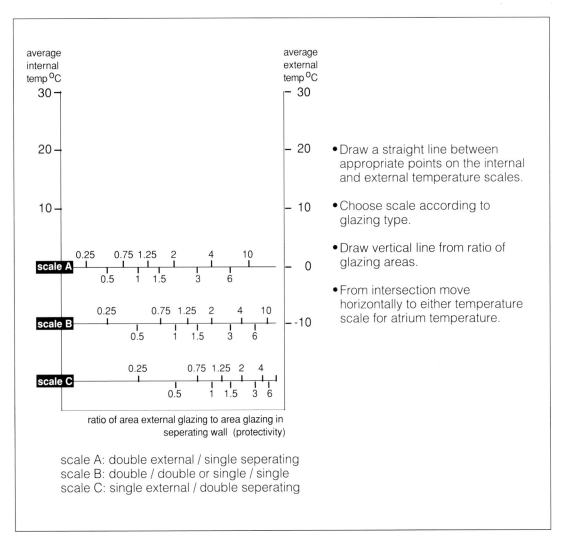

- Draw a straight line between appropriate points on the internal and external temperature scales.

- Choose scale according to glazing type.

- Draw vertical line from ratio of glazing areas.

- From intersection move horizontally to either temperature scale for atrium temperature.

ratio of area external glazing to area glazing in seperating wall (protectivity)

scale A: double external / single seperating
scale B: double / double or single / single
scale C: single external / double seperating

table B: Buffer Space Average Temperature Nomogram

3.0 LT CURVES, TABLES AND WORKSHEET

		A		B		C		D	
		x1	x2	x1	x2	x1	x2	x1	x2
1 N		0.62	0.71	0.71	0.77	0.65	0.72	1.27	1.39
2 N		0.62	0.84	0.82	1.01	0.69	0.84	1.24	1.50
3 N		0.47	0.64	0.60	0.75	0.57	0.71	1.02	1.27
4 N		0.39	0.60	0.60	0.80	0.49	0.62	1.01	1.34
5 N		0.35	0.49	0.47	0.67	0.41	0.54	0.60	0.89
6 N		0.62	0.95	0.84	1.22	0.71	0.99	1.09	1.57
7 N		0.56	0.75	0.67	0.90	0.65	0.81	1.01	1.34

Buffer space Thermal Savings (BTS) MWh/m y

orientation	UHA deg.	glazing ratio %	correction factors		
			lighting	heating	cooling
north	15 - 45	15.0	1.2	1.0	1.3
		30.0	1.3	1.0	0.9
		45.0	1.4	1.0	0.9
		60.0	1.4	1.0	0.8
	> 45	15.0	1.3	1.0	1.1
		30.0	1.6	1.0	0.8
		45.0	2.0	1.0	0.7
		60.0	2.1	1.0	0.6
east/west	15 - 45	15.0	1.3	1.0	1.2
		30.0	1.3	1.0	0.9
		45.0	1.3	1.1	0.9
		60.0	1.4	1.1	0.8
	> 45	15.0	1.3	1.1	1.1
		30.0	1.6	1.1	0.8
		45.0	2.0	1.1	0.7
		60.0	2.1	1.2	0.5
south	15 - 45	15.0	1.3	1.1	1.2
		30.0	1.3	1.2	0.9
		45.0	1.3	1.3	0.8
		60.0	1.3	1.4	0.8
	> 45	15.0	1.3	1.2	1.1
		30.0	1.7	1.3	0.8
		45.0	2.0	1.5	0.6
		60.0	2.1	1.6	0.5

Urban Horizon Factor (UHF)

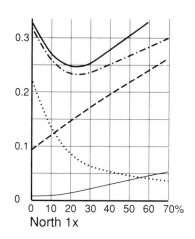

North 1x

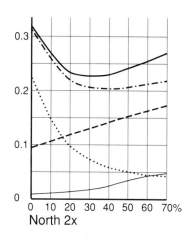

North 2x

MWh/m²y

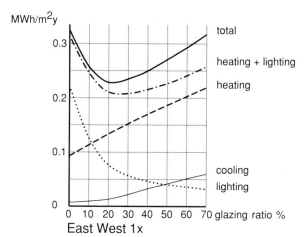

total

heating + lighting

heating

cooling

lighting

glazing ratio %

East West 1x

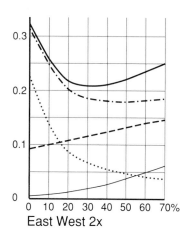

East West 2x

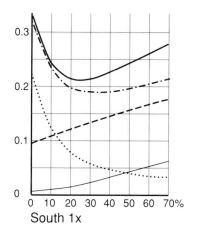

South 1x

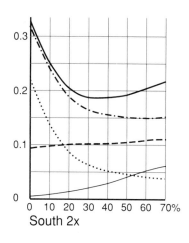

South 2x

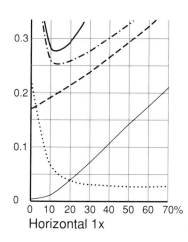

Horizontal 1x

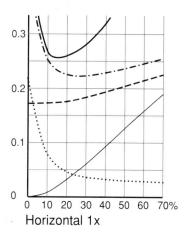

Horizontal 1x

	A x1	A x2	B x1	B x2	C x1	C x2	D x1	D x2
1 N	0.52	0.60	0.58	0.63	0.54	0.57	1.02	1.12
2 N	0.55	0.71	0.70	0.84	0.59	0.71	1.02	1.24
3 N	0.41	0.53	0.51	0.62	0.48	0.58	0.83	1.02
4 N	0.36	0.49	0.53	0.69	0.42	0.53	0.84	1.10
5 N	0.35	0.46	0.46	0.64	0.38	0.50	0.56	0.80
6 N	0.57	0.83	0.75	1.03	0.61	0.85	0.93	1.30
7 N	0.47	0.63	0.58	0.74	0.54	0.70	0.83	1.03

Buffer space Thermal Savings (BTS) MWh/m y

orientation	UHA deg.	glazing ratio %	correction factors lighting	heating	cooling
north	15 - 45	15.0	1.3	0.9	1.2
		30.0	1.3	0.9	0.7
		45.0	1.3	1.0	0.6
		60.0	1.4	1.1	0.6
	> 45	15.0	1.3	0.9	1.0
		30.0	1.6	0.8	0.9
		45.0	1.9	0.9	0.5
		60.0	2.1	0.9	0.5
east/west	15 - 45	15.0	1.2	1.2	1.1
		30.0	1.3	1.2	0.9
		45.0	1.4	1.3	0.8
		60.0	1.4	1.4	0.7
	> 45	15.0	1.3	1.2	0.9
		30.0	1.7	1.2	0.6
		45.0	2.1	1.4	0.4
		60.0	2.2	1.5	0.3
south	15 - 45	15.0	1.3	1.2	1.1
		30.0	1.3	1.5	0.9
		45.0	1.4	1.7	0.8
		60.0	1.5	1.8	0.8
	> 45	15.0	1.3	1.4	0.9
		30.0	1.7	1.7	0.7
		45.0	2.0	2.0	0.5
		60.0	2.3	2.3	0.4

Urban Horizon Factor (UHF)

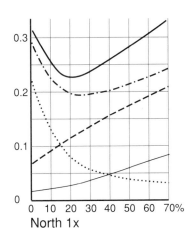

North 1x

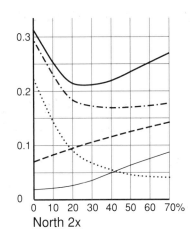

North 2x

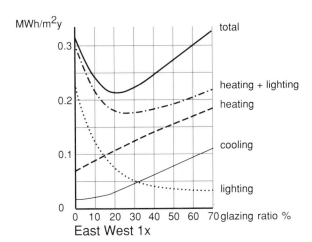

MWh/m²y

total

heating + lighting

heating

cooling

lighting

glazing ratio %

East West 1x

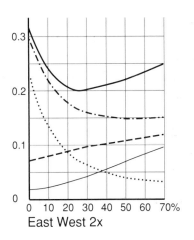

East West 2x

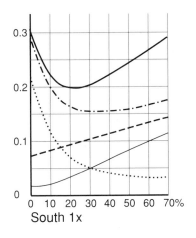

South 1x

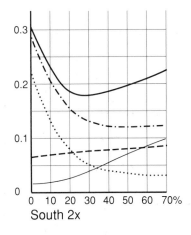

South 2x

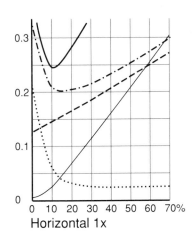

Horizontal 1x

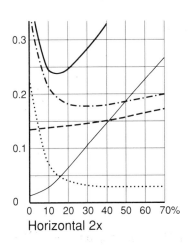

Horizontal 2x

	A x1	A x2	B x1	B x2	C x1	C x2	D x1	D x2
1 N	0.64	0.75	0.73	0.80	0.67	0.74	1.31	1.42
2 N	0.67	0.90	0.89	1.09	0.73	0.89	1.30	1.59
3 N	0.50	0.66	0.64	0.78	0.59	0.72	1.05	1.29
4 N	0.42	0.61	0.66	0.88	0.51	0.66	1.05	1.41
5 N	0.39	0.54	0.53	0.77	0.43	0.59	0.66	0.98
6 N	0.71	0.74	0.82	1.33	0.77	1.08	1.18	1.67
7 N	0.57	0.78	0.72	0.94	0.67	0.87	1.04	1.36

Buffer space Thermal Savings (BTS) MWh/m y

orientation	UHA deg.	glazing ratio %	correction factors lighting	heating	cooling
north	15 - 45	15.0	1.3	0.9	1.3
		30.0	1.3	1.0	0.9
		45.0	1.4	1.0	0.9
		60.0	1.4	1.0	0.8
	> 45	15.0	1.3	1.0	1.1
		30.0	1.7	1.0	0.8
		45.0	2.0	1.0	0.7
		60.0	2.2	1.1	0.5
east/west	15 - 45	15.0	1.2	1.0	1.1
		30.0	1.3	1.1	0.9
		45.0	1.3	1.1	0.8
		60.0	1.5	1.1	0.8
	> 45	15.0	1.3	1.1	1.1
		30.0	1.7	1.1	0.8
		45.0	2.0	1.2	0.6
		60.0	2.2	1.2	0.5
south	15 - 45	15.0	1.3	1.1	1.1
		30.0	1.3	1.2	0.9
		45.0	1.3	1.4	0.8
		60.0	1.5	1.4	0.8
	> 45	15.0	1.3	1.2	1.1
		30.0	1.7	1.4	0.8
		45.0	2.0	1.6	0.6
		60.0	2.3	1.7	0.5

Urban Horizon Factor (UHF)

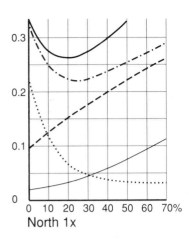

North 1x

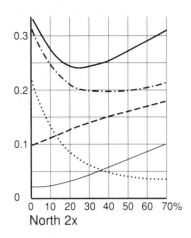

North 2x

MWh/m²y

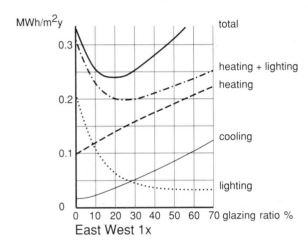

total

heating + lighting

heating

cooling

lighting

glazing ratio %

East West 1x

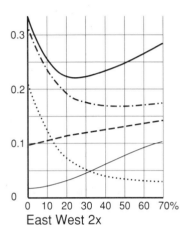

East West 2x

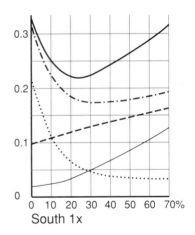

South 1x

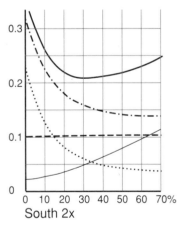

South 2x

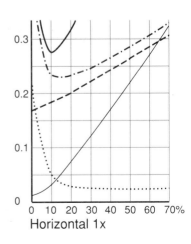

Horizontal 1x

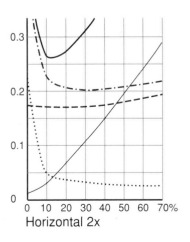

Horizontal 2x

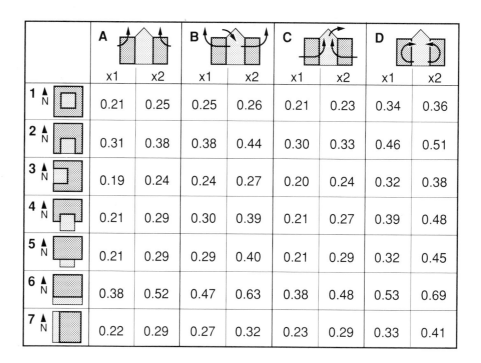

	A x1	A x2	B x1	B x2	C x1	C x2	D x1	D x2
1 N	0.21	0.25	0.25	0.26	0.21	0.23	0.34	0.36
2 N	0.31	0.38	0.38	0.44	0.30	0.33	0.46	0.51
3 N	0.19	0.24	0.24	0.27	0.20	0.24	0.32	0.38
4 N	0.21	0.29	0.30	0.39	0.21	0.27	0.39	0.48
5 N	0.21	0.29	0.29	0.40	0.21	0.29	0.32	0.45
6 N	0.38	0.52	0.47	0.63	0.38	0.48	0.53	0.69
7 N	0.22	0.29	0.27	0.32	0.23	0.29	0.33	0.41

Buffer space Thermal Savings (BTS) MWh/m y

orientation	UHA deg.	glazing ratio %	correction factors lighting	correction factors heating	correction factors cooling
north	15 - 45	15.0	1.3	0.9	1.2
		30.0	1.3	1.0	1.0
		45.0	1.5	1.0	0.9
		60.0	1.7	1.0	0.9
	> 45	15.0	1.4	0.9	1.1
		30.0	1.8	1.0	0.9
		45.0	2.2	1.1	0.8
		60.0	2.6	1.1	0.7
west/east	15 - 45	15.0	1.4	0.8	1.1
		30.0	1.3	1.1	1.0
		45.0	1.5	1.1	0.9
		60.0	1.6	1.3	0.9
	> 45	15.0	1.5	1.0	1.1
		30.0	1.8	1.3	0.9
		45.0	2.4	1.4	0.7
		60.0	2.5	1.5	0.6
south	15 - 45	15.0	1.4	1.0	1.1
		30.0	1.3	1.3	0.9
		45.0	1.5	1.3	0.9
		60.0	1.5	1.4	0.8
	> 45	15.0	1.5	1.5	1.1
		30.0	1.9	1.7	0.9
		45.0	2.5	1.8	0.7
		60.0	2.4	2.1	0.6

Urban Horizon Factor (UHF)

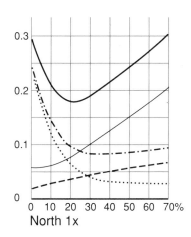

North 1x

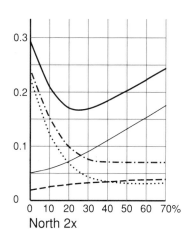

North 2x

MWh/m²y

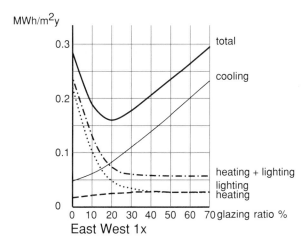

total

cooling

heating + lighting
lighting
heating

glazing ratio %

East West 1x

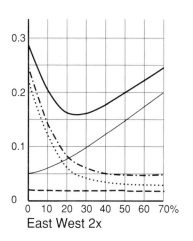

East West 2x

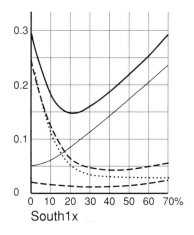

South1x

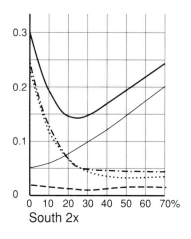

South 2x

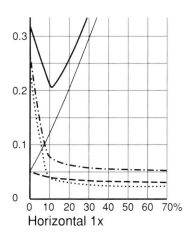

Horizontal 1x

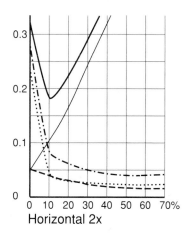

Horizontal 2x

LT WORKSHEET

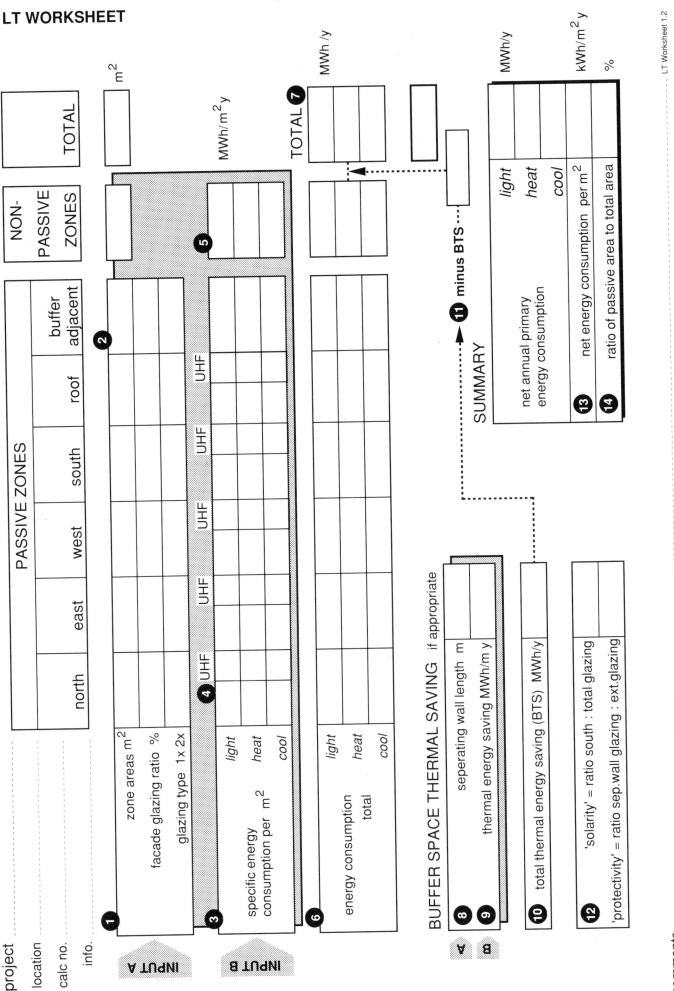

3.5 NOTES FOR THE LT METHOD WORKSHEET

These numbered notes correspond to the circled numbers appearing on the LT Worksheet. Note that the shaded boxes are inputs, whilst the white boxes are for calculated results. Type A inputs are from building data, whilst type B inputs are from LT curves or tables. To keep the number of digits small we use mega-watt hours (MWh) for energy units - 1MWh = 1000 kilo-watt hours (kWh)

1) Calculate passive and non-passive zone areas as shown in figure 3 and enter.   Enter facade glazing ratio (percentage glass of total wall area), and glazing type.

2) Enter buffer adjacent area if appropriate. Reduce zone depth according to buffer space characteristics as in table A. Glazing ratio and glazing type refers to separating wall.

3) Read specific energy consumption from LT Curves for appropriate location, orientation, and glazing type. For buffer adjacent zone use East/West curve irrespective of orientation.

4) Estimate urban horizon angle for each facade. From appropriate table determine the urban horizon factor (UHF) for appropriate orientation, urban horizon angle (UHA) and glazing ratio. There is no factor for rooflights.

5) Read specific energy consumption for non passive zone from LT curve at zero glazing ratio (not horizontal).

6) Multiply areas (1) by specific energy consumptions (3) and by UHF (4).

7) Sum energy consumption for all zones, by light, heat and cool. If there is a buffer-space, do not sum heat until BTS has been calculated (10).

8) For buffer space thermal savings enter total separating wall length multiplied by number of storeys, e.g. for 3 storey building with 30m of separating wall on plan, enter 90m.

9) From BTS tables choose appropriate buffer space type and ventilation mode and enter specific thermal saving (per metre separating wall) interpolating if necessary.

10) Multiply specific thermal saving by separating wall length and enter total buffer-space thermal saving (BTS).

11) Subtract BTS from heating and enter total at (7).

12) Solarity and protectivity are of interest but are not needed for the LT calculation.

13) Divide net total energy consumption by total floor area (usually excluding buffer space) and enter as kWh/m^2y.

14) Passive area ratio is of interest but is not essential for the LT calculation.

Computer Spreadsheets

Those who are familiar with using computer based spreadsheets will quickly appreciate that the LT Worksheet could be installed on one of these. This would greatly simplify the use of the LT Method and in particular would facilitate multiple calculations where a number of alternatives are being investigated, or the use of individual worksheets for each storey in order to accurately account for variation of urban horizon angle.

4.0 LT METHOD WORKED EXAMPLES

4.1 5 storey office building.

For the first example we are investigating a proposed office building with an L-shaped plan as shown in figure 3. The building is five storeys high and is on a site which is heavily overshadowed from the east, slightly overshadowed to the north and open to the south and west. The site location is Mid-European Coastal.

Passive zones

Figure 7 shows the original plan with dimensions and the areas of the zones indicated. In designating the 6m passive zones at outside corners we assume the best performer for light and heat, i.e. south rather than west or east. At the inside corner we have designated a west-facing passive zone although a small part of this is actually further than 6m away from the facade. This is to simplify calculation.

The top floor is all passive zone; we will light the centre of the plan with rooflights with a sidelit perimeter zone. There is a slight problem here because LT assumes zero heat losses through ceilings of sidelit spaces, but it will result in only small errors since heat losses are dominated by the area of glazing.

The areas are now calculated and set out in the working table, and then entered in the LT Worksheet.

The LT Curves

The next step is to enter the glazing ratios for the facades. Let us start off with 20% all round - maybe we will change it later.

It is possible that the fenestration has already been designed as part of the initial concept. If you are working from a preconceived glazing pattern, then remember that the glazing ratio refers to the actual area of glass not including glazing bars etc, as a ratio to the total wall area (not just opaque). The actual glazing area is often smaller than it might appear.

For cool climates, the LT Curves show that the overall energy performance for double glazing is better than single glazing (not so for Southern and Mediterranean), so let us choose double glazing. For the rooflights, the "Horizontal 2x" total curve shows a minimum at about 15% - we will choose that. Note that we are already using the curves to influence our choice here, whereas for the vertical glazing we are working from an initial proposal which might have originated from aesthetic considerations.

If the top floor is lit with monitor or other rooflight configurations with glazing tilted to more than 45° to horizontal, use the vertical glazing curve of appropriate orientation and UHF applied to the whole floor area.

Now marking off the glazing ratio on the appropriate graphs read off the specific energy consumption for light, heat and cool on the vertical axis. Try to work to an accuracy of about 5%. Enter the values on the Worksheet. For the Non-passive zone enter the value for Zero% glazing ratio - any sidelit curve - they are all the same for a given climate type.

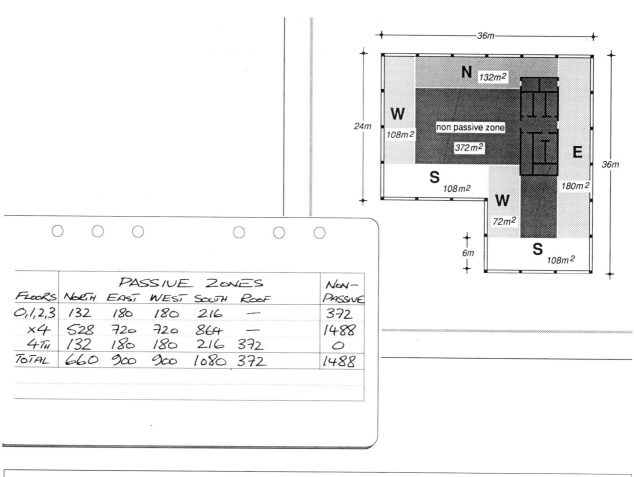

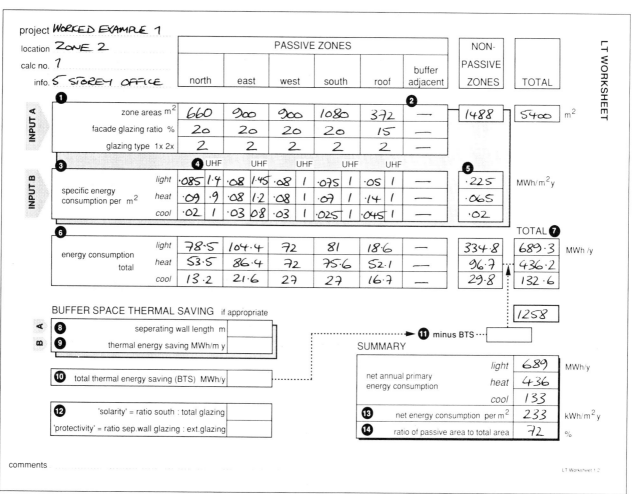

figure 7: Worked example no.1 - 5 storey office building.

The Urban Horizon Factor

From the site plan we have deduced (from a sketch section if necessary) that the UHA from the centre of the third storey is less than 15° for the south and west, between 15° and 45° for north, and greater than 45° for east. Reading off the Urban Horizon Factor table we can find the appropriate values of UHF, twelve values in all, and enter them in the Worksheet. This table is difficult to read - it helps to mask off all but the relevant part.

Now it is simply multiplying the areas by the appropriate specific energy consumption values and UHFs, adding across the page to give totals for lighting heating and cooling. In the summary box we enter the totals again and the specific energy consumption in kWh per m². It is also interesting to enter the ratio of the area of passive zone to the total floor area to indicate the "passivity" of the design.

4.2   Four storey building with atrium.

Passive zones

This building has a south-facing atrium as in figure 6, and is four storeys. It is on a "green field" site with no significant obstructions and so we do not have to apply UHFs. The climatic zone is again Mid-European Coastal.

The passive and non-passive zones are taken off the plan and set out in the working table as shown in figure 8. Note that the buffer adjacent zone is defined as 4.5m deep. With reference to Table B, we have assumed movable shading, clear glazing and light internal finishes. From the geometry of the atrium we can see that the effective sky angle is less than 45°.

Note also that the south-facing zone takes precedence over the buffer adjacent zone. The top floor is designated as all roof lit. The areas are entered on the worksheet.

We have proposed values for the glazing ratios as follows - north, east and west facing - 25%, south facing 40%, roof - 15%, and buffer-adjacent - 60%. All is double glazed except for the buffer adjacent (separating wall). This larger area of single glazing makes sense due to the protective effect of the atrium.

Specific energy consumption values are read off from the LT curves as before. Note that the buffer adjacent space is read off from the East/West curve irrespective of orientation (see note 3 in Section 3.5). The energy consumption for the zones is calculated, but before totalling the heating energy we have to calculate the buffer-space thermal savings, BTS.

Buffer-space thermal saving

First we measure the separating wall length, as indicated on figure 8 Since the plan is the same on all four floors, we simply multiply by four to get the total length. Note that this length together with an assumed storey height of 3m provides an area, which is a measure of the thermal coupling between the unheated atrium and the parent building. (If the storey height were different an allowance could be made - e.g. if the storey height were 4m the effective separating wall length should be increased by 4/3 to allow for this).

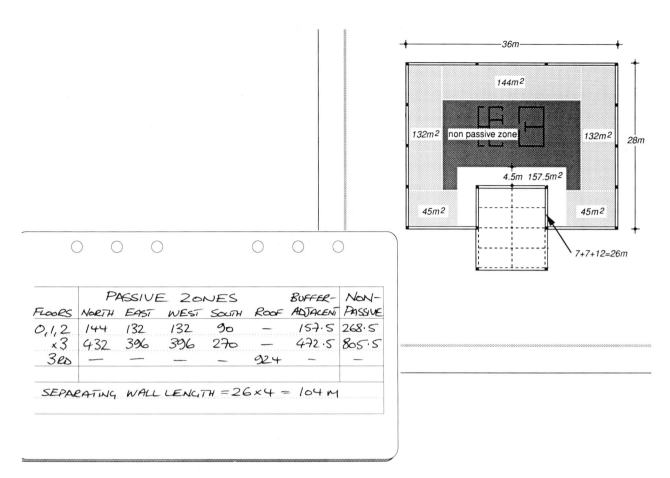

FLOORS	PASSIVE ZONES					BUFFER ADJACENT	NON-PASSIVE
	NORTH	EAST	WEST	SOUTH	ROOF		
0,1,2	144	132	132	90	—	157.5	268.5
×3	432	396	396	270	—	472.5	805.5
3RD	—	—	—	—	924	—	—

SEPARATING WALL LENGTH = 26 × 4 = 104 M

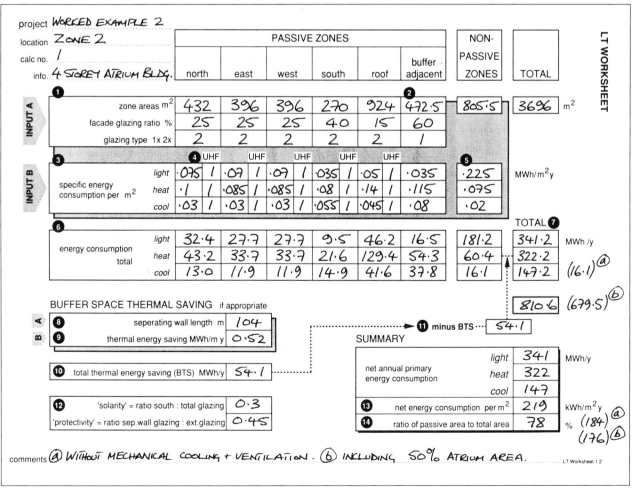

project WORKED EXAMPLE 2
location ZONE 2
calc no. 1
info. 4 STOREY ATRIUM BLDG.

LT WORKSHEET

	north	east	west	south	roof	buffer adjacent	NON-PASSIVE ZONES	TOTAL	
INPUT A — zone areas m²	432	396	396	270	924	472.5	805.5	3696	m²
facade glazing ratio %	25	25	25	40	15	60			
glazing type 1x 2x	2	2	2	2	2	1			

INPUT B — specific energy consumption per m² (MWh/m²y)

	north UHF	east UHF	west UHF	south UHF	roof UHF	buffer adjacent	NON-PASSIVE
light	.075 / 1	.07 / 1	.07 / 1	.035 / 1	.05 / 1	.035	.225
heat	.1 / 1	.085 / 1	.085 / 1	.08 / 1	.14 / 1	.115	.075
cool	.03 / 1	.03 / 1	.03 / 1	.055 / 1	.045 / 1	.08	.02

energy consumption total — TOTAL (MWh/y)

	north	east	west	south	roof	buffer adj	NON-PASSIVE	TOTAL	
light	32.4	27.7	27.7	9.5	46.2	16.5	181.2	341.2	MWh/y
heat	43.2	33.7	33.7	21.6	129.4	54.3	60.4	322.2	
cool	13.0	11.9	11.9	14.9	41.6	37.8	16.1	147.2	(16.1) ⓐ

810.6 (679.5) ⓑ

BUFFER SPACE THERMAL SAVING if appropriate

8 separating wall length m 104
9 thermal energy saving MWh/m y 0.52

10 total thermal energy saving (BTS) MWh/y 54.1

11 minus BTS ··· 54.1

12 'solarity' = ratio south : total glazing 0.3
'protectivity' = ratio sep.wall glazing : ext.glazing 0.45

SUMMARY

net annual primary energy consumption	light	341	MWh/y
	heat	322	
	cool	147	
13 net energy consumption per m²		219	kWh/m²y ⓐ
14 ratio of passive area to total area		78	% (184) ⓐ (176) ⓑ

comments ⓐ WITHOUT MECHANICAL COOLING + VENTILATION. ⓑ INCLUDING 50% ATRIUM AREA.

LT Worksheet 1 2

figure 8: Worked example no.2 - 4 storey building with atrium.

Now, referring to the Buffer-space Thermal Saving table for Mid-European Coastal, clearly the atrium is of Type 4 in the left hand column. We have already decided that the external glazing of the atrium is single, and that we will use it for ventilation pre-heating if possible as in Mode B. The Specific Buffer-space Thermal Saving (BTS) is 0.53 MWh/my. This is entered on the Worksheet.

Now we can simply multiply the BTS by the separating wall length to get the total Buffer-space Thermal Saving. Enter this in the buffer-space box and transfer it to the insert box in the main Worksheet, as indicated by the dotted line.

Two other values are requested - the Solarity Ratio and the Protectivity Ratio, although they are not actually required for the LT calculation. We can use the Protectivity Ratio to calculate the average atrium temperature in absence of solar radiation using the nomogram Table B

The final totals are calculated with the BTS deducted from the heating total. The Primary Energy Consumption per m^2 of 219 has been calculated without including the area of the atrium since it is assumed that this is not serviced. However an allowance for the potentially useful atrium space could be made in which case the specific energy consumption figure will be reduced further. This is a matter of choice but whatever is decided should be noted on the worksheet.

Comparison with non-passive building.

It is always difficult to know how to define the non-passive building with which to make the comparison. But supposing we consider a building of the same plan and section but with no atrium. There is 60% glazing all round (of single glazed tinted glass) but there are no photo-electric controls and we know that the electric lighting will be used all day. The whole building will be mechanically ventilated. This may sound a bit of a horror storey, but it is not at all unusual!

If we interpret this as a zero passive zone building, i.e. take the specific energy consumptions off the zero glazing ratio intercept, we shall get the correct value for lighting, but it will be an underestimate for heating and cooling since it assumes an opaque envelope with U-value 0.6. So we should use the actual glazing ratios for heating and cooling. Even then, the cooling will be calculated assuming optimum shading as described in para 1.2. In practice the cooling load could be at least twice as large due to direct radiation on east, south and west facades.

This "non-passive" option still utilises solar gains on the south facade. Furthermore, the LT Method has under estimated fan power and so the actual performance would probably be considerably worse.

The third worksheet is completed for this building and results in a Primary Energy Consumption of 427 kWh/m^{2}y, that is over twice as much as the passive atrium building above. It must be pointed out, however, that most of the difference is due to the non-utilization of daylight, rather than the presence of the atrium.

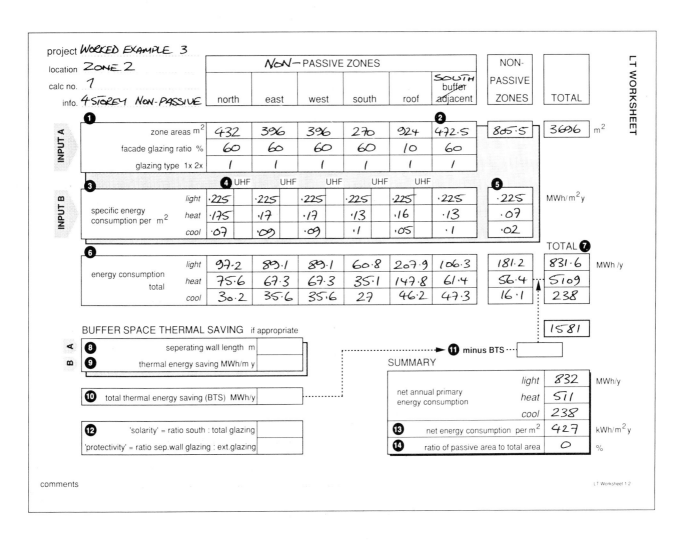

project _WORKED EXAMPLE 3_
location _ZONE 2_
calc no. _1_
info. _4 STOREY NON-PASSIVE_

LT WORKSHEET

	NON-PASSIVE ZONES						NON-PASSIVE ZONES	TOTAL
	north	east	west	south	roof	SOUTH buffer adjacent		
INPUT A								
zone areas m²	432	396	396	270	924	472.5	805.5	3696 m²
facade glazing ratio %	60	60	60	60	10	60		
glazing type 1x 2x	1	1	1	1	1	1		
INPUT B	UHF	UHF	UHF	UHF	UHF			
specific energy consumption per m² — light	.225	.225	.225	.225	.225	.225	.225	MWh/m²y
heat	.175	.17	.17	.13	.16	.13	.07	
cool	.07	.09	.09	.1	.05	.1	.02	
energy consumption total — light	97.2	89.1	89.1	60.8	207.9	106.3	181.2	831.6
heat	75.6	67.3	67.3	35.1	147.8	61.4	56.4	510.9
cool	30.2	35.6	35.6	27	46.2	47.3	16.1	238

TOTAL ❼ MWh/y

BUFFER SPACE THERMAL SAVING if appropriate

A ❽ seperating wall length m
B ❾ thermal energy saving MWh/m y

❿ total thermal energy saving (BTS) MWh/y

⓫ minus BTS ┈ 1581

⓬ 'solarity' = ratio south : total glazing
'protectivity' = ratio sep.wall glazing : ext.glazing

SUMMARY

net annual primary energy consumption	light	832	MWh/y
	heat	511	
	cool	238	
⓭ net energy consumption per m²		427	kWh/m²y
⓮ ratio of passive area to total area		0	%

comments

LT Worksheet 1.2

figure 9: Worked example no.3 - non-passive, non-atrium version of building example no.2.

5.0 DESIGN CHECKLIST

This checklist is not really part of the LT Method, but it is included here to remind the reader of all the other factors which will influence building performance - both in relation to energy and occupant comfort and well-being. It is not usually possible to deal quantitatively with these other factors at the level of simplicity of the LT Method. Thus, these areas tend to be dealt with descriptively and will be found elsewhere in this volume.

1) SITE
Site planning and layout, micro-climate, noise and pollution, landscape and planting, sunlight and daylight access.

2) BUILDING FORM
Plan shape, section shape, floor to ceiling height, proportion of passive zone, orientation, adjacent buildings.

3) BUILDING FABRIC
Construction and insulation of floor, walls, roof, glazing type and areas, thermal mass, cold bridges, vapour barriers.

4) DAYLIGHT AND SUNLIGHT
Glazing area, facade design, glazed area distribution, shading, external reflective surfaces, internal reflective surfaces, views, glare control.

5) PASSIVE SOLAR
Thermal stratergy relating to climate and building type, glazing area and orientation, distribution, thermal mass, solar ventilation pre-heat, effect on thermal comfort.

6) NATURAL VENTILATION
Prevailing wind conditions, building shape, pressure coefficients, stack effect, distribution of openings, noise and pollution, minimal ventilation, ventilation pre-heat, structural cooling, night ventilation, physiological cooling by air movement.

7) PREVENTION OF OVERHEATING / REDUCTION OF COOLING
Shading, building reflectance, insulation, thermal mass, ventilation gains, lighting gains, casual gains, comfort standards.

8) ARTIFICIAL LIGHTING
Sources, luminaires, levels and standards, task lighting, ambient lighting, integration with daylight, zones, glare and visual comfort, controls.

9) HEATING
Choice of fuel, distribution of plant, emitters, thermal comfort, controls, distribution and storage of hot water.

10) MECHANICAL SERVICING / AIR CONDITIONING
Centralized or distributed air-handling units, duct sizes, heating/cooling recovery and storage, integration with structure, refrigeration type, controls, comfort standards.

11) ATRIA / SUN SPACES / BUFFER SPACES
Glazing type and area, daylight level, effect on daylighting of adjacent rooms, reflected light, thermal conditions in winter and summer, ventilation pre-heat, heating, shading, ventilation, openable areas, comfort conditions, types of planting and requirements for heat and light.

12) ENVIRONMENTAL ISSUES
Materials' energy of production , avoidance of CFCs, avoidance of non-renewable timber, choice of fuels to minimise CO_2 and other pollutant production. Building health, avoidance of Sick Building Syndrome and Legionaires' disease etc.

NEW METHOD 5000

INTRODUCTION

The more exact analysis of the performance of passive solar buildings is difficult without computers. Some simplified or 'manual' methods can, however, be used for quick and approximate results. One such method is the *New Method 5000* which provides a procedure based on a set of blank forms to be filled out in sequence with appropriate hand calculations.

Manual methods are justified if their results are in reasonable agreement with those of more elaborate computer models, where those models themselves are properly validated.

There are various sources of error, however, in all predictions of thermal behaviour, whether by computer or manual methods. There is obvious scope for uncertainty in the fact that no building when constructed, corresponds exactly with its plans and specification, and that no material behaves *in situ* exactly as in the laboratory. In addition, we have no choice but to approximate the effects of infiltration, weather conditions and building use, as averages over a fairly long period.

The *New Method 5000* is used to predict the auxiliary heating required for any specified month. This is done by subtracting the useful heat gains (in *kWh* for the month) from the gross heat losses (in *kWh*) for the same month.

CLASSIFICATION OF POSSIBILITIES

It is convenient to divide passive solar buildings into two subsets- those heated by direct gain only, (Case A) which are very common, and those not relying on direct gain only, (Case B), of which there are many types.

For each subset, certain questions may have to be answered: for example, whether the insulation properties of the building are the same by night as by day; whether the thermostat setting is the same by night and by day; whether heating is intermittent, and whether the heated space is comprised of one or more zones (defined by different thermostat settings). Particular to Case B will be the question of which particular solar-gain devices, singly or in combination, are used, from the wide range of possibilities. It will be rare in practice for a building to use all the possible solar devices.

Heat losses are essentially the same, but for both cases two possibilities arise; -whether buffer spaces are, or are not used. The application of the Method can be resolved into a sequence of five stages (stages 1 through 5). We have, however, arranged the text in sections to make it easy to skip the discussion of miscellaneous passive-solar features in those cases where we are concerned with a direct-gain-only building. As a result, the individual steps are named according to the section headings and subheadings and not according to the numbering of the stages. The relationship can be seen overleaf and in the flow chart. The labelling of the forms (A1, A2 etc.) has not been changed from that in the 1986 edition.

APPLICATION OF THE METHOD

The method proceeds through the following stages:

SECTION I and II : Heat losses and Gains for all cases

STAGE 1:
Compute heat loss rate for the heated space *(kWh/month)*

STAGE 2:
Compute gross heat gain rate *(kWh/month)*

SECTION III : Net Heat Load

STAGE 3:
Compute useful heat gains *(kWh/month)*

STAGE 4:
Compute Auxiliary Heat required *(kWh/month)*

STAGE 5:
Check for Discomfort Conditions

The stages can be elaborated in the following steps:

STAGE 1 (Heat Loss Rate)
(a) Calculate Heat loss rate for exterior walls and roof *(W/K)*
(may be two values, for day and night, if $U_d \neq U_n$)
(b) Compute heat loss rate of exterior windows *(W/K)*
(may be two values, for day and night, if $U_d \neq U_n$)
(c) Compute heat-loss rate for floor slabs *(W/K)*
(normally $U_d = U_n$)   [Note 1]
(d) Compute heat-loss rate through buffer space, if any *(W/K)*
(e) Compute heat-loss rate from ventilation/infiltration *(W/K)*   [Note 2]
(f) Sum the losses (daily rates) *(W/K)*
(g) Convert heat-loss rates (average daily) to *monthly* heating load. *(kWh/month)*

(Daytime and night-time losses, if derived separately, are apportioned according to the relative duration of day and night insulation use over 24 hours) *(kWh/month)*

STAGE 2 (Heat Gain Rate)
(a) Compute direct solar gains *(kWh/day)*
(b) Enter casual heat gains Φ_i *(kWh/day)*
(c) through (i): Calculate all other solar gains apart from direct *(kWh/day)*
(i) Add all gains *(kWh/day)*

STAGE 3 (Calculate *useful* Gains)
(a) Compute fraction η of gains that is useful *($0 \leq \eta \leq 1$)*
(b) Multiply η by gains from all sources, and list *(kWh/day)*
(c) Sum useful gains for month *(kWh/month)*

STAGE 4 (Net Heat Demand)
(a) Calculate auxiliary heat demand by subtracting useful gains for month [3(c)] from monthly heating load [1(g)] *(kWh/month)*

STAGE 5 (Comfort Conditions)
(a) Estimate the number of hours (for the month) during which the predicted indoor temperature might exceed a specified criterion.

PROCEDURE
We divide buildings into two types
A-(Section I) those having direct-solar-gain features only; and B-(Section II), those having various passive-solar features in addition.

The treatment succeeding, deals first with analysis common to both types of building, after which one skips to Section III if dealing with a Direct-gain building, or proceeds to Section II otherwise. The essential difference lies in Step 2, which, for the general case permits detailed analysis [Step 2(c) through 2(i)] of the several passive solar features possible.

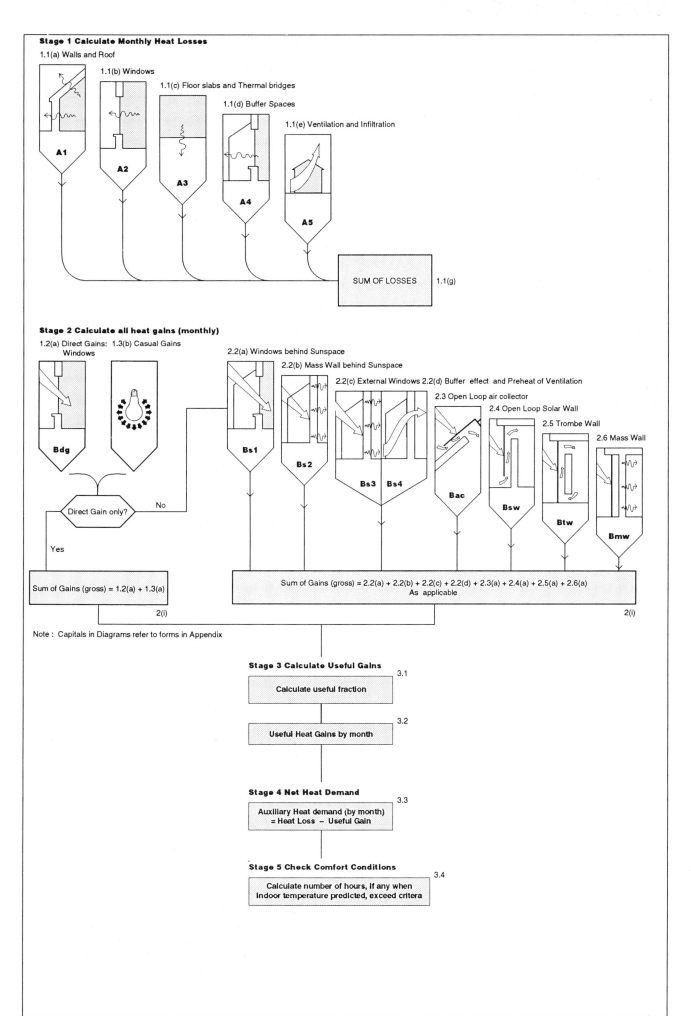

Stage 1 Calculate Monthly Heat Losses

1.1(a) Walls and Roof

1.1(b) Windows

1.1(c) Floor slabs and Thermal bridges

1.1(d) Buffer Spaces

1.1(e) Ventilation and Infiltration

A1

A2

A3

A4

A5

SUM OF LOSSES   1.1(g)

Stage 2 Calculate all heat gains (monthly)

1.2(a) Direct Gains: 1.3(b) Casual Gains
Windows

2.2(a) Windows behind Sunspace

2.2(b) Mass Wall behind Sunspace

2.2(c) External Windows 2.2(d) Buffer effect and Preheat of Ventilation

2.3 Open Loop air collector

2.4 Open Loop Solar Wall

2.5 Trombe Wall

2.6 Mass Wall

Bdg

Bs1

Bs2

Bs3   Bs4

Bac

Bsw

Btw

Bmw

Direct Gain only?

No

Yes

Sum of Gains (gross) = 1.2(a) + 1.3(a)

Sum of Gains (gross) = 2.2(a) + 2.2(b) + 2.2(c) + 2.2(d) + 2.3(a) + 2.4(a) + 2.5(a) + 2.6(a)
As applicable

2(i)

2(i)

Note : Capitals in Diagrams refer to forms in Appendix

Stage 3 Calculate Useful Gains

Calculate useful fraction   3.1

Useful Heat Gains by month   3.2

Stage 4 Net Heat Demand

**Auxiliary Heat demand (by month)
= Heat Loss − Useful Gain**   3.3

Stage 5 Check Comfort Conditions

**Calculate number of hours, if any when
Indoor temperature predicted, exceed critera**   3.4

SECTION I

DIRECT GAIN BUILDINGS

1.1

Calculating the Heat Load (Forms A1 through A7)

Except where solar devices contribute to the preheating of ventilation air, [paragraph 1(e), below], losses are computed in the same way for both TYPE A (Direct-Gain-only), and for TYPE B buildings.

1.1(a) Wall and Roof

Compute (for heated space) the wall and roof (conduction) losses: (Form A1). For buffer spaces or walls behind them see 1(d), below.

Where L_1 denotes the total heat-loss rate for the walls and roof, A the area of a wall or roof section, and U the corresponding thermal transmittance-

[Wall + Roof] conduction loss-rate, $L_1 = \sum (A \times U)$ W/K

Note: The method can distinguish between the daytime and night-time U-values , U_d and U_n, and hence between $\sum (U_d \times A)$ and $\sum (U_n \times A)$. The distinction applies to solar walls with night shutters or to parts separated from the heated volume, at night, by a shutter. For solar walls used to preheat outdoor air, mass walls, or Trombe walls, the U-value is calculated as follows:

$$U = 1/(r_g + r_a + r_{wall} + r_s) \qquad (W/m^2K)$$

where

r_g = resistance of glazing (0.0 for single glazing; 0.11 for double glazing) (m²K/W)

r_a = airspace resistance (0.16 if the wall has a non-selective surface, or 0.40 m²K/W if the wall has a selective coating)

r_{wall} = resistance of the wall itself-(if insulating shutters are used at night, add their resistance when computing U_n.)

r_s = the sum of internal and external surface resistance (usually = 0.17 m²K/W)

1.1(b) Windows

Compute heat-loss rate for exterior windows (Form A2) For buffer spaces or walls behind them see 1(d), below.

Where L_2 denotes the total heat-loss rate for the windows, A the area of a window section, and U the corresponding thermal transmittance-

Window conduction loss-rate, $L_2 = \sum (A \times U)$ (W/K)

Note: The method can distinguish between the daytime and night-time U-values , U_d and U_n, and hence between $\sum (U_d \times A)$ and $\sum (U_n \times A)$. This is to allow for the use of curtaining or insulated shutters, for example, at night. The calculation corresponds to that in 1(a), above, that is

$$U_{night} = 1/(r_g + r_a + r_{ni} + 0.17) \qquad (m^2K/W)$$

where r_{ni} = resistance of night insulation

1.1(c) Floor Slab

Compute heat-loss rate for solid floor slab and cold bridges:(Form A3)

Where L_3 denotes the total heat-loss rate for the exposed edges of the floor slab(s) in contact with the exterior, l the length of an edge, and k the corresponding transmission coefficient of heat loss (linear) – *[Note 1].*

Floor-slab conduction loss-rate, $L_3 = \sum (l \times k)$ (W/K)

Note: A difference is unlikely between daytime and night-time k.

1.1(d) Buffer Spaces

Compute heat-loss rate through buffer space, if any: (Form A4) *(W/K)*
If there is no buffer space, skip to 1(e)

Heat losses from each buffer space will have to be added to those above. (However, if one is adding a buffer space to an existing heated space whose heat losses are already calculated as if the buffer space did not exist, it is necessary to recalculate those heat losses. This is because the heated space will now be losing heat not directly to the exterior through that part of the envelope shared with the buffer space, but indirectly via the buffer space). The buffer space will lose heat to the exterior and gain heat from the interior. Using Form A4, proceed as follows but consider only surfaces on exterior envelope of heated space.

(i) Heat loss from buffer-space to outside air
For the elements separating the buffer space from the outdoors, calculate the conduction losses [as in forms A1, A2, A3];

If q is the rate of introducing outside air into the buffer space [m³/h], calculate ventilation loss rate [= 0.34q, in W/K]. Then, assuming the glazing transmittance differs between night and day, and the total daytime heat loss rate can be designated L_{bd}

$$L_{bd} = \sum (U \times A_w) + \sum (U_d \times A) + \sum (k \times l) + 0.34 \times q \ (W/K)$$

representing walls, glazing, slab edges, and ventilation where U_d is daytime transmittance of the glazing.

Similarly, night-time loss rate (L_{bn}) is –

$$L_{bn} = \sum (U \times A_w) + \sum (U_n \times A) + \sum (k \times l) + (0.34 \times q) \qquad (W/K)$$

where U_n is night-time transmittance of the glazing.
Now compute mean value, L_{bm}

$$L_{bm} = (L_{bd} + L_{bn})/2 \qquad (W/K)$$

If $U_d = U_n$,

$$L_{bm} = L_{bd} \text{ in first equation above.} \qquad (W/K)$$

(ii) Heat loss from heated space to buffer space.

On the lower part of the same form, compute these losses as in forms A1, A2, A3, but in respect of the elements separating the heated space from the buffer space. Total daytime heat loss to the buffer space is designated $\mathbf{L}_{bd}$

$$\mathbf{L}_{bd} = \Sigma(\mathbf{U} \times \mathbf{A}_w) + \Sigma(\mathbf{U}_d \times \mathbf{A}) \qquad (W/K)$$

Only glazing and opaque surfaces need be considered as edge losses are negligible, and ventilation losses do not apply.

For night-time, similarly –

$$\mathbf{L}_{bn} = \Sigma(\mathbf{U} \times \mathbf{A}_w) + \Sigma(\mathbf{U}_n \times \mathbf{A}) \qquad (W/K)$$

Now compute mean value, $\mathbf{L}_{bm}$

$$\mathbf{L}_{bm} = (\mathbf{L}_{bd} + \mathbf{L}_{bn})/2 \qquad (W/K)$$

If $\mathbf{U}_d = \mathbf{U}_n$
$\mathbf{L}_{bm} = \mathbf{L}_{bd}$ in first equation above. $\qquad (W/K)$

Heat-loss reduction coefficient of buffer space is:
$\mathbf{C}_{lb} = \mathbf{L}_{bm}/(\mathbf{L}_{bm} + \mathbf{L}_{bm})$

Daytime heat-loss rate from heated space to outdoors through buffer space is:
$\mathbf{L}_{bd} \times \mathbf{C}_{lb} \qquad (W/K)$

Night-time heat-loss rate from heated space to outdoors through the buffer space is:
$\mathbf{L}_{bn} \times \mathbf{C}_{lb} \qquad (W/K)$

The heat-loss rate from the heated space to the outdoors *via the buffer space* is the same as if the common surface (elements separating the heated space from the buffer space) was in direct contact with the exterior, but with its heat-loss rate $\mathbf{L}_b$ multiplied by the factor $\mathbf{C}_{lb}$. [$\mathbf{C}_{lb} < 1$]

1.1(e) Ventilation

Compute heat-loss rate from ventilation and infiltration (Form A5)

Ventilation heat-loss rate $= 0.34 \times \mathbf{q}_t \qquad (W/K)$

where $\mathbf{q}_t$ is m^3/h, and represents the total rate of outdoor air entering the heated space, including flowrate through buffer space, (and through heat-exchanger, or solar wall used to preheat ventilation air where not just direct-gain is used).

Net ventilation heat-loss rate

If the total flow rate of air through the heated space is $\mathbf{q}_t$ (m^3/h) from all sources, the gross heat-loss rate due to air would be $\mathbf{0.34} \times \mathbf{q}_t \qquad (W/K)$

This may be reduced by heat exchangers, buffer spaces or solar walls (Figures 1–4)

Reduction due to air to air heat exchanger
$\mathbf{R}_e = 0.34 \times \mathbf{q}_e \times \eta_e \qquad (W/K)$
where $\mathbf{q}_e$ is flow rate through exchanger [m^3/h]; η_e its efficiency

Reduction due to buffer space
$\mathbf{R}_b = 0.34 \times \mathbf{q}_b \times (1 - \mathbf{C}_{lb}) \qquad (W/K)$
where $\mathbf{q}_b$ is flow rate through through buffer space from outside to heated space [m^3/h]; $\mathbf{C}_{lb} =$ heat-loss reduction factor of buffer space (from Form A4)
Where there is a direct coupling of the buffer space with a heat exchanger, R_e and R_b become

$$\mathbf{R}_e = 0.34 \times \mathbf{q}_e \times \eta_e \qquad (W/K)$$
$$\mathbf{R}_b = 0.34 \times \mathbf{q}_b \times (1 - \mathbf{C}_{lb}) \times (1 - \eta_e) \qquad (W/K)$$

Reduction due to presence of an open-loop solar wall:
Reduction of ventilation heat-loss
$\mathbf{R}_{sw} = 0.34 \times \mathbf{q}_{sw} \times \eta_{sw} \qquad (W/K)$
where $\mathbf{q}_{sw}$ is flow rate of outside air introduced through solar wall [m^3/h], and η_{sw} efficiency of wall (obtainable from Figure 5)

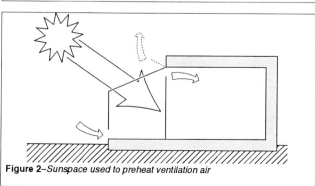

① Outdoor
② From heated space
③ To heated space

Figure 1–*Reduction due to air-to-air heat exchanger*

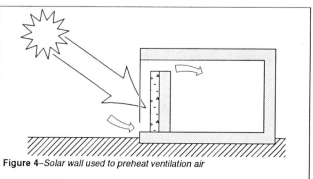

Figure 3–*Sunspace preheating coupled with an air-to-air heat exchanger*

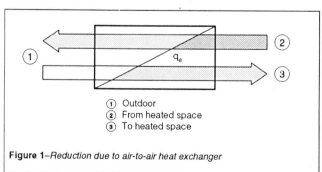

Figure 2–*Sunspace used to preheat ventilation air*

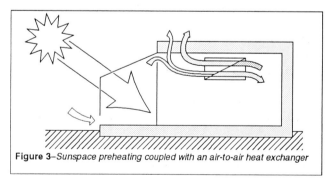

Figure 4–*Solar wall used to preheat ventilation air*

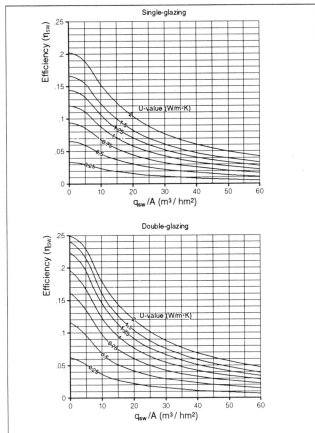

Figure 5 – *Values of η_{sw} as function of (q_{sw}/A) for different U values of wall, and single or double glazing. These graphs have been derived using formulae given by the C.S.T.B. [10]*

Total ventilation heat-loss rate reduction

$$R_t = R_e + R_b + R_{sw} \qquad (W/K)$$

Net heat-loss rate from ventilation

$$= 0.34 \times q_t - R_t$$

1.1(f) Summation
Sum the loss rates [Form A6] *W/K*
List the heat-loss rates from all sources, a, b, c, d, e (above), differentiating between day and night if appropriate. Compute grand total and calculate each source as a percentage of the total.

Note: In *New Method 5000* a different thermostat temperature can be set for day and night. Compute heating load with the high thermostat setting t_t. Savings due to the lowering of the thermostat setting are computed elsewhere. (See Appendix IV.3b)

1.1(g) Heating Load
Convert heat-loss rates to monthly heating load
(kWh/month)
The (average) total daily heat loss rate is calculated by taking the total day loss rate (*W/K*) and multiplying by the fraction [H/24] of the day the day-insulation is used. To this is added the night-loss rate multiplied by the fraction of night-insulation hours, [(24-H)/24]. (*W/K*)
To convert this to a quantity of heat for the relevant month obliges us first to integrate the temperature deficiency over time (one month). The deficiency is the difference between the average temperature for the

month outdoors (t_o), and that required indoors (ie, thermostat setting, t_t). The summation is the 'accumulated temperature, which, when multiplied by the number of days in the month, has units *degree-days,* or *K-days*. This is then multiplied by 24, for consistency, giving *K-hours* [units:*Kh*]

The total heat-loss rate [unit: W/K] multiplied by degree-hours [unit *Kh*] produces the total quantity of heat loss for the month., that is, *Wh* per month-
Divide by 1000 to give *kWh/month*

1.2

Solar gains (Forms Bdg through B)

1.2(a) Direct Gains
Calculate Direct solar gains (Form Bdg)
Direct gains are calculated by entering the window areas in Form Bdg and reading in the average daily heat transmitted (**E**), in *kWh/m²* for the orientation and tilt of each window. Each value of **E** is multiplied by the appropriate window area and the results summed. The result is the total solar heat admitted per day, and this, multiplied by the number of days in the month gives the total direct solar heat gain (gross) for the month. (*kWh*)

Calculate the solar energy transmitted into the building through the windows as follows:

for each window $\Phi_{dg} = E \times A \times m \times C_c \times S_f \times C_f$
(kWh/day)

where
E = *solar energy transmitted per m² of glazing per day*
A = *gross window area m²*
m = *effective area of window (fraction)*
C_c = *transmission factor for net curtains.(If no curtaining, C_c = 1; a typical value for net curtains is C_c= 0.93)*
S_f = *shading fraction (ie, fraction of window effective after allowing for shading and obstruction) [Note 3].*
C_f = *fraction of solar radiation incident on floor to be allowed for, according to the degree of subfloor insulation and floor absorptance. See Figure 7.*

The solar energy contributed by all windows including those facing north, is summed.

1.3

Calculating Internal Gains

1.3(a) Internal Gains
Calculate average daily casual heat gains (gross), from Section 4.5. The amounts are dependent on the occupancy of the building and on the way hot water, electric lighting and appliances are used. The daily value is multiplied by the number of days in the month. (*kWh*)
Estimates of casual heat gains in the typical house vary widely due to differences in the ways buildings are used and managed.

Proceed now to SECTION III, if the building uses Direct-Gain only; Proceed to SECTION II for buildings with additional Passive-Solar features.

BUILDINGS WITH PASSIVE-SOLAR FEATURES

In the general case a building may, in addition to windows, rely on one or more passive-solar features, the gains from which may be calculated in the following steps before summation in form B.

2.1   Windows (Form Bdg, previous)

2.2   Sunspaces (Form Bs1, Bs2, Bs3, Bs4)

2.3   Air collectors (Form Bac)

2.4   Solar Walls (Form Bsw)

2.5   Trombe walls (Form Btw)

2.6   Mass Wall (Form Bmw)

We summarise the different ways solar gain is effected: Windows (direct gain); Sunspaces; Open loop air collector (used to preheat ventilation air); Sunspace to preheat ventilation air; Open-loop solar wall (used to preheat ventilation air); Trombe wall; Mass wall

2.1

Direct gain in Type B building

Calculate direct gain from windows as for Type A step 1.2(a) previous (Form Bdg).

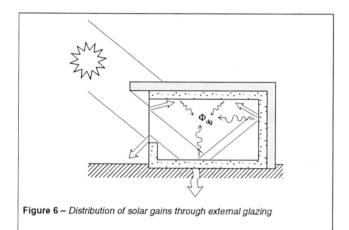

Figure 6 – Distribution of solar gains through external glazing

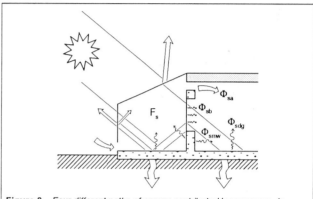

Figure 8 – Four different paths of energy contributed by sunspace to heated space

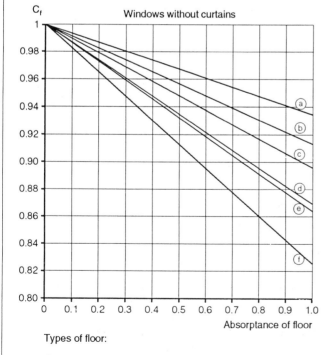

Types of floor:

(a) Carpeted slab floor on slightly vented crawl space
(b) As a but not carpeted
(c) Carpeted slab floor on vented crawl space

(d) Carpeted slab floor on soil
(e) As c but not carpeted
(f) As d but not carpeted

Figure 7–Values of solar loss factor

Solar gains from sunspace

In a sunspace solar gain arises in 4 possible ways (Fig.8).

1. By direct gain through windows between heated space and sunspace
2. By use of a mass wall in the sunspace
3. By a buffer effect whereby air in the sunspace heated by the sun reduces the heat loss of the heated space.
4. By preheating ventilation air for the heated space.

Total sunspace energy = $\Phi_{sdg} + \Phi_{smw} + \Phi_{sb} + \Phi_{sa}$

2.2(a) : Internal Windows

Gain through windows separating sunspace from heated space.

Read, from tables, **E**, the transmitted solar energy per m^2 in *kWh/day*, for slope and orientation of each window. For each **E**, multiply by area and τ_s

($\tau_s = \tau_g \times m_s$, with $\tau_g = 0.79$ for single-glazed sunspace, and 0.63 for double-glazed sunspace and m_s representing ratio of sunspace glass, net of mullions, frames etc, to total area of transparent sunspace envelope). Φ_{sdg} is computed for each window and summed.

$$\Phi_{sdg} = E \times \tau_s \times A \times m \times S_f$$

where

A = *the area of the window (m^2) between the sunspace and heated space including mullions.*

m = *the ratio between areas of glass and total area of the opening between the sunspace and the heated space (as in Form Bdg), 'm' accounts for mullions and frames.*

S$_f$ = *the shading factor as a fraction. If the roof of the sunspace, or any of the walls is opaque, S$_f$ must take into account the reduction, caused by these obstructions in the energy transmitted into the heated space .*

Since Φ_{sdg} is likely to be small when compared to Φ_{dg}, we can ignore the reduction caused by indoor net curtains and energy losses through the ground.

2.2(b) : Mass Wall

Gain from mass wall in sunspace (Φ_{mw})

For each uninsulated mass wall separating heated space and sunspace

$$\Phi_{smw} = 0.11 \times U \times \alpha \times E_i \times \tau_s \times A \times S_f \quad (kWh/day)$$

where

U = *the U-value of the wall (W/m^2 K). For example a concrete mass wall 20cm thick has a U-value of 3 W/m^2 K*

α = *the absorptance of the wall (= 0.9 for a matt black wall)*

E$_i$ = *the incident solar energy, in kWh/m^2 day read directly from tables for the orientation and slope of the wall considered*

τ_s = *the overall solar transmittance of the envelope of the sunspace (see Form Bs1)*

A = *the total area of the wall, (m^2)*

S$_f$ = *the shading factor, as a fraction, to take into account external obstructions or shading by overhangs or fins within the sunspace or the sunspace envelope*

2.2(c) : External Windows

Solar energy entering sunspace (E_s)

For each glazed area of the envelope of the sunspace,: compute energy transmitted

$$E_s = E \times S_f \times A \times m \quad (kWh/day)$$
where

E = *transmitted solar energy (kWh/m^2 day), read directly from tables for the orientation and slope and nature of the glazing considered*

S$_f$ = *the shading factor of the glazed area to take into account shades or site obstructions*

A = *overall area of the sunspace glazed surfaces (m^2).*

m = *ratio of glass to overall area of glazed surface to take account of mullions and frames (a typical value of m = 0.85)*

The transmitted energy through all sunspace glazing is then summed.

2.2(d) : Sunspace buffer-effect (Φ_{sb}), and ventilation preheat (Φ_{sa})

First compute energy F_s trapped in the sunspace

$$F_s = (a_1 \times E_s) - (a_2 \times \Phi_{sdg}) - \Phi_{smw} \quad (kWh/day)$$

where E_s, Φ_{sdg}, Φ_{smw} are the totals appearing at the bottom of Forms Bs3, Bs1, Bs2, respectively.

a_1 and a_2 are two coefficients which take into account solar energy losses by multiple reflections within the sunspace, and solar energy losses through the ground of the sunspace. They have been computed for different geometries and average climatic conditions using a detailed method [11]. Values for a_1 and a_2 are given in Table 1 as a function of the geometry of the sunspace, glazing type, absorptance of floor and floor insulation.

(i) Once F_s has been computed, two monthly mean temperatures can be computed:

The monthly mean temperature of the sunspace without solar gains: t_{sng}

$$t_{sng} = ((t_o \times L_{bm}) + (t_t \times L_{bm}))/(L_{bm} + L_{bm})$$

where

t_o = *the monthly mean outdoor temperature, specified in Form A7*

t_t = *the thermostat set temperature of the heated space,*
(Form A7)

L_{bm} = *the mean heat loss from heated space to buffer*
(Form A4)

L_{bm} = *the mean heat loss from buffer space to outdoors*
(Form A4)

The monthly mean temperature of the sunspace with solar gains: t_s

$$t_s = t_{sng} + (F_s / (0.024 \times (L_{bm} + L_{bm}))) \qquad [°C]$$

t_s is a good indication of the usability of the sunspace.

(ii). Solar gains from the buffer effect (Φ_{sb}) and ventilation of the sunspace (Φ_{sa}) are then very simply computed:

$$\Phi_{sb} = (1 - C_{lb}) \times F_s \qquad (kWh/day)$$

$$\Phi_{sa} = R_b \times F_s / L_{bm}$$

where
C_{lb} = *heat loss reduction factor of the buffer space (Form A4)*
R_b = *reduction of the ventilation heat losses due to a buffer space (Form A5).*

Method 5000 does not take into account thermocirculation between the sunspace and the heated space. This is because in most European climates such a scheme is not efficient, the sunspace rarely reaching high enough temperatures, during the heating season, to induce significant warm air currents.

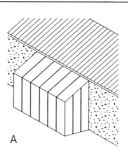
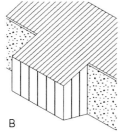
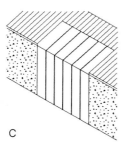

A

B

C

Table 1 – *Types of sunspace for which factors a_1 and a_2 are given on Page 325*

Solar Gain from Open-loop Air Collector (Φ_{ac})

The gains from preheating the ventilation air with an air collector are computed as follows:

$$\Phi_{ac} = \eta_{ac} \times (E / \tau_n) \times S_f \times A \times m \qquad (kWh/day)$$

where
η_{ac} = *the efficiency of the air collector for the flowrate considered and 'average' climatic conditions (fraction): ratio of the energy collected by the air to the incident solar energy. A thorough study has shown that for open-loop air collectors the efficiency was mostly dependent on the flowrate per unit area, rather than on climatic conditions. Efficiency curves extracted from this study are given in Figures 10 to 13. E is the average solar energy transmitted by single glazing in kWh/m^2 day read directly from the tables for the orientation and slope of the air collector considered. τ_n is the transmittance factor of single glazing at normal incidence; $\tau_n = 0.85$*

S_f = *the shading factor, as a fraction. If the collector is roof-mounted S_f will normally be 1.*
A = *the collector area, (m^2), and*
m = *the ratio between glazed area and collector area to take into account the collector frame and mullions.*

Solar Gain from Open-loop Solar Wall (Φ_{sw})

In the open-loop solar wall the ventilation air is heated between the glazing and the wall (painted black) before being introduced into the heated space.
Solar gains arise in two ways: from the ventilation air; and from conduction through the wall

Solar gains from the ventilation air: Φ_{asw}

$$\Phi_{asw} = F \times R_{sw} \times (r_{int} / A) \times C \qquad (kWh/day)$$

where
F = *the solar energy absorbed by the wall (kWh/day). F is computed in the same way as for external glazing (Form Bdg) except that the energy is multiplied by α, the absorptance of the wall.*
$F = E \times S_f \times A \times m \times \alpha$
R_{sw} = *reduction of ventilation heat loss computed in Form A5*   (W/k)
r_{int} = *the resistance from absorber to indoor (m^2K/W)*
r_{int} = $r_{wall} + 0.11$
(r_{wall} = *resistance of the wall itself*)

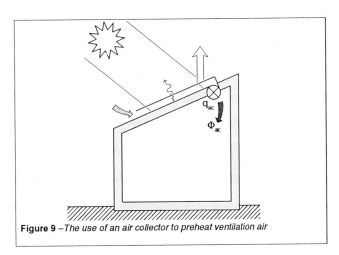

Figure 9 –*The use of an air collector to preheat ventilation air*

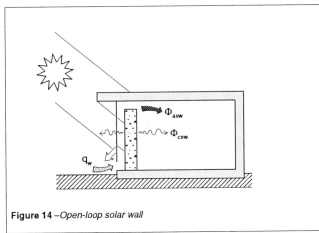

Figure 14 –*Open-loop solar wall*

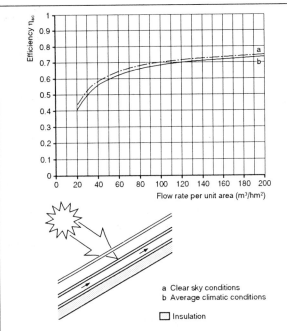

Figure 10 –*Efficiency curves for open-loop air collector where air circulation is through the absorber. Matt black absorber with clear glass*

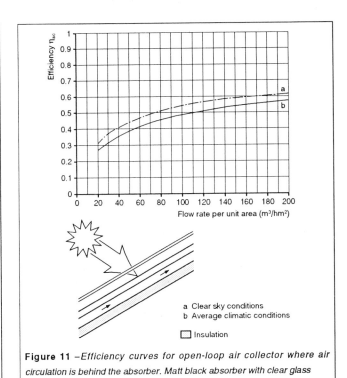

Figure 11 –*Efficiency curves for open-loop air collector where air circulation is behind the absorber. Matt black absorber with clear glass*

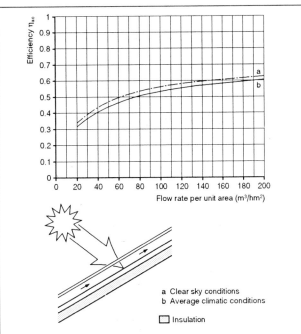

Figure 12 –*Efficiency curve for open-loop air collector where air circulation is between the glazing and the absorber. Matt black absorber with clear glass*

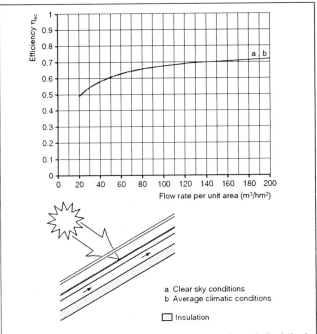

Figure 13 –*Efficiency curve for open-loop air collector where air circulation is behind the absorber. Black absorber with selective coating and clear glass*

A = *the area of the wall (m^2)*

C = *a factor which has the following values according to the position of the insulation of the wall:*
insulation on inside surface: **C**= 0.90
insulation on outside wall: **C**= 0.86

Solar gains from conduction through the wall: (Φ_{csw})

$$\Phi_{csw} = \textbf{F x U x r}_{ext} \textbf{ x C} \quad (kWh/day)$$

where

F = *the absorbed solar energy. See computation described above.*

U = *the U-value of the wall plus its glazing. This has been computed in Form A1.*

$\textbf{r}_{ext}$ = *the resistance from absorber to outdoor ($m^2 \cdot k/W$)*

$\textbf{r}_{ext}$ = *$0.22 + r_g$*

$\textbf{r}_g$ = *the resistance of the glazing:*
$\textbf{r}_g$= *0 for single glazing*
$\textbf{r}_g$= *0.11 for double glazing*

C *is as in the formula of Φ_{asw} above.*

Finally, solar gains from the solar wall are:

$$\Phi_{sw} = \Phi_{asw} + \Phi_{csw} \quad (kWh/day)$$

2.5

Solar Gain from Trombe Wall (Φ_{tw})

In a Trombe Wall the indoor air thermosiphons between the glazing and the absorber. At night the vents are closed to prevent reverse thermo-circulation.
Solar gains are computed as follows:

$$\Phi_{tw} = \textbf{F x C} \quad (kWh/day)$$
where

F = *the energy absorbed by the wall (kWh/day). As with a solar wall, **F** is equal to:*
$\textbf{E x A x S}_f \textbf{x m x } \alpha$

E = *the transmitted solar energy per m^2 of glazing (kWh/m^2 .day). This is read directly from the tables for the orientation, slope and nature of the wall.*

$\textbf{S}_f$ = *the shading coefficient of the wall, as a fraction*

A = *the area of the wall (m^2)*

m = *the ratio of glazed area to total area of wall, to take account of mullions.*

α = *the absorptance of the wall (α = 0.90 for a wall painted matt black).*

and **C** *is an efficiency factor read from the following table. **C** depends on the nature of the glazing (single or double glazing), the emissivity of the absorber, and the presence of night insulation.*

Nature of glazing	Emissivity of absorber	Night insulation	C
S.G	0.9	No	0.46
S.G	0.9	Yes	0.58
D.G	0.9	No	0.66
D.G	0.9	Yes	0.76
S.G	0.1	No	0.65
S.G	0.1	Yes	0.76
D.G	0.1	No	0.77
D.G	0.1	Yes	0.85

Table 2: – *values of the efficiency factor for various types of Trombe* Wall

2.6

Solar Gain from Mass Wall (Φ_{mw})

A mass wall is simply an unvented Trombe Wall. Solar gains result from conduction of solar energy through the wall. Solar gains are computed as follows:

$$\Phi_{mw} = \textbf{F x U x r}_{ext}$$
where

F = *the energy absorbed by the wall (kWh/day), computed in exactly the same way as for Trombe walls and solar walls. (forms Btw and Bsw).*

U = *the U-value of the wall plus its glazing, (W/m^2K), computed in Form A1.*

$\textbf{r}_{ext}$ = *the resistance from absorber to outdoors, (m^2K/W)*
$\textbf{r}_{ext}$ = *$0.06 + \textbf{r}_g + \textbf{r}_a$*

$\textbf{r}_g$ = *resistance of glazing (0 for single, 0.11 for double glazing) (m^2K/W)*

$\textbf{r}_a$ = *air space resistance (m^2K/W)*
$\textbf{r}_a$ = *0.16 m^2 K/W if the absorber is not selective*
$\textbf{r}_a$ = *0.40 m^2 K/W if the absorber has a selective coating.*

When night insulation is used, the above formula is replaced by:

$$\Phi_{mw} = \textbf{F x } ((0.7 \textbf{ x U}_d \textbf{ x r}_{ed}) + (0.3 \textbf{ x U}_n \textbf{ x r}_{en}))$$

where
$\textbf{U}_d$ *and* $\textbf{U}_n$ *are the day and night-time U-values of the wall plus its glazing;*
$\textbf{r}_{ed}$ *and* $\textbf{r}_{en}$ *are the day and night-time thermal resistances from absorber to outdoors;*
$\textbf{U}_n$ *and* $\textbf{r}_{en}$ *take into account the resistance of the night insulation:*

$$1/\textbf{U}_n = 1/\textbf{U}_d + \textbf{r}_{ni}$$
$$\textbf{r}_{en} = \textbf{r}_{ed} + \textbf{r}_{ni}$$

where $\textbf{r}_{ni}$ = *resistance of night insulation.*

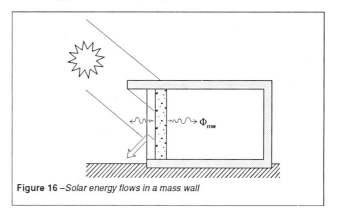

Figure 16 –*Solar energy flows in a mass wall*

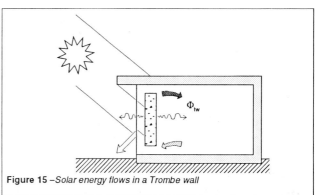

Figure 15 –*Solar energy flows in a Trombe wall*

SECTION III

DERIVING AUXILIARY HEATING LOAD

3.1

Sum of all Solar and Internal Gains

Add all the solar gains contributed by the different components.

Where there are sunspaces, the value of Φ_s is computed month by month:

$$\Phi_s = \Phi_{sdg} + \Phi_{smw} + \Phi_{sb} + \Phi_{sa}$$

For each type of component, values of Φ entered are the values appearing at the bottom of the corresponding forms. When several forms are needed for the same type of component the sum of all subtotals is entered in Form B.

The bottom line of Form B is for the total amount of solar gains (kWh/day):

$$\Phi_{solar} = \Phi_{dg} + \Phi_s + \Phi_{ac} + \Phi_{sw} + \Phi_{tw} + \Phi_{mw}$$

(Some of the terms in the above formula can be zero)

3.2

Calculating useful fraction of all gains

Compute fraction η of gains that is useful $(0 \leq \eta \leq 1)$

The useful fraction η of gross solar and casual gains is computed (Figures 17, 18 and 19) from a knowledge of the building's thermal inertia classification and the gains/load ratio of the building. The thermal inertia classification is derived from the ratio $\mathbf{I}$ (useful thermal mass/floor area). (kg/m^2)

Proceed as follows

(a)
First compute the total amount of energy contributed daily to the heated space.

$$\Sigma\Phi = \Phi_i + \Phi_{solar} \qquad (kWh/day)$$

where Φ_i = the average daily 'raw' internal gains from occupancy. This value should be specific to the country and type of house considered.

(b)
Then compute the useful thermal mass (inertia) per unit of heated floor area (I). The method used is the one given by C.S.T.B.

For each wall or partition wall, compute the useful thermal mass per unit area of wall.
When there is no specific layer of insulation*, one takes half the total mass of the material. In all cases the useful mass per unit area is limited to 150 kg/m^2. If a higher figure is found, assume that it is 150 kg/m^2.

For walls or floors in contact with the ground or a crawl space, one assumes 150 kg/m^2 if the wall or floor has no layer of insulation. When there is a layer of insulation one takes the mass per unit area of the material located between the layer of insulation and the heated space.

For walls or floors in contact with another heated space one takes the mass per unit area of the material located between the insulation and the heated space. If there is no insulation layer, half the total mass per unit area is taken. In all cases the limit is 150 kg/m^2.

For partition walls within the heated space, take the total mass of the wall per unit area, but not exceeding the value of 300 kg/m^2.

The useful thermal mass to consider is illustrated in Figure 17.

Multiply the useful thermal mass per unit area of wall by the area of the wall. This is done for all walls, partition walls, floors, etc., in the heated space. The figures obtained are then summed. This yields the total useful thermal mass of the heated space.

Finally, this figure is divided by the heated floor area. Then one gets $\mathbf{I}$ in kg/m^2.

Depending on the value of $\mathbf{I}$, the building falls within one of the 5 categories ** shown in Figure 18. For each category, an average value of the main time constant τ of the building, characteristic of its thermal inertia, is given. A way to calculate a more accurate value of τ is provided in Section 4.3(a)

---

* In this definition of I, a layer of insulation means a layer of material having a conductivity less than 0.12 $W/m.K$, the resistance of the layer being greater than 0.5 m^2K/W.

** Categories 2 to 4 are the categories defined by C.S.T.B. [13]

(c)
the total monthly solar and internal gains are:

$$\mathbf{N} \, \Sigma\Phi$$

where
$\Sigma\Phi$ = the total amount of solar and internal gains (kWh/day), computed in part (a), above
$\mathbf{N}$ = the number of days per month

The useful monthly gains are given by the formula:

$$\Sigma Q = \eta \, \mathbf{N} \, \Sigma\Phi$$

where
η = the utilisation factor (fraction) taken from the curves in Figure 19, for each category of thermal inertia.

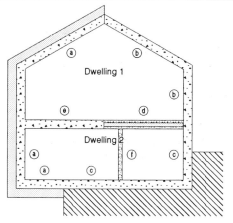

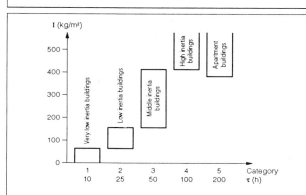

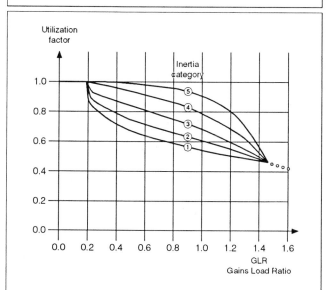

(a) mass with external insulation -use all the mass-maximum 150 kg/m² of floor, wall or ceiling.

(b) mass with no external insulation -use half the total mass- maximum 150 kg/m² of floor or wall area.

(c) mass in contact with the ground -use 150 kg/m² of floor or wall area.

(d) insulated mass between two heated dwellings -use all mass between insulation and dwelling considered- maximum 150 kg/m² of floor, wall or ceiling.

(e) uninsulated mass between two heated dwellings -use half the total mass- maximum 150 kg/m² of floor, wall or ceiling.

(f) partition walls within heated space -use all the mass- maximum 300 kg/m² of partition.

Figure 17 –*Mass to be considered in calculating 'I''*

Figure 18 –*Building category for different thermal mass and 'I''*

Figure 19 –*Efficiency curves: Utilization factor versus GLR for each inertia category*

Category   Building Type

5   **Apartment building**
    I close to 400 kg/m^2 or greater:– τ is close to 200h
4   **High-inertia houses**
    I 400 kg/m^2 or greater: – τ is about 100h
3   **Medium -inertia houses**
    I between 150 and 400 kg/m^2 – τ is about 50 h
2   **Low-inertia houses**
    I between 60 and 150 kg/m^2– τ is around 25 h
1   **Very-low-inertia houses**
    I less than 60 kg/m^2– τ is around 10 h

The abscissa GLR of the curves in Figure 19 is calculated by the formula:

$$GLR = \frac{N \sum \Phi}{Q_{ng}}$$

where

Q_{ng} = *the monthly loads without gain calculated in Form A7*

The different curves in figure 19 are drawn using the following correlations,

$$\eta = 1 - (0.581 - \frac{0.0957}{GLR}) e^{- (0.0279 - 0.0195\, GLR)\, \tau}$$

where

GLR = *Gains Load Ratio calculated above*

τ = *Main time constant of the building whose value is deduced from Figure 18 or calculated as explained in Section 4.3(a)*

The use of the previous correlation is limited by the following constraints:

When η yields a value of $\sum Q$ greater than Q_{ng}, one limits $\sum Q$ to Q_{ng}. The auxiliary heating load is then zero and η can be recomputed from the formula:

$$\eta = Q_{ng} / N \sum \Phi$$

When η calculated with the above formula is greater than 1.0, η is limited to 1.0

When **GLR** is greater than 1.4, the correlation is no longer valid and the curves become imprecise. But these points correspond to very large gains and small loads, and the accuracy of the method is not affected. Nevertheless, circles have been drawn to indicate that the utilization factor η must decrease when **GLR** increases.

These circles correspond in fact to periods when the building is uncomfortable: heat losses may be increased by increasing the ventilation rate, thus decreasing the **GLR** value. For the analysis of comfort conditions in each period, see Stage 5 or Section 3.4.

Auxiliary Heating Load

The auxiliary heating load is simply the difference between the heating demand without gains and the useful solar and internal gains:

$$Q_{aux} = Q_{ng} - \Sigma Q \qquad (kWh/month)$$

To investigate the proportion of useful gains each passive solar component contributes, the following formula is used:

$$Q = \Phi \times \eta \times N \qquad (kWh/month)$$

where

Φ = *the 'raw' gains computed in one of the B-forms: the contributed energy by the component considered. (kWh/day)*
η = *the utilization factor computed in 3.2 (c)*
N = *the number of days per month.*

Note: To assess the net contribution of the passive solar component one has to subtract the heat loss of the component from the useful gain.

It may be useful to compute the annual auxiliary heating load, Q_{auxtot} and to derive the ratios:

$$Q_{auxtot}/A_h$$
auxiliary heating demand per unit of heated floor area

$$(kWh/m^2).$$

$$Q_{auxtot}/(A_h \times DD_a)$$
auxiliary heating demand per unit of heated floor area and per degree-day

$$(kWh/m^2K.day).$$

where
DD_a = *annual degree-days (Form A7)*

The **B**-value is used to characterize the auxiliary heating demand of a dwelling:

$$B = (Q_{auxtot} \times G)/Q_{ng\,tot} \qquad (W/m^3K)$$

where

G = *coefficient of annual heating load without gains (Form A7)*
$Q_{ng\,tot}$ = *annual heating load without gains (Form A7)*

Checking comfort Conditions

To check the risk that by designing so as to minimise the heating load, we might produce uncomfortably hot conditions in the heated space, we can use the following method to predict the number of hours per day when the indoor temperature will rise above a given value.

First, compute the average monthly indoor temperatures without heating, t_{wh}

$$t_{wh} = t_o + (t_t - t_o)\,GLR$$

where
t_o = *the average monthly outdoor temperature*
t_t = *the thermostat set point temperature*
GLR = *Gains Load Ratio calculated in forms C*

Then, the average indoor temperature is deduced from:

$$t_i = t_{wh} + (t_t - t_o)(1 - \eta\,GLR)$$

where
η = *is the utilization factor calculated in forms C*

Then, depending on the inertia category **I**, use one of the graphs in Figure 20 to 23.

Each graph indicates, for a given value of t_i, the number of hours per day during which the indoor temperature will rise above a specified temperature **t**. If this number is too high a shading device can be used. If this device reduces the 'raw' solar gains with a factor S_f then the temperature without heating reduces with $\Delta t_{wh} = S_f\,\Phi_{solar}/LL$. So if in the same month $Q_{aux} = 0$ before and after using the device then the average monthly indoor temperature decreases with the same amount since:
$Q_{aux} = 0;\; \Delta t_i = \Delta t_{wh}$

The curves were developed using the computer results which also provide the correlations of Method 5000*. Since the code used cannot predict indoor temperature with full accuracy, the method used here should be considered as one which gives only an estimate of mean comfort conditions. One must also bear in mind that winter occupancy patterns are implicitly assumed in this method: the house is assumed to be vented at a fixed rate, specified in the computation of Q_{aux} with no opening of windows.

In reality, during the summer when windows are opened, ventilation is dependent on the temperature difference between indoors and outdoors, and a fixed rate cannot be assumed..

To investigate the effect of movable shading devices, calculate the solar gains with the shading coefficient (S_f) corresponding to the device. Then Q_{aux} and t_i are computed. With the new values of t_i, one then reads n(t).

Example

Consider a housing unit in a condominium located at Carpentras (Southern France). The characteristics are:

G-value: 0.60 $W/m^3.K$
Heated volume: 250 m^3, so **LL** = 3.6 $kWh/dayK$

Direct gain with 15 m^2 of south facing double glazed windows.

m = 0.8 $\qquad$ t_i = 19°C

For May: t_i = t_{wh} = 25.7°C (the auxiliary heating load is zero). The graph in Figure 23 indicates that for t_i = 25.7°C, n(24) = 18h. The indoor temperature will rise above 24°C for an average of 18 hours per day, in May.

Since this comfort level is unacceptable, the effect of

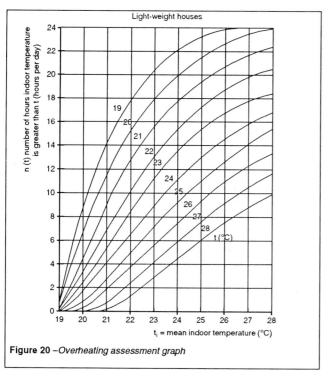

Figure 20 –*Overheating assessment graph*

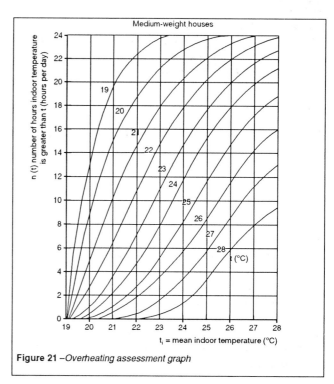

Figure 21 –*Overheating assessment graph*

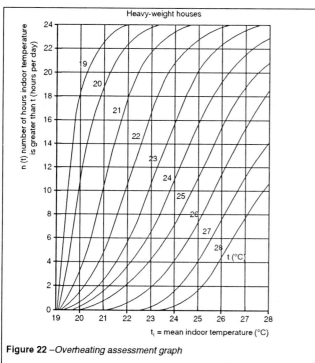

Figure 22 –*Overheating assessment graph*

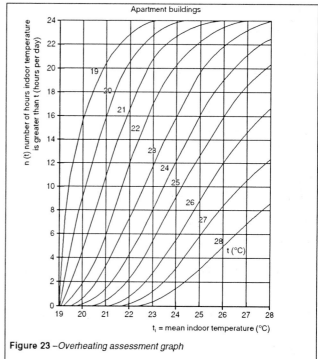

Figure 23 –*Overheating assessment graph*

movable shading devices is assessed. It is assumed that these will provide 60% shading in May and 40% in October.

	May	**Oct**
Δt_{wb} (oC)	3.5	2.3
Q_{aux}	0	0
t_i	22.2	21.0

The auxiliary heating load is still zero but, this time, we have n (24) < 6h. The comfort level is acceptable, without a reduction in the annual performance of the dwelling.

This example indicates one of the advantages of this method. It is possible to assess monthly comfort conditions and the effect of shading devices.

* The thermostat set temperature assumed for these curves is 19°C. If any other thermostat set temperature is used, the curves will be valid fot t_i and t greater than 22°C.

Designation		Jan	Feb	Mar	Ap	May	Oct	Nov	Dec
t_o	(°C)	4.9	6.1	9.7	12.5	16.3	14.0	9.1	5.7
LL	(kWh/K.day)	3.6	3.6	3.6	3.6	3.6	3.6	3.6	3.6
Φ_{solar}	(kWh/day)	23.8	24.5	26.5	24.4	20.9	20.6	22.8	21.0
t_{wb}	(°C)	15.1	16.5	20.7	22.9	25.7	23.3	19	15.1
Q_{aux}	(kWh)	446	293	43	0	0	0	94	458
t_i	(°C)	19.3	19.4	21.1	22.9	25.7	23.3	19.9	19.2

APPENDIX

 Calculation Forms

FORM A1 HEAT-LOSS RATE FOR EXTERIOR WALLS AND ROOF

Element	Area m²	Day U_d W/m²K	Night U_n W/m²K	HEAT LOSS Day U_d X A W/K	HEAT LOSS Night U_n X A W/K
TOTALS (WALLS/ROOF)					

Heat loss = U-Value x area

FORM A2 HEAT-LOSS RATE FOR EXTERIOR WINDOWS

TOTALS (WINDOWS)					

Heat loss = U-Value x area

FORM A3 HEAT-LOSS RATE FROM FLOOR SLABS AND THERMAL BRIDGES

Floor slab or thermal bridge element	Length m	k-value W/mK	Heat loss W/K
TOTALS (SLABS & BRIDGES)			

Heat loss = k-Value x length

Element	Area m²	Day U_d W/m²K	Night U_n W/m²K	HEAT LOSS Day $U_d \times A$ W/K	HEAT LOSS Night $U_n \times A$ W/K
SOUTH WALL	17.88	0.43	0.43	7.69	7.69
NORTH WALL	26.50	0.43	0.43	11.39	11.39
WEST WALL	15.67	0.43	0.43	6.73	6.73
CEILING	104	0.22	0.22	22.88	22.88
DOOR	2.20	3.33	3.33	7.32	7.32
TOTALS (WALLS/ROOF)	166.25			56.01	56.01

FORM A1 HEAT-LOSS RATE FOR EXTERIOR WALLS AND ROOF

EXAMPLE

SOUTH GLAZING	6.40	3.3	0.7	21.12	4.48
NORTH GLAZING	3.12	3.3	0.7	10.30	2.18
WEST GLAZING	4.20	3.3	0.7	13.86	2.94
TOTALS (WINDOWS)	13.72			45.28	9.60

FORM A2 HEAT-LOSS RATE FOR EXTERIOR WINDOWS

EXAMPLE

Floor slab or thermal bridge element	Length m	k-value W/mK	Heat loss W/K
FLOOR	30.8	1.2	36.96
TOTALS (SLABS & BRIDGES)	30.8		36.96

FORM A3 HEAT-LOSS RATE FROM FLOOR SLABS AND THERMAL BRIDGES

EXAMPLE

FORM A4 HEAT-LOSS RATE FROM BUFFER SPACES

(a) Loss from Buffer Space to Exterior [L_b]   Buffer Space Name:_____

Element separating buffer-space from outdoors	HEAT LOSS OF WALLS AND WINDOWS			HEAT LOSS FROM THERMAL BRIDGES		AIR CHANGE	TOTALS	
	U_{day}	U_{night}	Area	k	length	q	day	night
	W/m^2K	W/m^2K	m^2	W/mK	m	m^3/h	W/K	W/K
TOTALS							$L_{bd} =$	$L_{bn} =$
Mean Value of $L_b = (L_{bd} + L_{bn})/2$							$L_{bm} =$	

(b) Loss from Interior to Buffer Space [L_h]

TOTALS							$L_{hd} =$	$L_{hn} =$
Mean Value of L_h: $L_{hm} = (L_{hd} + L_{hn})/2$							$L_{hm} =$	

Mean heat loss reduction factor of buffer space:

$C_{lb} = L_{bm} / (L_{hm} + L_{bm}) =$ _____

Daytime heat loss from heated space to exterior through buffer space:

$L_{hd} \times C_{lb} =$ _____ (W/K)

Nightime heat loss from heated space to exterior through buffer space:

$L_{hn} \times C_{lb} =$ _____ (W/K)

FORM A5 NET HEAT-LOSS FROM INFILTRATION AND VENTILATION

Reduction due to a heat echanger:

$R_e = 0.34 \times q_e \times \eta_e$

Reduction due to air circulation through a buffer space:

- with no heat exchanger:

$R_b = 0.34 \times q_b \times (1-C_{lb})$

- with buffer space coupled to heat exchanger:

$R_b = 0.34 \times q_b \times (1-C_{lb}) \times (1-\eta_e)$

Reduction due to air circulation in open loop solar wall:

$R_{sw} = 0.34 \times q_{sw} \times \eta_{sw}$

TOTAL VENTILATION HEAT LOSS REDUCTION:

$R_t = R_e + R_b + R_{sw} =$ _____ W/K

NET HEAT LOSS FROM VENTILATION:

$0.34 \times q_t - R_t =$ _____ W/K

FORM A5 NET HEAT-LOSS FROM INFILTRATION AND VENTILATION
Total Rate of introduction of outside air $q_t =$ _____ m^3/h

Element reducing heat loss	m^3/h			R_e W/K	R_b W/K	R_{sw} W/K
Heat exchanger	$q_e =$	$\eta_e =$				
Buffer space 1	$q_b =$	$\eta_e =$ *	$C_{lb} =$			
Buffer space 2	$q_b =$	$\eta_e =$	$C_{lb} =$			
Buffer space 3	$q_b =$	$\eta_e =$	$C_{lb} =$			
Buffer space 4	$q_b =$	$\eta_e =$	$C_{lb} =$			
Open loop solar wall	$q_{sw} =$	$\eta_{sw} =$				

* if space coupled to heat exchanger

(a) Loss from Buffer Space to Exterior [L_b] Buffer Space Name: CONSERVATORY

Element separating buffer-space from outdoors	HEAT LOSS OF WALLS AND WINDOWS			HEAT LOSS FROM THERMAL BRIDGES		AIR CHANGE	TOTALS	
	U_{day}	U_{night}	Area	k	length	q	day	night
	W/m²K	W/m²K	m²	W/mK	m	m³/h	W/K	W/K
ROOF	1.0	1.0	6.42				6.42	6.42
GLAZING	5.0	5.0	17.98				89.90	89.90
FLOOR				1.75	7.19		12.58	12.58
AIR CHANGE						90	30.60	30.60
TOTALS							L_{bd} = 139.5	L_{bn} = 139.5
Mean Value of $L_b = (L_{bd} + L_{bn})/2$							L_{bm} =	139.5

(b) Loss from Interior to Buffer Space [L_h]

WALL	0.42	0.42	1.63				0.69	0.69
WINDOW	2.75	1.70	6.60				18.15	11.22
EDGE WALL/FLOOR				1.20	3.2		3.84	3.84
TOTALS							L_{hd} = 22.68	L_{hn} = 15.75
Mean Value of L_h : $L_{hm} = (L_{hd} + L_{hn})/2$							L_{hm} =	19.22

FORM A5 NET HEAT-LOSS FROM INFILTRATION AND VENTILATION
Total Rate of introduction of outside air q_t = ___186___ m³/h

Element reducing heat loss	m³/h			R_e	R_b	R_{sw}
				W/K	W/K	W/K
Heat exchanger	q_e = 180	η_e = 0.60		36.72		
Buffer space 1	q_b = 90	η_e = * 0.60	c_{lb} = 0.88		1.48	
Buffer space 2	q_b =	η_e =	c_{lb} =			
Buffer space 3	q_b =	η_e =	c_{lb} =			
Buffer space 4	q_b =	η_e =	c_{lb} =			
Open loop solar wall	q_{sw} =	η_{sw} =				
TOTALS					1.48	

* if space coupled to heat exchanger

FORM A4 HEAT-LOSS RATE FROM BUFFER SPACES

EXAMPLE

FORM A4 HEAT-LOSS RATE FROM BUFFER SPACES
(a) Loss from Buffer Space to Exterior [L_b]   Buffer Space Name: GARAGE

Element separating buffer-space from outdoors	HEAT LOSS OF WALLS AND WINDOWS			HEAT LOSS FROM THERMAL BRIDGES		AIR CHANGE	TOTALS	
	U_{day}	U_{night}	Area	k	length	q	day	night
	W/m²K	W/m²K	m²	W/mK	m	m³/h	W/K	W/K
SOUTH WALL	2.47	2.47	5.3				13.09	13.09
EAST WALL	2.47	2.47	20.0				49.4	49.4
SOUTH DOOR	3.33	3.33	4.4				14.65	14.65
NORTH WALL	2.47	2.47	3.1				7.66	7.66
NORTH DOOR	3.33	3.33	2.2				7.33	7.33
CEILING	5.18	5.18	24.0				124.32	124.32
FLOOR				1.75	14		24.50	24.50
AIR CHANGE						30	10.24	10.24
TOTALS							L_{bd}= 251.2	L_{bn} = 251.2
Mean Value of $L_b = (L_{bd} + L_{bn})/2$							L_{bm} =	251.2

(b) Loss from Interior to Buffer Space [L_h]

WEST WALL	0.42	0.42	20.0				8.40	8.40
EDGE WALL/FLOOR				1.20	8		9.60	9.60
TOTALS							L_{hd}= 18.0	L_{hn} = 18.0
Mean Value of L_h : $L_{hm} = (L_{hd} + L_{hn})/2$							L_{hm} =

FORM A5 NET HEAT-LOSS FROM INFILTRATION AND VENTILATION

EXAMPLE

FORM A6 SUM OF LOSSES

From Form	Source of heat loss	Area m²	DAY LOSS RATE W/K	DAY LOSS RATE % of total	NIGHT LOSS RATE W/K	NIGHT LOSS RATE % of total
A1	Wall and Roof					
A2	External Windows					
A3	Floor Slab etc...					
A4	Buffer Space 1					
	Buffer Space 2					
	Buffer Space 3					
	Buffer Space 4					
	Total from heated space to exterior thru buffer spaces					
	Total heat loss by transmission: A1+A2+A3+A4					
A5	Net heat loss from ventilation					
TOTAL HEAT-LOSS RATE				100		100

FORM A7 MONTHLY HEATING LOAD WITHOUT GAINS

Daytime heat loss = _____ W/K

Night-time heat loss = _____ W/K

Thermostat set temperature:

$t_t =$ _____ °C

Heated volume: $V_h =$ _____ m³

$DD_m = (t_t - t_o) \times N$

$LL = (H \times$ daytime heat loss $+ (24 - H) \times$ night-time heat loss $/ 1000$

$Q_{ng} = LL \times DD_m$

$G = Q_{ngtot} / (0.024 \times V_h \times DD_a) =$ _____ (W/m³.K)

Climate of:

Heating Season: Month Number of days N	OCT 31	NOV 30	DEC 31	JAN 31	FEB 28	MAR 31	APR 30	MAY 31	TOTAL (243)
Average exterior temp t_o (°C)									
Degree days DD_m (K-days)									
Length of day H (hrs)									
Heat loss per day LL (kWh/Kday)									
Load Q_{ng} (kWh/mo)									

FORM Bdg DIRECT SOLAR GAINS

Element	Glazing Properties			OCT	NOV	DEC	JAN	FEB	MAR	APR	MAY
	Area A =	Orientation	E kWh/m² day								
	m = C_c =	Tilt	S_f								
	C_f =	SG/DG Glazing	Φ_{dg} kWh/day								
	Area A =	Orientation	E kWh/m² day								
	m = C_c =	Tilt	S_f								
	C_f =	SG/DG Glazing	Φ_{dg} kWh/day								
	Area A =	Orientation	E kWh/m² day								
	m = C_c =	Tilt	S_f								
	C_f =	SG/DG Glazing	Φ_{dg} kWh/day								
	Area A =	Orientation	E kWh/m² day								
	m = C_c =	Tilt	S_f								
	C_f =	SG/DG Glazing	Φ_{dg} kWh/day								
TOTAL	Area:	Φ_{dg} kWh/day									

$\Phi_{dg} = E \times A \times m \times C_c \times S_f \times C_f$

From Form	Source of heat loss	Area m²	DAY LOSS RATE W/K	DAY LOSS RATE % of total	NIGHT LOSS RATE W/K	NIGHT LOSS RATE % of total
A1	Wall and Roof	166.25	56.01	27.9	56.01	35.3
A2	External Windows	13.72	45.28	22.6	9.60	6.0
A3	Floor Slab etc...	30.8	36.96	18.5	36.96	23.3
A4	Buffer Space 1 CONSERVATORY	19.93			13.84	
A4	Buffer Space 2 GARAGE	16.74			16.74	
A4	Buffer Space 3					
A4	Buffer Space 4					
A4	Total from heated space to exterior thru buffer spaces		36.73	18.3	30.64	19.3
	Total heat loss by transmission: A1+A2+A3+A4		174.98	87.4	133.21	84.0
A5	Net heat loss from ventilation		25.04	12.6	25.04	15.9
	TOTAL HEAT-LOSS RATE		200	100	158	100

Heating Season: Month Number of days N	OCT 31	NOV 30	DEC 31	JAN 31	FEB 28	MAR 31	APR 30	MAY 31	TOTAL (243)
Average exterior temp t_a (°C)	11.5	6.9	3.9	3.1	4.0	7.7	10.5	14.0	
Degree days DD_m (K-days)	233	363	468	493	420	350	255	155	2737
Length of day H (hrs)	10.7	9.2	8.4	8.7	10.1	11.7	13.4	14.8	
Heat loss per day LL (kWh/Kday)	4.24	4.18	4.15	4.16	4.22	4.29	4.36	4.42	
Load Q_{ng} (kWh/mo)	987	1518	1942	2051	1772	1502	1111	685	11568

Element	Glazing Properties			OCT	NOV	DEC	JAN	FEB	MAR	APR	MAY
SOUTH GLAZING	Area A = 6.40	Orientation SOUTH	E kWh/m² day	1.70	1.07	0.87	0.98	1.37	1.69	1.74	1.64
	m = 0.70 C_c = 1	Tilt 90°	S_f	/	/	/	/	/	/	/	/
	C_f = 0.88	SG/DG Glazing	Φ_{dg} kWh/day	6.70	4.22	3.43	3.86	5.40	6.66	6.86	6.47
WEST GLAZING	Area A = 4.20	Orientation WEST	E kWh/m² day	0.98	0.49	0.34	0.40	0.73	1.21	1.74	2.09
	m = 0.70 C_c = 1	Tilt 90°	S_f	/	/	/	/	/	/	/	/
	C_f = 0.88	SG/DG Glazing	Φ_{dg} kWh/day	2.54	1.27	0.88	1.03	1.89	3.13	4.50	5.41
NORTH GLAZING	Area A = 3.12	Orientation NORTH	E kWh/m² day	0.58	0.31	0.22	0.26	0.46	0.81	1.32	1.76
	m = 0.70 C_c = 1	Tilt 90°	S_f	/	/	/	/	/	/	/	/
	C_f = 0.88	SG/DG Glazing	Φ_{dg} kWh/day	1.11	0.60	0.42	0.50	0.88	1.56	2.54	3.38
	Area A =	Orientation	E kWh/m² day								
	m = C_c =	Tilt	S_f								
	C_f =	SG/DG Glazing	Φ_{dg} kWh/day								
TOTAL	Area:	Φ_{dg} kWh/day		10.35	6.09	4.73	5.39	8.17	11.35	13.90	15.26

FORM Bmw SOLAR GAINS FROM MASS WALL(Φ_{mw})

Element	Glazing Properties			OCT	NOV	DEC	JAN	FEB	MAR	APR	MAY
	Area A = m²	Orientation	E kWh/m² day								
	α =	Tilt									
	m =	SG/DG	S$_f$								
	U-value day night		F kWh/day								
	r$_{ext}$ day night										
	Weighted average:		Φ_{mw} kWh/day								
	Area A = m²	Orientation	E kWh/m² day								
	α =	Tilt									
	m =	SG/DG	S$_f$								
	U-value day night		F kWh/day								
	r$_{ext}$ day night										
	Weighted average:		Φ_{mw} kWh/day								
TOTALS	Area:		Φ_{mw} kWh/day								

$F = E \times A \times S_f \times m \times \alpha$

$\Phi_{mw} = F \times ((0.7 \times U_{day} \times r_{ext\ day}) + (0.3 \times U_{night} \times r_{ext\ night}))$

U_{day} and U_{night} are read from Form A1

$r_{ext} = 0.06 + r_g + r_a\ (\ + r_{night\ insul.})$

selective coating: $r_a = 0.40\,m^2.K/W$
non-selective coating: $r_a = 0.16\,m^2.K/W$

$r_g = 0.$ for single-glazing

$r_g = 0.11$ for double-glazing

FORM Btw SOLAR GAINS FROM TROMBE WALL(Φ_{tw})

Element	Glazing Properties			OCT	NOV	DEC	JAN	FEB	MAR	APR	MAY
	Area A = m²	Orientation	E kWh/m² day								
	α =	Tilt									
	m =	SG/DG	S$_f$								
	Selective coating ? Y/N		F kWh/day								
	Night insulation ? Y/N										
	Efficiency C =		Φ_{tw} kWh/day								
	Area A = m²	Orientation	E kWh/m² day								
	α =	Tilt									
	m =	SG/DG	S$_f$								
	Selective coating ? Y/N		F kWh/day								
	Night insulation ? Y/N										
	Efficiency C =		Φ_{tw} kWh/day								
TOTALS	Area:		Φ_{tw} kWh/day								

$F = E \times A \times S_f \times m \times \alpha$

$\Phi_{tw} = C \times F$

Element	Wall Properties			OCT	NOV	DEC	JAN	FEB	MAR	APR	MAY
	Area m² =	Orientation	E kWh/m² day								
	α =	Tilt	S_t								
	m =	SG/DG Glazing	F kWh/day								
	R_{sw} = W/K	r_{int} = m²K/W	Φ_{asw} kWh/day								
	U = W/m²K	r_{ext} = m²K/W	Φ_{csw} kWh/day								
	Area m² =	Orientation	E kWh/m² day								
	α =	Tilt	S_t								
	m =	SG/DG Glazing	F kWh/day								
	R_{sw} = W/K	r_{int} = m²K/W	Φ_{asw} kWh/day								
	U = W/m²K	r_{ext} = m²K/W	Φ_{csw} kWh/day								
	Area m² =	Orientation	E kWh/m² day								
	α =	Tilt	S_t								
	m =	SG/DG Glazing	F kWh/day								
	R_{sw} = W/K	r_{int} = m²K/W	Φ_{asw} kWh/day								
	U = W/m²K	r_{ext} = m²K/W	Φ_{csw} kWh/day								
TOTALS	Area:		Φ_{asw} kWh/day								
			Φ_{csw} kWh/day								

FORM Bsw SOLAR GAINS FROM OPEN-LOOP SOLAR WALLS (Φ_{sw})

$F = E \times S_f \times A \times m \times \alpha$

$\Phi_{asw} = F \times R_{sw} \times (r_{int} / A) \times c$

$\Phi_{csw} = F \times U \times r_{ext} \times c$

U read from Form A1
R_{sw} read from Form A5

r_{int} = resistance of wall itself + 0.11
r_{ext} = resistance of glazing + 0.22

c = 0.90 for internal insulation
c = 0.86 for external insulation

Area of collector :   A = _____ m²
Flowrate   : q_{ac} = _____ m³/h
Efficiency   : η_{ac} = _____
Orientation   : _____
Tilt   : _____ m = _____

		OCT	NOV	DEC	JAN	FEB	MAR	APR	MAY
E	kWh/m²day								
S_t									
Φ_{ac}	kWh/day								

FORM Bac SOLAR GAINS FROM OPEN-LOOP AIR COLLECTER (Φ_{ac})

Area of collector: A = _____ m²

Flowrate: q_{ac} = _____ m³/h

Efficiency: η_{ac} = _____

Orientation: _____

Tilt: _____ m = _____

$\Phi_{ac} = \eta_{ac} \times (E / 0.85) \times S_f \times A \times m$

FORM B SUM OF SOLAR GAINS (Φ_{solar})

Form	TYPE OF SOLAR GAINS (kWh/day)		OCT	NOV	DEC	JAN	FEB	MAR	APR	MAY	YEARLY TOTALS* kWh
	Number of days/month		31	30	31	31	28	31	30	31	243
Bdg	External windows Φ_{dg}										
Bs1		Φ_{sdg}									
Bs2	SUNSPACES	Φ_{smw}									
Bs4		Φ_{sb}									
		Φ_{ss}									
	Sub-totals for sunspaces Φ_s										
Bac	Air collector Φ_{ac}										
Bsw	SOLAR WALLS	Φ_{asw}									
		Φ_{csw}									
	Sub-totals for solar walls Φ_{sw}										
Btw	Trombe walls Φ_{tw}										
Bmw	Mass walls Φ_{mw}										
TOTALS	Φ_{solar}										

* The yearly totals in kWh are:

$$\Phi_{yearly} = \sum_{i=1}^{12} \Phi_i \times N_i$$

(N_i = number of days in month i)

FORM Bs1 SOLAR GAINS FROM WINDOWS BETWEEN SUNSPACE AND HEATED SPACE (Φ_{sdg})

Name of Sunspace: _____

Overall solar transmittance of sunspace τ_s = _____

Element	Glazing Properties			OCT	NOV	DEC	JAN	FEB	MAR	APR	MAY
	Area A =	γ	E kWh/m² day								
	m =	Tilt	S_t								
		SG/DG Glazing	Φ_{sdg} kWh/day								
	Area A =	γ	E kWh/m² day								
	rn =	Tilt	S_t								
		SG/DG Glazing	Φ_{sdg} kWh/day								
	Area A =	γ	E kWh/m² day								
	m =	Tilt	S_t								
		SG/DG Glazing	Φ_{sdg} kWh/day								
	Area A =	γ	E kWh/m² day								
	m =	Tilt	S_t								
		SG/DG Glazing	Φ_{sdg} kWh/day								
TOTAL	Area:	Φ_{sdg} kWh/day									

$\Phi_{sdg} = E \times \tau_s \times A \times m \times s_t$

Form	TYPE OF SOLAR GAINS (kWh/day)		OCT	NOV	DEC	JAN	FEB	MAR	APR	MAY	YEARLY TOTALS* kWh
	Number of days/month		31	30	31	31	28	31	30	31	243
Bdg	External windows Φ_{dg}		10.35	6.09	4.73	5.39	8.17	11.35	13.90	15.26	2288
Bs1	SUNSPACES	Φ_{sdg}	3.32	2.58	2.33	2.40	3.15	2.75	2.73	3.34	
Bs2		Φ_{smw}									
Bs4		Φ_{sb}	1.54	0.82	0.59	0.71	1.14	1.80	2.27	2.31	
		Φ_{ea}	0.98	0.52	0.38	0.45	0.72	1.15	1.45	1.47	
	Sub-totals for sunspaces Φ_{s}		5.84	3.92	3.30	3.56	5.01	5.70	6.45	7.12	1243
Bac	Air collector Φ_{ac}										
Bsw	SOLAR WALLS	Φ_{ssw}									
		Φ_{csw}									
	Sub-totals for solar walls Φ_{sw}										
Btw	Trombe walls Φ_{tw}										
Bmw	Mass walls Φ_{mw}										
TOTALS Φ_{solar}			16.19	10.01	8.03	8.95	13.18	17.05	20.35	22.38	3530

Name of Sunspace: CONSERVATORY

Overall solar transmittance of sunspace τ_s = 0.63

Element	Glazing Properties			OCT	NOV	DEC	JAN	FEB	MAR	APR	MAY
SOUTH WINDOW	Area A = 6.60	γ SOUTH	E kWh/m² day	1.70	1.07	0.87	0.98	1.37	1.69	1.74	1.64
	m = 0.70	Tilt 90°	S_f	0.67	0.83	0.92	0.84	0.79	0.56	0.54	0.70
		SG/DG Glazing	Φ_{sdg} kWh/day	3.32	2.58	2.33	2.40	3.15	2.75	2.73	3.34
	Area A =	γ	E kWh/m² day								
	m =	Tilt	S_f								
		SG/DG Glazing	Φ_{sdg} kWh/day								
	Area A =	γ	E kWh/m² day								
	m =	Tilt	S_f								
		SG/DG Glazing	Φ_{sdg} kWh/day								
	Area A =	γ	E kWh/m² day								
	m =	Tilt	S_f								
		SG/DG Glazing	Φ_{sdg} kWh/day								
TOTAL	Area:		Φ_{sdg} kWh/day	3.32	2.58	2.33	2.40	3.15	2.75	2.73	3.34

FORM Bs2 SOLAR GAINS FROM MASS WALLS WITHIN A SUNSPACE (Φ_{smw})

Name of Sunspace:_____

Overall solar transmittance of sunspace $\tau_s =$ _____

Element	Area m²	Glazing Properties		OCT	NOV	DEC	JAN	FEB	MAR	APR	MAY
	A =	Orientation	E* kWh/m²day								
		Tilt	S_t								
		U-value =	F kWh/day								
		α =	Φ_{smw} kWh/day								
	A =	Orientation	E kWh/m²day								
		Tilt	S_t								
		U-value =	F kWh/day								
		α =	Φ_{smw} kWh/day								
	A =	Orientation	E kWh/m²day								
		Tilt	S_t								
		U-value =	F kWh/day								
		α =	Φ_{smw} kWh/day								
	A =	Orientation	E kWh/m²day								
		Tilt	S_t								
		U-value =	F kWh/day								
		α =	Φ_{smw} kWh/day								
TOTALS	Area:	Φ_{smw} kWh/day									

Solar energy absorbed by mass wall:

$F = E_i \times \tau_s \times A \times S_f \times \alpha$

Gains from mass wall:

$\Phi_{smw} = 0.11 \times U \times F$

Incident solar energy values read from tables in The European Solar Radiation Atlasses Vol. 1 and 2.

FORM Bs3 SOLAR ENERGY ENTERING THE SUNSPACE E_s

Name of Sunspace:_____

$E_s = E \times S_t \times A \times m$

Element	Area m²	Glazing Properties		OCT	NOV	DEC	JAN	FEB	MAR	APR	MAY
	A =	Orientation	E kWh/m²day								
		Glazing SG/DG	S_t								
		Tilt : m =	E_s kWh/day								
	A =	Orientation	E kWh/m²day								
		Glazing SG/DG	S_t								
		Tilt : m =	E_s kWh/day								
	A =	Orientation	E kWh/m²day								
		Glazing SG/DG	S_t								
		Tilt : m =	E_s kWh/day								
	A =	Orientation	E kWh/m²day								
		Glazing SG/DG	S_t								
		Tilt : m =	E_s kWh/day								
TOTALS	Area:	E_s kWh/day									

Name of Sunspace: _____
Overall solar transmittance of sunspace τ_s = _____

Element	Area m²	Glazing Properties			OCT	NOV	DEC	JAN	FEB	MAR	APR	MAY
	A =	Orientation		E* kWh/m²day								
		Tilt		S_l								
		U-value =		F kWh/day								
		α =		Φ_{smw} kWh/day								
	A =	Orientation		E kWh/m²day								
		Tilt		S_l								
		U-value =		F kWh/day								
		α =		Φ_{smw} kWh/day								
	A =	Orientation		E kWh/m²day								
		Tilt		S_l								
		U-value =		F kWh/day								
		α =		Φ_{smw} kWh/day								
	A =	Orientation		E kWh/m²day								
		Tilt		S_l								
		U-value =		F kWh/day								
		α =		Φ_{smw} kWh/day								
TOTALS	Area:	Φ_{smw} kWh/day										

Name of Sunspace: _CONSERVATORY_
$E_s = E \times S_l \times A \times m$

Element	Area m²	Glazing Properties			OCT	NOV	DEC	JAN	FEB	MAR	APR	MAY
SOUTH GLAZING	A = 8.23	Orientation	SOUTH	E kWh/m²day	2.07	1.29	1.04	1.18	1.66	2.10	2.20	2.11
		Glazing SG/DG	SG	S_l	/	/	/	/	/	/	/	/
		Tilt : 90 m = 0.8		E_s kWh/day	13.63	8.49	6.85	7.77	10.93	13.83	14.48	13.89
EAST GLAZING	A = 4.88	Orientation	EAST	E kWh/m²day	1.21	0.61	0.43	0.51	0.90	1.49	2.14	2.56
		Glazing SG/DG	SG	S_l	/	/	/	/	/	/	/	0.88
		Tilt : 90 m = 0.8		E_s kWh/day	4.72	2.38	1.68	1.99	3.51	5.82	8.35	8.79
WEST GLAZING	A = 4.88	Orientation	WEST	E kWh/m²day	1.21	0.51	0.43	0.51	0.90	1.49	2.14	2.56
		Glazing SG/DG	SG	S_l	/	/	/	/	/	/	/	0.97
		Tilt : 90 m = 0.8		E_s kWh/day	4.72	2.38	1.68	1.99	3.51	5.82	8.35	9.70
	A =	Orientation		E kWh/m²day								
		Glazing SG/DG		S_l								
		Tilt : m =		E_s kWh/day								
TOTALS	Area: 17.99	E_s kWh/day			23.07	13.25	10.21	11.75	17.95	25.47	31.18	32.38

FORM Bs4 SOLAR HEAT GAINS FROM SUNSPACE BUFFER EFFECT (Φ_{sb}) AND VENTILATION PREHEAT (Φ_{sa})

Sunspace:_____

L_{bm} = _____ W/K L_{hm} = _____ W/K C_{lb} = _____

a_1 = _____ a_2 = _____ R_b = _____ W/K

DESIGNATION		OCT	NOV	DEC	JAN	FEB	MAR	APR	MAY
E_s	kWh/day								
Φ_{sdg}	kWh/day								
Φ_{smw}	kWh/day								
F_s	kWh/day								
t_o ℃ Average outdoor temp									
t_{sng}	kWh/day								
t_s ℃ Mean temp of sunspace									
Φ_{sb}	kWh/day								
Φ_{sa}	kWh/day								

E_s values are totals of Form Bs3
Φ_{sdg} values are totals of Form Bs1
Φ_{smw} values are totals of Form Bs2

$F_s = (a_1 \times E_s) - (a_2 \times \Phi_{sdg}) - \Phi_{smw}$

$t_{sng} = ((t_o \times L_{bm}) + (t_t \times L_{hm})) / (L_{bm} + L_{hm})$

$t_s = t_{sng} + (F_s / (0.024 \times (L_{bm} + L_{hm})))$

$\Phi_{sb} = (1 - C_{lb}) \times F_s$

$\Phi_{sa} = R_b \times F_s / L_{hm}$

FORM C1 USEFUL GAINS AND AUXILIARY HEATING LOAD

Raw internal gains: Φ_i = _____ kWh/day

Inertia category: I = _____

Main time constant: τ = _____ h

		OCT	NOV	DEC	JAN	FEB	MAR	APR	MAY	YEARLY TOTALS
Number of days	N	31	30	31	31	28	31	30	31	243
Daily solar gains Φ_{solar} kWh/day										
Daily total gains $\Sigma\Phi$ kWh/day										
Monthly total gains $N\Sigma\Phi$ kWh/month										
Heating loads without gains Q_{ng} kWh/month										
Gains load ratio GLR										
Utilization factor η										
Useful gains ΣQ kWh/month										
Auxiliary heating load Q_{aux} kWh/month										

$\Sigma\Phi = \Phi_{solar} + \Phi_i$

$GLR = N\Sigma\Phi / Q_{ng}$

$\Sigma Q = \eta\, N\Sigma\Phi$

$Q_{aux} = Q_{ng} - \Sigma Q$

$B = (Q_{auxtot} \times G / Q_{ngtot}) =$ _____ W / m^3.K

Q_{auxtot} and Q_{ngtot} are sums of Q_{aux} and Q_{ng} for the whole heating season.

FORM C2 SPLITTING OF USEFUL GAINS

		OCT	NOV	DEC	JAN	FEB	MAR	APR	MAY	YEARLY TOTALS
Number of days	N	31	30	31	31	28	31	30	31	243
Useful internal gains Q_i kWh/month										
Useful solar direct gains Q_{dg} kWh/month										
Useful solar gains from sunspace Q_s kWh/month										
Useful solar gains from air collector Q_{ac} kWh/month										
Useful solar gains from solar walls Q_{sw} kWh/month										
Useful solar gains from Trombe walls Q_{tw} kWh/month										
Useful solar gains from mass walls Q_{mw} kWh/month										

$Q_i = \eta\, N\, \Phi_i$
$Q_{dg} = \eta\, N\, \Phi_{dg}$
$Q_s = \eta\, N\, \Phi_s$
$Q_{ac} = \eta\, N\, \Phi_{ac}$
$Q_{sw} = \eta\, N\, \Phi_{sw}$
$Q_{tw} = \eta\, N\, \Phi_{tw}$
$Q_{mw} = \eta\, N\, \Phi_{mw}$

Sunspace: CONSERVATORY

$L_{bm} =$ __139.5__ W/K $L_{hm} =$ __19.2__ W/K $C_{ib} =$ __0.88__

$a_1 =$ __0.68__ $a_2 =$ __0.89__ $R_b =$ __1.48__ W/K

DESIGNATION		OCT	NOV	DEC	JAN	FEB	MAR	APR	MAY
E_s	kWh/day	23.07	13.25	10.21	11.75	17.95	25.47	31.18	32.38
Φ_{sdg}	kWh/day	3.32	2.58	2.33	2.40	3.15	2.75	2.73	3.34
Φ_{smw}	kWh/day	/	/	/	/	/	/	/	/
F_s	kWh/day	12.73	6.71	4.87	5.85	9.40	14.87	18.77	19.05
t_o °C Average outdoor temp		11.5	6.9	3.9	3.1	4.0	7.7	10.5	14.0
t_{eng}	kWh/day	12.4	8.4	5.7	5.0	5.8	9.1	11.5	14.6
t_s °C Mean temp of sunspace		15.7	10.2	7.0	6.5	8.3	13.0	16.4	19.6
Φ_{sb}	kWh/day	1.54	0.81	0.59	0.71	1.14	1.80	2.27	2.31
Φ_{sa}	kWh/day	0.98	0.52	0.38	0.45	0.72	1.15	1.45	1.47

FORM Bs4 SOLAR HEAT GAINS FROM SUNSPACE BUFFER EFFECT (Φ_{sb}) AND VENTILATION PREHEAT (Φ_{sa})

EXAMPLE

Raw internal gains: $\Phi_i =$ __13.5__ kWh/day

Inertia category: $I =$ __3__

Main time constant: $\tau =$ __50__ h

FORM C1 USEFUL GAINS AND AUXILIARY HEATING LOAD

EXAMPLE

		OCT	NOV	DEC	JAN	FEB	MAR	APR	MAY	YEARLY TOTALS
Number of days	N	31	30	31	31	28	31	30	31	243
Daily solar gains Φ_{solar}	kWh/day	16.2	10.0	8.0	8.95	13.2	17.1	20.3	22.4	
Daily total gains $\Sigma\Phi$	kWh/day	29.7	23.5	21.5	22.5	26.7	30.6	33.8	35.9	
Monthly total gains $N\Sigma\Phi$	kWh/month	921	705	667	696	748	949	1014	1113	6813
Heating loads without gains Q_{ng}	kWh/month	987	1518	1942	2051	1772	1502	1111	685	11568
Gains load ratio GLR		0.93	0.46	0.34	0.34	0.42	0.63	0.91	1.62	
Utilization factor η		0.71	0.86	0.90	0.90	0.87	0.80	0.71	0.37	
Useful gains ΣQ	kWh/month	654	606	600	626	651	759	720	412	5028
Auxiliary heating load Q_{aux}	kWh/month	333	912	1342	1425	1121	743	391	273	6540

FORM C2 SPLITTING OF USEFUL GAINS

EXAMPLE

		OCT	NOV	DEC	JAN	FEB	MAR	APR	MAY	YEARLY TOTALS
Number of days	N	31	30	31	31	28	31	30	31	243
Useful internal gains Q_i	kWh/month	297	348	377	377	329	335	288	155	2506
Useful solar direct gains Q_{dg}	kWh/month	228	157	132	150	199	281	296	175	1618
Useful solar gains from sunspace Q_s	kWh/month	129	101	92	99	122	141	137	82	903
Useful solar gains from air collector Q_{ac}	kWh/month									
Useful solar gains from solar walls Q_{sw}	kWh/month									
Useful solar gains from Trombe walls Q_{tw}	kWh/month									
Useful solar gains from mass walls Q_{mw}	kWh/month									

FORM C3

Thermostat set point: $t_t =$ _____ °C

	OCT	NOV	DEC	JAN	FEB	MAR	APR	MAY
Average outdoor temp. t_o °C								
Average indoor temp. without heating t_{wh} °C								
Average indoor temp. with heating t_i °C								

$t_{wh} = t_o + (t_t - t_o)\,GLR$

$t_i = t_{wh} + (t_t - t_o)(1 - \eta GLR)$

FORM E GAINS WITH INTERMITTENT HEATING

$t_t =$ _____ °C

$t_t^i =$ _____ °C

$d =$ _____ h

$I =$ _____ $\tau =$ _____ h

	OCT	NOV	DEC	JAN	FEB	MAR	APR	MAY	YEARLY TOTALS
Number of days N	31	30	31	31	28	31	30	31	243
Daily heat loss LL kWh/dayK									
Maximum intermittent heating gain Q_{ih}^{max} kWh									
Gains load ratio for intermit. heating GLR_{ih} kWh									
Intermittent heating recovery factor θ									
Useful intermittent heating gain Q_{ih} kWh									
Auxiliary heating load with intermit. heating $Q_{aux,ih}$ kWh									

$Q_{ih}^{max} = N(t_t - t_t^e)\,LL \times d/24$

$GLR_{ih} = Q_{ih}^{max} / Q_{ng}$

$\theta = f(GLR_{ih}, \tau)$

$Q_{ih} = \theta\,Q_{ih}^{max}$

$Q_{aux,ih} = Q_{aux} - Q_{ih}$

FORM D1 MAIN TIME CONSTANT – EXTERNAL WALLS

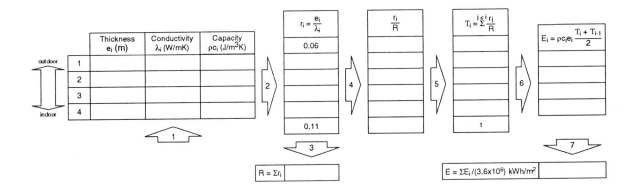

FORM D2 MAIN TIME CONSTANT – INTERNAL WALLS

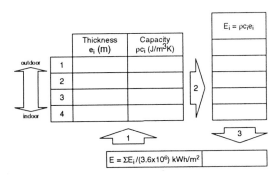

Thermostat set point: $t_t =$ ___19___ °C

		OCT	NOV	DEC	JAN	FEB	MAR	APR	MAY
Average outdoor temp. t_o	°C	11.5	6.9	3.9	3.1	4.0	7.7	10.5	14.0
Average indoor temp. without heating t_{wh}	°C	18.5	12.5	9.0	8.5	10.3	14.8	18.2	22.1
Average indoor temp. with heating t_i	°C	21.0	19.8	19.5	19.5	19.8	20.4	21.2	24.1

$t_t =$ ___19___ °C

$t_t^l =$ ___15___ °C

$d =$ ___10___ h

$I =$ ___3___ $\tau =$ ___50___ h

		OCT	NOV	DEC	JAN	FEB	MAR	APR	MAY	YEARLY TOTALS
Number of days	N	31	30	31	31	28	31	30	31	243
Daily heat loss LL	kWh/dayK	4.24	4.18	4.15	4.16	4.22	4.29	4.36	4.42	
Maximum intermittent heating gain Q_{ih}^{max}	kWh	219	209	214	215	197	222	218	228	1722
Gains load ratio for intermit. heating GLR_{ih}	kWh	0.22	0.14	0.11	0.11	0.11	0.15	0.20	0.33	
Intermittent heating recovery factor θ		0	0.25	0.33	0.35	0.33	0.21	0	0	
Useful intermittent heating gain Q_{ih}	kWh	0	52	71	75	65	47	0	0	310
Auxiliary heating load with intermit. heating $Q_{aux, ih}$	kWh	333	860	1271	1350	1056	696	391	273	6230

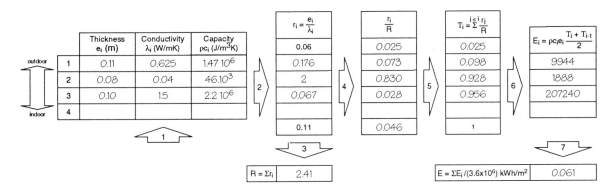

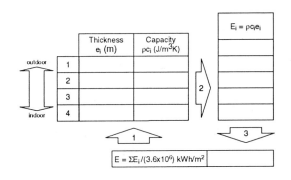

4.2

Worked Example

The following example is provided to illustrate the method and the way the forms are used.

Single-family house. Heated floor area: 104 m^2. Heated volume: $260\,m^3$
I-value category: 3 (concrete floor, light-weight partition walls, internal insulation). High Level of insulation.

Mechanical ventilation with air to air heat exchanger. Flow-rate: $180\,m^3/h$.

Sunspace coupled to the mechanical ventilation system with $90\,m^3/h$ pre-heated in the sunspace. Sunspace has $6.4\,m^2$ of floor area and $18\,m^2$ of vertical single glazing.

Double glazed external windows with night insulation: $6.4m^2$ on the sough façade, $4.2m^2$ on the west façade, $3.1m^2$ on the north façade.

Calculations are done with the climatic data of Bourges.

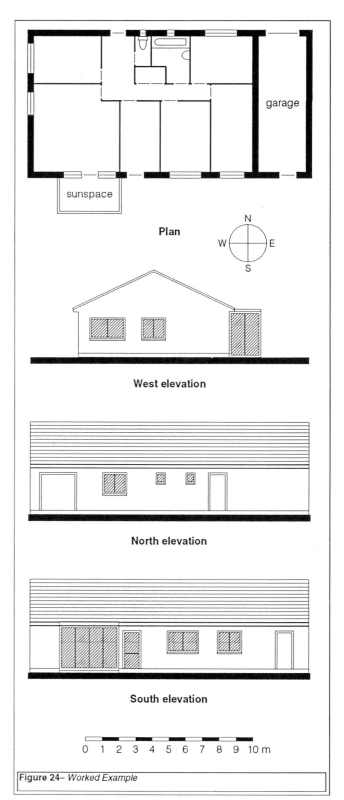

Figure 24– Worked Example

4.3

Special Cases

(a); Calculation of the main time constant

This calculation procedure applies only to a complete building or to the zones of a building having a uniform indoor temperature. The limits of the building are walls, roof, etc in contact with the outside. Zones are delimited by walls, ceilings, etc in contact with the outside, or by other zones of different temperature (which can be considered as "external walls"). The calculation of the main time constant τ is made when the building (or zone) is in a particular thermal state (although the result is usable in any state); the indoor temperature is fixed at 1°C, with the other ones (outside or other-zone temperatures) fixed at zero, and the building or zone presumed to be in a thermal steady state.

We use two forms: form D1 for external walls separating the building from its environment, and form D2 for internal walls. These forms are used to calculate the total energy stored (for a difference of $1K$ between indoor and outdoor temperature) in the walls constituting the building. One form must be filled for each kind of wall. The total energy is equal to the sum of the energies stored in external and internal walls:

$$E = \sum_{i=1}^{N_i} E_i A_i + \sum_{e=1}^{N_e} E_e A_e$$

where
$N_i E_i$ and A_i = the number, energy and area of internal walls
$N_e E_e$ and A_e = the number, energy and area of external walls

The value of the time constant τ is:

$$\tau = \frac{E \times 24}{LL} \qquad (b)$$

where
E = the total energy stored in the zone or building (kWh/K)
LL = the daily heat loss coefficient $(kWh/dayK)$ [Form A 7]

320

If the calculation is for the entire building, the value of **LL** can be taken from Form A7; if the calculation is made for a zone, the **LL** value of the zone can be calculated using the same forms A1 to A7 as for an entire building.

Use of Form D1

Form D1 is filled from left to right and from top to bottom. The order of the calculations is indicated by Numbers 1 to 7.

1. The values of thickness e_i, thermal conductivity λ_i and the specific heat capacity ρc_i of each layer i of the wall are put in the first box, beginning with the more external layers.

2. The second box is filled with the value of the thermal resistance of each layer $r_i = e_i / \lambda_i$ where e_i and λ_i are taken from the same row of the first box. The first and last row correspond to the surface resistances on both sides of the wall. [Values 0.11 and 0.06 m^2K/W respectively, can be used for the internal and external surface resistances].

3. The sum of all the resistances is the value of the wall resistance.

4. Each row of the third box is equal to the ratio of the resistance of the corresponding layer to the total resistance of the wall

5. Each value of the fourth box is equal to the sum of all the rows of the third column which are above or at the same level as the current row

6. The fifth box contains the value of the energy stored in each layer, per square meter. This value is equal to the product of the thickness e_i, the specific heat capacity ρc_i, and the mean temperature in the layer. This temperature is the mean between the values which are in the previous column, in the same row and in the above one.

7. The sum of the last column is the total energy stored in each square meter of wall (the factor represents 3.6×10^6 is a unit change from J/m^2K to kWh/m^2K)

Use of Form D2

The calculations in this box are faster.
Form D2 is used from left to right and from top to bottom. The order of the calculations is indicated by numbers 1, 2 and 3.

1. The values of thickness e_i, the specific heat capacity ρc_i of each layer of the wall are written in the first box, beginning by the more external layers.

2. The second box is filled with the value of the energy stored in each layer, per square meter. This value is equal to the product of the thickness e_i, and the specific heat capacity ρc_i (all points of internal walls are at 1°C).

3. The sum of the last column is the total energy stored in each square meter of wall.

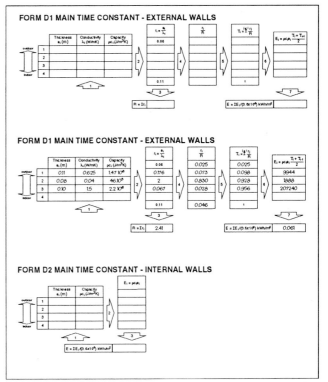

Example

For a wall in contact with the outside, using:
brick 11 cm thick (λ = 0.625 W/mK; ρc = 1.47 × 10^6 J/m^3K)
insulation 8 cm thick (λ = 0.04 W/mK; ρc = 46000 J/m^3K)
concrete 10 cm thick (λ = 1.5 W/mK; ρc = 2.2 10^6 J/m^3K)
from the outside to the inside –

Form D1 would be filled as follows (See page 319):

(b): Intermittent heating

Allowing a lower thermostat set point during some periods of the day decreases the auxiliary heating demand. The following method enables one to calculate heating loads when there is intermittent heating. This simplified method is based on the assumption that the recovery of solar and internal gains and intermittent heating savings are not linked. The intermittent heating gain is calculated by multiplying a utilization factor by the maximum theoretical gain.

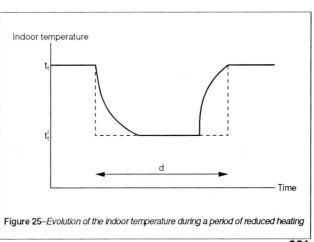

Figure 25–*Evolution of the indoor temperature during a period of reduced heating*

321

The energy saving per day provided by intermittent heating has a maximum value which is given by the following formula:

$$\Phi^{max}_{ih} = LL \, (t_t - t_t^l) \times d/24$$

where

LL = daily heat loss per K temperature difference (kWh/Kday)

t_t = usual thermostat set temperature used in the previous calculations (°C)

t_t^l = low thermostat temperature during the period of reduced heating (°C)

d = duration of the period of reduced heating (h) [Figure 25]

It is assumed that only one period of reduced heating occurs each day and that this period has the same characteristics during the whole heating season. The maximum energy saving per month is:

$$Q^{max}_{ih} = N\Phi^{max}_{ih}$$

The "Gains Load Ratio" due to intermittent heating is then:

$$GLR_{ih} = \frac{Q^{max}_{ih}}{Q_{ng}}$$

And the utilization factor θ of this energy saving can be deduced from the following empirical correlation:

$$\theta = \frac{58.47 - (303.93)\, GLR_{ih}}{58.47 - (303.93)\, GLR_{ih} + \tau}$$

where

GLR_{ih} = Gains Load Ratio related to intermittent heating

τ = main time constant of the building or zone whose value is taken from Figure 18 or calculated in SECTION 4.3(a) The energy saving is then:

$$Q_{ih} = \theta \, Q^{max}_{ih}$$

and the auxiliary heating load is:

$$Q_{aux, ih} = Q_{aux} - Q_{ih}$$

where

Q_{aux} = auxiliary heating demand with constant thermostat set temperature

When the GLR_{ih} value is greater than 0.19, θ is supposed to be equal to zero. This case corresponds to high outdoor temperature and low indoor temperature set back. The savings by intermittent heating are then negligible.

It must be noted that d is the length of the period during which an indoor temperature lower than the usual one is allowed, not the one during which the heating system is not functioning.

(c): Multizoning

A building in which various zones have very different temperature behaviour must be described as a multizone building. An extended use of "New Method 5000" can be made to assess the response of each zone in terms of auxiliary monthly heating loads and monthly mean indoor temperature. The method will be explained only for a two-zone building. The method can be easily generalized to N-zone buildings, but the calculations would have to be made with a computer.

The principle is to use the monozone method described previously for each zone of the building. Let us consider the input data.–

Firstly, the mean indoor temperature of Zone 1 without heating is given by:

$$t^1_{wh} = \frac{\sum\Phi^1}{(LL^1 + LL^{12})} + \frac{LL^1 \, t_o + LL^{12} \, t^2_{wh}}{(LL^1 + LL^{12})}$$

where

$\sum\Phi^1$ = daily gains of Zone 1 (kWh/day)

LL^1 = daily heat loss coefficient from Zone 1 to outside (kWh/Kday)

LL^{12} = daily heat exchange coefficient between Zones 1 and 2 (kWh/Kday)

t_o = mean outdoor temperature (°C)

t^2_{wh} = mean temperature without heating of Zone 2 (°C)

The permutation of indices 1 and 2 would give the respective expression of t^2_{wh}. The value of LL^1, LL^2 and LL^{12} can be calculated as explained in Section 1.1 for a monozone building. The system of equations relative to Zones 1 and 2 with two unknown variables can be solved by methods which provide explicit expressions of the solution (for two zones, such a resolution is faster but for more than two zones, the iterative technique is appropriate if done by hand):

1. An initial value of t_o is given to t^1_{wh} and t^2_{wh}

2. A new value of t^1_{wh} and t^2_{wh} can then be calculated with the previous expressions

3. These new values are used as initial ones for step 2

4. The iterations are made until new values become very close to the previous ones

Secondly, the auxiliary heating loads and the mean indoor temperature are calculated using similar iterative technique. Assuming that the mean indoor temperature of Zone 2 is known, the loads without gain of Zone 1 are:

$$Q^1_{ng} = N \, LL^1 \, (t^1_t - t_o) + N \, LL^{12} \, (t^1_t - t^2_i)$$

where

t^2_i = mean indoor temperature of Zone 2 (with gains) and then the Gains Load Ratio of zone 1 is:

$$GLR^1 = \frac{N\Sigma\Phi^1}{Q^1_{ng}}$$

where

$N\Sigma\Phi^1$ = total monthly gains of Zone 1 calculated as shown in Section 2

Q^1_{ng} = monthly loads without heating calculated previously

After deducing the value of the main time constant of zone 1 from Figure 18 or from Section 4.3(a), the curves from Figure 19 give the value of the utilization factor η^1, and the auxiliary heating loads are, – in Zone 1:

$$Q^1_{aux} = Q^1_{ng} - \eta^1 N\Sigma\Phi^1$$

Finally, the mean indoor temperature of Zone 1 is expressed as:

$$t^1_i = t^1_{wb} + \frac{Q^1_{aux}}{N(LL^1 + LL^{12})} + \frac{LL^{12}}{(LL^1 + LL^{12})}(t^2_i - t^2_{wb})$$

Here, an iterative method must be used to solve the problem:

1. Initial values of t^1_i and t^2_i are taken equal to t^1_{wb} and t^2_{wb}

2. First values of Q^1_{ng} and Q^2_{ng} are calculated with the previous equations

3. Then, GLR^1, GLR^2, Q^1_{aux} and Q^2_{aux} are evaluated

4. New values of t^1_i and t^2_i are finally deduced from the last equation

5. These new values are used as initial one in Step 2 above until they remain invariant under iteration.

NB: The correlation used in this extended use has not been developed for multizone purposes; this extended use would have to be more rigorously validated. However, this extension provides information which is not otherwise available, and the sensitivity of the results to changes of the input parameters is certainly good.

4.4

Table of Materials Properties

See Appendix 9

4.5

Table of Internal Gains

Source	Occupants	Lighting	Appliances	Cooking	Hot water	Total
Billington	4.8	1.8	7.8	8.0	15.0	37.4
Wolf	5.4	>	12.0	<	—	17.4
Brundrett	6.0	>	12.0	<	11.0	29.0
Heap	4.0	1.4	3.4	3.4	3.4	15.6
Siviour and Haslett	4.67	2.7	2.7	4.1	4.3	18.5
Searle (H)*	5.48	2.5	> 6.22	<	4.7	18.9
(L)	3.97	2.17	> 4.84	<	3.7	14.68
L.P.B.**						
Liège (H)*	3.8	0.8	> 5.6	<	1.1	11.3
(L)	3.5	1.0	> 3.5	<	0.6	8.6
Siviour	5.0	4.0	4.0	4.0	5.0	22.0
(Summer)	>		12.0	<	5.0	15.0
Thermal Insulation Laboratory	6.98	>	11.23	<	3.06	21.0 ***

*	H: Heavy occupation profile L: Light occupation profile
**	Low levels of appliance and hot water use
***	Total energy consumption for hot water supply is 10.2 but only 30% incidental gains is assumed.

Average daily incidental heat gains from inhabitant, appliances and hot water for a three-bedroom town or row house in winter, and summer if indicated (kWh/day). [19, 20, and 22 to 29].

	kWh/day
Occupants	4.0
Lighting	1.5
Appliances and Cooking	6.5
Hot Water	3.0
Total	15.0

Incidental heat gains per day. Representative European values for a three-bedroomed house constructed from data in the table above.

Notes

Notes: 1(a) Heat is not lost from a solid ground floor uniformly over its area, but principally through its exposed edges, and secondarily by contact with the ground below. Two floors of the same superficial area and composition can have different heat-loss rates by virtue of different ratios of perimeter to area. For analytical convenience we can use a notional U-value, as if it were uniform for the entire floor . Its value, obtained from tables, will have been determined experimentally [from a range of sizes and plan shapes, taking account also of whether insulation is also applied at the edges (vertical) or below the slab(horizontal)] This approach allows one to treat the floor losses in principle as those of the roof and walls, that is by multiplying area by U-value in Form A1

1(b) An alternative is to use linear-based transmission coefficients (units W/mK), also available from tables, applied to each exposed metre of floor slab, as in Form A3. This approach might be more convenient where a buffer space is attached to a heated space and shares a common edge.

2. It would be logical to consider not just a change in insulation properties at night, but also a change in ventilation rates, and hence ventilation loss-rates, both in heated spaces and in buffer spaces.

3. The use of shading masks will indicate how far obstructions or shades might reduce the radiation incident on a window, and hence the amount transmitted to the inside. However, in respect of direct solar radiation, the sunpaths and the aperture masks are normally shown on a cylindrical rather than an equal-area projection. As a result, occlusion-patches of equal area on the sunpath projection may not represent the same loss of radiation, apart from the influence the sun's position on the diagram has on the intensity of incident radiation.

References

[1] Produced for the CEC, DGXII by Prof. J.K. Page, Dept. of Building Science, University of Sheffield.

[2] "Tables of Temperature, Relative Humidity, Precipitation and Sunshine for the World", Part III Europe and the Azores, Meteorological Office, 1982.

[3] "ESP Manual", Abacus, Strathclyde University, Glasgow, UK.

[4] "Handbook of Air Conditioning, Heating and Ventilating", Strock, C, Koral, R.L., Industrial Press, New York, 1965.

[5] CIBS Guide, Section A3, Chartered Institute of Building Services.

[6] "Handbook of Chemistry and Physics", based on data from Coblentz, Cammerer and Drysdale, Department of Scientific and Industrial Research.

[7] "Solar Energy Thermal Processes", Duffie, J.A., Beckmann, W.A., John Wiley & Sons Inc., 1974.

[8] Règles TH-K-77, Centre Scientifique et Technique du Bâtiment, France, November 1977 - Revised April 1982.

[9] ASHRAE Handbook 1981, Fundamentals, Chapter 25: Heating Load.

[10] Règles TH-G-77, Centre Scientifique et Technique du Bâtiment, France, November 1977 - Revised April 1982.

[11] "Methode 5000", Claux, P., Franca, J.P., Gilles, R., Pesso, A., Pouget, A., Raoust, M., PYC Edition, France, December 1982.

[12] "Modélisation de Capteurs Solaires à Air Assurant un Préchauffage de l'Air Neuf", Raoust, M., Research Report, Direction de la Construction, France, 1980.

[13] Règles THB 82: "Calcul du Coefficient Voluminique des Besoins de Chauffage des Logements", Centre Scientifique et Technique du Bâtiment, France, November 1977 - Revised April 1982.

[14] "Analyse et réduction modales d'un modèle de comportement thermique de bâtiment", Lefebvre, G, thèse de doctorat de l'Université Paris VI, November 1987.

Factors a_1 and a_2 for various types of attached sunspace (read in conjunction with Table 1, Page 297)

	Type of sunspace		Sunspace with insulated floor*		Sunspace with uninsulated floor			
					light coloured ($\alpha = 0.3$)		dark coloured ($\alpha = 0.8$)	
A	Attached sunspace with 4 collecting		S.G	D.G	S.G	D.G	S.G	D.G
	surfaces: roof	a_1	.65	.69	.63	.67	.59	.63
	front and sides.	a_2	.85	.87	.85	.87	.85	.87
B	Attached sunspace with 3 collecting		S.G	D.G	S.G	D.G	S.G	D.G
	surfaces: front	a_1	.70	.74	.68	.71	.65	.68
	and sides.	a_2	.89	.91	.89	.91	.89	.91
C	Integrated or attached sunspace		S.G	D.G	S.G	D.G	S.G	D.G
	with 2 collecting	a_1	.87	.90	.84	.87	.80	.82
	surfaces: roof and front.	a_2	.87	.90	.87	.90	.87	.90

TABLE of SYMBOLS and GLOSSARY

Symbol Unit Meaning

A m^2
Area of each solar aperture; total opening including mullions (window, air collector, Trombe wall &c

A_e, A_i m^2
Area of external and internal walls of zone or building

A_b m^2
Heated floor area

A_w m^2
Area of each opaque wall

a_1, a_2   (fraction)
Factors taking into account solar energy losses by multiple reflections within a sunspace, and solar-energy losses through the ground of a sunspace

B W/m^3K
Coefficient of annual heating load [B-value]

C   (fraction)
Efficiency factor for Trombe Wall

C_c   (fraction)
Effective solar-energy transmission factor of net curtains

C_f   (fraction)
Solar loss factor of an uninsulated floor

C_{lb}   (fraction)
Heat-loss reduction factor of a buffer space, mean diurnal values

c   (fraction)
Correction factor for the solar gain of an open-loop solar wall

DD_a   K-days
Annual degree-days for the thermostat set temperature

DD_m   K-days
Monthly degree-days for the thermostat set temperature

d b
Duration of night set-back in temperature

E kWb/m^2day
Trasmitted solar energy per m^2 (read from the tables)

E_i kWb/m^2day
Incident solar energy per m^2 (read from the tables)

E_e, E_i kWb/m^2
Energy per sq metre stored in external and internal walls

E_s kWb/day
Transmitted solar energy through each glazed surface of the envelope of a sunspace

e_i m
Thickness of a layer i

F kWb/day
Solar energy absorbed by a wall

F_s kWb/day
Solar energy collected by a sunspace contributing to its elevation in temperature.

G W/m^3K
Coefficient of annual heating load without gains (G-value); Volumetric heat-loss rate

GLR   (fraction)
Gains/Load Ratio

H b
Number of hours per day during which movable insulation is removed (for the month considered)

I kg/m^2
Useful thermal mass in the heated space per unit of heated floor area

k W/mK
Transmission heat-loss coefficient for a floor slab (per m of perimeter), or for thermal bridges, (per m of edge).

L_{bd} W/K
Total daytime heat loss coefficient from the buffer space to outdoors

L_{bm} W/K
Mean heat-loss coefficient for heat loss from buffer space to outdoors

L_{bn} W/K
Total night-time heat-loss coefficient for heat loss from buffer space to outdoors

L_{bd} W/K
Total daytime heat-loss coefficient for heat loss from heated space to buffer space

L_{bm} W/K
Mean heat-loss coefficient for heat loss from heated space to buffer space

L_{bn} W/K
Total night-time heat-loss coefficient for heat loss from heated space to buffer space

LL $kWb/Kday$
Daily heat loss per K temperature difference

LL_i $kWb/dayK$
Daily heat-loss coefficient from zone i to outdoor

LL_{ij} $kWb/dayK$
Daily heat exchange coefficient between zones i and j

l m
Length of 'edge' separating heated space from outdoor environment

m   *(fraction)*
Ratio of glazing area to total area of opening, to take account of window frames, mullions etc

m_s *(fraction)*
Ratio of total area of sunspace glass to the total area of the transparent envelope

N   *(integer)*
Number of days in particular month

N_e, N_i *(integer)*
Number of external and internal walls of a zone or building

$N\Sigma\Phi$ *kWh/mo*
Monthly total gains

$n(t)$ *h/day*
Number of hours that the indoor temperature is greater than t

ΣQ *kWh/mo*
Monthly useful solar and internal gains

Q_{aux} *kWh/mo*
Monthly auxiliary heating load

$Q_{aux\,tot}$ *kWh/a*
Annual auxiliary heating load

Q_{ng} *kWh/mo*
Monthly heating load without solar and internal gains

$Q_{ng\,tot}$ *kWh/a*
Annual heating load without gains

q *m³/h*
Air-flow rate of outdoor air entering a buffer space

q_b *m³/h*
Air-flow rate through the buffer space, from the outdoor to the heated space

q_e *m³/h*
Air-flow rate through heat exchanger

q_{sw} *m³/h*
Air-flow rate through an open-loop solar wall

q_t *m³/h*
Total flow rate of outside air entering heated space

R_e *W/K*
Reduction of ventilation heat loss due to an air to air heat exchanger

R_b *W/K*
Reduction of ventilation heat loss due to a buffer space

R_{sw} *W/K*
Reduction of ventilation heat loss due to an open-loop solar wall

R_t *W/K*
Total reduction of ventilation heat loss

r_a *m²K/W*
Thermal resistance of an air space (Airspace resistance)

r_{ext} *m²K/W*
Thermal resistance from absorber to outdoors (r_{ed} daytime value; r_{en}, night-time value)

r_g *m²K/W*
Thermal resistance of glazing

r_{int} *m²K/W*
Thermal resistance from absorber to indoor

r_{ni} *m²K/W*
Thermal resistance of night insulation

r_s *m²K/W*
Sum of internal and external surface resistance

r_w *m²K/W*
Thermal resistance of walls

S_f *(fraction)*
Shading factor (If no shading:, $S_f = 1$)

t_i *°C*
Monthly mean indoor temperature

t_{wh} *°C*
Indoor air temperature without heating

t_o *°C*
Monthly mean outdoor temperature

t_s *°C*
Monthly mean temperature inside a sunspace, taking solar gains into account

t_{sng} *°C*
Monthly mean temperature inside a sunspace without solar gains

t_t *°C*
Thermostat set temperature

U   *W/m²K*
Thermal transmittance of a wall or window(U-value).

U_d *W/m²K*
Daytime U-Value

U_n,   *W/m²K*
Night-time U-Value

V_b *m³*
Heated volume

α *(fraction)*
Absorptance of wall or floor

η *(fraction)*
Monthly utilization factor of solar and internal gains

η_{ac} *(fraction)*
Efficiency of an open-loop air collector

η_e *(fraction)*
Efficiency of an air-to-air heat exchanger

η_{sw} *(fraction)*
Efficiency of an open-loop solar wall

λ_i *W/mK*
Thermal conductivity of material of layer i

$\sum\Phi$ *kWh/day*
Daily amount of solar and internal gains contributed to the heated space

Φ_{ac} *kWh/day*
Solar gains from an open-loop air collector

Φ_{asw} *kWh/day*
Solar gains from ventilation air in an open-loop solar wall

Φ_{csw} *kWh/day*
Solar gains from conduction through the wall of an open-loop solar wall

Φ_{sw} *kWh/day*
Total Solar gains from an open-loop solar wall

Φ_{dg} *kWh/day*
Solar gains from direct gain

Φ_{sdg} *kWh/day*
Solar gains from windows between the sunspace and the heated space

Φ_{smw} *kWh/day*
Solar gains from mass walls within a sunspace

Φ_{sa} *kWh/day*
Solar gains from the ventilation of a sunspace

Φ_{sb} *kWh/day*
Solar gains from the buffer effect of the sunspace

Φ_s *kWh/day*
Total solar gains of a sunspace

Φ_{tw} *kWh/day*
Solar gains from a Trombe wall

Φ_{mw} *kWh/day*
Solar gains from a mass wall

Φ_i *kWh/day*
Daily amount of internal gains (occupants and equipment)

Φ_{solar} *kWh/day*
Total solar energy contributed to the heated space

ρc_i *J/m³K*
Thermal capacity of material of layer *i*

τ_g *(fraction)*
Solar transmittance of glazing

τ_s *(fraction)*
Overall solar transmittance of the envelope of a sunspace ($\tau_g \times \mathbf{m}_s$)

τ *h*
Main time constant of a building (or zone)

328

ACKNOWLEDGEMENTS

Method 5000 was originally released in France in 1982 for the Architectural Design competition called *5000 Maisons Solaires*. The original authors are P Claux, J P Franca, R Gilles, A Pesso, A Pouget, and M Raoust. G Lefebvre, using the results of his doctoral thesis, revised the method in 1990 to take into account a calculated time constant, intermittent heating and multi-zoning. M Raoust is responsible for producing the English version originally published in the Preliminary Edition of the *European Passive Solar Handbook*, for the present revision *New Method 5000* incorporating Lefebvre's additions, and for the computer version of the method.

Editing work on this English edition was carried out by the Energy Research group.

DAYLIGHT FACTOR METER

A15.1
PRINCIPLES

The Daylight Factor (DF) is defined by the following -

$$DF = (I_i / I_o) \times 100\%$$

where I_i is the illumination inside at the point of interest and I_o is the illumination outside due to the unobstructed sky.

This important parameter is dependent upon room geometry and window design, room finishes, external obstructions, etc. From a known DF, the designer can assess if there is likely to be sufficient daylight for the use of the room by comparing with recommended values, and can check the quality of the daylight from the variation of the DF across the room. Also, if the hourly sky luminance for the site is available, the energy saving of the daylight, by the avoidance of use of artificial light, can be evaluated.

The DF can be calculated for rooms of simple geometry, but for more complex lighting features such as light wells, light shelves, atria etc, physical models are often used. These are illuminated under a real sky, or an artificial sky and the illuminance levels measured inside and outside the model with an electrical Luxmeter. The device described here is a low cost alternative to this.

The DF Meter, (fig 1&2), consists in effect of two little white rooms, the *measuring room* and the *sky room* connected by a *dark corridor*. The measuring room is illuminated with light falling onto a *diffusing window* in the top of the measuring room, which in a 1:20 model is in the working plane - 1m above floor level. The sky room is illuminated with light from the sky passing through a *diffusing skylight* and down the *light tower*.

Through a hole in the sky room wall, the *occulus*, you can look down the corridor to the measuring room. However, half way down there is a *mirror* across the lower half of the corridor allowing a view of the sky room also. Look above the mirror and you see into the measuring room; look at the mirror and you see into the sky room. The two room surfaces appear vertically adjacent.

The Daylight Factor in a room is always much less than 100% and so to get a brightness match we have to reduce the light in the sky room. The light tower already reduces this to about 25% and the remainder of the reduction is carried out by the *sliding shutter*. The higher the value of DF in the room, the less the shutter will have to be closed. When a brightness match is made, the DF can be read off from the position of the shutter.

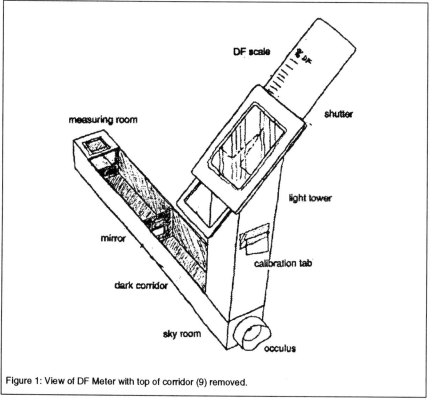

Figure 1: View of DF Meter with top of corridor (9) removed.

CONSTRUCTION

The meter is constructed from parts cut from 1mm thick white card. It is best to mark out the card on a drawing board with a knife or very fine pencil, taking dimensions from the drawing (fig 3). It might be possible to use an enlarged photocopy direct onto card or onto paper, but distortion may occur. The paper will have to be stuck to card using a strong spray adhesive. Card with a single black side could be used, or black paper for the appropriate areas.

You will also need to obtain the small mirror cut from thin glass with one good clean edge, and two pieces of drawing film for the diffusers.

Cutting out and Assembly

Cut out the parts with an accuracy of at least 0.3mm and with a clean square edge. The components have been set out to minimise the number of separate cuts. Punch or burn the hole in part (7). Save the square, cut from the hole in part (8).

For assembly, refer to fig (3) and the long section fig(2).

Start with the rooms and the corridor. The base (3) is marked with the positions of parts (4), (5), (6) and (7). Glue them into position making sure they are square in plan and section. PVA glue is best but needs support during drying. You may prefer to use a contact adhesive. Joints can finally be taped to protect the cut edges. Parts (5) and (6) have their white side to the measuring room. Glue on side pieces (1) and (2). Complete the measuring room with (8) and add two landing pieces

(10) and (11) inside the top of the corridor at each end. The small diffuser should be glued over the square hole in (8).

Now position the mirror with the reflective side to the sky room, so that looking through the occulus the reflection of the hole is just not visible at the edge of the mirror. The top of the mirror should be horizontal and if the frame (6) is vertical and square, the mirror should fill the lower half of the aperture. Inspect the mirror edges and position the best edge upwards. The mirror may be stuck onto the frame with glue, double sided tape, or 'Blue-Tack'. The latter will permit easy alignment. Blacken all edges and surfaces other than actually in the rooms.

The top (9) may now be glued on, black side down, and the tower base (16), (17), (18) and (19) glued into the sky room, white side in. Note that these parts should protrude 32mm above the top edge of the room.

The light tower is made from parts (12), (13), (14) and (15). Reinforce the joints at the base with some tape. If everything is accurate, the tower should fit snugly on the base in any of four orientations. You may need to chamfer the corners of the base with a blade or sandpaper. If it is loose, use four map pins. Blacken all edges of the joint to prevent stray light.

The skylight is made from parts (20) and (21), and spacers (23) and (24)., The glueing must be clean and the spacers positioned right at the edge. The shutter (24), which must be cut out very accurately, can have sharp edges removed and the sliding surfaces

rubbed with candle wax. The scale should be photocopied at a magnification that gives a length of 109 mm, and positioned so that when the shutter is fully open it indicates 25%.

The top edges of (14) and (15) the tower should be bevelled to take the sloping skylight . The skylight can now be glued to the tower.

Stick the large diffuser over the skylight. Add a shaped piece of cardboard tube around the occulus. Check that all joints are light proof. It is useful to draw a scale on the side of the meter in units of 50 mm from the centre of the measuring room. This is a metre scale for a 1:20 model. The DF meter is now finished except for calibration.

Preferably, the light tower is attached to the corridor so that the skylight faces the same direction as the main windows of the model room being tested, but it must not be obstructed by the model.

PRACTICE

Before the final calibration it is a good idea to practice obtaining a brightness match. Outside on an overcast day, insert the meter into a model, as in fig (4) and obtain a match by moving the shutter. You may find that the measuring room appears as a slightly different colour from the sky room, This is due to the internally reflected component from the model. This effect can almost be eliminated by almost closing the eyelids - reducing the amount of light to the eye suppresses colour vision.

Make sure that you wait long enough for your eye to adapt to the low level of light in the model - a black patch over the spare eye may help! You may also find it helps to rock the eye up and down, scanning the edge between the two surfaces. This seems to accentuate any slight difference in brightness.

Your performance can now be tested. Fix the measuring room in one position in the model, say about 3m back from the window, and adjust the shutter to give a match. Make a pencil mark where the shutter emerges from the skylight. Repeat the operation several times; when your marks are within 1mm of each other, you are ready to calibrate.

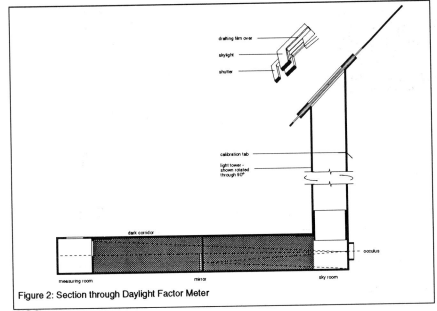

drafting film over

skylight

shutter

calibration tab

light tower - shown rotated through 90°

occulus

dark corridor

measuring room

mirror

sky room

Figure 2: Section through Daylight Factor Meter

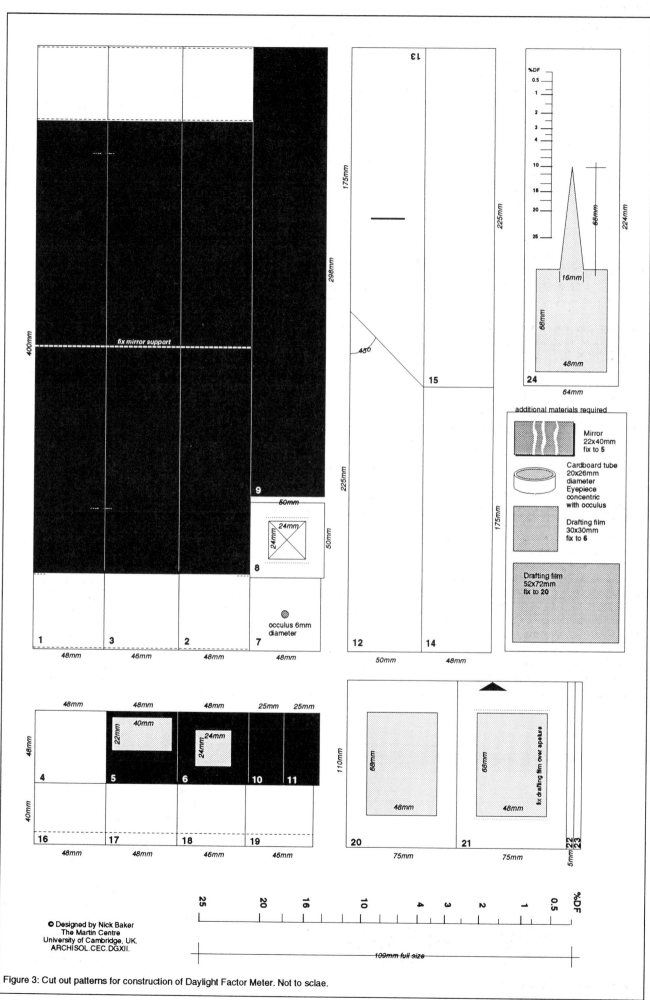

Figure 3: Cut out patterns for construction of Daylight Factor Meter. Not to sclae.

A15.4

CALIBRATION

With the shutter fully open, the skylight and the tower reduce the light in the sky room to a little over 25% of the illuminance due to the sky. The scale provided in this kit indicates the DF as a function of shutter position; the DF is exactly 25% when the shutter is fully open. So before positioning this scale, the meter must be "trimmed" to this condition.

Find the square piece of card that you cut out of part (8) - it is probably in the bin or on the floor - and accurately cut into four pieces across the diagonals. With the model fixed to the table under an overcast sky (not in the model), place the three triangular pieces over the measuring window so that they obscure exactly 75% of the aperture. Make sure that the shutter is accurately fully open.

It is clear that we have now reduced the illumination of the measuring room to 25%. If all is well, the sky room should appear just a little brighter.

Assuming that it does, now push a strip of stiff white paper through the calibration slot in (15) and use your newly developed optical skill to obtain a really good match. Try moving the meter to face different parts of the sky, and move the triangular pieces around on the window. When you are sure of the match, fold the trimming strip down and secure with tape.

If you find the sky room already dimmer, something has gone a little wrong. There are two possible cures - shorten the tower by 10 mm or so, or reduce the transmission of the measuring room diffuser by spots of drawing ink or permanent spirit pen.

A15.5

USE OF THE DF METER

The main disadvantage of the DF meter is its size. Clearly the model must be designed with a 50mm x 50mm aperture in the side to allow the meter to be inserted. This should be a good fit to reduce extraneous light. Do not forget that the model should have the realistic reflectances on the inside - colours do not matter but the degree of lightness or darkness must be roughly correct. Ideally the outside of the part of the meter which goes into the model should be a medium grey. Also for the final DF values an allowance must be made for the reduction of light due to glazing and glazing bars, not normally present in models. Information on this can usually be found in text books.

It is useful to use the DF meter to measure the minimum Daylight Factor, and the Daylight Factor Distribution. Recommended values according to room use are generally available for the former. The distribution has attracted more interest recently, since it has been found that rooms with a DF ratio of greater than 10 often require supplementary artificial lighting at higher levels than normal night time levels. This is to compensate for very bright daylight levels in part of the room, leaving the remainder of the room looking gloomy. Thus good daylighting design is concerned with obtaining sufficient light furthest away from the window without providing too much light close to the window. Advanced daylighting devices such as light shelves, are specifically aimed at overcoming this problem.

The design of the Daylight Factor Meter is by Nick Baker of the Martin Centre for Architectural and Urban Studies University of Cambridge, 6 Chaucer Rd, Cambridge, UK.
It was originally developed for the EC Architectural
Competition "Working in the City" coordinated by the Energy Research Group, University College Dublin within The SOLINFO programme of the CEC DGXII.

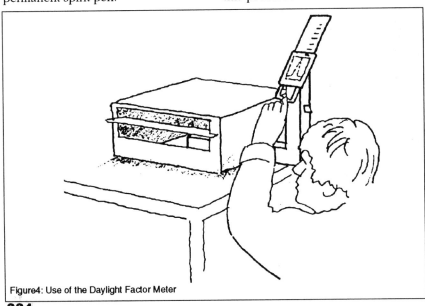

Figure4: Use of the Daylight Factor Meter

Pmv program

Computer program for calculation of predicted mean vote (PMV) and predicted percentage of dissatisfied (PPD)

The following BASIC program computes the PMV and the PPD in accordance with the International Standard, ISO 7730 [1]. PMV is the mean vote of a large group of persons on the following scale:

+3	hot
+2	warm
+1	slightly warm
0	neutral
-1	slightly cool
-2	cool
-3	cold

The PMV predicts the thermal sensation for the entire body. It does not predict local discomfort, i.e. compliance with the limits for thermal nonuniformity should be checked separately. The calculation is based on the following variables:

Variables Symbols in program	
Clothing, clo	CLO
Metabolic rate, met	MET
External work, met	WME
Air temperature, °C	TA
Mean radiant temperature, °C	TR
Relative air velocity, m/s	VEL
Relative humidity, %	RH
Water vapour pressure, Pa	PA

An example of the output follows the program listing:

```
10      'Computer program (BASIC) for calculation of
20      'Predicted Mean Vote (PMV) and Predicted Percentage of Dissatisfied
        (PPD)
30      'in accordance with International Standard, ISO 7730
40      CLS:PRINT"DATA ENTRY" :'data entry
50      INPUT " Clothing (clo)"; CLO
60      INPUT " Metabolic rate (met)"; MET
70      INPUT " External work, normally around 0 (met)"; WME
80      INPUT " Air temperature ( C )"; TA
90      INPUT " Mean radiant temperature ( C )"; TR
100     INPUT " Relative air velocity (m/s)"; VEL
110     PRINT " ENTER EITHER RH OR WATER VAPOUR PRESSURE BUT
        NOT BOTH"
120     INPUT " Relative humidity ( % )"; RH
130     INPUT " Water vapour pressure    ( Pa)"; PA
140     DEF FNPS(T)=EXP(16.6536 - 4030.183/(T+235)) :'saturated vapour
        pressure,KPa
150     IF PA=0 THEN PA=RH*10*FNPS(TA) :'water vapour pressure, Pa
160     ICL = .155 * CLO:'thermal insulation of the clothing in m2K/W
170     M   = MET * 58.15 :'metabolic rate in W/m2
180     W   = WME * 58.15 :'external work in W/m2
190     MW  = M - W :'internal heat production in the human body
200     IF ICL < .078 THEN FCL = 1 + 1.29 * ICL ELSE FCL=1.05 + .645*ICL
        :'clothing area factor
210     HCF=12.1*SQR(VEL):'heat transf. coeff. by forced     convection
220     TAA = TA + 273 :'air temperature in Kelvin
230     TRA = TR + 273 :'mean radiant temperature in Kelvin
240     '————CALCULATE SURFACE TEMPERATURE OF CLOTHING BY
        ITERATION————
250     TCLA = TAA + (35.5-TA) / (3.5*(6.45*ICL+.1)) :'first guess for surface
        temperature of clothing
260     P1 = ICL * FCL :'calculation term
270     P2 = P1 * 3.96 :'calculation term
280     P3 = P1 * 100 :'calculation term
290     P4 = P1 * TAA :'calculation term
300     P5 = 308.7 - .028 * MW + P2 * (TRA/100)^4 :'calculation term
310     XN = TCLA / 100
```

```
320     XF = XN
330     N=0:'N: number of iterations
340     EPS = .00015:'stop criteria in iteration
350     XF=(XF+XN)/2
360     HCN=2.38*ABS(100*XF-TAA)^.25 :'heat transf. coeff. by natural
        convection
370     IF HCF>HCN THEN HC=HCF ELSE HC=HCN
380     XN=(P5+P4*HCP2*XF^4)/(100+P3*HC)
390     N=N+1
400     IF N > 150 THEN GOTO  550
410     IF ABS(XN-XF)>EPS GOTO  350
420     TCL=100*XN-273:'surface temperature of the clothing
430     '——————————————HEAT LOSS COMPONENTS———————
        ——————————
440     HL1 = 3.05*.001*(5733-6.99*MW-PA):'heat loss diff. through skin
450     IF MW > 58.15 THEN HL2 = .42 * (MW-58.15)
        ELSE HL2 = 0! :'heat loss by sweating(comfort)
460     HL3 = 1.7 * .00001 * M * (5867-PA) :'latent respiration heat loss
470     HL4 = .0014 * M * (34-TA) :'dry respiration heat loss
480     HL5=3.96*FCL*(XN^4- (TRA/100)^4) :'heat loss by radiation
490     HL6 = FCL * HC * (TCL-TA) :'heat loss by convection
500     '———————————————— CALCULATE PMV AND PPD———————————
        ———
510     TS = .303 * EXP(-.036*M) + .028 :'thermal sensation trans coeff
520     PMV = TS * (MW-HL1-HL2-HL3-HL4-HL5-HL6) :'predicted mean vote
530     PPD=100-95*EXP(-.03353*PMV^4.2179*PMV^2):'predicted percentage
        dissat.
540     GOTO 570
550     PMV=999999!
560     PPD=100
570     PRINT:PRINT"OUTPUT" :'output
580     PRINT " Predicted Mean Vote (PMV): " ;:PRINT USING "##.#"; PMV
590     PRINT " Predicted Percent of Dissatisfied (PPD):";:PRINT
        USING"###.#"; PPD
600     PRINT: INPUT "NEXT RUN (Y/N)" ; R$
610     IF (R$="Y" OR R$="y") THEN RUN
620     END
```

A16.2

EXAMPLE

DATA ENTRY

Clothing
(clo)? 1.0
Metabolic rate
(met)? 1.0
External work, normally around 0
(met)? 0
Air temperature   (C)?
 23.3
Mean Radiant Temperature (C)?
 23.3
Relative air velocity (m/s)?
0.1

ENTER EITHER RH OR WATER
VAPOUR PRESSURE BUT NOT BOTH
Relative humidity (%)?
50
Water vapour pressure (Pa)?

OUTPUT
Predicted Mean Vote (PMV):
0.0
Predicted Percent of Dissatisfied (PPD):
5.0

NEXT RUN (Y/N)?

REFERENCE

[1] ISO 7730 (International Organization for Standardization), Moderate Thermal Environments - Determination of the PMV and PPD indices and specification of the conditions for thermal comfort, Geneva, 1984.

INDEX